Lecture Notes in Computer Science　　9659

Commenced Publication in 1973
Founding and Former Series Editors:
Gerhard Goos, Juris Hartmanis, and Jan van Leeuwen

More information about this series at http://www.springer.com/series/7409

Bin Cui · Nan Zhang · Jianliang Xu
Xiang Lian · Dexi Liu (Eds.)

Web-Age Information Management

17th International Conference, WAIM 2016
Nanchang, China, June 3–5, 2016
Proceedings, Part II

 Springer

Editors
Bin Cui
Peking University
Beijing
China

Xiang Lian
University of Texas Rio Grande Valley
Edinburg, TX
USA

Nan Zhang
The George Washington University
Washington, D.C.
USA

Dexi Liu
Jiangxi University of Finance and
    Economics
Nanchang
China

Jianliang Xu
Hong Kong Baptist University
Kowloon Tong, Hong Kong
SAR China

ISSN 0302-9743   ISSN 1611-3349   (electronic)
Lecture Notes in Computer Science
ISBN 978-3-319-39957-7   ISBN 978-3-319-39958-4   (eBook)
DOI 10.1007/978-3-319-39958-4

Library of Congress Control Number: 2016940123

LNCS Sublibrary: SL3 – Information Systems and Applications, incl. Internet/Web, and HCI

Printed on acid-free paper

This Springer imprint is published by Springer Nature
The registered company is Springer International Publishing AG Switzerland

Preface

This volume contains the proceedings of the 17th International Conference on Web-Age Information Management (WAIM), held during June 3–5, 2016, in Nanchang, Jiangxi, China. As a flagship conference in the Asia-Pacific region focusing on the research, development, and applications of Web information management, its success has been witnessed through the previous conference series that were held in Shanghai (2000), Xi'an (2001), Beijing (2002), Chengdu (2003), Dalian (2004), Hangzhou (2005), Hong Kong (2006), Huangshan (2007), Zhangjiajie (2008), Suzhou (2009), Jiuzhaigou (2010), Wuhan (2011), Harbin (2012), Beidahe (2013), Macau (2014), and Qingdao (2015). With the fast development of Web-related technologies, we expect that WAIM will become an increasingly popular forum to bring together outstanding researchers in this field from all over the world.

This high-quality program would not have been possible without the authors who chose WAIM for disseminating their contributions. Out of 249 submissions to the research track and 17 to the demonstration track, the conference accepted 80 research papers and eight demonstrations. The contributed papers address a wide range of topics, such as big data analytics, data mining, query processing and optimization, security, privacy, trust, recommender systems, spatial databases, information retrieval and Web search, information extraction and integration, data and information quality, distributed and cloud computing, among others.

The technical program of WAIM 2016 also included two keynote talks by Profs. Beng Chin Ooi (National University of Singapore) and Yanchun Zhang (Victoria University, Australia), as well as three talks in the Distinguished Young Lecturer Series by Profs. Tingjian Ge (University of Massachusetts at Lowell), Hua Lu (Aalborg University), and Haibo Hu (Hong Kong Polytechnic University). We are immensely grateful to these distinguished guests for their invaluable contributions to the conference program.

A conference like WAIM can only succeed as a team effort. We are deeply thankful to the Program Committee members and the reviewers for their invaluable efforts. Special thanks to the local Organizing Committee headed by Guoqiong Liao and Xiaobing Mao. Many thanks also go to our workshop co-chairs (Shaoxu Song and Yongxin Tong), proceedings co-chairs (Xiang Lian and Dexi Liu), DYL co-chairs (Hong Gao and Weiyi Meng), demo co-chairs (Xiping Liu and Yi Yu), publicity co-chairs (Ye Yuan, Hua Lu, and Chengkai Li), registration chair (Yong Yang), and finance chair (Bo Shen). Last but not least, we wish to express our gratitude for the hard work of our webmaster (Bo Yang), and for our sponsors who generously supported the smooth running of our conference.

We hope you enjoy the proceedings WAIM 2016!

June 2016

Zhanhuai Li
Sang Kyun Cha
Changxuan Wan
Bin Cui
Nan Zhang
Jianliang Xu

Organization

Organizing Committee

Honor Chair

Qiao Wang — Jiangxi University of Finance and Economics, China

General Co-chairs

Zhanhuai Li	Northwestern Polytechnical University, China
Sang Kyun Cha	Seoul National University, Korea
Changxuan Wan	Jiangxi University of Finance and Economics, China

PC Co-chairs

Bin Cui	Peking University, China
Nan Zhang	George Washington University, USA
Jianliang Xu	Hong Kong Baptist University, SAR China

Workshop Co-chairs

Shaoxu Song	Tsinghua University, China
Yongxin Tong	Beihang University, China

Proceedings Co-chairs

Xiang Lian	The University of Texas Rio Grande Valley, USA
Dexi Liu	Jiangxi University of Finance and Economics, China

DYL Series Co-chairs (Distinguished Young Lecturer)

Hong Gao	Harbin Institute of Technology, China
Weiyi Meng	SUNY Binghamton, USA

Demo Co-chairs

Xiping Liu	Jiangxi University of Finance and Economics, China
Yi Yu	National Institute of Informatics, Japan

Publicity Co-chairs

Ye Yuan	Northeastern University, China
Hua Lu	Aalborg University, Denmark
Chengkai Li	The University of Texas at Arlington, USA

Local Organization Co-chairs

Xiaobing Mao	Jiangxi University of Finance and Economics, China
Guoqiong Liao	Jiangxi University of Finance and Economics, China

Registration Chair

Yong Yang	Jiangxi University of Finance and Economics, China

Finance Chair

Bo Shen	Jiangxi University of Finance and Economics, China

Web Chair

Bo Yang	Jiangxi University of Finance and Economics, China

Steering Committee Liaison

Weiyi Meng	SUNY Binghamton, USA

CCF DBS Liaison

Xiaofeng Meng	Renmin University of China, China

Program Committee

Alex Thomo	University of Victoria, Canada
Anirban Mondal	Xerox Research Centre India, India
Baihua Zheng	Singapore Management University, Singapore
Baoning Niu	Taiyuan University of Technology, China
Byron Choi	Hong Kong Baptist University, SAR China
Carson Leung	University of Manitoba, Canada
Ce Zhang	Stanford University, USA
Chengkai Li	The University of Texas at Arlington, USA
Chih-Hua Tai	National Taipei University, China
Cuiping Li	Renmin University of China, China
David Cheung	The University of Hong Kong, SAR China
Dejing Dou	University of Oregon, USA
De-Nian Yang	Academia Sinica, Taiwan, China

Dongxiang Zhang	National University of Singapore, Singapore
Feida Zhu	Singapore Management University, Singapore
Feifei Li	University of Utah, USA
Fuzhen Zhuang	ICT, Chinese Academy of Sciences, China
Gang Chen	Zhejiang University, China
Gao Cong	Nanyang Technological University, Singapore
Giovanna Guerrini	Università di Genova, Italy
Guohui Li	Huazhong University of Science and Technology, China
Guoliang Li	Tsinghua University, China
Guoqiong Liao	Jiangxi University of Finance and Economics, China
Haibo Hu	Hong Kong Polytechnic University, SAR China
Hailong Sun	Beihang University, China
Hiroaki Ohshima	Kyoto University, Japan
Hongyan Liu	Tsinghua University, China
Hongzhi Wang	Harbin Institue of Technology, China
Hongzhi Yin	The University of Queensland, Australia
Hua Lu	Aalborg University, Denmark
Jae-Gil Lee	Korea Advanced Institute of Science and Technology, Korea
Jeffrey Xu Yu	Chinese University of Hong Kong, SAR China
Jiaheng Lu	Renmin University of China
Jianbin Qin	The University of New South Wales, Australia
Jianbin Huang	Xidian University, China
Jiannan Wang	University of Berkerley, USA
Jie Shao	University of Electronic Science and Technology of China, China
Jinchuan Chen	Renmin University of China, China
Jingfeng Guo	Yanshan University, China
Jiun-Long Huang	National Chiao Tung University, Taiwan, China
Jizhou Luo	Harbin Institue of Technology, China
Ju Fan	National University of Singapore, Singapore
Junfeng Zhou	Yanshan University, China
Junjie Yao	East China Normal University, China
Ke Yi	Hong Kong University of Science and Technology, SAR China
Kun Ren	Yale University, USA
Kyuseok Shim	Seoul National University, South Korea
Lei Zou	Peking University, China
Leong Hou U	University of Macau, SAR China
Lianghuai Yang	Zhejiang University of Technology, China
Lidan Shou	Zhejiang University, China
Lili Jiang	Max Planck Institute for Informatics, Germany
Ling Chen	University of Technology, Sydney, Australia
Luke Huan	University of Kansas, USA
Man Lung Yiu	Hong Kong Polytechnic University, SAR China
Muhammad Cheema	Monash University, Australia

Peiquan Jin	University of Science and Technology of China, China
Peng Wang	Fudan University, China
Qi Liu	University of Science and Technology of China, China
Qiang Wei	Tsinghua University, China
Qingzhong Li	Shandong University, China
Qinmin Hu	East China Normal University, China
Quan Zou	Xiamen University, China
Richong Zhang	Beihang University, China
Rui Zhang	The University of Melbourne, Australia
Rui Chen	Samsung Research America, USA
Saravanan Thirumuruganathan	Qatar Computing Research Institute, Qatar
Senjuti Basu Roy	University of Washington, USA
Shengli Wu	Jiangsu University, China
Shimin Chen	Chinese Academy of Sciences, China
Shinsuke Nakajima	Kyoto Sangyo University, Japan
Shuai Ma	Beihang University, China
Sourav Bhowmick	National Taiwan University, China
Takahiro Hara	Osaka University, Japan
Taketoshi Ushiama	Kyushu University, Japan
Tingjian Ge	University of Massachusetts Lowell, USA
Wang-Chien Lee	Penn State University, USA
Wei Wang	University of New South Wales, Australia
Weiwei Sun	Fudan University, China
Weiwei Ni	Southeast University, China
Wen-Chih Peng	National Chiao Tung University, Taiwan, China
Wenjie Zhang	University of New South Wales, Australia
Wolf-Tilo Balke	TU-Braunschweig, Germany
Wookey Lee	Inha University, Korea
Xiang Lian	The University of Texas Rio Grande Valley, USA
Xiangliang Zhang	King Abdullah University of Science and Technology, Saudi Arabia
Xiaochun Yang	Northeast University, China
Xiaofeng Meng	Renmin University of China, China
Xiaohui Yu	Shandong University, China
Xiaokui Xiao	Nanyang Technological University, Singapore
Xifeng Yan	University of California at Santa Barbara, USA
Xin Lin	East China Normal University, China
Xin Cao	Queen's University Belfast, UK
Xin Wang	Tianjin University, China
Xingquan Zhu	Florida Atlantic University, USA
Xuanjing Huang	Fudan University, China
Yafei Li	Henan University of Economics and Law, China
Yang Liu	Shandong University, China
Yanghua Xiao	Fudan University, China
Yang-Sae Moon	Kangwon National University, Korea

Contents – Part II

Privacy and Trust

Detecting Anomalous Ratings Using Matrix Factorization
for Recommender Systems.................................... 3
   Zhihai Yang, Zhongmin Cai, and Xinyuan Chen

A Novel Spatial Cloaking Scheme Using Hierarchical Hilbert Curve
for Location-Based Services................................. 15
   Ningning Cui, Xiaochun Yang, and Bin Wang

Efficient Privacy-Preserving Content-Based Image Retrieval in the Cloud ... 28
   Kai Huang, Ming Xu, Shaojing Fu, and Dongsheng Wang

Preserving the d-Reachability When Anonymizing Social Networks 40
   Xiangyu Liu, Jiajia Li, Dahai Zhou, Yunzhe An, and Xiufeng Xia

Personalized Location Anonymity - A Kernel Density Estimation Approach... 52
   Dapeng Zhao, Jiansong Ma, Xiaoling Wang, and Xiuxia Tian

Detecting Data-model-oriented Anomalies in Parallel Business Process 65
   Ning Yin, Shanshan Wang, Hongyan Li, and Lilue Fan

Learning User Credibility on Aspects from Review Texts 78
   Yifan Gao, Yuming Li, Yanhong Pan, Jiali Mao, and Rong Zhang

Detecting Anomaly in Traffic Flow from Road Similarity Analysis........ 92
   Xinran Liu, Xingwu Liu, Yuanhong Wang, Juhua Pu,
   and Xiangliang Zhang

Query Processing and Optimization

PACOKS: Progressive Ant-Colony-Optimization-Based Keyword Search
over Relational Databases 107
   Ziyu Lin, Qian Xue, and Yongxuan Lai

Enhanced Query Classification with Millions of Fine-Grained Topics 120
   Qi Ye, Feng Wang, Bo Li, and Zhimin Liu

A Hybrid Machine-Crowdsourcing Approach for Web Table Matching
and Cleaning 132
   Chunhua Li, Pengpeng Zhao, Victor S. Sheng, Zhixu Li, Guanfeng Liu,
   Jian Wu, and Zhiming Cui

An Update Method for Shortest Path Caching with Burst Paths Based
on Sliding Windows   145
   Xiaohua Li, Ning Wang, Kanggui Peng, Xiaochun Yang, and Ge Yu

Low Overhead Log Replication for Main Memory Database System   159
   Jinwei Guo, Chendong Zhang, Peng Cai, Minqi Zhou, and Aoying Zhou

Diversification of Keyword Query Result Patterns.   171
   Cem Aksoy, Ananya Dass, Dimitri Theodoratos, and Xiaoying Wu

Efficient Approximate Substring Matching in Compressed String   184
   Yutong Han, Bin Wang, and Xiaochun Yang

Top-K Similarity Search for Query-By-Humming   198
   Peipei Wang, Bin Wang, and Shiying Luo

Social Media

Restricted Boltzmann Machines for Retweeting Behaviours Prediction.   213
   Xiang Li, Lijuan Xie, Yong Tan, and Qiuli Tong

Cross-Collection Emotion Tagging for Online News   225
   Li Yu, Xue Zhao, Chao Wang, Haiwei Zhang, and Ying Zhang

Online News Emotion Prediction with Bidirectional LSTM   238
   Xue Zhao, Chao Wang, Zhifan Yang, Ying Zhang, and Xiaojie Yuan

Learning for Search Results Diversification in Twitter.   251
   Ying Wang, Zhunchen Luo, and Yang Yu

Adjustable Time-Window-Based Event Detection on Twitter   265
   Qinyi Wang, Jieying She, Tianshu Song, Yongxin Tong, Lei Chen,
   and Ke Xu

User-IBTM: An Online Framework for Hashtag Suggestion in Twitter   279
   Jia Li and Hua Xu

Unifying User and Message Clustering Information for Retweeting
Behavior Prediction.   291
   Bo Jiang, Jiguang Liang, Ying Sha, Lihong Wang, Zhixin Kuang, Rui Li,
   and Peng Li

KPCA-WT: An Efficient Framework for High Quality Microblog
Extraction in Time-Frequency Domain.   304
   Min Peng, Xinyuan Dai, Kai Zhang, Guanyin Zeng, Jiahui Zhu,
   Shuang Ouyang, Qianqian Xie, and Gang Tian

Big Data Analytics

Active Learning Method for Constraint-Based Clustering Algorithms   319
  Lijun Cai, Tinghao Yu, Tingqin He, Lei Chen, and Meiqi Lin

An Effective Cluster Assignment Strategy for Large Time Series Data   330
  Damir Mirzanurov, Waqas Nawaz, JooYoung Lee, and Qiang Qu

AdaWIRL: A Novel Bayesian Ranking Approach for Personal Big-Hit
Paper Prediction   342
  Chuxu Zhang, Lu Yu, Jie Lu, Tao Zhou, and Zi-Ke Zhang

Detecting Live Events by Mining Textual and Spatial-Temporal Features
from Microblogs   356
  Zhejun Zheng, Beihong Jin, Yanling Cui, and Qiang Ji

A Label Correlation Based Weighting Feature Selection Approach
for Multi-label Data   369
  Lu Liu, Jing Zhang, Peipei Li, Yuhong Zhang, and Xuegang Hu

Valuable Group Trajectory Pattern Mining Directed by Adaptable Value
Measuring Model   380
  Xinyu Huang, Tengjiao Wang, Shun Li, and Wei Chen

DualPOS: A Semi-supervised Attribute Selection Approach for Symbolic
Data Based on Rough Set Theory   392
  Jianhua Dai, Huifeng Han, Hu Hu, Qinghua Hu, Jinghong Zhang,
  and Wentao Wang

Semi-supervised Clustering Based on Artificial Bee Colony Algorithm
with Kernel Strategy   403
  Jianhua Dai, Huifeng Han, Hu Hu, Qinghua Hu, Bingjie Wei,
  and Yuejun Yan

Distributed and Cloud Computing

HMNRS: A Hierarchical Multi-source Name Resolution Service
for the Industrial Internet   417
  Yang Liu, Guoqiang Fu, and Xinchi Li

Optimizing Replica Exchange Strategy for Load Balancing
in Multienant Databases   430
  Teng Liu, Qingzhong Li, Lanju Kong, Lei Liu, and Lizhen Cui

ERPC: An Edge-Resources Based Framework to Reduce Bandwidth Cost
in the Personal Cloud   444
  Shaoduo Gan, Jie Yu, Xiaoling Li, Jun Ma, Lei Luo, Qingbo Wu,
  and Shasha Li

Multidimensional Similarity Join Using MapReduce   457
  Ye Li, Jian Wang, and Leong Hou U

Real-Time Logo Recognition from Live Video Streams Using an Elastic
Cloud Platform   469
  Jianbing Ding, Hongyang Chao, and Mansheng Yang

Profit Based Two-Step Job Scheduling in Clouds   481
  Shuo Zhang, Li Pan, Shijun Liu, Lei Wu, and Xiangxu Meng

A Join Optimization Method for CPU/MIC Heterogeneous Systems   493
  Kailai Zhou, Hong Chen, Hui Sun, Cuiping Li, and Tianzhen Wu

GFSF: A Novel Similarity Join Method Based on Frequency Vector   506
  Ziyu Lin, Daowen Luo, and Yongxuan Lai

Erratum to: Web-Age Information Management (Part I and II) E1
  Bin Cui, Nan Zhang, Jianliang Xu, Xiang Lian, and Dexi Liu

Demo Papers

SHMS: A Smart Phone Self-health Management System Using Data Mining. . .   521
  Chuanhua Xu, Jia Zhu, Zhixu Li, Jing Xiao, Changqin Huang,
  and Yong Tang

MVUC: An Interactive System for Mining and Visualizing Urban
Co-locations   524
  Xiao Wang, Hongmei Chen, and Qing Xiao

SNExtractor: A Prototype for Extracting Semantic Networks from Web
Documents   527
  Chi Zhang, Yanhua Wang, Chengyu Wang, Wenliang Cheng,
  and Xiaofeng He

Crowd-PANDA: Using Crowdsourcing Method for Academic Knowledge
Acquisition.   531
  Zhaoan Dong, Jiaheng Lu, and Tok Wang Ling

LPSMon: A Stream-Based Live Public Sentiment Monitoring System   534
  Kun Ma, Zijie Tang, Jialin Zhong, and Bo Yang

DPBT: A System for Detecting Pacemakers in Burst Topics.   537
  Guozhong Dong, Wu Yang, Feida Zhu, and Wei Wang

CEQA - An Open Source Chinese Question Answer System Based
on Linked Knowledge   540
  Zeyu Du, Yan Yang, Qinming Hu, and Liang He

OSSRec:An Open Source Software Recommendation System Based
on Wisdom of Crowds.   544
    Mengwen Chen, Gang Yin, Chenxi Song, Tao Wang, Cheng Yang,
    and Huaimin Wang

Author Index   549

Contents – Part I

Data Mining

More Efficient Algorithm for Mining Frequent Patterns with Multiple
Minimum Supports   3
   Wensheng Gan, Jerry Chun-Wei Lin, Philippe Fournier-Viger,
   and Han-Chieh Chao

Efficient Mining of Uncertain Data for High-Utility Itemsets 17
   Jerry Chun-Wei Lin, Wensheng Gan, Philippe Fournier-Viger,
   Tzung-Pei Hong, and Vincent S. Tseng

An Improved HMM Model for Sensing Data Predicting in WSN 31
   Zeyu Zhang, Bailong Deng, Siding Chen, and Li Li

eXtreme Gradient Boosting for Identifying Individual Users Across
Different Digital Devices 43
   Rongwei Song, Siding Chen, Bailong Deng, and Li Li

Two-Phase Mining for Frequent Closed Episodes 55
   Guoqiong Liao, Xiaoting Yang, Sihong Xie, Philip S. Yu,
   and Changxuan Wan

Effectively Updating High Utility Co-location Patterns in Evolving Spatial
Databases 67
   Xiaoxuan Wang, Lizhen Wang, Junli Lu, and Lihua Zhou

Mining Top-*k* Distinguishing Sequential Patterns with Flexible Gap
Constraints 82
   Chao Gao, Lei Duan, Guozhu Dong, Haiqing Zhang, Hao Yang,
   and Changjie Tang

A Novel Chinese Text Mining Method for E-Commerce Review Spam
Detection 95
   Xiu Li and Xinwei Yan

Spatial and Temporal Databases

Retrieving Routes of Interest Over Road Networks 109
   Wengen Li, Jiannong Cao, Jihong Guan, Man Lung Yiu,
   and Shuigeng Zhou

Semantic-Aware Trajectory Compression with Urban Road Network   124
    Na Ta, Guoliang Li, Bole Chen, and Jianhua Feng

Discovering Underground Roads from Trajectories Without Road Network . . .   137
    Qiuge Song, Jiali Mao, and Cheqing Jin

Ridesharing Recommendation: Whether and Where Should I Wait?   151
    Chengcheng Dai

Keyword-aware Optimal Location Query in Road Network   164
    Jinling Bao, Xingshan Liu, Rui Zhou, and Bin Wang

Point-of-Interest Recommendations by Unifying Multiple Correlations   178
    Ce Cheng, Jiajin Huang, and Ning Zhong

Top-*k* Team Recommendation in Spatial Crowdsourcing   191
    *Dawei Gao, Yongxin Tong, Jieying She, Tianshu Song, Lei Chen,
    and Ke Xu*

Explicable Location Prediction Based on Preference Tensor Model   205
    Duoduo Zhang, Ning Yang, and Yuchi Ma

Recommender Systems

Random Partition Factorization Machines for Context-Aware
Recommendations   219
    *Shaoqing Wang, Cuilan Du, Kankan Zhao, Cuiping Li, Yangxi Li,
    Yang Zheng, Zheng Wang, and Hong Chen*

A Novel Framework to Process the Quantity and Quality of User Behavior
Data in Recommender Systems   231
    Penghua Yu, Lanfen Lin, and Yuangang Yao

RankMBPR: Rank-Aware Mutual Bayesian Personalized Ranking for Item
Recommendation   244
    *Lu Yu, Ge Zhou, Chuxu Zhang, Junming Huang, Chuang Liu,
    and Zi-Ke Zhang*

Unsupervised Expert Finding in Social Network for Personalized
Recommendation   257
    *Junmei Ding, Yan Chen, Xin Li, Guiquan Liu, Aili Shen,
    and Xiangfu Meng*

An Approach for Clothing Recommendation Based on Multiple Image
Attributes   272
    *Dandan Sha, Daling Wang, Xiangmin Zhou, Shi Feng, Yifei Zhang,
    and Ge Yu*

SocialFM: A Social Recommender System with Factorization Machines 286
  Juming Zhou, Dong Wang, Yue Ding, and Litian Yin

Identifying Linked Data Datasets for sameAs Interlinking Using
Recommendation Techniques 298
  *Haichi Liu, Ting Wang, Jintao Tang, Hong Ning, Dengping Wei,
  Songxian Xie, and Peilei Liu*

Query-Biased Multi-document Abstractive Summarization via Submodular
Maximization Using Event Guidance........................... 310
  Rui Sun, Zhenchao Wang, Yafeng Ren, and Donghong Ji

Graph Data Management

Inferring Diffusion Network on Incomplete Cascade Data 325
  Peng Dou, Sizhen Du, and Guojie Song

Anchor Link Prediction Using Topological Information in Social Networks ... 338
  Shuo Feng, Derong Shen, Tiezheng Nie, Yue Kou, and Ge Yu

Collaborative Partitioning for Multiple Social Networks with Anchor Nodes ... 353
  Fenglan Li, Anming Ji, Songchang Jin, Shuqiang Yang, and Qiang Liu

A General Framework for Graph Matching and Its Application in Ontology
Matching 365
  Yuda Zang, Jianyong Wang, and Xuan Zhu

Internet Traffic Analysis in a Large University Town: A Graphical
and Clustering Approach 378
  Weitao Weng, Kai Lei, Kuai Xu, Xiaoyou Liu, and Tao Sun

Conceptual Sentence Embeddings 390
  Yashen Wang, Heyan Huang, Chong Feng, Qiang Zhou, and Jiahui Gu

Inferring Social Roles of Mobile Users Based on Communication
Behaviors... 402
  Yipeng Chen, Hongyan Li, Jinbo Zhang, and Gaoshan Miao

Sparse Topical Coding with Sparse Groups 415
  *Min Peng, Qianqian Xie, Jiajia Huang, Jiahui Zhu, Shuang Ouyang,
  Jimin Huang, and Gang Tian*

Information Retrieval

Differential Evolution-Based Fusion for Results Diversification of Web
Search 429
  Chunlin Xu, Chunlan Huang, and Shengli Wu

BMF: An Indexing Structure to Support Multi-element Check   441
   Chenyang Xu, Qin Liu, and Weixiong Rao

Efficient Unique Column Combinations Discovery Based on Data
Distribution   454
   Chao Wang, Shupeng Han, Xiangrui Cai, Haiwei Zhang,
   and Yanlong Wen

Event Related Document Retrieval Based on Bipartite Graph   467
   Wenjing Yang, Rui Li, Peng Li, Meilin Zhou, and Bin Wang

SPedia: A Semantics Based Repository of Scientific Publications Data   479
   Muhammad Ahtisham Aslam and Naif Radi Aljohani

A Set-Based Training Query Classification Approach for Twitter Search   491
   Qingli Ma, Ben He, Jungang Xu, and Bin Wang

NBLucene: Flexible and Efficient Open Source Search Engine   504
   Zhaohua Zhang, Benjun Ye, Jiayi Huang, Rebecca Stones, Gang Wang,
   and Xiaoguang Liu

Context-Aware Entity Summarization   517
   Jihong Yan, Yanhua Wang, Ming Gao, and Aoying Zhou

Erratum to: BMF: An Indexing Structure to Support Multi-element Check . . .   E1
   Chenyang Xu, Qin Liu, and Weixiong Rao

Author Index   531

Privacy and Trust

Detecting Anomalous Ratings Using Matrix Factorization for Recommender Systems

Zhihai Yang, Zhongmin Cai$^{(\boxtimes)}$, and Xinyuan Chen

Ministry of Education Key Lab for Intelligent Networks and Network Security,
Xi'an Jiaotong University, Xi'an 710049, China
zmcai@mail.xjtu.edu.cn

Abstract. Personalization recommendation techniques play a key role in the popular E-commerce services such as Amazon, TripAdvisor and etc. In practice, collaborative filtering recommender systems are highly vulnerable to "shilling" attacks due to its openness. Although attack detection based on such attacks has been extensively researched during the past decade, the studies on these issues have not reached an end. They either extract extra features from user profiles or directly calculate similarity between users to capture concerned attackers. In this paper, we propose a novel detection technique to bypass these hard problems, which combines max-margin matrix factorization with Bayesian non-parametrics and outlier detection. Firstly, mean prediction errors for users and items are calculated by utilizing trained prediction model on test sets. And then we continue to comprehensively analyze the distribution of mean prediction errors of items in order to reduce the scope of concerned items. Based on the suspected items, all anomalous users can be finally determined by analyzing the distribution of mean prediction error on each user. Extensive experiments on the MovieLens-100K dataset demonstrate the effectiveness of the proposed method.

1 Introduction

Personalization recommendation technique has become a crucial component in existing E-commerce services such as Amazon, TripAdvisor etc. Recommender systems make rating predictions for users' potential preferences based on currently observed preferences and their relations with others' on currently unrated items [9,24]. However, recommender systems especially for collaborative filtering recommender systems (CFRSs) are highly vulnerable to "profile injection" attacks or "shilling" attacks [1,10,14,18,22,26]. It is common occurance that attackers promote (called push attack) or demote (called nuke attack) target items with the maximum or minimum rating to achieve their attack intentions. In addition, detecting weaknesses in products and services on real-world datasets attracted much attentions [2,3]. Therefore, it is imperative to construct an effective detection technique to defend and remove such attacks.

Detection methods based on such attacks have received much attentions, however, they either used classification methods based on extracted features to

© Springer International Publishing Switzerland 2016
B. Cui et al. (Eds.): WAIM 2016, Part II, LNCS 9659, pp. 3–14, 2016.
DOI: 10.1007/978-3-319-39958-4_1

spot anomalous users [1,5,8,26] or directly calculated similarity of user to distinguish between attackers and genuine users. In reality, it is difficult to extract extra features from user profiles (containing attack profiles and genuine profiles) in compared with previous methods [15,16]. Furthermore, calculating similarity between users to capture abnormal users is unsubstantial on large-scale datasets (with thousands of users), although it is effective to find out concerned attackers in some extent [8]. In contrast of the aforementioned issues, we consider a matrix factorization based method to bypass the problems of feature extraction and similarity calculation.

In this paper, we present a stepwise detection method which consists of rating prediction by exploiting matrix factorization and outlier detection based on mean prediction errors of users and items. Since extracting extra features from user profiles is really difficult in compared with the existing methods, we employ a fast max-margin matrix factorization method [21] to predict users' potential preferences. It is noteworthy that using matrix factorization not only can bypass feature extraction problem but also can be favorable to reveal the difference between predicting rating and actual rating. Based on prediction results on test sets, we further calculate mean prediction error (MPE) of each user and item, respectively. Due to attackers focus on target items with the same attack intention (push or nuke target items), the predicting ratings on corresponding target items generally deviate from actual ratings, which may leads to higher MPEs on these target items in compared with other items. Therefore, it is easy to capture the concerned target items that show abnormally high MPE in the distribution of item. To incorporate the MPE distribution of user into outlier detection, we utilize the captured target items to reduce detection scope for spotting concerned attackers. What's even more important, calculating MPE for each user just focuses on the concerned target items rather than total items, due to the fact that the MPEs of attackers show higher and consistent in compared with genuine users'. Extensive experiments on MovieLens dataset demonstrate the effectiveness of the proposed method in diverse attack models.

The main contributions of this paper can be summarized as follows:

- We propose a novel detection method to spot anomalous ratings by combing max-margin matrix factorization with Bayesian nonparametrics and outlier detection, which bypasses the problems of feature extraction and similarity calculation.
- Based on prediction results, mean prediction errors for users and items are respectively calculated and abnormal users are further spotted by comprehensively considering both the distributions of mean prediction errors of user and item.
- Extensive experiments on the MovieLens-100K dataset demonstrate the effectiveness of the proposed method.

The rest of this paper is organized as follows: Section 2 reviews related work. Attack models will be introduced in Sect. 3. In Sect. 4, we detail the proposed method. Experimental results will be reported and analyzed in Sect. 5. Finally, we conclude the paper with a brief summary and predict the direction of future work.

2   Related Work

Spotting anomalous ratings in recommender systems attracted much attentions. In the following, we just discuss methods related to this work. Su et al. [13] proposed a spreading similarity algorithm for detecting groups of similar attackers. Then, [1,5,8,15,16,26] presented different features directly extracted from user profiles to capture abnormal users. Mehta et al. [6] proposed an unsupervised detection method based on principal component analysis and performed well against shilling attacks. The motivation behind this method is that attackers have higher similarity (i.e., using Pearson Correlation coefficient) each other and as well as having high similarity with a large number of genuine users. Zhang et al. [23] proposed an online method, HHT-SVM, to detect "profile injection" attacks by combining Hilbert-Huang transform (HHT) and support vector machine (SVM). After that, Zhou et al. [27–29] presented detection method for detecting group attack profiles, which analyzes target items to find out anomalous ratings. Roughly speaking, on the one hand, it is difficult to extract more features from user profiles. On the other hand, directly calculating similarity between users is not a wise choice, especially for large-scale datasets, although it is useful to distinguish between attacker and genuine user. Based on these challenges, we aim at constructing a detection method to bypass the two challenges and effectively defend abnormal ratings.

3   Attack Models

The attackers have different attack intentions to bias the recommendation results to achieve their benefits, which demotes or promotes the target items with the lowest or highest rating [1,4,7,8,15,24]. The general form of an attack profile is described in Table 1. The details of the four sets of items are briefly described as follows:

I_T: A set of target items with singleton or multiple items, called single-target attack or multi-target attack. The rating is $\gamma(i_j^T)$, generally rated the maximum or minimum value in the entire profiles.

I_S: The set of selected items with specified rating by the function $\sigma(i_k^S)$;

I_F: A set of filler items, received items with randomly chosen by the function $\rho(i_l^F)$;

I_N: A set of items with no ratings;

In this work, we utilize 8 different attack models, the details of each attack model will be introduced in Sect. 5.

4   The Proposed Approach

The proposed method consists of two stages: the stage of rating prediction and the stage of outlier detection. In this section, we firstly introduce the mechanism of rating prediction based on matrix factorization, which aims to induce a trained prediction model. Then, we utilize the trained model to examine on test sets

Table 1. General form of attack profiles

I_T		I_S			I_F			I_N		
i_1^T	$\cdots$	i_j^T	i_1^S	$\cdots$	i_k^S	i_1^F	$\cdots$	i_l^F	i_1^N	$\cdots$	i_v^N
$\gamma(i_1^T)$	$\cdots$	$\gamma(i_j^T)$	$\sigma(i_1^S)$	$\cdots$	$\sigma(i_k^S)$	$\rho(i_1^F)$	$\cdots$	$\rho(i_l^F)$	$null$	$\cdots$	$null$

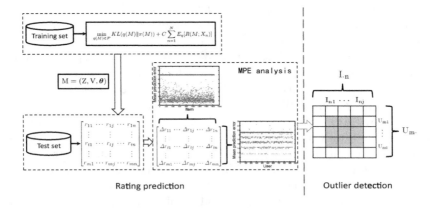

Fig. 1. The framework of the proposed approach.

and further calculate prediction errors (as shown in Fig. 1). By calculating mean prediction errors (MPE) for users and items, suspected items with higher mean prediction error can be observed from MPE distribution of item. Based on the suspected items, we further focus on abnormal error distribution of users in order to spot all anomalous users who simultaneously rated the suspected items with the maximum (for push attack) or minimum (for nuke attack) rating. In the end, detected attackers can be directly returned.

4.1 Rating Prediction via Matrix Factorization

Matrix factorization has been used to collaborative predictions [19–21]. Inspired from these studies, we exploit the advantages of matrix factorization to predict ratings for target users on target items. In the stage of rating prediction, we employ a fast max-margin matrix factorization method which combines maximum a posteriori (MAP) estimation and Bayesian nonparametrics to construct a novel interpretation of max-margin matrix factorization (M^3F) [21].

Given a set of training data $X = \{X_n\}_{n=1}^N$, a probabilistic extension to the original deterministic risk minimization problem (Ω, R) [21] is described as follows:

$$\min_{q(M) \in P} KL(q(M) \| \pi(M)) + C \sum_{n=1}^N E_q[R(M; X_n)] \tag{1}$$

where $M = (U, V, \boldsymbol{\theta})$, $q(M)$ presents a posterior distribution derived from the generic inference procedure [21], $\boldsymbol{\theta} = (\theta_1, \cdots, \theta_{L-1})^\top$, $X_n = (x_n, y_n)$, the prediction rule is $Y_{ij} = \max\{r|U_iV_j^\top \geq \theta_{r-1}, r = 1, \cdots, L\}$, L is the maximum rating. C denotes a balancing factor. In addition, P denotes a space of valid probability distributions, $q(M)$ denotes a posterior distribution derived from out generic inference procedure. $KL(q(M)\|\pi(M))$ is the KL-divergence and $E_q[R(M; X_n)] = \int \log L(M|X_n)q(M)dM$.

By following IBP (Indian buffet process) [21] to solve problem (1), we replace U by Z and substitute it to corresponding equations such as $M = (Z, V, \boldsymbol{\theta})$. Thereupon, we have the normalized dele-prior as follows:

$$\pi(Z, V, \boldsymbol{\theta}) = \text{IBP}(Z|\alpha) \cdot \Pi_{j=1}^M N(V_j|0, \sigma^2 I) \cdot \Pi_{i=1}^N N(\boldsymbol{\theta}_i|\rho, \varsigma^2 I) \tag{2}$$

where we set $\pi(Z, V, \boldsymbol{\theta})$ equal to $\hat{p}_0(Z, V, \boldsymbol{\theta})$.

Regularizer and loss fully specified, a probabilistic formulation of MAP estimation can be described as follows:

$$\dot{p}_0(Z, V, \boldsymbol{\theta}) = e^{-\frac{1}{2\sigma^2}(\|Z\|_F^2 + \|V\|_F^2) - \frac{1}{2\varsigma^2}\sum_{i=1}^N \|\theta_i - \rho\|_2^2} \tag{3}$$

$$\dot{L}(Z, V, \boldsymbol{\theta}|(i, j), Y_{ij}) = e^{-C\sum_{r=1}^{L-1} h_\ell(T_{ij}^r(\theta_{ir} - Z_iV_j^\top))} \tag{4}$$

Ultimately, the prediction rule is accordingly changed to $Y_{ij} = \max\{r|Z_iV_j^\top \geq \theta_{r-1}, r = 1, \cdots, L\}$ for user i on item j (in a rating matrix $ZV^\top$). By solving the following optimization problem, we can obtain the trained binary matrix Z and V and directly use them on test sets.

$$\min_{q(Z,V,\boldsymbol{\theta})\in P} KL(q(Z, V, \boldsymbol{\theta})\|\pi(Z, V, \boldsymbol{\theta})) + C\sum_{n=1}^N E_q[R(Z, V, \boldsymbol{\theta}; X_n)] \tag{5}$$

4.2   Outlier Detection

In the detection stage, we firstly calculate predicted ratings using the trained prediction model on test sets and analyze the distribution of Mean Prediction Error (MPE) [21] for users and items. Then, all detected users will be directly returned by comprehensively analyzing both the MPE distribution of user and item (as shown in Algorithm 1). Due to attackers focus on target items to achieve their intention, the ratings of prediction generally deviate from actual ratings on the target items. For this observation, we aim at analyzing the distribution of MPE of items in order to capture suspected items as illustrated in Fig. 2. In practice, attackers focus on target items many times to promote or demote the same target items, therefore, the MPEs of these target items very likely reach higher than other items that just genuine users focused on. To find out the suspected items as far as possible, we firstly calculate MPE for each item and select the top-N items which have higher MPE than an empirical threshold ε_i as the suspected items. Based on these suspected items, we continue to calculate MPE for each user and spot concerned attackers who have high MPEs and simultaneously rated these target items with the maximum or minimum rating. In Algorithm 1, k is the number of latent factors in line 2.

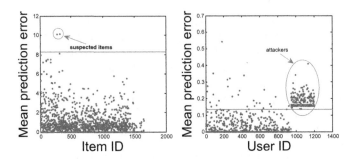

Fig. 2. The distributions of mean prediction error for items and users.

5 Experiments and Analysis

In this section, we conduct a series of experiments on the MovieLens-100K dataset in 8 different attack models and compare the effectiveness of the proposed method with benchmark methods. Finally, various experimental results are briefly analyzed.

5.1 Experimental Data and Settings

Dataset. In our experiments, we utilize the MovieLens-100K[1] dataset as the data set describing the behaviors of genuine users in recommender system. It consists of 100,000 ratings on 1682 movies by 943 raters and each rater had to rate at least 20 movies. All ratings are in the form of integral values between minimum value 1 and maximum value 5. The minimum score means the rater distastes the movie, while the maximum score means the rater enjoyed the movie. Attack profiles are generated by using corresponding attack models (as shown in Table 2) with diverse attack size[2] {1.1 %, 6.4 %, 11.7 %, 17.0 %, 22.3 %, 27.6 %} and filler sizes[3] {1.2 %, 4.2 %, 7.3 %, 10.3 %, 13.3 %, 16.4 %}. After that, the generated attack profiles directly inserted into genuine profiles to construct final user profiles (consisting of attack profiles and genuine profiles). Therefore, we have 288 (8 × 6 × 6) experimental datasets including 8 attack models, 6 different attack sizes and 6 different filler sizes.

Attack Profiles. To conduct experimental data, we introduce 8 attack models to generate attack profiles (as shown in Table 2). Let $N(r, \sigma^2)$ denotes the Gaussian distribution with mean r and standard deviation σ, the details of each attack model are described as follows:

---

[1] http://grouplens.org/datasets/movielens/.

[2] The ratio between the number of attackers and genuine users.

[3] The ratio between the number of items rated by user u and the number of entire items in the recommender systems.

Algorithm 1. Detecting abnormal users.

Require:
    Rating matrix of test set M_T;
    Binary matrix Z and matrix V;
    Thresholds ε_u and ε_i;
Ensure:
    Detected result DR;
1: $I_e = $ null, $U_e = $ null;
2: $E_{mn} = M_T - Z_{mk}V_{nk}^{\top}$; // All ratings with the lowest or highest score are considered.
3: **for** each item $i \in [1, n]$ **do**
4:     $e_{\cdot i} = \frac{\sum_{t=1}^{m} |e_{ti}|}{m}$;
5:     **if** $e_{\cdot i} > \varepsilon_i$ **then**
6:         $I_e \leftarrow i$;
7:     **end if**
8: **end for**
9: **for** each user $u \in [1, m]$ **do**
10:     $e_{u\cdot} = \frac{\sum_{t=1}^{N_{I_e}} |e_{ut}|}{N_{I_e}}$;
11:     **if** $e_{u\cdot} > \varepsilon_u$ **then**
12:         $U_e \leftarrow u$;
13:     **end if**
14: **end for**
15: $DR = U_e$;
16: **return** DR;

---

(1) Average attack: $I_S = \phi$ and $\rho(i) \sim N(\overline{r_i}, \overline{\sigma_i}^2)$, where $\overline{r_i}$ denotes the average rating of item i over all the users that rated this item, and $\overline{\sigma_i}$ denotes the standard deviation of ratings of item i over all the users who have rated this item [25].

(2) Bandwagon (average) attack: I_S contains a set of popular items. And then, we use these items as I_S, $\sigma(i) = r_{max}$ or r_{min} (push or nuke) and $\rho(i) \sim N(\overline{r_i}, \overline{\sigma_i}^2)$ [17].

(3) Segment attack: I_S contains a set of segmented items. And then, we use these items as I_S, $\sigma(i) = r_{max}$ or r_{min} (push or nuke) and $\rho(i) = r_{min}$ or r_{max} (push or nuke) [4].

(4) Reverse Bandwagon attack: I_S contains a set of unpopular items, $\sigma(i) = r_{min}$ or r_{max} (push or nuke) and $\rho(i) \sim N(\overline{r}, \overline{\sigma}^2)$ [4].

(5) Love/Hate attack: $I_S = \phi$ and $\rho(i) = r_{min}$ or r_{max} (push or nuke) [4].

(6) AOP attack: A simple and effective strategy to obfuscate the Average attack is to choose filler items with equal probability from the top x % of most popular items rather than from the entire collection of items [12].

(7) PIA-AS attack: The top-N items with the highest aggregate similarity (AS) scores become the selected set of power items. This method requires at least 5 users who have rated the same item i and item j [12].

(8) PUA-AS attack: The top 50 users with the highest Aggregate Similarity scores become the selected set of power users. This method requires at least 5 co-rated items between users u and v and does not use significance weighting [11].

Table 2. Attack models summary.

Attack models	I_S Items	Rating	I_F Items	Rating	I_N	I_T Push or nuke
Average	Null		Randomly chosen	Normal dist item mean around	Null	r_{max} or r_{min}
Bandwagon (average)	Popular items	r_{max} or r_{min}	Randomly chosen	Normal dist item mean around	Null	r_{max} or r_{min}
Segment	Segmented items	r_{max} or r_{min}	Randomly chosen	r_{min} or r_{max}	Null	r_{max} or r_{min}
Reverse Bandwagon	Unpopular items	r_{min} or r_{max}	Randomly chosen	System mean	Null	r_{max} or r_{min}
Love/Hate	Null		Randomly chosen	r_{min} or r_{max}	Null	r_{max} or r_{min}
AOP	Null		x% popular items, ratings set with normal dist around item mean.		Null	r_{max} or r_{min}
PIA-AS	Power items, ratings set with normal dist around item mean.		Null		Null	r_{max} or r_{min}
PUA-AS	Copy ratings and items from power user profiles.		Null		Null	r_{max} or r_{min}

Experimental Setting and Evaluation Metrics. As in Xu et al. [19–21], we utilize the all-but-one protocol to generate training sets and test sets. Note that, we just consider *weak* generalization, which indicates all users contribute to the learning of the latent factors. We randomly select 800 users from the constructed user profiles for *weak* and repeat the random selection thrice. To evaluate prediction results, we exploit mean prediction error (MPE) or mean absolute error (MAE). In addition, the standard measurements of precision and recall are considered to evaluate experimental results. The metrics are respectively described as follows:

$$\text{MPE}_i = \frac{\sum_{u \in U} |p_{ui} - r_{ui}|}{|U|}, \text{MPE}_u = \frac{\sum_{i \in I} |p_{ui} - r_{ui}|}{|I|} \tag{6}$$

$$\text{Recall} = \frac{\#\text{True Positives}}{\#\text{True Positives} + \#\text{False Negatives}} \tag{7}$$

$$\text{Precision} = \frac{\#\text{True Positives}}{\#\text{True Positives} + \#\text{False Positives}} \tag{8}$$

where I and U are item set and user set, respectively. r_{ui} is actual rating on item i rated by user u and p_{ui} is predictive rating on item i rated by user u. *#False Positives* denotes the number of genuine profiles that are misclassified, *#True Positives* denotes the number of attack profiles correctly identified as attacks, and *#False Negatives* denotes the number of attack profiles that were misclassified [1]. Do not forget that algorithm is strongly dependent on the characteristics of the dataset, so the results obtained in the study may not coincide with those obtained with data from other domains.

Table 3. Precision and recall of the proposed method with diverse sample sizes, where attack size is 17.0 % and filler size is 4.2 %.

N_w	Precision	Recall
400	0.76225	0.49656
500	0.78857	0.54625
600	0.82861	0.67125
700	0.85172	0.78375
800	0.93945	0.88125

5.2 Experimental Results Analysis

Just as aforementioned settings of experiments, we randomly select "weak" data to construct the training and test datasets. Different sizes of selected samples may perform different experimental results, therefore, we generate experiments to examine the impact of sample size in order to choose a comparatively reasonable value. Take AOP attack for example, Table 3 shows the precision and recall with diverse numbers of sample size N_w. We can clearly see that the highest precision and recall corresponds to 800 samples. It means that N_w equal to 800 is a relatively reasonable choice for constructing the "weak" dataset.

For evaluating detection performance, we utilize the metrics of precision and recall. To look at each of these metrics with respect to attack identification, we are interested in how well the detection methods spot attacks. In this paper, we compare the performance of the proposed method with four alternative methods including GM3F-based, iBPM3F-based, iPM3F-based and M3F-based [21]. As shown in Fig. 3(a), the proposed method significantly outperforms all alternative methods. One observation is that both the precision and recall of the proposed method present increasing with the increasing of attack sizes.

To examine the effectiveness of the proposed method in other attacks, we conduct a list of experiments by using 7 different attack models with diverse attack sizes as illustrated in Fig. 3(b). It is noteworthy that recently published attacks such as PIA-AS and PUA-AS, can be partly detected. With the attack sizes increasing, both the precision and recall of the proposed method gradually increasing. In addition, the other observation is that the larger attack sizes, the higher precision and recall. These results might be indicated that more attackers can be sampled with the attack profiles increased in the sampling stage.

6 Conclusion and Further Discussion

"Shilling" attacks and "profile injection" attacks are the main threats in CFRSs. These attack profiles have a good probability of being similar rating details to a large number of genuine profiles in order to make them hard to be detected. In this paper, we developed a new detection approach to spot anomalous users by combining max-margin matrix factorization with Bayesian nonparametrics

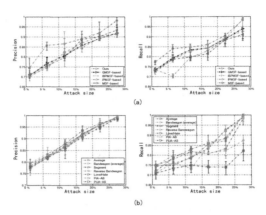

Fig. 3. (a) Precision and recall of alternative methods in AOP attack when filler size is 1.2 % and attack size varies. (b) Precision and recall of the proposed method in 7 different attacks when filler size is 1.2 % and attack size varies.

and outlier detection. Furthermore, matrix factorization based detection method successfully bypasses the challenges of feature extraction and similarity calculation while maintaining an acceptable detection results. Mean prediction errors (MPE) of items and users are respectively calculated by utilizing trained prediction model on test set. Then, comprehensively analyzing the MPE distribution of item is applied to find out anomalous items. Based on the suspected items, the MPE of each user can be directly calculated to reveal abnormal distribution. Finally, users with abnormal distribution can be returned as concerned attackers. Extensive experiments on MovieLens-100K dataset have demonstrated the effectiveness of the proposed method. It is noteworthy that recently published attacks such as PIA-AS, PUA-AS, can be partly detected by the proposed method.

In the future work, we will further investigate the MPE distributions of users and items to improve detection performance. Several crucial issues should be considered: (a) Are there any other better evaluations to measure prediction results? (b) How to well evaluate the distribution of MPE for improving discrimination between users or items? Besides, detecting anomalous ratings in real-world datasets including Amazon, TripAdvisor etc., will be indispensable point in our further work.

Acknowledgments. The research is supported by NSFC (61175039, 61221063), 863 High Tech Development Plan (2012AA011003), Research Fund for Doctoral Program of Higher Education of China (20090201120032), International Research Collaboration Project of Shaanxi Province (2013KW11) and Fundamental Research Funds for Central Universities (2012jdhz08).

References

1. Burke, R., Mobasher, B., Williams, C.: Classification features for attack detection in collaborative recommender systems. International Conference on Knowledge Discovery and Data Mining, pp. 17–20 (2006)
2. Gnnemann, N., Gnnemann, S., Faloutsos, C.: Robust multivariate autoregression for anomaly detection in dynamic product ratings. In: Proceeding WWW 2014 Proceedings of the 23rd International Conference on World Wide Web, pp. 361–372 (2014)
3. Gnnemann, S., Gnnemann, N., Faloutsos, C.: Detecting anomalies in dynamic rating data: a robust probabilistic model for rating evolution. In: KDD 2014, pp. 841–850 (2014)
4. Gunes, I., Kaleli, C., Bilge, A., Polat, H.: Shilling attacks against recommender systems: a comprehensive survey. Artif. Intell. Rev. pp. 1–33 (2012)
5. Mehta, B.: Unsupervised shilling detection for collaborative filtering. Assoc. Adv. Artif. Intell. (2007)
6. Mehta, B., Hofmann, T., Fankhauser, P.: Lies and propaganda: detecting spam users in collaborative filtering. In: IUI 2007: Proceedings of the 12th International Conference on Intelligent User Interfaces, pp. 14–21 (2007)
7. Mehta, B., Nejdl, W.: Unsupervised strategies for shilling detection and robust collaborative filtering. User Model. User-Adap. Inter. **19**, 65–97 (2009)
8. Morid, M., Shajari, M.: Defending recommender systems by influence analysis. Information Retrieval **17**, 137–152 (2014)
9. Saleh, A., Desouky, A., Ali, S.: Promoting the performance of vertical recommendation systems by applying new classification techniques. Knowl.-Based Syst. **75**, 192–223 (2015)
10. Savage, D., Zhang, X., Yu, X., Chou, P.L., Wang, Q.: Detection of opinion spam based on anomalous rating deviation. Expert Syst. Appl. **42**(22), 8650–8657 (2015)
11. Seminario, C., Wilson, D.: Assessing impacts of a power user attack on a matrix factorization collaborative recommender system. In: Florida Artificial Intelligence Research Society Conference (2014)
12. Seminario, C., Wilson, D.: Attacking item-based recommender systems with power items. In: ACM Conference on Recommender Systems, pp. 57–64 (2014)
13. Su, X., Zeng, H., Chen, Z.: Finding group shilling in recommendation system. In: WWW, pp. 960–961 (2005)
14. Wang, Y., Zhang, L., Tao, H., Wu, Z., Cao, J.: A comparative study of shilling attack detectors for recommender systems. In: The 12th International Conference on Service Systems and Service Management (ICSSSM), pp. 1–6 (2015)
15. Williams, C., Mobasher, B., Burke, R.: Defending recommender systems: detection of profile injection attacks. SOCA **1**, 157–170 (2007)
16. Williams, C., Mobasher, B., Burke, R., Bhaumik, R.: Detecting profile injection attacks in collaborative filtering: a classification-based approach. In: Advances in Web Mining and Web Usage Analysis, pp. 167–186 (2007)
17. Wu, Z., Wang, Y., Cao, J.: A survey on shilling attack models and detection techniques for recommender systems. Sci. China **59**(7), 551–560 (2014)
18. Xia, H., Fang, B., Gao, M., Ma, H., Tang, Y., Wen, J.: A novel item anomaly detection approach against shilling attacks in collaborative recommendation systems using the dynamic time interval segmentation technique. Inf. Sci. **306**, 150–165 (2015)

19. Xu, M., Zhu, J.: Discriminative infinite latent feature models. In: Proceedings of the 1st IEEE China Summit and International Conference on Signal and Information Processing (ChinaSIP 2013), pp. 184–188 (2013)

20. Xu, M., Zhu, J., Zhang, B.: Nonparametric max-margin matrix factorization for collaborative prediction. In: Advances in Neural Information Processing Systems (NIPS 2012), pp. 64–72 (2012)

21. Xu, M., Zhu, J., Zhang, B.: Fast max-margin matrix factorization with data augmentation. In: Proceedings of the 30th International Conference on Machine Learning, vol. 28, 3, pp. 978–986 (2013)

22. Zhang, F., Zhou, Q.: A meta-learning-based approach for detecting profile injection attacks in collaborative recommender systems. J. Comput. $7(1)$, 226–234 (2012)

23. Zhang, F., Zhou, Q.: HHT-SVM: an online method for detecting profile injection attacks in collaborative recommender systems. Knowl.-Based Syst. (2014)

24. Zhang, X., Lee, T.M.D., Pitsilis, G.: Securing recommender systems against shilling attacks using social-based clustering. J. Comput. Sci. Technol. **28**, 616–624 (2013)

25. Zhang, Z., Kulkarni, S.: Graph-based detection of shilling attacks in recommender systems. In: IEEE International Workshop on Machine Learning for, Signal Processing, pp. 1–6 (2013)

26. Zhang, Z., Kulkarni, S.R.: Detection of shilling attacks in recommender systems via spectral clustering. In: International Conference on Information Fusion, pp. 1–8 (2014)

27. Zhou, W., Koh, Y.S., Wen, J.H., Burki, S., Dobbie, G.: Detection of abnormal profiles on group attacks in recommender systems. In: Proceedings of the 37th International ACM SIGIR Conference on Research & Development in Information Retrieval, pp. 955–958 (2014)

28. Zhou, W., Wen, J., Koh, Y.S., Alam, S., Dobbie, G.: Attack detection in recommender systems based on target item analysis. In: International Joint Conference on Neural Networks (IJCNN) (2014)

29. Zhou, W., Wen, J., Koh, Y.S., Xiong, Q., Gao, M., Dobbie, G., Alam, S.: Shilling attacks detection in recommender systems based on target item analysis. PloS one (2015)

A Novel Spatial Cloaking Scheme Using Hierarchical Hilbert Curve for Location-Based Services

Ningning Cui[✉], Xiaochun Yang, and Bin Wang

School of Computer Science and Engineering, Northeastern University,
Liaoning 110819, China
willber1988@163.com,
{yangxc,binwang}@mail.neu.edu.cn

Abstract. With the rapid development of positioning and wireless technologies, Location-based Services (LBS) appear anywhere in our daily life. Though LBS brings a great convience to users, the location privacy is vulnerable in many ways (e.g., untrusted LBS Server). To address privacy issues, we propose *HHScloak*, a novel Hierarchical Hilbert Curve Spatial Cloaking algorithm to effectively achieve k-anonymity for mobile users in LBS. Different from existing methods, we take Average Query Density (AQD) into consideration, and generate the Anonymity Set (AS) which satisfies reciprocity and uniformity. Based on the hierarchical method and optimal splitting bucket strategy, our scheme provides a larger cloaking region which guarantees privacy level. Security analysis and experimental evaluation prove that *HHScloak* scheme can perform more efficiently and effectively than other methods.

1 Introduction

With the fast development of the wireless positioning technologies, and the appearing of the mobile devices (e.g., smart phone, panel PC, etc.), more and more location-based services are widely used by people. But the user information may be divulged by untrusted Location-based Provider (LSP) and other compromised entities. In this case, the user suffers from a threat to privacy (e.g., LSP may get the information and location of the user, and then the LSP can infer users hobbies, religious belief and political activities, etc.). At present, Gruteser et al. [1] first introduce k-anonymity into location-based privacy. Cloaking-based location privacy preserving mechanisms are mainly divided into two aspects [2]. One is Trusted Third-party Server (TTP), and the other is Peer to Peer (P2P). But TTP is easy to become a single point of failure and bottleneck of system performance and P2P is vulnerable by colluding attack and is difficult to satisfy privacy level for generating a too small Cloaking Region (CR).

The work is partially supported by the NSF of China for Key Program (No. 61532021), the NSF of China for Outstanding Young Scholars (No. 61322208), and the NSF of China (Nos. 61173031, 61572122, 61272178).

B. Cui et al. (Eds.): WAIM 2016, Part II, LNCS 9659, pp. 15–27, 2016.
DOI: 10.1007/978-3-319-39958-4_2

According to the aforementioned problems, we propose a novel hierarchical Hilbert Curve spatial k-anonymity algorithm that called *HHScloak* algorithm. Different from existing methods, *HHScloak* takes relevant features (e.g., reciprocity, uniformity) of Anonymity Set into consideration, and assigns equal degree of anonymity to each subregion to resist against positions clustering in a single subregion in candidate set. What is more important is the *HHScloak* scheme can generate a more larger Cloaking Region to realize k-anonymity well. The main contributions are as follows.

a. We conduct a novel system architecture which is based on a cooperative group containing an Auxiliary Server and mobile users.
b. We describe an attack model: Subregion Probability Attack Model. Proposed properties (e.g., reciprocity, uniformity) of Anonymity Set can resist to subregion probability attack well.
c. We propose a novel Hierarchical Hilbert Spatial k-anonymity algorithm. We take Average Query Density into consideration, and construct hierarchical index to fill each layer with Hilbert Curve respectively.

The rest of the paper is organized as follows. We give an overview of related work in Sect. 2. We present preliminaries in Sect. 3. *HHScloak* algorithm is proposed in Sect. 4. The security analysis and experimental evaluation results are shown in Sects. 5 and 6 respectively. Section 7 represents conclusion of our paper.

2 Related Work

In order to protect users location privacy, researchers have proposed a great deal of methods which are based on k-anonymity in LBS. These methods are mainly classified into TTP [3–6,11] and P2P [7–10,12] model.

On the one hand, based on TTP, Bamba et al. [3,4] proposed a *Privacy grid*, according to privacy preference profile (P3P) model, which can satisfy users personal measures (e.g., location k-anonymity and location l-diversity). Through a dynamic bottom-up or top-down spatial grid cloaking algorithm, *Privacy grid* makes sure it has a higher anonymization success rate than Quad Grid Cloaking algorithm. Kalnis et al. [5] proposed Hilbert Cloak (*HC*) algorithm who is the first person to introduce the Hilbert Curve to achieve k-anonymity. *HC* yields a mininal Anonymization Spitial Region (ASR) to achieve k-anonymity which satisfies reciprocity. Lee et al. [6] proposed a Cloaking Region generating algorithm by storing adjacent grid information which is not considered by Hilbert Curve. This method may discard the extention of Cloaking Region, but it may lead to all selected locations clustering in a small region to reveal users real location.

On the other hand, based on P2P, Ghinita et al. [7] proposed a *Prive* algorithm. It creates a distributed B^+ tree based on user information, and generates a k-anonymity region by the order of Hilbert Curve in B^+ tree. Lu et al. [8] proposed the Privacy-Area Dummy (*PAD*) location generating approach, which is generated by a virtual grid or a virtual circle. *PAD* algorithm generates a bigger CR which can be controlled for location k-anonymity. Based on side information and entropy metric, Niu et al. [9,10] proposed a privacy-aware location k-anonymity called Dummy Location Selection (*DLS*) and *enhanced DLS*, and also proposed a Fine-Grained Spatial Cloaking Scheme (*FGcloak*). Based on query distribution, *FGcloak* modifies the standard Hilbert Curve to finer grains which has higher query distribution.

3 Preliminaries

3.1 System Architecture

In this paper, a novel system architecture is proposed, as shown in Fig. 1. System contains an Auxiliary Server and mobile users form a cooperative group through WiFi/Bluetooth/3G/4G Ad Hoc network. Mobile users communicate with LBS Server through similar network. Each mobile user exchanges information with Auxiliary Server periodically, and only sends delayed positions to Auxiliary Server without anything else. Auxiliary Server maintains a location table, which stores user positions and query quantities in each grid (e.g., dividing the entire map into n × n grids). Mobile users send a query $Q = (Q_{id}, Q_{POI}, AS, CR,$ Others) to LBS Server, where Q_{id} represents query *id*, which can uniquely identify a query; Q_{POI} represents query interest content; AS represents Anonymity Set; CR represents cloaking region which contains a real user and k-1 dummy users; Others represent time, etc. LBS Server receives queries, deals with them, and returns query results to users.

3.2 Basic Concept

Definition 1 (Average Query Density). *Given a map G, dividing G into $n \times n$ grids, for any grid $G_{ij}, 1 \leq i \leq n, 1 \leq j \leq n, P_{ij}, U_{ij},$ and D_{ij} respectively represents Query Probability, User Density and Average Query Density of G_{ij}, and*

$$P_{ij} = \frac{Count_{ij}(Q_u)}{\sum_{i=1}^{n}\sum_{j=1}^{n} Count_{ij}(Q_u)} \tag{1}$$

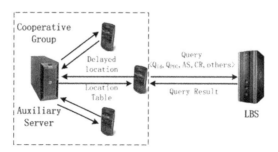

Fig. 1. System architecture

$$U_{ij} = \frac{Count_{ij}(U)}{\sum_{i=1}^{n} \sum_{j=1}^{n} Count_{ij}(U)} \tag{2}$$

where $Count_{ij}(Q_u)$ and $Count_{ij}(U)$ respectively represents the number of user querys and the number of users in G_{ij}

$$D_{ij} = \frac{P_{ij}}{U_{ij}} \tag{3}$$

Definition 2 (Subregion Probability). *Given a region grid C_i, which is the one of i-th layers of the map division. The real user locates in the grid whose AQD is D_u. In C_i, n_i represents the number of grids which has the same AQD with D_u, and SP_i represents subregion probability,*

$$SP_i = \frac{n_i}{N} \tag{4}$$

where N represents all grids which has the same AQD with D_u.

3.3   Attack Model

Generally, different attackers have different background knowledge, and have different abilities to attack. At present, the common attack models including colluding attack model and inference attack model (see Sect. 5). In this paper, we propose a *Subregion Probability Attack* (SPA) model.

Subregion Probability Attack Model. Based on positions distributing in Anonymity Set, for any layer of map division, if existing a subregion probability is much more bigger than others, then attackers can obtain the region of clustering positions and filter out positions in other subregions. What is more, attackers can infer Minimum Containing Anonymity Region (MCAR) of AS, and this disgrades the level of privacy preserving. MCAR represents minimum region which contains maximum subregion probability of AS, and then attackers can further infer the real position.

3.4   Basic Idea and Motivation

Property 1 (Reciprocity) [5]. *Given a mobile user U, U sends a query $Q = (Q_{id}, Q_{POI}, AS, CR, Others)$ to LBS Server, with a cooperative group, and U generates an AS. If AS satisfies: (i) AS contains U and other k-1 mobile users; (ii) with the same anonymity degree k, each user in AS generates the same AS. Then we call AS satisfies reciprocity.*

For example, *Hilbert Cloak* [5] satisfies reciprocity, but *Interval Cloak* [2] and *Casper* [13] algorithms do not satisfy reciprocity. However, *Hilbert Cloak* still exists a few problems, (i) Kalnis et al. [5] indicated that two adjacent points in the transformed space are likely to be close in the original space. Therefore it may lead that the CR is too small to reveal position information. (ii) It does not take AQD into consideration. As shown in Fig. 2, the map is divided into 8×8 grids, and different grid types represent different AQD, generating two anonymity sets AS_1 (e.g., yellow region) and AS_2 (e.g., red region). In the case of the similar size of CR, with the background knowledge of AQD, attackers may filter out the positions which has low AQD, therefore this way disgrades privacy preserving degree. However, different positions have the same AQD in AS_2, and attackers can not distinguish the real position from dummy positions according to background knowledge. Therefore it realizes location-privacy preserving.

Based on aforementioned issues, we propose a *Basic* method. As shown in Fig. 3. The real user u_8 locates in subregion II where the AQD of the grid is D_u. *Basic* method selects the grids which contains red point having the same AQD with u_8 and randomly selects a position in these grids as dummy position. *Basic* regards the selected positions $\{u_1, u_2, \cdots, u_{19}, u_{20}\}$ as a candidate set. According to anonymity degree k, Basic divides positions in candidate set into buckets, and based on relative location of the real user in bucket, *Basic* selects dummy locations in each bucket which corresponds to relative location of the real user. The anonymity set satisfies reciprocity and generates a more larger CR. For $k = 10$, generating an AS is $\{u_2, u_4, u_6, u_8, u_{10}, u_{12}, u_{14}, u_{16}, u_{18}, u_{20}\}$. Though *Basic* realizes k-anonymity well, it still exists a certain problem.

Property 2 (Uniformity). *Given a region grid C_i, which is the one of i-th layers of the map division (see Sect. 4.1). C_i is divided into four subregions $C_{i1}, C_{i2}, C_{i3},$ and C_{i4}. As the anonymity degree k' of C_i, if each division of the map satisfies, (i) $\mid C_i \mid = \sum_{j=1}^{4} \mid C_{ij} \mid = k', 1 \leqslant j \leqslant 4, \mid C_i \mid$ represents anonymity degree of the corresponding region; $\mid C_{ij} \mid$ represents anonymity degree of j-th subregion of C_i. (ii) $0 \leqslant \parallel C_{ij} \mid - \mid C_{ij'} \parallel \leqslant \sigma, 1 \leqslant j' \leqslant 4$. Then we call AS satisfies uniformity.*

As shown in Fig. 3, the anonymity set of subregion I is u_2, and the anonymity set of subregion II is $\{u_4, u_6, u_8, u_{10}, u_{12}\}$. Note that, the subregion probability of AS_{II} is 5 times bigger than AS_I, that is AS does not satisfy uniformity. Attackers can do subregion probability attack through background knowledge, filter out

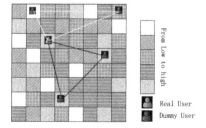

Fig. 2. Problem of existing approach (Color figure online)

Fig. 3.   Basic   method (Color figure online)

positions in other regions and generage MCAR (e.g., red rectangle region) where have a higher probability containing the real position. To address this problem, a hierarchical spatial cloaking algorithm is proposed.

4 Hierarchical Hilbert Curve Spatial Cloaking Scheme

4.1 Hierarchical Algorithm Based on Grid

In hierarchical algorithm, the entire map is divided into grids of size $2^n \times 2^n$, n is the system-difined parameter, as shown in Fig. 4 (e.g., $n = 4$). In the system, Auxiliary Server collects users position information, calculates statistic information in the grid, and updates the location table. Location table contains many attributes like grid id, grid position, grid AQD and user position, which is the form of $([i, j], [\langle X_s, Y_s \rangle, \langle X_t, Y_t \rangle], D_{ij}, [u_k, \langle x_k, y_k \rangle])$. Grid position denotes coordinates of points of left-bottom and right-up. User position denotes a list of k positions which have the most queries.

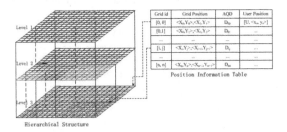

Fig. 4. Hierarchical structure

The hierarchical algorithm is as Algorithm 1. First, through the mapping methods, the algorithm maps the real location into a grid, matchs grid id where the real user located in with grid id in location table, and obtains AQD D_u which the grid id is matchable (Line 1). The first layer division (Line 3), the algorithm divides the map region G of $2^n \times 2^n$ into four subregions G_i of $2^{(n-1)} \times 2^{(n-1)}$, $1 \leqslant i \leqslant 4$, and calculates quantity of grids n_i which has the same D_u with the grid where the real user located in and corresponding subregion probability SP_i (Line 4). If the results satisfy $0 \leqslant | SP_i - SP_{i'} | \leqslant \delta_1, i, i' \in [1, 4]$, and $i \neq i', \delta_1 \in [0, 1]$, then the algorithm fills the map region of $2^n \times 2^n$ in Hilbert Curve directly (Lines 6–7). Otherwise the algorithm does the second layer division, and calculates corresponding n_{ij} and SP_{ij} in each subregion $G_{ij}, 1 \leqslant j \leqslant 4$ of region G_i in sequence. If $0 \leqslant | SP_{ij} - SP_{ij'} | \leqslant \delta_2, j, j' \in [1, 4]$, and $j \neq j', \delta_2 \in [0, 1]$, then the algorithm fills the corresponding subregion of $2^{(n-1)} \times 2^{(n-1)}$ in Hilbert Curve directly. Otherwise the algorithm continues to judge subregion of region G_{ij}. If the number of divided layers h is equal with h_{max}, then the hierarchical algorithm terminates (Lines 9–12).

Algorithm 1. Hierarchical Algorithm

Data: Map G, Location Table T, Threshold Set $\{\delta_i\}$, h_{max}
Result: Hierarchical Index
1 Calculate grid id and AQD D_u where the real user located in ;
2 Hierarchical_algorithm (G, T, δ_i, h, D_u);
3 divide the map region G of $2^n \times 2^n$ into four subregions G_i of $2^{(n-1)} \times 2^{(n-1)}$;
4 calculate the corresponding subregion probability SP_i ;
5 **for** *judge subregion grid in sequence* **do**
6   **if** *difference of arbitrary two subregion probabilities is no more than δ_i* **then**
7   fill the region in Hilbert curve directly;
8   **else**
9   **if** $h < h_{max}$ **then**
10   Hierarchical_algorithm $(G_i, T, \delta_{(i+1)}, h+1, D_u)$;
11   **else**
12   break;

13 **return** *Hierarchical Index*

4.2   Hierarchical Hilbert Filling Curve Anonymization Algorithm

According to hierarchical algorithm (as shown in Fig. 5), we can easily obtain Hilbert filling Curve of subregion I, II, III and IV (called H_1, H_2, H_3 and H_4). Dummy positions selection algorithm is described as follows. Specifically, we first divide the degree of anonymity k into each subregion, that is subregion degree of anonymity is $k_{sub} = \lfloor k/4 \rfloor$. If k can not be divided exactly, as subregion probability order from big to small, we make $k\%4$ subregion degree of anonymity plus 1, that is $k_{sub} = \lfloor k/4 \rfloor + 1$. As shown in Fig. 5, the subregion degree of anonymity of I, II, III and IV is 2, 3, 2 and 3, the anonymity set satisfies uniformity. Then, as Basic approach, we split the candidate set of each subregion into corresponding k_{sub}-bucket on Hilbert Curve of H_1, H_2, H_3 and H_4. Based on relative location of the real user in bucket, we select dummy locations in each bucket which corresponds with relative location of the real user. As shown in Fig. 6, $k = 5$, respectively, subregion degree of anonymity is 1, 2, 1 and 1, the relative bucket position of the real user u_3 is 3, then Basic selects dummy positions in each bucket which has relative position of 3.

However, if predefined k is 10, as Basic approach, subregion degree of anonymity is 2, 3, 2 and 3, and the relative position of the real user u_3 in bucket is 3. Note that we can not find relative position of 3 in each bucket of subregions of I and III, so the anonymity set can not satisfy reciprocity.

Based on aforementioned issue, we propose an optimal splitting bucket strategy. As Definition 1, we can know the density in each grid is uniform distribution,

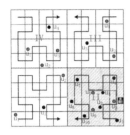

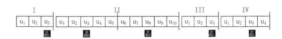

Fig. 5. Hierarchical Hilbert Curve

Fig. 6. Spliting of Hierarchical Hilbert Curve

and after dividing a grid into four sub-grids, each sub-grid has the same AQD with the grid. Specifically, we first calculate the maximum number of grids n_{max} in all candidate sets of subregion dummy positions. Then, as the order of Hilbert filling Curve, in each candidate set of subregion dummy positions except the one which contains n_{max}, we subdivide each grid until the number of grids in each candidate set of subregion dummy position is similar to n_{max}. As shown in Fig. 7, subregion II contains the maximum number of grids n_{max} (e.g., $n_{max} = 10$), and subregion I contains the number of candidate grid n (e.g., $n = 3$). From u_1 in subregion I, we subdivide the grid which contains u_1 into u_{11}, u_{12}, u_{13} and u_{14}, then subdivide the grid which contains u_2, until subregion I containing the number of candidate grids n (e.g., $n = 9$) is similar to n_{max}. Subregion III and IV continue to subdivide grids in sequence like I. After subdivision, the number of candidate grids in each subregion is 9, 10, 9 and 10, based on the formula of k_{sub}, we can obtain that each subregion degree of anonymity is 2, 3, 2 and 3. As shown in Fig. 7, We split each subregion positions of candidate set into buckets. The relative position of the real user u_3 in bucket is 3, then we select dummy positions in each bucket which has relative position of 3.

In reality, selecting dummy positions in a grid is random, but it is probable to select a position where there is no user or no LBS. In this paper, we provide a replacement scheme using a structure of cooperative group. Each user cooperates with other users who are in the same cooperative group to maintain a location table, and user position denotes a list of k positions which have the most queries in a period. If the grid in candidate set has not been subdivided, then the scheme

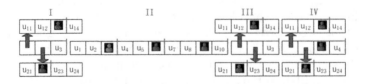

Fig. 7. Optimal splitting bucket scheme

Algorithm 2. Optimal Spliting Bucket Dummy Selecting Algorithm

---

Data: Hierarchical Index; Anonymity degree k; AQD D_u
Result: Anonymity Set
1   Dummy_Position Selection (Hierarchical Index,k, D_u);
2   Calculate the maximum number of grids n_{max} in all subregion dummy position candidate sets;
3   **for** *each subregion except containing n_{max}* **do**
4   **for** *candidate grids as Hilbert Curve order in subregion* **do**
5   **if** *the number of grids in subregion $n \cong n_{max}$* **then**
6   break;
7   **else**
8   subdivide the grid, $n = n + 3$;

9   Calculate k_{sub} in each subregion;
10   Split each subregion positions of candidate set into buckets;
11   Calculate dummy positions, and obtain Anonymity Set;
12   **return** *Anonymity Set*

---

directly selects the position which has the most queries as dummy position in the grid. Otherwise, the scheme selects the position from top-k positions which has the most queries in corresponding sub-grid as dummy position. Though this approach costs more storage space and communication, it guarantees the selected dummy position is a real query position.

5   Security Analysis

In this section, we analyse the security of the proposed *HHScloak* algorithm.

Resistance Against Colluding Attack. Generally, colluding attack means that attackers compromise with other collaborative users or Auxiliary Server, even compromise with LBS Server, and obtain the real user location information.

Theorem 1. *HHScloak is resistant against colluding attack.*

Proof. We prove this theorem in two aspects. First, we analyse the system architecture. Our paper organizes a novel cooperative group structure. Each user does not exchange information with each other, even though attackers compromising with users can not get the real user location information. Moreover, Auxiliary Server only stores users delayed positions periodically, and does not know current user position. Second, aspect of selecting dummy positions. Our proposed method is based on a location table, and attackers can compromise with Auxiliary Server to get the table. But the positions in AS have the same AQD, and attackers can not filter out dummy positions from background knowledge, that is the probability of recognizing real position is no more than $1/k$.

Resistance Against Inference Attack. Inference attack means attackers possess certain knowledge (e.g., historical position, query content, anonymity algorithm, etc.). Based on the submitted AS, attackers infer users real position. Generally, attackers compromise with LBS Server to get background knowledge.

Theorem 2. *HHScloak is resistant against inference attack.*

Proof. Based on background knowledge, attackers obtain submitted AS, and execute anonymity algorithm at each position of AS in sequence to distinguish different positions. Our proposed *HHScloak* k-anonymity algorithm satisfies reciprocity, that is any position in AS generates the same AS. Attackers can not distinguish the real position from other k-1 dummy positions.

Resistance Against Subregion Probability Attack. (*SPA* see Sect. 3.3)

Theorem 3. *HHScloak is resistant against subregion probability attack.*

Proof. The reason of being vulnerable to *subregion probability attack* is positions in AS clustering in a small subregion, which results in the *subregion probability* is much more bigger than other subregions. Attackers may filter out positions in other subregions based on MCAR. Our proposed *HHScloak* algorithm satisfies uniformity, that is each *subregion probability* is similar, and will not appear that existing a *subregion probability* in AS is much more than others. Therefore, attackers can not distinguish the real position through *subregion probability attack*.

6   Performance Evaluations

In our experiments, we use the map of city Aalborg which contains 129680 mobile users in the size of 60 km × 60 km. We divide the map into 64 × 64 grids that the size of each grid is 937.5 m × 937.5 m. In every minute, we randomly choose 10 % mobile users to send location-based queries to LBS Server and at the same time the position table T stores every delayed query record. After 2 h simulation, the Auxiliary Server records all queries which only contains users delayed location information from the selected users, and exchanges the table T with every user (we assume the position information in table T does not change any more in the period).

We set several parameters to evaluate the performance of anonymity approachs. Specifically, k denotes anonymity degree (from 3 to 30 in evaluation) and h denotes the maximum of hierarchy, which are determined by users. Cloaking Region shows the minimum area contains all k-1 dummy positions and the real position. Moreover we use Entropy and Variance to evaluate the uncertainty of AS. Entropy and Variance are defined respectively as follows.

$$Entropy = -\sum_{i=1}^{k} \frac{D_i}{\sum D_i} \times \log_2 \frac{D_i}{\sum D_i} \tag{5}$$

$$Variance = \sum_{i=1}^{k}(D_i - \mu)^2 \qquad (6)$$

where D_i denotes the AQD of the grid where the selected user located in. μ denotes mean value of AQD in AS.

We analyse the performance of *HHScloak* from two aspects. First, we compare our proposed *HHScloak* with several recently proposed methods. *Hilbert* denotes Kalnis et al. [5] proposed Hilbert Cloak to achieve k-anonymity. *FGcloak* represents Niu B et al. [9] proposed a fine-gained modified Hilbert Cloak to achieve k-anonymity. The *Optimal* represents ideal condition to complete the k-anonymity. Second, we analyse the relationship between parameter h and the other parameters.

6.1 K vs Cloaking Region, Entropy and Variance

The results are respectively shown in Fig. 8(a) represents k vs Cloaking Region, we can see that the *optimal* scheme achieves ideal result no matter what the value of k is, its Cloaking Region reachs the entire map (size of $3.6 \times 10^9 m^2$). *Hilbert* scheme can not reach a large region, and its maximum is still very small because the approach splits candidate users into buckets in a restricted region (adjacent region). Based on fine-gained Hilber Curve, *FGcloak* can reach a larger region than aforementioned two schemes, but it does not take subregion information of every layer into consideration and just considers the side information of individual grids. Observe that no matter what the value of k is, *HHScloak* can achieve a more larger Cloaking Region, even $k = 3$. Figure 8(b) represents k vs Entropy, Entropy represents uncertainty of AS, the larger value of Entropy is, the higher of uncertainty of AS is, and AS can achieve a better k-anonymity. We know that the same with *Optimal* scheme, *HHScloak* and *Basic* have the equal value of AQD in each grid in AS. So the value of Entropy in these three schemes is $\log_2 k$. However, in *Hilbert* scheme and *FGcloak* scheme, the value of AQD in each grid in AS is uncertain and has lower probability to achieve optimal condition. Figure 8(c) represents k vs Variance, similar to Entropy, Variance also represents uncertainty of AS by the way of statistics. But contrast to Entropy, the lower value of Variance is, the higher of uncertainty of AS is. Note that the mean value of AQD of *HHScloak* and *Basic* scheme in AS is equal to *Optimal*

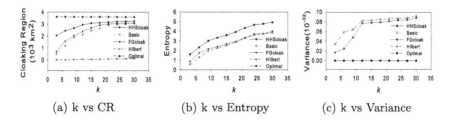

(a) k vs CR   (b) k vs Entropy   (c) k vs Variance

Fig. 8. The effect of HHScloak algorithm, h = 3

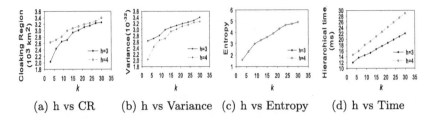

Fig. 9. The effect of h in HHScloak algorithm

condition in a large extent, so the value of Variance tends to be zero, but in *Hilbert* and *FGcloak* scheme, the difference between mean value and them is obvious. Along with the increasing value of k, the uncertainty of different grids become more distinct and the curve unexpectedly rise up.

6.2 H Vs Cloaking Region, Entropy, Variance and Hierarchical Time

The relationship between these several parameters in *HHScloak* is shown in Fig. 9. We respectively set $h = 3$ and $h = 4$ to evaluate the effect of different number of layers. Figure 9 represents when h is 3 or 4, the system creates a hierarchical structure of 3 layers or 4 layers and h is 3 which the minimum subregion is 16×16 and h is 4 which the minimum subregion is 8×8. Then we can note that when h is 4, it has a higher resolution of Hilbert Curve traversal. As show in Fig. 9(a), when h is 4, the Cloaking Region is larger than h is 3, because h is 4 has more exact division. But along with the increasing value of h, the cost becomes higher. Figure 9(d) shows that h is 3 has a lower cost of creating hierarchical time than h is 4. Figure 9(b) and (c) shows that the hierarchical influence on Variance and Entropy is very unconspicuous. Though the curve of h is 3 and h is 4 in Fig. 9(b) is different, the magnitude is 10^{-32} that we can ignore the influence.

7 Conclusion

In this paper, we studied the problem of *subregion probability attack*. Based on this problem, we proposed a novel Hierarchical Hilbert Curve Spatial Cloaking Scheme which satisfies reciprocity and uniformity. Meanwhile we also proposed a optimal splitting bucket strategy during selecting dummy positions. Aforementioned works are based on a system structure of cooperative group which overcomes the demerits of TTP and P2P. The security analysis and experiment results have proved our proposed method can achieve an effective performance.

Acknowledgments. The work is partially supported by the NSF of China for Key Program (No. 61532021), the NSF of China for Outstanding Young Scholars (No. 61322208), and the NSF of China (Nos. 61173031, 61572122, 61272178).

References

1. Gruteser, M., Grunwald, D.: Anonymous usage of location-based services through spatial and temporal cloaking. In: International Conference on Mobile Systems, Applications, and Services (MobiSys), pp. 31–42. CA (2003)
2. Rao, U.P., Girme, H.: A novel framework for privacy preserving in location based services. In: Fifth International Conference on Advanced Computing and Communication Technologies (ACCT), pp. 272–277. IEEE (2015)
3. Bamba, B., Liu, L.: PRIVACYGRID: Supporting Anonymous Location Queries in Mobile Environments. J. GEORGIA INST OF TECH ATLANTA COLL OF, COMPUTING, pp. 7–17 (2007)
4. Bamba, B., Liu, L., Pesti, P., et al.: Supporting anonymous location queries in mobile environments with privacygrid. In: International Conference on World Wide Web, pp. 237–246. ACM (2008)
5. Kalnis, P., Ghinita, G., Mouratidis, K., et al.: Preventing location-based identity inference in anonymous spatial queries. IEEE Trans. J. Knowl. Data Eng. **19**, 1719–1733 (2007)
6. Lee, H.J., Hong, S.T., Yoon, M., et al.: A new cloaking algorithm using Hilbert curves for privacy protection. In: ACM Sigspatial International Workshop on Security and Privacy in GIS and LBS, pp. 42–46. ACM (2010)
7. Ghinita, G., Kalnis, P., Skiadopoulos, S.: PRIVE: anonymous location-based queries in distributed mobile systems. In: WWW 2007: International Conference on World Wide Web, pp. 371–380. IEEE (2007)
8. Lu, H., Jensen, C.S., Yiu, M.L.: PAD: privacy-area aware, dummy-based location privacy in mobile services. In: The Seventh ACM International Workshop on Data Engineering for Wireless and Mobile Access, pp. 16–23. ACM (2008)
9. Niu, B., Li, Q., Zhu, X., et al.: A fine-grained spatial cloaking scheme for privacy-aware users in Location-Based Services. In: Computer Communication and Networks (ICCCN), pp. 1–8. IEEE (2014)
10. Niu, B., Li, Q., Zhu, X., et al.: Achieving k-anonymity in privacy-aware location-based services. In: INFOCOM, pp. 754–762. IEEE (2014)
11. Gkoulalas-Divanis, A., Kalnis, P., Verykios, V.S.: Providing K-Anonymity in location based services. J. ACM SIGKDD Explor. Newsl. **12**, 3–10 (2010)
12. Hossain, A., Hossain, A.A., Jang, S.J., et al.: Privacy-aware cloaking technique in location-based services. In: IEEE First International Conference on Mobile Services, pp. 9–16. IEEE (2012)
13. Mokbel, M.F., Chow, C.Y., Aref, W.G.: The new Casper: query processing for location services without compromising privacy. In: The 32nd International Conference on Very Large Data Bases, pp. 763–774. VLDB (2006)

Efficient Privacy-Preserving Content-Based Image Retrieval in the Cloud

Kai Huang, Ming Xu$^{(\boxtimes)}$, Shaojing Fu, and Dongsheng Wang

College of Computer, National University of Defense Technology, Changsha, China
{kai.huang,xuming}@nudt.edu.cn, shaojing1984@yahoo.cn,
wdsh2011@gmail.com

Abstract. Cloud storage systems are increasingly being used to host personal or organizational data of critical importance, especially for image data that needs more storage space than ordinary data. While bringing in much convenience, existing cloud storage solutions could seriously breach the privacy of users. Encryption before outsourcing images to the cloud helps to protect the privacy of the data, but it also brings challenges to perform image retrieval over encrypted data. To address this issue, considerable amount of searchable encryption schemes have been proposed in the literature. However, most existing schemes are either less secure or too computation and communication intensive to be practical. In this paper, we propose an efficient privacy-preserving content-based image retrieval scheme. We first convert the high-dimensional image descriptors to compact binary codes, and then adapt the asymmetric scalar-product-preserving encryption (ASPE) to ensure the confidentiality of the sensitive images. The security analysis and experiments show the security, accuracy and efficiency of our proposed scheme.

1 Introduction

With the popularity of digital cameras and smart phones and the development of mobile internet, which makes the acquisition and transmission of image data cheaper and faster, the amount of image data has experienced exponential growth over the past decade. For example, to improve the management of urban flows and allow for real time responses to challenges, many governments are making efforts to build the smart cities, and the most important part of which is the monitoring and surveillance system generating countless images day and night. Meanwhile, image data is usually of high resolution and therefore needs more storage space than ordinary data. Consequently, the large amount of images are far beyond the storage and processing capacity of a small company or a resource-constained organization.

Advances in cloud computing have prompted many customers to outsource their storage and computing needs. By moving their image data to the cloud, customers can avoid the costs of building and maintaining a private storage infrastructure. However, such image data as monitoring and surveillance images usually contains plenty of confidential or sensitive information. Outsourcing the

© Springer International Publishing Switzerland 2016
B. Cui et al. (Eds.): WAIM 2016, Part II, LNCS 9659, pp. 28–39, 2016.
DOI: 10.1007/978-3-319-39958-4_3

data to the cloud means that the data is outside its control and could potentially be granted to untrusted parties. This poses a dramatic threat to the privacy of users. To mitigate this threat, more and more users tend to encrypt their image data before outsourcing. While preserving privacy of the image data on the cloud, it brings new challenges to the image retrieval problem, that is, the problem of searching for images in the encrypted domain.

Recently, numerous schemes have been designed to implement text document search over the encrypted data, which perform either Boolean search [1,2] to find the exact match between the query strings and the index strings or similarity search [3,4] to return index strings that are similar to the query strings. Although they can be applied to keyword-based image retrieval, they are not suitable for privacy-preserving content-based image retrieval (PCBIR) which needs to calculate the distances among the encrypted high dimensional vectors. Therefore, Lu et al. [5]proposed several randomization techniques to protect the image features but approximately preserve the distance between the randomized feature vectors. Although the proposed randomization techniques are computationally efficient and can obtain high search accuracy, such transformation is shown to be not secure in practice due to the distance-preserving property. The weakness of distance-preserving encryption comes from the fact that the distance information could be recovered from the encrypted databases [6]. To enhance security, the semantically secure homomorphic encryption schemes such as Paillier and Gentry, which allow computations in the encrypted domain directly, have attracted the most attention in the literature [7–9]. Through homomorphic encryption, the similarity computation can be outsourced to the cloud without exposing the data and at the same time the search accuracy can be retained. Unfortunately, the currently established homomorphic encryption techniques are too computation and communication intensive to be practical.

To address this issue, in this paper, we propose an efficient privacy-preserving content-based image retrieval scheme which allows users to retrieve the encrypted images outsourced to the cloud server. We first adopt the BPBC (bilinear projection-based binary codes) [10] to convert the high-dimensional VLAD descriptors [11] to compact binary codes. The similarity between the images is subsequently measured through the asymmetric distance between the reduced query vector and the compact points in the database. Then we adapt the asymmetric scalar-product-preserving encryption to design PCBIR to achieve the privacy requirements in the cloud environment. We implement PCBIR over real image repositories and the experiment results show that PCBIR achieves high search accuracy with low overhead.

This paper makes the following contribution: (1) we propose an efficient construction for privacy-preserving content-based image retrieval over encrypted data; (2) we formally prove the security of our proposed scheme; (3) we experimentally show that our scheme provides increased performance and lower overhead.

The rest of this paper is organized as follows: Section 2 introduces the formalization of the problem and necessary preliminaries. In Sect. 3, we describe the detail of our proposed PCBIR. Security analysis and performance evaluation are presented in Sects. 4 and 5 respectively. We conclude the paper in Sect. 6.

2   Problem Formulation

2.1   System and Threat Model

In our scheme, we consider three main entities: the cloud server, the image data owner and the image data user, as shown in Fig. 1. Generally, images are outsourced to the cloud server by the owner, and the authorized users are allowed to access the images and retrieve the required ones. For privacy protection, the owner should first generate secret keys and encrypt the images before outsourcing. To facilitate the image retrieval, the owner should also construct the secure indexes which are submitted to the cloud along with the encrypted data. Any user who has the secret keys is permitted to access and retrieve the required image data. The secret keys can be obtained through the state-of-the-art access control policy and key distribution protocol. During the retrieval process, the authorized user first generates a secure trapdoor according to a query image. Then, he sends the trapdoor to the cloud server to perform the image retrieval over the encrypted data. Upon receiving the retrieval request from the user, the cloud server returns the required images to the user according to the similarity of the images.

Similar to previous works [12–14] found in the literature, we consider an honest-but-curious cloud environment in our threat model, which means that the cloud server honestly follows the designated protocol while curiously infers some private information of interest based on the data stored and processed on it. Meanwhile, both the authorized and unauthorized data user are semi-trusted, since they may eavesdrop on other users' image data.

2.2   Design Goals

Our goal is to design a secure and well functioning image retrieval scheme over encrypted data. Specifically, we have the following goals:

Image Similarity Retrieval. Our primary goal is to support the similarity retrieval of the images, that is to allow the users to retrieve not only the duplicate images but also the most similar images that have common objects or backgrounds.

Privacy-Preserving. The main concern in our design is to protect the privacy of the image data and prevent the cloud server or the malicious users from learning additional information of the owner, which includes the original image data, the index and the query image data.

Efficiency. Compared with existing privacy-preserving image retrieval schemes, our design should take into full account the communication and computation overhead and achieve an effective but efficient privacy-preserving image retrieval scheme.

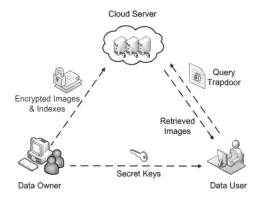

Fig. 1. System Model of Our Scheme

Adaptability. Our design should be applicable to different image representations derived from other feature extraction techniques according to the user requirements.

2.3 Preliminaries

In the following, we present some basic terminologies and algorithms that are adopted in our scheme.

Vector of Locally Aggregated Descriptors (VLAD). It is a first order extension to the popular Bag-of-Visual-Words (BoVW) model which is local features based technique for content-based image retrieval [11] and is considered to be a reliable approach to bridge the gap between low-level visual features and high-level image contents. For each image, the local features such as SIFT [15] are extracted and encoded to the corresponding visual words. Different from the BoVW model, which involves simply counting the number of descriptors associated with each cluster in the visual vocabulary and creating a histogram vector for each set of descriptors from an image, VLAD accumulates the residual of each descriptor with respect to its assigned cluster and can achieve better discrimination for the classifier.

Specifically, let $C = \{c_1, c_2, \cdots, c_k\}$ be the k clusters generated through k-means. Each d-dimensional local descriptor x from an image is associated with its nearest cluster. Then for each cluster c_i, we accumulate the difference $x - c_i$, where $c_i = NN(x)$. Representing the VLAD vector for each image by v, we have $v_{ij} = \sum_{x|x=NN(c_i)} (x_j - c_{ij})$, for $i = 1, 2, \cdots, k$ and $j = 1, 2, \cdots, d$. Often, the vector v will be power-normalized or L_2 normalized.

Then, Euclidean distance can be adopted to compute the distance of the VLAD vectors and find the k-nearest neighbor (kNN) of the query image.

Secure kNN Computation. In kNN computation, distance-preserving transformation is proved to be not secure and exact distance computation is not necessary in practice. The asymmetric scalar-product-preserving encryption (ASPE) is distance-recoverable and supports secure and accurate kNN query computation. For details, please refer to [6]. In this paper, we adapt the ASPE to implement the privacy-preserving content-based image retrieval.

3 Our Proposed Scheme

In this section, we present the design of our privacy-preserving content-based image retrieval scheme based on the secure kNN computation. To improve the search efficiency and accuracy, we first describe a method that converts the high-dimensional descriptors to compact binary codes and an asymmetric distance measure. Then we present the details of our privacy-preserving content-based image retrieval scheme.

3.1 BPBC Based Vector Quantization

As introduced in Sect. 2.3, an image can be represented as a $(d * k)$-dimensional vector using VLAD. Then, Euclidean distance between the VLAD descriptors is used for the similarity search. Note that the descriptors should be high-dimensional when the number of clusters k is large and computation over them becomes quite expensive. There are a number of methods to quantize high-dimensional descriptors, including Locality Sensitive Hashing (LSH), Spectral Hashing (SH), and Product Quantization (PQ) [16]. Considering the trade-off between search efficiency and storage requirement [10], in our scheme, we employ the bilinear projection-based binary codes (BPBC) to convert VLAD descriptors into compact binary codes.

For a descriptor $p \in \mathbb{R}^D$, it can be reshaped into a matrix $P \in \mathbb{R}^{d_1 \times d_2}$, $D = d_1 d_2$. Then, two random orthogonal matrices $R_1 \in \mathbb{R}^{d_1 \times d_1}$ and $R_2 \in \mathbb{R}^{d_2 \times d_2}$ are applied to P:

$$H(P) = vec(sgn(R_1^T P R_2))$$

where vec($\cdot$) denotes column-wise concatenation and $H(P)$ is the generated D-dimensional binary string.

Further, reduced-dimension codes can be learned through dimensionality reduction. Random orthogonal projection matrices $R_1 \in \mathbb{R}^{d_1 \times c_1}$ and $R_2 \in \mathbb{R}^{d_2 \times c_2}$ are utilized to produce codes of size $c = c_1 \times c_2$, where $c_1 < d_1$ and $c_2 < d_2$. In this way, we can convert the D-dimensional descriptors of the images to reduced c-dimensional binary codes.

3.2 Asymmetric Distance Based Similarity Measure

At retrieval time, the cloud server determines the similarity between the points in the database and the query point through distance computation. Although both the query image and the images in the cloud can be represented by compact binary codes, to improve the search accuracy, we use asymmetric distance in our scheme. That is, the points in the database are quantized but the query point is not. For the database points $p \in \{-1, +1\}^c$ and a query point $q \in \mathbb{R}^c$, the Euclidean distance between p and q is:

$$d(p, q) = \sqrt{\|p\|_2^2 + \|q\|_2^2 - 2p^T q}$$

Since $p \in \{-1, +1\}^c$, we have $\|p\|_2^2 = c$. And q is on the query side, thus we only need to compute $p^T q$.

In the following, we will adopt the secure inner product scheme to protect the points p in the database and the query point q without revealing the exact distance information between p and q.

3.3   The Privacy-Preserving CBIR

The process of our scheme can be divided into four phases: (1) Key generation phase which is designed to generate the needed secret keys for encryption and decryption of the images; (2) System setup phase which is conducted by the image data owner to generate not only the encrypted image sets that will be outsourced to the cloud server but also the secure indexes of the images; (3) Trapdoor generation phase where the user generates query trapdoors and initiates search sessions with the cloud server; (4) Retrieval phase which is performed by the cloud server to search for the required images and return them to the user in a ranked manner.

Accordingly, our scheme consists of a tuple of four basic polynomial time algorithms as follows:

K ← KeyGen(1^λ) : given the input security parameter λ, output a set of secret keys K.

I ← BuildIndex(Δ, K) : given the image dataset Δ and the secret key K, output the secure index I.

T ← TrapdoorGen(Q, K) : given a query image Q and the secret key K, output the query trapdoor T.

R ← Search(T, I) : given the secure trapdoor T and the index I, output the matched result R.

To enhance the security, we perform a pseudo-random permutation $\pi(\cdot)$ on the binary codes of the data points $p \in \{-1, +1\}^c$. Moreover, we add r dummy dimensions to extend the c-dimensional permuted data points to $(c+r)$-bit. The details are as follows:

Key Generation Phase. The image data owner first runs the KeyGen algorithm to generate the secret keys. The KeyGen algorithm takes as input a security parameter λ and outputs a set of secret keys $K = \{M_1, M_2, S, sk\}$, where M_1 and M_2 are two $(c+r) \times (c+r)$ invertible matrices, S is a $(c+r)$-dimensional bit vector which functions as a splitting indicator, and sk is a symmetric secret key to encrypt the original image data set.

System Setup Phase. Before outsourcing, the owner represents the images as compact binary codes through aforementioned BPBC. Then the owner generates the searchable secure index using the BuildIndex algorithm. We first perform a pseudo-random permutation $\pi(\cdot)$ on p. After that, the permutated vector $\bar{p}$ is extended into $\tilde{p}$ that is $(c + r)$-bit. By choosing t out of r dummy dimensions, the corresponding entries in $\tilde{p}$ are set to 1. And $\tilde{p}$ is further split into two d-dimensional points $\tilde{p}_a$ and $\tilde{p}_b$ in such a way: for $j = 1$ to $c + r$, if the j-th bit of S is 1, $\tilde{p}_a[j]$ and $\tilde{p}_b[j]$ are set to two random numbers so that their sum is equal to $\tilde{p}[j]$, that is $\tilde{p}[j] = \tilde{p}_a[j] + \tilde{p}_b[j]$; if the j-th bit of S is 0, $\tilde{p}_a[j]$ and $\tilde{p}_b[j]$ are set to the same value as $\tilde{p}[j]$, that is $\tilde{p}[j] = \tilde{p}_a[j] = \tilde{p}_b[j]$. Finally, the split data vector pair $\tilde{p}_a$ and $\tilde{p}_b$ is encrypted using M_1^T and M_2^T as $\hat{p}_a = M_1^T \tilde{p}_a$, $\hat{p}_b = M_2^T \tilde{p}_b$. After that, we can outsource the encrypted image data set and the encrypted indexes to the cloud server.

Trapdoor Generation Phase. In order to retrieve the required image data, the user runs the TrapdoorGen algorithm to generate a query trapdoor. Since we use the asymmetric distance, the query image is represented as a reduced c-dimensional descriptor q, which is permuted to be $\bar{q}$ with the same pseudo-random permutation $\pi(\cdot)$. And $\bar{q}$ is extended into $\tilde{q}$ that is $(c + r)$-dimensional. For $j = c + 1$ to $c + r$, the entry in $\tilde{q}$ is set to a random number ε_i. Then $\tilde{q}$ is split in the same way as the index except that the split process is the opposite. That is, for $j = 1$ to $c + r$, if the j-th bit of S is 0, $\tilde{q}[j]$ is split into $\tilde{q}_a[j]$ and $\tilde{q}_b[j]$ so that $\tilde{q}[j] = \tilde{q}_a[j] + \tilde{q}_b[j]$; if the j-th bit of S is 1, $\tilde{q}_a[j]$ and $\tilde{q}_b[j]$ are both set to $\tilde{q}[j]$. The split query vectors $\tilde{q}_a$ and $\tilde{q}_b$ are then encrypted using M_1^{-1} and M_2^{-1} as $\hat{q}_a = M_1^{-1}\tilde{q}_a$, $\hat{q}_b = M_2^{-1}\tilde{q}_b$. Finally, the user submits the trapdoor $\{\hat{q}_a, \hat{q}_b\}$ to the cloud server to search for the required image data.

Image Retrieval Phase. After receiving the trapdoor from the user, the cloud server runs the Search algorithm to find out the most similar results against q. Here, we adopt the Euclidean distance to compute similarity between images. Then, for p and q, we have:

$$d(\hat{p}, \hat{q}) = \hat{p}_a^T \hat{q}_a + \hat{p}_b^T \hat{q}_b = \tilde{p}_a^T \tilde{q}_a + \tilde{p}_b^T \tilde{q}_b = \tilde{p}^T \tilde{q} + \Sigma_{j \in \hat{t}} \varepsilon_j$$

where $\hat{t}$ is the set of the t selected dummy dimensions, and it is different for each index point. In this way, the cloud server can determine the similarity between the query point and the points in the database. And the algorithm selects the top-k secure indexes that have the k biggest matching degree against q and gathers them into a result set R. Finally, the cloud server returns the user with a set of image data that are associated with indexes in R.

4   Security Analysis

In this section, we analyze the security strength of our proposed scheme.

In our scheme, in order to ensure the security of the data stored in the cloud server, the original images are encrypted by standard symmetric encryption algorithm which is semantically secure. As long as the secret key is kept confidential, it can be assumed computationally difficult for the cloud server to infer the original image data. Thus the privacy of the original images are properly protected.

However, during the interaction process between the cloud server and the users, the cloud server can build up access pattern and search pattern based on the recorded trapdoors and search results. Explicitly, access pattern includes the search results corresponding to the query trapdoors, while search pattern includes information describing which queries contain the similar image features.

Therefore, our scheme should guarantee that no additional information beyond the access pattern and the search pattern be leaked to the attackers. We will follow the simulation based adaptive security definition, extended from searchable symmetric encryption(SSE), to investigate the security of our scheme.

Theorem 1. *Our scheme is secure under the known-plaintext model.*

Before proving the Theorem 1, we introduce some notions used in [17].

- *History* is an image data set Δ, an index set I built from Δ and a set of queries $Q = (q_1, \cdots, q_k)$ submitted by the users, denoted as $H = (\Delta, I, Q)$. H is the plaintext knowledge of the attacker.
- *View* is the encrypted form of H under some secret key sk, denoted as $V(H) = (E_{sk}(\Delta), E_{sk}(I), E_{sk}(Q))$. Note that the attacker can see $V(H)$ from the cloud server as its ciphertext knowledge.
- *Trace* is the information that the scheme may leak about the history H, denoted as $\tau(H) = \{|E_{sk}(I_i)|, \alpha(H), \sigma(H)\}$, where $|E_{sk}(I_i)|$ is the bit length of the encrypted index; $\alpha(H)$, denoted as the access pattern induced by H, is a tuple $\alpha(H) = (R(q_1), \cdots, R(q_k))$; $\sigma(H)$, denoted as the search pattern induced by H, is a symmetric binary matrix such that $\sigma(H)_{i,j} = 1$, for $1 \leq i, j \leq k$, if q_i and q_j have the similar features, and 0 otherwise.

As defined in [17], given two histories that have the same trace, i.e. $\tau(H') = \tau(H)$, if there is no P.P.T adversary that can distinguish the two views of them, the scheme will be non-adaptively semantically secure.

Proof: Given a history H, there exists a polynomial-time simulator $\mathcal{S}$ that takes $\tau(H)$ as input and outputs a view $V(H') = \{E_{sk'}(\Delta'), E_{sk'}(I'), E_{sk'}(q')\}$ indistinguishable from an attacker's view. It runs as follows:

- $\mathcal{S}$ initializes an image data set Δ' indistinguishable with real Δ.
- $\mathcal{S}$ generates a visual vocabulary V' indistinguishable with real V.
- $\mathcal{S}$ builds up an index set I' based on V'. I' is indistinguishable with real I.
- $\mathcal{S}$ randomly generates the secret keys: $sk' = \{M_1', M_2', S'\}$ to encrypt I'.
- $\mathcal{S}$ generates a random string to simulate the query trapdoor q' and then operates a random oracle to point at random selected encrypted indexes in I' and reveal the identical identifiers from $\tau(H)$. The identical number of simulated index set randomly assigned from I' are considered to be the simulated results.
- $\mathcal{S}$ sets $V(H') = \{E_{sk'}(\Delta'), E_{sk'}(I'), E_{sk'}(q)\}$.

The correctness of such construction is easy to demonstrate by querying encrypted q' over encrypted I'. The secure index $E(I')$ and the trapdoor $E(q')$ generate the same trace as the one that the attacker has. Since the image vector is first pseudorandomly permuted with a pseudorandom permutation (PRP) and then encrypted by the secure kNN, due to the properties of the cryptographic primitive (PRP) and the security of kNN, we claim that no P.P.T adversary can distinguish $\{E_{sk}(I), E_{sk}(q)\}$ from $\{E_{sk'}(I'), E_{sk'}(q')\}$. Since Δ is encrypted under standard symmetric encryption such as AES which is considered as pseudorandom and Δ' is randomly generated, the adversary cannot distinguish $E_{sk}(\Delta)$ from $E_{sk'}(\Delta')$. Thus, there is no P.P.T adversary that can distinguish $V(H')$ from $V(H)$ although their underlying histories have the same trace.

Moreover, Yao et al. [18] point out that the secure kNN scheme is susceptible under the chosen plaintext attack. That is, the server can construct d linear equations to derive the coordinates of the data point p through d times observation of the query points and their encryption. However, on the one hand, by using the random dummy dimensions [3], it is not applicable for the cloud server to acquire enough plaintext query information, i.e., the query image vector; on the other hand, the vector is still high-dimensional in our scheme and is randomly permuted before encrypted by the secure kNN method. It is really hard for the attacker to construct such linear equations. Therefore, the attack scenario described in [18] will not make sense.

5   Experimental Evaluation

In this section, we evaluate the performance of our scheme on a real-world image database: Corel Image Set [19], which contains 1000 color images grouped into 10 categories. This database has been used as ground-truth for evaluating color image retrieval and image annotation [5]. We build up our system prototype using the Python language and the computer vision library OpenCV 2.4.11. We deploy the cloud on a server with Intel E5-2620 CPU and 120G RAM, and deploy the user client on a PC with Intel i7-3930 CPU and 16G RAM. The performance of our scheme is evaluated in terms of the search accuracy and efficiency in the process of index construction, trapdoor generation and search operation.

5.1   Search Accuracy

We use the precision and recall rate to evaluate the search accuracy of our proposed scheme. Precision is the percentage of retrieved relevant images over all returned results; recall is the proportion of retrieved relevant images over all the relevant results for the query in the original image data set. Denote TP as true positive, FP as false positive and FN as false negative, then the precision can be calculated as $Precision = \frac{TP}{TP+FP}$, and the recall can be calculated as $Recall = \frac{TP}{TP+FN}$.

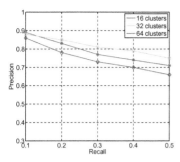
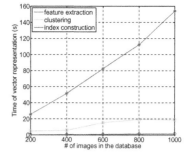

Fig. 2. Search precision and recall with different clusters

Fig. 3. Time of index construction before encryption

Since VLAD is parametrized by a single parameter k [11], the number of visual words, we first conduct our experiment to verify that excellent results can be obtained even with a relatively small number of k. The existing tool, vlfeat [20], is used to describe the images with SIFT and VLAD descriptors. We choose $k = 16, 32$ and 64 respectively and conduct 50 encrypted queries over the secure indexes. Figure 2 illustrates that comparable search accuracy can be obtained with the three different k values. Since the accuracy depends largely on the techniques such as feature extraction, image representation, and vector quantization, we can utilize more complex image representation techniques combining various features to obtain higher search accuracy.

5.2 Search Efficiency

Index Construction. The index construction process includes feature extraction, clustering, vector representation, vector quantization, and vector encryption. Figure 3 shows that the feature extraction consumes most of the time before encryption and outsourcing compared with clustering and vector representation. Figure 4 shows that the encryption over the descriptors in our scheme is a negligible overhead. Note that the time-consuming parts are the one-off operations and are therefore affordable for the data owner.

Trapdoor Generation. For a query request, the data user needs to instruct the same index as the outsourced images and generates the query trapdoor. The whole process is similar to the index construction, which consists of feature extraction, vector representation and vector encryption. Here, the vector representation is based on the previous clustering step. Through our 50 times of experiment, the average time consumption in the process of trapdoor generation is 1.837 s, which is affordable on the user's side.

Search Operation. After receiving the query trapdoor, the cloud server searches for the top-k images upon the secure index, which includes computing the inner product for all the encrypted indexes in the image data set.

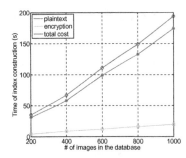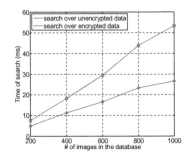

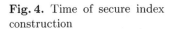

Fig. 4. Time of secure index construction

Fig. 5. Time of search over unencrypted and encrypted index

As shown in Fig. 5, image retrieval over the encrypted index consumes comparable time to the unencrypted index and the search time grows linearly with the size of the file set. This is intuitive since we just perform exhaustive computation in our experiment and the search process needs to go over all the files in the dataset before the cloud server can get the final result. Even so, the search can be done within milliseconds, which is comparable to the image retrieval schemes for plaintext.

6   Conclusion

In this paper, we proposed a privacy-preserving content-based image retrieval scheme, which allows the data owner to outsource the images to the untrusted cloud without revealing the privacy of the images. To improve the security and the efficiency, we combined the secure kNN scheme with the BPBC based vector quantization and asymmetric distance based similarity measure. The security analysis and experiment shows the security and efficiency of our scheme. For the future work, we will extend our scheme to support much more complicated application requirements.

Acknowledgment. The work is supported by the National Natural Science Foundation of China under Grant Nos. 61379144 and 61572026.

References

1. Ning, C., Cong, W., Ming, L., Ren, K., Lou, W.: Privacy-preserving multi-keyword ranked search over encrypted cloud data. IEEE Trans. Parallel Distrib. Syst. **25**(1), 222–233 (2014)
2. Song, D.X., Wagner, D., Perrig, A.: Practical techniques for searches on encrypted data. In: IEEE Symposium on Security & Privacy, pp. 44–55 (2000)
3. Sun, W., Wang, B., Cao, N., Li, M., Lou, W., Hou, Y.T., Li, H.: Privacy-preserving multi-keyword text search in the cloud supporting similarity-based ranking. IEEE Trans. Parallel Distrib. Syst. **25**(11), 71–82 (2013)

4. Wang, B., Yu, S., Lou, W., Hou, Y.T.: Privacy-preserving multi-keyword fuzzy search over encrypted data in the cloud. In: 2014 Proceedings IEEE INFOCOM, pp. 2112–2120 (2014)
5. Lu, W., Varna, A.L., Wu, M.: Confidentiality-preserving image search: a comparative study between homomorphic encryption and distance-preserving randomization. Access IEEE **2**, 125–141 (2014)
6. Wong, W.K., Cheung, D.W.l., Kao, B., Mamoulis, N.: Secure knn computation on encrypted databases. In: Proceedings of the 2009 ACM SIGMOD International Conference on Management of Data, pp. 139–152 (2009)
7. Chu, W.T., Chang, F.C.: A privacy-preserving bipartite graph matching framework for multimedia analysis and retrieval. In: Proceedings of the 5th ACM on International Conference on Multimedia Retrieval, pp. 243–250 (2015)
8. Zhang, L., Jung, T., Liu, C., Ding, X.: Pop: Privacy-preserving outsourced photo sharing and searching for mobile devices. In: 2015 IEEE 35th International Conference on Distributed Computing Systems (ICDCS), pp. 308–317 (2015)
9. Cui, H., Yuan, X., Wang, C.: Harnessing encrypted data in cloud for secure and efficient image sharing from mobile devices. In: 2015 IEEE Conference on Computer Communications (INFOCOM) (2015)
10. Gong, Y., Kumar, S., Rowley, H., Lazebnik, S.: Learning binary codes for high-dimensional data using bilinear projections. In: 2013 IEEE Conference on Computer Vision and Pattern Recognition (CVPR), pp. 484–491, June 2013
11. Jégou, H., Douze, M., Schmid, C., Pérez, P.: Aggregating local descriptors into a compact image representation. In: 2010 IEEE Conference on Computer Vision and Pattern Recognition (CVPR), pp. 3304–3311 (2010)
12. Ferreira, B., Rodrigues, J., Leitão, J., Domingos, H.: Privacy-preserving content-based image retrieval in the cloud. Eprint Arxiv (2014)
13. Xia, Z., Zhu, Y., Sun, X., Qin, Z.: Towards privacy-preserving content-based image retrieval in cloud computing. IEEE Trans. Cloud Comput. 1–11 (2015). doi:10.1109/TCC.2015.2491933
14. Yuan, J., Yu, S., Guo, L.: Seisa: secure and efficient encrypted image search with access control. In: 2015 IEEE Conference on Computer Communications (INFOCOM), pp. 2083–2091 (2015)
15. Lowe, D.G.: Distinctive image features from scale-invariant keypoints. Int. J. Comput. Vis. **60**(2), 91–110 (2004)
16. Jégou, H., Douze, M., Schmid, C.: Product quantization for nearest neighbor search. IEEE Trans. Pattern Anal. Mach. Intell. **33**(1), 117–128 (2010)
17. Curtmola, R., Garay, J., Kamara, S., Ostrovsky, R.: Searchable symmetric encryption: improved definitions and efficient constructions. In: Proceedings of CCS 2006, pp. 79–88 (2006)
18. Yao, B., Li, F., Xiao, X.: Secure nearest neighbor revisited. In: 2013 IEEE 29th International Conference on Data Engineering (ICDE), pp. 733–744 (2013)
19. Wang, J.Z., Li, J., Wiederholdy, G.: Simplicity: semantics-sensitive integrated matching for picture libraries. IEEE Trans. Pattern Anal. Mach. Intell. **23**(9), 947–963 (2001)
20. Vedaldi, A., Fulkerson, B.: VLFeat: an open and portable library of computer vision algorithms (2008). http://www.vlfeat.org/

Preserving the *d*-Reachability When Anonymizing Social Networks

Xiangyu Liu$^{(\boxtimes)}$, Jiajia Li, Dahai Zhou, Yunzhe An, and Xiufeng Xia

School of Computer Science, Shenyang Aerospace University,
Shenyang 110136, China
liuxy@sau.edu.cn

Abstract. The goal of graph anonymization is avoiding disclosure of privacy in social networks through graph modifications meanwhile preserving data utility of anonymized graphs for social network analysis. Graph reachability is an important data utility as reachability queries are not only common on graph databases, but also serving as fundamental operations for many other graph queries. However, the graph reachability is severely distorted after anonymization. In this work, we study how to preserve the *d*-reachability of vertices when anonymizing social networks. We solve the problem by designing a *d*-reachability preserving graph anonymization (*d*-RPA for short) algorithm. The main idea of *d*-RPA is to find a subgraph that preserves the *d*-reachability, and keep it unchanged during anonymization. We show that *d*-RPA can efficiently find such a subgraph and anonymize the releasing graph with low information loss. Extensive experiments on real datasets illustrate that anonymized social networks generated by our method can be used to answer *d*-reachable queries with high accuracy.

1 Introduction

Social networks usually contain individuals' sensitive information, which makes it an important concern to avoid compromising individual privacy when releasing social network data. Graph anonymization, as an valid method for privacy protection, has been extensively studied in the past few years [3,7,11–13]. One of the fundamental issues when anonymizing social network data is avoiding disclosure of individuals' sensitive information while still permitting certain analysis and queries on the network.

The *d*-reachability is an important graph query as *d*-reachable queries are not only common on graph databases, but they also serve as fundamental operations for many other graph queries [1,6]. For instance, SNS (Social Networking Services) generally support queries on the relationship path between two users within a threshold length. In practice, the reachability of vertices is nontrivially distorted due to graph anonymization [8].

The work is partially supported by the National Natural Science Foundation of China (Nos. 61502316, 61502317).

B. Cui et al. (Eds.): WAIM 2016, Part II, LNCS 9659, pp. 40–51, 2016.
DOI: 10.1007/978-3-319-39958-4_4

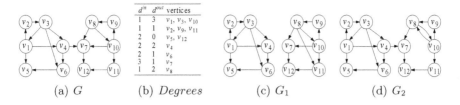

d^{in}	d^{out}	vertices
1	3	v_1, v_3, v_{10}
1	1	v_2, v_9, v_{11}
2	0	v_5, v_{12}
2	2	v_4
2	1	v_6
3	1	v_7
1	2	v_8

(a) G (b) $Degrees$ (c) G_1 (d) G_2

Fig. 1. A toy microblog network G and its tow k-degree anonymized versions.

Figure 1(a) describes a toy microblog network G and the degrees of vertices are illustrated in Fig. 1(b), where d^{in} and d^{out} refer to the in-degree and out-degree of a vertex, respectively. A directed edge (u, v) indicates that u is a follower of v. Assuming the adversary has acquired that Bob has two followers and follows two other individuals, i.e. $d^{in}(Bob) = 2$ and $d^{out}(Bob) = 2$, it is not difficult for the adversary to re-identify vertex v_4 in G as Bob with 100% confidence, since there is only one vertex having the same degree (including in-degree and out-degree) as Bob.

To avoid vertex re-identification using degrees as background knowledge, k-degree anonymization has been proposed in [7] and the general idea is to modify the network so that for each vertex v, there exist at least $k - 1$ other vertices having the same degrees as v, thus the probability of the adversary re-identifying the identity of a vertex would not be larger than $\frac{1}{k}$. For instance, G_1 and G_2 in Fig. 1 are two anonymized versions for G. It could be easily verified that both G_1 and G_2 satisfy 2-degree anonymity. We say $\langle u, v \rangle$ is a d-reachable pair, if vertex v is reachable from u within d hops. Specifically, when $d = 2$, compared with G there are four 2-reachable pairs lost in G_1, which are $\langle v_1, v_7 \rangle$, $\langle v_3, v_7 \rangle$, $\langle v_4, v_7 \rangle$ and $\langle v_4, v_{12} \rangle$; $\langle v_9, v_{11} \rangle$ is a new 2-reachable pair in G_1. Correspondingly, all 2-reachable pairs in G are preserved in G_2, and no 2-reachable pair is newly introduced. Let $R_d(G)$ denote the set of all d-reachable pairs in G. For simplicity, we define that a vertex is d-reachable from itself. For graph G and its k-anonymized version G_k, we use similarity function Sim_d to measure the utility of G_k on d-reachable query, which is calculated as:

$$Sim_d(G, G_k) = \frac{|R_d(G) \cap R_d(G_k)|}{|R_d(G) \cup R_d(G_k)|} \qquad (1)$$

Higher $Sim_d(G, G_k)$ implies better utility of G_k on d-reachable queries. By employing Eq. 1, we get $Sim_2(G_1, G) = 0.9$ and $Sim_2(G_2, G) = 1.0$, indicating that the 2-reachability in G is better preserved in G_2 than in G_1.

When anonymizing graph G, in order to preserve the d-reachability of vertieces in G, we hope to find an anonymized version G_k, such that $R_d(G_k)$ and $R_d(G)$ are as close as possible.

2   Related Work

In previous work, privacy attacks in social networks are mainly classified into two categories, including *vertex re-identification* attacks and *link re-identification* attacks.

In vertex re-identification attacks (a.k.a. identity disclosure), an adversary identifies the identities of vertices in the published network using the subgraphs associated with target individuals as background knowledge. Liu et al. in [7] propose k-degree anonymity to prevent from privacy attacking using vertex degree as adversary knowledge. Zhou et al. in [12] protect identity privacy through anonymizing the 1-neighborhood subgraph of each vertex. To resist subgraph based vertex re-identification attacks, Hay in [3] propose to cluster vertices into super nodes, thus vertices in a super node are indistinguishable from each other. Zou et al. in [13] present a privacy preserving model K-Automorphism for protecting identity privacy. In link re-identification (a.k.a. link disclosure) attacks, an adversary aims at identifying sensitive relationships among individuals in social network. Ying et al. in [10] study graph randomization through adding/removing and switching edges randomly while preserving the spectrum of the network. Liu et al. in [9] study protecting link inference attacks. Fard et al. in [2] propose subgraph randomization to protect sensitive relationships for directed social network graphs. Besides protecting identity and link privacy, Yuan et al. in [11] give a solution to satisfy different requirements on privacy protecting level.

Liu et al. in [8] preserve the reachability of anonymized graphs, which is similar to our work. However, we study the problem of preserving the d-reachability of anonymized graphs, which is a general case of [8].

3   Preliminaries and Problem Definition

We represent a social network as a directed graph $G(V, E)$ with $|V| = n$ and $|E| = m$, where V is the vertex set, and $E \subseteq V \times V$ is the edge set. We also use $V(G)$ and $E(G)$ to denote the vertex set and edge set of G. We use (u, v) to represent the edge from vertex u to vertex v. Specifically, we say (u, v) is an *out-edge* of u and an *in-edge* of v. Correspondingly, vertex v is an *out-neighbor* of u, and u is an *in-neighbor* of v. We use $d^{in}(u)/d^{out}(u)$ to denote the *in/out-degree* of u, i.e., the number of edges coming to or out of u. For vertex $u \in V$, we represent the degree of u in the form of $(d^{in}(u), d^{out}(u))$. When adding an edge (u, v) into E, we say that we link u to v.

In this work, we assume that the adversary uses both the in-degrees and out-degrees as background knowledge to reveal the identities of individuals. Based on the privacy model of k-*degree anonymity* in [7], we formally present our privacy model in social networks.

Definition 1 (k-DEGREE ANONYMITY). *Given graph $G(V, E)$ and integer k, if $\forall v \in V$, there exist $m(m \geq k - 1)$ other vertices $u_1, u_2, \ldots u_m$ that satisfy $d^{in}(v) = d^{in}(u_i)$ and $d^{out}(v) = d^{out}(u_i)(1 \leq i \leq m)$, we say graph G is k-degree anonymous.*

For instance, both G_1 and G_2 in Fig. 1 are 2-degree anonymous graphs. In this paper, we focus on anonymizing directed graphs with d-reachability preservation. With no difficulty our approach can be applied in undirected graphs, for each edge can be regarded as an edge with bi-directions.

Definition 2 (d-REACHABILITY). *Given graph $G(V, E)$ and vertices $u, v \in V$, we say vertex v is d-reachable from vertex u if there is a path P starting from u and ending at v meanwhile the length of P satisfies $|P| \le d$, and we define that $\langle u, v \rangle$ is a d-reachable pair of G, denoted as $u \xrightarrow{d} v$. The d-reachability of graph G refers to the set of all d-reachable pairs in G, denoted as $R_d(G)$.*

Given graph $G(V, E)$, integers k and d, we expect to construct a k-degree anonymous graph $G_k(V_k, E_k)$ through a set of graph-modification operations on G such that G_k preserves the d-reachability of G as much as possible. Moreover, we restrict the vertex-modification operations to vertex additions, i.e. $V \subseteq V_k$. We formally propose the d-REACHABILITY PRESERVING GRAPH ANONYMIZATION problem as follows.

Problem 1. (d-REACHABILITY PRESERVING GRAPH ANONYMIZATION). Given graph $G(V, E)$, integers k and d, find a k-degree anonymous graph $G_k(V_k, E_k)$ such that anonymization cost $C(G, G_k)$ is minimized, where $C(G, G_k)$ consists of three parts:

(1) the cost of d-reachability;
(2) the cost of vertex-modifications;
(3) the cost of edge-modifications.

Notice that we evaluate the anonymization cost $C(G, G_k)$ in three parts: d-reachability, vertex-modifications and edge-modifications. In particular, we use $C_r(G, G_k) = 1 - \frac{|R_d(G) \cap R_d(G_k)|}{|R_d(G) \cup R_d(G_k)|}$ to evaluate the anonymization cost on d-reachability. The anonymization cost of vertex and edge modifications can be quantified by the number of modified vertices and edges.

Theorem 1. *The problem of d-REACHABILITY PRESERVING GRAPH ANONYMIZATION is NP-hard.*

The proof of Theorem 1 is by reducing the NP-complete problem of k-DIMENSIONAL PERFECT MATCHING [4]. Limited by space, we omit the details here.

4   Generate a d-Reachability Preserved Graph

Given graph $G(V, E)$, in order to preserve the d-reachability of G in its anonymized version G_k, we hope to find a subgraph that preserves the d-reachable pairs of G, and keep it unchanged during the anonymization. Based on such intuitions, we propose the definition of d-reachability preserved graph and design an efficient algorithm to generate it.

Definition 3 (d-REACHABILITY PRESERVED GRAPH). *Given graph $G(V, E)$ and integer d, let $G'(V', E')$ be a subgraph of G with $V' = V$ and $E' \subseteq E$. For each d-reachable pair $\langle u, v \rangle \in R_d(G)$, if $u \xrightarrow{d} v$ still holds in G', we say G' is a d-reachability preserved graph (d-RPG for short) of G, denoted as $G' \prec_d G$.*

For instance, G_3 in Fig. 2(a) is a 3-RPG of G in Fig. 1(a), i.e. $G_3 \prec_3 G$. A special case is $G_{|V|}$ in Fig. 2(b), which is a $|V|$-RPG of G and preserves all reachable pairs. Notice that there is always a *feasible* method to find a d-RPG G_d, that is, let G_d preserve all edges presented in G. It is obvious that some edges are redundant for preserving d-reachability. For the sake of anonymization, we hope to find the minimal d-RPG.

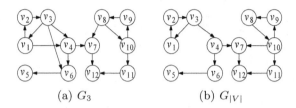

(a) G_3 (b) $G_{|V|}$

Fig. 2. Reachability preserved graphs of G in Fig. 1(a).

Definition 4 (MINIMAL d-RPG). *Given graph $G(V, E)$, we say G_d is a minimal d-reachability preserved graph of G if and only if $G_d \prec_d G$ and there is no subgraph G' of G_d satisfying $G' \prec_d G$.*

Given G in Fig. 1(a), it is easily verified that G_3 in Fig. 2(a) is a minimal 3-RPG. Here we use *minimal* in Definition 4, which refers to a property of the set, not the size of the set.

Theorem 2. *Given graph $G(V, E)$, finding the minimal d-reachability preserved graph of G is NP-hard.*

Proof. The proof is constructed by reducing the NP-hard DIRECTED HAMIL-TONIAN CIRCUIT [5] problem to the FINDING MINIMAL d-RPG problem. Limited by space, we omit the details here.

Given graph $G(V, E)$ and integer d, a straightforward method for generating a minimal d-RPG is described as follows. We firstly initialize G_d with G. For each edge (u, v) in G_d, we check whether it could be removed. When removing edge (u, v) from G_d, there would be two situations: (1) $R_d(G_d)$ keeps unchanged; (2) $R_d(G_d)$ loses some d-reachable pairs. If $R_d(G_d)$ does not change after removing (u, v), (u, v) is removable and definitely removed from G_d; otherwise, edge (u, v) should be kept in G_d. We perform such checking and removing operation for each edge in G_d, and the final G_d is a minimal d-RPG of G. However, the time complexity would be $O(nm(n+m))$, which is impractical for real social networks.

Instead, as shown in Algorithm 1, we design an efficient algorithm to generate a d-RPG. The basic idea is as follows. For each vertex $v \in V(G_d)$, we propose an heuristic method to make v reach its d-reachable neighbors within d hops by inserting fewer edges (lines 2–10). We evaluate each path p with $S(\Delta R, p) = \frac{|\Delta R \cap V(p)|}{|E(p) - E_d(p)|}$ (line 8), where $V(p)$ and $E(p)$ refer to the amount of vertices and edges included in p, respectively. Obviously, higher $S(\Delta R, p)$ indicates inserting

---

Algorithm 1. Generate a d-Reachability Preserved Graph

Input: Graph $G(V, E)$ and integer d
Output: a d-RPG G_d

1   $E_d \leftarrow \emptyset$;
2   **for** *each* $v \in V$ **do**
3     $R_d(v) \leftarrow$ the d-reachable vertices of v in $G(V, E)$;
4     $R'_d(v) \leftarrow$ the d-reachable vertices of v in $G_d(V, E_d)$;
5     $\Delta R \leftarrow R_d(v) - R'_d(v)$;
6     $P \leftarrow$ paths starting from v with length$\leq d$;
7   **while** $\Delta R \neq \emptyset$ **do**
8       $p' \leftarrow p \in P$ with maximal $S(\Delta R, p)$;
9       $\Delta R \leftarrow \Delta R - V(p')$;
10      $E_d \leftarrow E_d \cup E(p')$;

11 **return** $G_d(V, E_d)$;

---

edges on path p could make v reach more d-reachable neighbors by adding fewer edges. It requires $O(n + m)$ time to achieve the d-reachable neighbors for each vertex v, thus the time complexity of Algorithm 1 is $O(n(n + m))$. In practical applications, the d value is usually small ($d \leq 6$), and the execution time of achieving the d-reachable neighbors is much less than $O(n + m)$, which makes the algorithm practical for real social networks.

5   Reachability Preserving Graph Anonymization

In Algorithm 2, we introduce our d-reachability preserving graph anonymization (d-RPA for short) algorithm. Given graph $G(V, E)$ and its d-RPG G_d, the main idea of the algorithm is to keep G_d unchanged during the anonymization in order to preserve the d-reachability of the releasing graph.

We first outline the d-RPA algorithm. Different from performing modifications on G directly, the algorithm iteratively set the labels of vertices and edges in G during the anonymization (lines 1–11). The labels are initialized based on G_d (line 1). In the anonymization, the algorithm first selects an "unanonymized" vertex as $Seed$ (line 3). The cost parameters α and β are the weights specified by users for evaluating anonymization cost on edge modifications and d-reachability, respectively. Based on $Seed$, the algorithm tries to find a vertex set V_A of size k (line 4), in which the vertices have close degrees. Vertex $Seed$ is labeled as "postprocessing" if such V_A is not found (lines 5–6). The algorithm calculates an optimal degree (d^{in}, d^{out}) (line 8) and anonymizes each vertex in V_A into degree (d^{in}, d^{out}) (lines 9–10). A vertex is labeled as "anonymized" after anonymization. The algorithm stops the iteration when the number of unanonymized vertices is less than $2k - 1$ (line 11), and anonymizes the remaining vertices labeled "unanonymized" and "postprocessing" (line 12). When all vertices are labeled "anonymized", the algorithm generates the releasing graph G_k based on the labels of vertices and edges (line 13). The details of each step will be discussed shortly.

Algorithm 2. d-Reachability Preserving Graph Anonymization

   Input: Graph $G(V, E)$, d-RPG G_d, integer k, cost parameters α and β
   Output: k-degree anonymous graph $G_k(V_k, E_k)$
1 Initialize labels of vertices and edges in G based on G_d;
2 **repeat**
3      $Seed \leftarrow SearchSeedVertex(G)$;
4      $V_A \leftarrow AnonymizationVertexSet(G, Seed, k, \alpha, \beta)$;
5   **if** $|V_A| < k$ **then**
6   Set $Seed$ as "postprocessing";
7   **else**
8         $(d^{in}, d^{out}) \leftarrow OptimalDegree(V_A, \alpha, \beta)$;
9   **for** each vertex $v \in V_A$ **do**
10           $AnonymizeVertex(v, d^{in}, d^{out})$;
11 **until** $UnAnonymized(G) < 2k - 1$;
12 Anonymize the "unanonymized" and "postprocessing" vertices in G;
13 Generate G_k based on G;
14 **return** G_k;

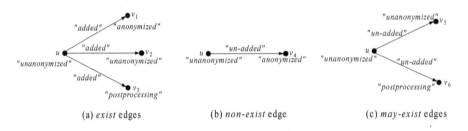

(a) *exist* edges   (b) *non-exist* edge   (c) *may-exist* edges

Fig. 3. Out-edge categories determined by labels.

The labels for edges include "added" and "un-added". For vertex v, there exist three labels to be assigned, including "anonymized," "unanonymized" and "postprocessing". When initializing the labels using G_d (line 1), all vertices in G are labeled with "unanonymized", edges presented in G_d are labeled with "added" and other ones are labeled with "un-added". For simplicity of discussion, given unanonymized vertex u, we classify the out-edges of u into three categories: *exist*, *non-exist*, and *may-exist*. Figure 3 shows the categories of out-edges, which are determined by the assigned labels. The in-edges are classified similarly.

The algorithm anonymizes vertex u into the specified degree (d^{in}, d^{out}) (line 10). Let $d^{out}_{min}(u)$ be the number of *exist* out-edges of u, and $d^{out}_{max}(u)$ be the number of *exist* and *may-exist* out-edges of u. The specified d^{in} and d^{out} satisfy $d^{in} \geq d^{in}_{min}(u)$ and $d^{out} \geq d^{out}_{min}(u)$. We introduce the anonymization of the out-degree, and the process of anonymizing in-degree is similar and omitted here. If $d^{out}_{min}(u) \leq d^{out} \leq d^{out}_{max}(u)$, the algorithm randomly selects $d^{out} - d^{out}_{min}(u)$ *may-exist* out-edges of u to label with "added"; otherwise, besides setting all *may-exist* out-edges of u as "added," the algorithm inserts $d^{out} - d^{out}_{max}(u)$ vertices and links u to each vertex.

The cost of anonymizing u into degree (d^{in}, d^{out}) is $Cost(u, d^{in}, d^{out}) = Cost_{in}(u, d^{in}) + Cost_{out}(u, d^{out})$, where $Cost_{in}(u, d^{in})/Cost_{out}(u, d^{out})$ refers to the cost of anonymizing the in/out-degree of u. Specifically, $Cost_{in}(u, d^{in})$ is formalized as:

$$Cost_{in}(u, d^{in}) = \begin{cases} \alpha(d^{in}_{max}(u) - d^{in}), d^{in}_{min}(u) \le d^{in} \le d^{in}_{max}(u) \\ \beta(d^{in} - d^{in}_{max}(u))|Ngr_{d-1}(u)|, d^{in} > d^{in}_{max}(u) \end{cases}$$

where $Ngr_{d-1}(u)$ refers to the $(d-1)$-neighborhood of vertex u. When $d^{in}_{min}(u) \le d^{in} \le d^{in}_{max}(u)$, $Cost_{in}(u, d^{in})$ measures the cost due to removing edges. When $d^{in} > d^{in}_{max}(u)$, $Cost_{in}(u, d^{in})$ measures the cost of newly introduced d-reachable pairs due to insertions of vertices and edges. The out-degree anonymization cost $Cost_{out}(u, d^{out})$ is calculated similarly.

When searching for the *Seed* (line 3), Algorithm 2 selects the unanonymized vertex $u \in V$ with maximal $d^{in}_{max}(u) + d^{out}_{max}(u)$. Thus, vertices with high degrees would be anonymized with priority. We adopt such anonymization strategy because it is relatively easy to anonymize vertices with low degrees.

Algorithm 3. AnonymizationVertexSet

Input: Graphs $G(V, E)$, seed vertex *Seed*, integer k, and cost parameters α and β

Output: Anonymization vertex set V_A

1   $d^{in} \leftarrow d^{in}_{max}(Seed), d^{out} \leftarrow d^{out}_{max}(Seed)$;
2   $V_A \leftarrow \emptyset$;
3   $VertexList \leftarrow$ Valid unanonymized vertices in V;
4   Ascending sort $VertexList$ based on $Cost(v, d^{in}, d^{out})$;
5   **repeat**
6     $v \leftarrow VertexList.head$, remove v from $VertexList$;
7   **if** v has no edges linking with vertices in V_A **then**
8        $V_A \leftarrow V_A \cup \{v\}$;
9   **until** $|V_A| = k$ *or* $VertexList = \emptyset$;
10   **return** V_A;

Given the *Seed*, Algorithm 3 generates vertex set V_A, which contains k unanonymized vertices. Vertex list $VertexList$ stores the valid unanonymized vertices in V (line 3), which are the ones satisfying the lower bounds, i.e. d^{in} for in-degree and d^{out} for out-degree. Given d^{in} and d^{out}, the algorithm ascending sorts the vertices in $VertexList$ based on the anonymization cost (line 4). The algorithm iteratively select vertex from $VertexList$ and add it to V_A until V_A contains k vertices or $VertexList$ is empty (lines 5–9). Notice that if vertex v has edges linking with other vertices in V_A, v would not be added to V_A (line 7), which ensures that anonymizing v would not affect the degrees of other vertices in V_A. For a seed vertex , there are at most n vertices to be checked for V_A. Thus, the time complexity of Algorithm 3 is $O(n)$.

Given vertex set V_A, Algorithm 2 calculates an optimal degree (d^{in}, d^{out}) (line 8). Following the degree lower bounds specified by V_A, d^{in} and d^{out} satisfy $d^{in} \in [\max_{u \in V_A}(d^{in}_{min}(u)), \max_{u \in V_A}(d^{in}_{max}(u))]$ and $d^{out} \in [\max_{u \in V_A}(d^{out}_{min}(u)), \max_{u \in V_A}(d^{out}_{max}(u))]$. We choose the (d^{in}, d^{out}) that minimizes $\sum_{u \in V_A} Cost(u, d^{in}, d^{out})$. In Algorithm 2, when the iteration stops, some unanonymized vertices need postprocessing anonymization (line 12). We map each vertex into a 2-dimension space based on degree and cluster them into several groups of size $\geq k$. For each group, specify (d^{in}, d^{out}) with the maximal d^{in}_{max} and d^{out}_{max} of vertices included, and anonymize through insertions of vertices and edges. Obviously, the newly added vertices have degree $(1,0)$ or $(0,1)$, the quantity of which is much larger than the anonymity parameter k and we safely mark them as "anonymized". When generating G_k (line 13), $V(G_k)$ and $E(G_k)$ are initialized with $V(G)$ and edges labeled "added" in $E(G)$, respectively. As the iteration in Algorithm 2 requires $O(2n)$ running time and the number of iterations is at most $\frac{n}{k}$, the time complexity of Algorithm 2 is $O(\frac{n^2}{k})$.

6   Experimental Evaluation

In this section, we provide extensive experiments to evaluate our methods. We have used two real social network datasets, which are both directed graphs[1]: HepTh and Epinions. Table 1 presents the statistics of these two datasets.

Table 1. Statistics of datasets

	HepTh	Epinions
Number of vertices	27770	75879
Number of edges	352807	508837
Maximum in-degree	2414	3035
Maximum out-degree	562	1801
Average in/out-degree	12.70	6.71
Number of triangles	1478735	1624481
Average clustering coefficient	0.3120	0.1378
Diameter (longest shortest path)	13	14

We implemented our d-RPA algorithm. For comparison, we modified and implemented the anonymization algorithm in [7] and named the modified version as Ngr-Degree. An anonymization cost function is introduced in [7], which measures the similarity between two vertices on their neighborhoods. We employed the same function in Ngr-Degree except that we used the degrees of vertices instead of the neighborhoods as the metric.

---

[1] Available at http://snap.stanford.edu/data/.

We evaluate the running time, the anonymization cost, the graph structural properties, and the quantities of added vertices, where we set $k = 5, 10, 15, 20, 25$, $d = 2, 4, 6, |V|$, $\alpha = 40$ and $\beta = 1$ by default. All programs are implemented in Java and performed on a 2.33 GHz Intel Core 2 Duo CPU with 4 GB DRAM running the Windows XP operating system.

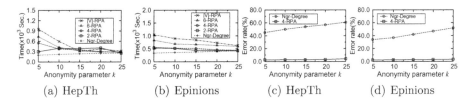

(a) HepTh   (b) Epinions   (c) HepTh   (d) Epinions

Fig. 4. The runtime of graph anonymization and error rates of *d*-reachable queries.

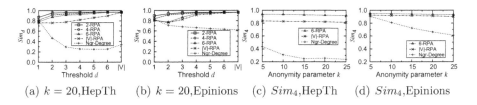

(a) $k = 20$,HepTh   (b) $k = 20$,Epinions   (c) Sim_4,HepTh   (d) Sim_4,Epinions

Fig. 5. The *d*-reachability similarity of anonymized graphs.

Figure 4(a) and (b) show the runtime of graph anonymization. It can be observed that Ngr-Degree costs less runtime than *d*-RPA, as Ngr-Degree does not consider preserving the reachability during the anonymization. Generally, it requires less runtime for *d*-RPA as k increases, which accords with the time complexity of *d*-RPA. For the same k value, it costs *d*-RPA more runtime as d increases.

We show the error rates of *d*-reachable queries in Fig. 4(c) and (d). We randomly select 100000 vertex pairs in anonymized graphs and perform 4-reachable queries on them. We define $error\ rate = \frac{\#correct\ answers}{\#total\ queries}$, where *correct answer* implies a *d*-reachable query on a vertex pair achieves the same result in both the original graph and the anonymized one. Ngr-Degree gets higher error rates than 4-RPA, which are above 43 % and 34 % in both datasets. With the increment of k, the error rate of Ngr-Degree gets higher while 4-RPA is still lower than 4.2 %, indicating that anonymized social networks generated by *d*-RPA can be used to answer *d*-reachable queries with high accuracy.

We evaluate the *d*-reachability similarity of anonymized graphs. Figure 5(a) and (b) show the results on HepTh and Epinions for different d values when $k = 20$. Our *d*-RPA gets higher Sim_d than Ngr-Degree, indicating that *d*-RPA preserves the *d*-reachability better. Overall, the Sim_d of *d*-RPA is over 0.8 by average. With the increment of d, the Sim_d of *d*-RPA gets increased and the one of Ngr-Degree decreases. Figure 5(c) and (d) show the 4-reachability similarities

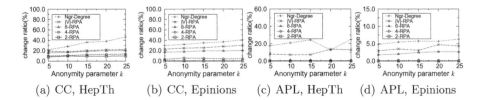

(a) CC, HepTh   (b) CC, Epinions   (c) APL, HepTh   (d) APL, Epinions

Fig. 6. The change ratios of graph structures.

for different k values. Generally, the Sim_4 obtained by d-RPA keeps stable as k increases while Ngr-Degree gets decreased.

We also examine anonymized graphs on graph structural properties. We consider two graph structural properties, *Clustering Coefficient* and *Average Path Length*. We use `Change ratio` $= |P_o - P_a|/|P_o|$ to evaluate the property change ratio, where P_o and P_a refer to the property values of the original graph and the anonymized version, respectively. Figure 6(a) and (b) show the change ratios of *Clustering Coefficient*, while the results of *Average Path Length* are shown in Fig. 6(c) and (d). Generally, the change ratios of d-RPA are lower than Ngr-Degree. Such observations could be explained by the fact that d-RPA preserves the essential edges for maintaining the connections between vertices in a social network graph, which retains the structure of the anonymized graph. As k increases, the change ratios of d-RPA keeps unchanged, indicating that preserving d-reachability is fundamental for retaining graph structure properties.

Table 2. The number of added fake vertices

	HepTh					Epinions				
k	5	10	15	20	25	5	10	15	20	25
2-RPA	54	70	98	113	134	160	181	207	221	232
4-RPA	46	53	84	91	106	132	146	169	181	208
6-RPA	31	47	64	73	83	108	127	146	155	172

Table 2 shows the number of fake vertices added by d-RPA when $d = 2, 4, 6$. As d keeps stable, the number of fake vertices gets increased as k increases. With the same k value, the algorithm inserts less fake vertices when d gets higher. In the anonymized graph, the ratio of fake vertices is quite small. For instance, when $k = 5$ and $d = 4$, the number of fake vertices in HepTh is 46 while the quantity of vertices with degree (1,0) or (0,1) is 1268, thus it is impossible for the adversary to infer privacy by removing these fake vertices.

7   Conclusions

In this paper, we propose the problem of preserving d-reachability when anonymizing social networks. We solve the problem by designing a d-reachability preserving graph anonymization (d-RPA for short) algorithm. The main idea of d-RPA is to find a subgraph (i.e. d-RPG) that guarantees the d-reachability, and keep it unchanged during anonymization. Firstly, we present an efficient algorithm to generate such a d-RPG, which is practical for real social networks. Then, based on the d-RPG, we anonymize the releasing graph with low information loss. Our extensive empirical studies illustrate that anonymized social networks generated by d-RPA can be used to answer d-reachable queries with high accuracy.

References

1. Cheng, J., Shang, Z., Cheng, H., Wang, H., Yu, J.: K-reach: who is in your small world. In: VLDB 2012 (2012)
2. Fard, A.M., Wang, K., Yu, P.S.: Limiting link disclosure in social network analysis through subgraph-wise perturbation. In: EDBT 2012 (2012)
3. Hay, M., Miklau, G., Jensen, D., Towsley, D., Weis, P.: Resisting structural re-identification in anonymized social networks. In: VLDB 2008, pp. 102–114 (2008)
4. Hazan, E., Safra, S., Schwartz, O.: On the complexity of approximating k-dimensional matching. In: Arora, S., Jansen, K., Rolim, J.D.P., Sahai, A. (eds.) RANDOM 2003 and APPROX 2003. LNCS, vol. 2764, pp. 83–97. Springer, Heidelberg (2003)
5. Karp, R.: Reducibility among combinatorial problems. In: Jünger, M., Liebling, T.M., Naddef, D., Nemhauser, G.L., Pulleyblank, W.R., Reinelt, G., Rinaldi, G., Wosley, L.A. (eds.) 50 Years of Integer Programming 1958–2008, pp. 219–241. Springer, Heidelberg (2010)
6. Li, M., Gao, H., Zou, Z.: K-reach query processing based on graph compression. J. Softw. **25**(4), 797–812 (2014)
7. Liu, K., Terzi, E.: Towards identity anonymization on graphs. In: SIGMOD 2008, pp. 93–106 (2008)
8. Liu, X., Wang, B., Yang, X.: Efficiently anonymizing social networks with reachability preservation. In: CIKM 2013, pp. 1613–1618 (2013)
9. Liu, X., Yang, X.: Protecting sensitive relationships against inference attacks in social networks. In: Lee, S., Peng, Z., Zhou, X., Moon, Y.-S., Unland, R., Yoo, J. (eds.) DASFAA 2012, Part I. LNCS, vol. 7238, pp. 335–350. Springer, Heidelberg (2012)
10. Ying, X., Wu, X.: Randomizing social networks: a spectrum preserving approach. In: SDM 2008, pp. 739–750 (2008)
11. Yuan, M., Chen, L., Yu, P.: Personalized privacy protection in social networks. In: VLDB 2010, pp. 141–150 (2010)
12. Zhou, B., Pei, J.: Preserving privacy in social networks against neighborhood attacks. In: ICDE 2008, pp. 506–515 (2008)
13. Zou, L., Chen, L., Özsu, M.: K-automorphism: a general framework for privacy preserving network publication. In: VLDB 2009, pp. 946–957 (2009)

Personalized Location Anonymity - A Kernel Density Estimation Approach

Dapeng Zhao[1], Jiansong Ma[1], Xiaoling Wang[1(✉)], and Xiuxia Tian[2]

[1] Shanghai Key Laboratory of Trustworthy Computing,
Institute for Data Science and Engineering, East China Normal University,
Shanghai, China
xlwang@sei.ecnu.edu.cn
[2] College of Computer Science and Technology,
Shanghai University of Electric Power, Shanghai, China

Abstract. In recent years, the problem of location privacy protection in location-based service (LBS) has drawn a great deal of researchers' attention. However, the existing technologies of location privacy protection rarely consider the personal visit probability and other side-information, which are likely to be exploited by attackers. In order to protect the users' location privacy more effectively, we propose a Personal Location Anonymity (PLA) combining side-information to achieve k-anonymity. On the offline phase, we utilize Kernel Density Estimation (KDE) approach to obtain the personal visit probability for each cell of space according to a specific users' visited locations. On the online phase, the dummy locations for each user's query can be selected based on both the entropy of personal visit probability and the area of Cloaking Region (CR). We conduct extensive experiments on the real dataset to verify the performance of privacy protection degree, where the privacy properties are measured by the location information entropy and the area of CR.

Keywords: Location-based services · Location privacy · k-anonymity · Cloaking region

1 Introduction

Location-based services (LBSs) have been wide-spread adopted. While users benefit from the convenience of LBSs, a lot of individual information is exposed to untrusted LBS service providers.

In many typical LBS usage scenarios, such as Dianping (www.dianping.com), users wish to get electronic coupons and discount by login information. However, users who send their locations to an untrusted LBS server do not want to expose his/her locations, which may be misused to harm his/her privacy. Once the untrusted LBS providers collect the users' location information, they can infer other information of users and release users' data to the third party, which will lead personal information is threatened.

© Springer International Publishing Switzerland 2016
B. Cui et al. (Eds.): WAIM 2016, Part II, LNCS 9659, pp. 52–64, 2016.
DOI: 10.1007/978-3-319-39958-4_5

To protect individual privacy, a lot of models are proposed. Location perturbation [1] and obfuscation [2,3] are the most popular and acceptable among these models. The main idea of most of the existing models is to reduce the spatial resolution of the user's exact location and achieve k-anonymity metric. And a lot of models exploit trusted-party structure to decrease client-side costs. When a user submits LBS queries with his exact location to the trusted-party anonymity server, the server hides the user's exact location by either building a bigger Cloaking Region (CR) [4] covering many other users or constructing a dummy location set [5] including the user's exact location and a lot of other locations geographically. Hence, the LBS service providers cannot determine what is the user's exact location. However, the data in the trusted third-party server faces certain risk, and hackers can capture all the users' data by breaking through the trusted server. Moreover, most of the existing approaches assumed adversary has no side-information, such as the probability that a user visits a location, which relates to locations and the semantics of geographical environment. Therefore, some improper dummy locations cover unlikely locations, such as lakes and mountains, which can be easily filtered out by the adversary. Even though, a few literatures [5] consider the side-information, researchers take the side-information as universal model for all users. The query probability is the same value in *DLS* [5] and enhanced-*DLS* [5] algorithm for all users, if location is the same one. As we know, influenced by the geographical environment and individual preference, the visited locations and movement trajectories for everyone are personalized. For instance, indoor persons tend to visit POIs around their living areas while outdoorsy persons prefer traveling around the world to explore new POIs. Therefore, the visit probability at the same location is different for different persons. From above analysis, existing models are difficult to guarantee the desired k-anonymity with expectations effectively.

To address above issues, we propose PLA, a personal location anonymity, to protect users' location privacy. Different from the existing approaches, PLA takes the individual geographical influence into account and selects dummy location according to the side-information, namely visited locations. The visit probability of a location is related to the places he visited, which is personal. Based on the finding, we model personal visit probability with kernel density estimation for each user. To enhance real-time, our scheme is divided into two phases: offline and online. In the offline phase, we utilize Kernel Density Estimation (KDE) to obtain the personal visit probability for each cell of space according to the a certain user's visited locations. In the online phase, the user firstly inputs his real location and parameter k. Then, we select $k-1$ dummy locations via maximizing location information entropy and the area of CR. Finally, we construct locations anonymity set as the user's location. And PLA makes sure large entropy and big area of CR to protect users' privacy.

The main contributions of this paper can be summarized as follows:

1. We propose a two-phases (online/offline) privacy protection framework to implement location protection, and this approach is efficient to save the anonymity time.

2. The personal location privacy is defined by KDE. And then, the dummy selection problem is formulated as Multiple-Criteria Decision Making (MCDM) model and a heuristic approach is used to solve this optimal problem.
3. Some preliminary experiments are conducted to evaluate the performance of our proposed scheme.

The rest of the paper is organized as follows. In Sect. 2, we review the work of location privacy protection. We present the relevant preliminaries of this paper in Sect. 3. The personal privacy protection approach PLA is proposed in Sect. 4. Section 5 presents the preliminary performance evaluation results. Finally, the paper is concluded in Sect. 6.

2   Related Work

For the existing techniques proposed to preserving users' location privacy, we categorize them into three main classes, i.e. spatial cloaking technique, pseudonyms and dummy location technique, cryptography technique and differential privacy technique.

Spatial cloaking technique [6] is a popular approach to protect location privacy. Gruteser M et al. [7] first observed the problem of location sensitive information disclosure and proposed spatial and temporal cloaking approach to protect privacy information. ICliqueCloak [3] was proposed to protect location privacy against location-dependent attacks. $SALS$ [8] enables users to cooperate with each other, in a Peer-to-Peer (P2P) way, to generate the cloaking zones based on the semantic context of the geographical locations.

Mix-zone [9,10] is one representative of Pseudonyms and dummy location techniques, which enables users only to change their pseudonyms inside a special region where users do not report the exact locations. [11] exploited dummy locations to achieve anonymity without employing anonymizer. PAD [12] approach is capable of offering privacy-region guarantees and uses so-called dummy locations that are deliberately generated according to either a virtual grid or circle. Niu et al. [5] considered query probability into account, and proposed DLS and enhanced-DLS algorithms to protect location privacy.

Cryptography technique [13] is also one of the main approaches to protect location privacy. Ghinita et al. [14] proposed a novel framework to support private location dependent queries, based on the theoretical work on Private Information Retrieval PIR. The PLAM framework [15] utilized a privacy-preserving request aggregation protocol with k-Anonymity and l-Diversity properties to keep users' location privacy. However, the calculating cost of cryptography technique is greater than other techniques.

Differential privacy technique [16,17] is a new technique to protect location privacy. Andrs et al. [16] presented a mechanism for achieving geo-indistinguishability by adding controlled random noise to the users' location with differential privacy.

From the analysis above, most of the existing approaches assumed adversary has no side-information and didn't take users' visit characteristics into consideration. So the approaches are difficult to make sure the users' privacy effectively.

3 Preliminaries

Our scheme assumes a client/server architecture without a third party anonymity server. Moreover, we focus on protecting users' location privacy when they issue snapshot queries in LBS which are expected to occur frequently in practice.

3.1 Query Model in LBS and Definition

In this paper, we divide the space into $n \times n$ cells and the user's location is indicated as a cell which he is in. For a location, namely a cell, we use the mean visit probability of five points in the cell to replace the personal visit probability of the cell for users. As shown in the Fig. 1, space is divided into a 4×4 $cells = \{c_1, c_2, \cdots, c_{16}\}$ and the selected five points set for c_6 is $P = \{P_1, P_2, P_3, P_4, P_5\}$ which includes four corners and a core point.

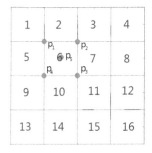

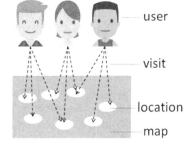

Fig. 1. Grid   **Fig. 2.** Users' visited locations

A query which a user sends to LBS service provider is presented as $Q = (uid, K, C)$, where uid is the user's identification and K denotes the keywords of the query which are not the focal point of research in this paper. The $C = \{c_1, c_2, \cdots, c_k\}$ denotes the anonymity set which includes the user's current location and $k - 1$ dummy locations.

Definition 1. *The distance between two cells is the Euclidean distance between the centers of them.*

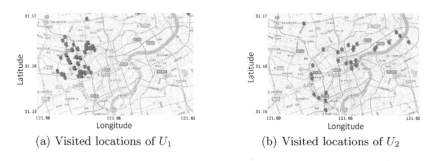

(a) Visited locations of U_1 (b) Visited locations of U_2

Fig. 3. Distributions of personal visited locations

3.2   The Personal Visit Preference

As we know, users' visit behaviors on locations are influenced by geographical factors and the users' preferences. In this paper, we aim to exploit the personal visit relation between users and locations to select dummy locations. In reality, each user's visit behavior should be unique and personal under the geographical influence as showed in Fig. 2.

Real-world motivating examples. To observe a certain user's unique visit preference, a spatial analysis is conducted on a real data set in Shanghai collected from China Telecommunications, which is one of the biggest communication enterprises. Figure 3 depicts the characteristics of the visited locations of two different users, U_1 and U_2. Obviously, the two users' visited locations can be affected by their visit preferences. Due to geographical influence and limit, each user's visit probability on the same location is different. So it is undesirable to model visit preference [5] as a universal distribution. It is not difficult to find that the distribution over the distance between every pair of visited locations by the same user is also unique.

3.3   Kernel Density Estimation of Distance Distribution

The experimental results (depicted in Sect. 3.2) inspire us to study the personal visit preference by distance distribution. Concretely, we model the personal distribution of the distance between any pair of locations visited by the user using KDE [18], since it can be used with arbitrary distributions and without the assumption that the form of the distance distribution is known. Modeling the personal distance distribution consists of two steps: distance sample collection and distance distribution estimation.

Distance sample collection. We can acquire a distance sample for a user by computing the Euclidean distance between every pair of locations that is randomly selected from the user's visited locations in reality. Cold-start is not a serious problem because a user can input some of his visited locations when he uses this service firstly.

Distance distribution estimation. Let D be the distance sample for a certain user, which is drawn from the distribution with an unknown density f. Its kernel density estimator $\hat{f}$ over distance d using D is given by:

$$\hat{f}(d) = \frac{1}{|D|h} \sum_{d' \in D} K\left(\frac{d - d'}{h}\right) \tag{1}$$

where $K(.)$ is the kernel function and h is a smoothing parameter, called the bandwidth. In this paper, we apply the most popular normal kernel:

$$K(x) = \frac{1}{\sqrt{2\pi}} e^{-\frac{x^2}{2}} \tag{2}$$

and the optimal bandwidth [19]:

$$h = \left(\frac{4\hat{\sigma}^5}{3|D|}\right)^{1/5} \approx 1.06\hat{\sigma}|D|^{-1/5} \tag{3}$$

where $\hat{\sigma}$ is the standard deviation of the sample in D.

3.4   Metrics for Location Privacy

In this paper, we use entropy [5] and the area of CR [3] to measure the level of privacy. Entropy is used as the location privacy metric. The general form of location information entropy is widely accepted in the location privacy research community. It is used to measure the uncertainty in distinguishing the real location of an individual from all the candidates. To compute the entropy, each possible location has a probability which is visited by the user, denoted by p_i, and the sum of all probabilities p_i is 1. The entropy H identifying the real location in the candidate set is defined as:

$$H(x) = -\sum_{i=1}^{k} p_i \cdot \log_2 p_i \tag{4}$$

As we know, the theoretical maximum entropy is $log_2 K$ when all the K possible locations have the same probability $\frac{1}{K}$.

The other metric for location privacy is the area of CR. The area of CR covered by all k locations tends to be greater when individual requires higher level of privacy. As we know, the area of polygon is difficult, so we use approximate approach to replace it. Intuitively, the sum of the distances between pairs of locations in anonymity set C can be used to measure the area of CR, which is $\sum_{i \neq j} d(c_i, c_j)$, where $d(c_i, c_j)$ denotes the distance between cell c_i and c_j in anonymity set C.

3.5   Attack Model

Any party can be a potential attacker who wants to gain sensitive information about a particular user. We consider two kinds of adversaries: weak adversary and strong adversary. Any party can be a weak adversary if he obtains the current queries from the particular user and a little visited positions by monitoring his communications. On the other hand, a strong adversary gains the user's current queries and a lot of his visited locations. Additionally, he knows the location privacy protection mechanism used in the system. Based on these information, he tries to infer the user's real location. Therefore, the probability that the strong adversary distinguishes the real location from anonymity set is larger than that of the weak adversary. In this paper, we only take strong adversary into account.

4   Personal Dummy Location Anonymity

Our objective is to generate a dummy location set based on the model of personal visit probability to achieve k-anonymity. To ensure high privacy level, dummies whose personal visit probability is similar to that of the user's current cell should be selected. In this section, we propose a personal dummy location anonymity scheme to satisfy the goal, which includes two parts: calculating personal visit probability of all cells offline and selecting dummy location set online.

4.1   Personal Visit Probability of Cell (PVPC) – Offline

We model personal distance distribution for all users based on KDE and then calculate the visit probability of any point in the space based on personal distance distribution. We propose a method based on KDE to derive the probability of a user U visiting a cell c_j given the user's set of sample visited locations $L = \{l_1, l_2, \cdots, l_n\}$. Firstly, we compute the distance of every pair of locations in L and P_k which is depicted by Sect. 3.1, $k \in \{1, 2, \cdots, 5\}$ as follows:

$$d_{ik} = \text{distance}(l_i, P_k), \forall l_i \in L \tag{5}$$

Each d_{ik} is then used to derive a probability based on Equation (1) as follows:

$$\hat{f}(d_{ik}) = \frac{1}{|D|h} \sum_{d' \in D} K\left(\frac{d_{ik} - d'}{h}\right) \tag{6}$$

The probability which U visits a point P_k in cell c_j can be obtained by taking the mean probability as follows:

$$p(P_k) = \frac{1}{n} \sum_{i=1}^{n} \hat{f}(d_{ik}) \tag{7}$$

Finally, the probability of U visiting a grid c_j can be calculated by taking the mean probability as follows:

$$p(c_j) = \frac{1}{5} \sum_{k=1}^{5} p(P_k) \tag{8}$$

Eventually, we can exploit the visited locations to calculate the personal visit probability of every cell for all users.

Algorithm 1. PVPC (Personal Visit Probability of Cell)

---

1 Input: a user's visited locations $L = \{l_1, l_2, \cdots, l_1\}$.
2 Output: p_i denoting the user's visit probability of $cell_i$
3 **for** *each $cell_i$ in the space* **do**
4 | Initialize points set $P = \{P_1, P_2, P_3, P_4, P_5\}$ containing four corners and core of $cell_i$
5 | **for** *each point P_k in P* **do**
6 | | Calculate the distance d_{ik} between l_i and P_k;
7 | | Obtain the probability $\hat{f}(d_{ik})$ of deriving d_{ik} via Equation (6);
8 | Calculate the probability of visit P_k via Equation (7);
9 | Calculate the visit probability of $cell_i$ via Equation (8);

4.2 Anonymity Set Selection(ASS) – Online

Since our PLA algorithm takes two criteria (namely location information entropy and the area of CR) into consideration, the anonymity set selection problem can be formulated as MCDM model. Let $C^* = \{c_1, c_2, \cdots, c_k\}$ denote the location anonymity set in LBS. The MCDM can be described as:

$$Max\{-\sum_{i=1}^{k} P_i \cdot \log P_i, \sum_{k \neq j} d(c_i, c_j)\}. \tag{9}$$

where $c_i, c_j \in C^*$, p_i and p_j denote the personal visit probabilities of c_i and c_j respectively.

It is difficult to select a anonymity set to satisfy all criteria simultaneously in MCDM. In this paper, we take heuristic solution [5] to solve the problem. We can take two steps as follows: (1) maximize entropy; (2) maximize the area of CR, and the concrete algorithm, namely *ASS*, is depicted by Algorithm 2.

Maximize Entropy: Firstly, a user in LBS inputs the parameter k which is closely related to k-anonymity and the real location c. Secondly, the user needs to read all the personal visit probabilities which are computed by *PVPC* and then sorts all cells by the personal visit probabilities. In the sorted list, the user chooses $4k$ candidates containing the $2k$ cells before c and the $2k$ cells after c to guarantee that all visit probability is as similar as possible to the visit probability of the user's current location. After that, m dummy location sets $C = \{C_1, C_2, \ldots, C_m\}$ is selected from $4k$ candidates, each with $2k$ cells. For every set, real location and $2k - 1$ dummy location are contained. The $j^{th}(j \in [1, m])$ set can be denoted as $C_j = \{c_{j1}, c_{j2}, \ldots, c_{ji}, \ldots, c_{j2k}\}$. According to the personal visit probabilities of the included cells, the normalized visit probabilities can be denoted as $p_{j1}, p_{j2}, \ldots, p_{ji}, \ldots, p_{j2k}$ and computed by:

$$P_{ji} = \frac{q_{ji}}{\sum_{i=1}^{2k} q_{ji}}. \tag{10}$$

where q_{ji} is the personal visit probability of the cell c_{ji} and $\sum_{i=1}^{2k} P_{ji} = 1$. Therefore, the entropy H_j of anonymity set C_j can be computed by:

$$H_j = -\sum_{i=1}^{2k} P_{ji} \cdot \log P_{ji} \tag{11}$$

Finally, the location set $C' = \{c'_1, c'_2, \cdots, c'_i, \cdots, c'_{2k}\}$ whose entropy is maximum in C is selected. And it can be described as:

$$C' = argmax H_j. \tag{12}$$

Maximize the Area: In this section, we select k location set $C^* = \{c_1, c_2, \cdots, c_k\}$ which contains the user's real location and $k-1$ dummy locations from C' according to $\sum_{i \neq j} d(c_i, c_j)$. It can be denoted as:

$$C^* = argmax\{\sum_{k \neq j} d(c_i, c_j)\}. \tag{13}$$

Firstly, we construct location set $C^* = \emptyset$. And the user's real location is put into C^* and deleted from C'. Then, the next location c^* is selected to be put into C^* and to be deleted from C', when $\sum_{c_j \in C^*} d(c^*, c_j)$ is greatest. As lines 9–12 in Algorithm 2 depicted, ASS repeats this step $k-1$ times. Eventually, the anonymity set C^* is selected successfully.

Algorithm 2. ASS (Anonymity Set Selection)

---

1 Input: p_i denoting the user's visit probability of $cell_i$; the user's real cell c; parameter k based on k-anonymity; the number of sets m.

2 Output: anonymity set C^*

3 Choose $4k$ dummy candidates including $2k$ cells before and $2k$ cells after the user's real cell in the sorted list;

4 Construct m location sets $C = \{C_1, C_2, \cdots, C_m\}$, each C_j contains a real cell and $2k - 1$ other cells randomly selected from $4k$ dummy candidates;

5 **for** each C_j **do**

6 $\quad\lfloor$ Calculate entropy H_j via Equation (11);

7 $C' = argmax H_j$

8 Initialize $C^* = \{c\}$ and $C' = C' - c$;

9 **for** $i = 0 : k - 1$ **do**

10 $\quad$ **if** c^* in C' and $\sum_{c_j \in C} d(c^*, c_j)$ **then**

11 $\quad\quad$ $C^* = C^* + c^*$;

12 $\quad\quad$ $C' = C' - c^*$;

13 Return C^*;

---

5   Performance Evaluation

5.1   Experiment Setup

Experiments are implemented in java and run on a desktop PC with a Quad AMD 3.2 GMHz processor and 8GB main memory. We use the publicly available large-scale real check-in data sets as the sample of users' visited locations and the data sets were crawled from Gowalla [20] to sample the distance for each user. The statistics of the data sets are shown in Table 1.

Moreover, we select $40\,km \times 40\,km$ region in America and divide it into 80×80 sells. The personal visit probability of each cell based on individual check-in data is calculated by KDE.

Table 1. Statistic of the gowalla data set

Number of users	196,591
Number of locations (POIs)	1,280,969
Number of check-ins	6,442,890
Avg. No. of visited POIs per user	37.18
Avg. No. of check-ins per location	3.11

We evaluate the performance of PLA by comparing three algorithms, namely dummy [11], DLS[5] and enhanced-DLS[5]. The dummy selection scheme in [11] is taken as the baseline, which randomly chooses dummy locations to protect location privacy based on random walk. The DLS and enhanced-DLS approaches are the dummy selection approaches proposed in [5], which take query probability into account to achieve k-anonymity. As we know, the location information entropy of the optimal scheme is $log_2 k$ in theory.

5.2 Evaluation Results

In the process of performance evaluation, we take location information entropy and the sum of distance as metrics. Real-time is also a important trait in LBS and we measure it using anonymity cost.

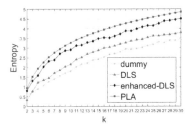

Fig. 4. Entropy vs. k

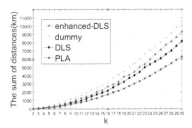

Fig. 5. The sum of distances vs. k

Entropy vs. k: As we know, the location information entropy of anonymous set indicates the protection strength. Larger information entropy implies that the attackers are more uncertain about the user's real location and higher privacy level. As Fig. 4 depicts, the location information entropies of all approaches increase along with k. It's not surprising to find dummy is terrible since it does not consider any side-information. It is not difficult to find that our scheme *PLA* is optimal and provides better protection than *DLS* and enhanced-*DLS* at all settings.

Area of CR vs. k: The area of CR covered by all k locations is a fine metric to evaluate privacy. Under the premise of ensuring a large entropy, we hope to obtain a bigger area. The sum of distances of each location pair in anonymity set indicates the area well. In Fig. 5, we measure the area of CR by the sum of distances of each location pair in anonymity set. With k increasing, we observe that the sum of the distances is rising. We can observe that PLA is slightly short of the other schemes. It is reasonable since users' visit locations are limited by geographical environment and these locations are local to a certain extent.

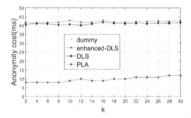

Fig. 6. Anonymity times vs. k

Anonymity times vs. k: We proposed two-phases (online/offline) privacy protection framework to save the anonymity time and enhance real-time in this paper. From Fig. 6, the anonymity costs for algorithms hardly increased with k increasing. The anonymity cost of dummy is close to zero since it doesn't take any side-information and generates dummy location set randomly. The performances of *DLS* and enhanced-*DLS* are similar, and both of them is terrible. Benefiting from two-phases (online/offline) privacy protection framework, our scheme is close to dummy and obtains a better performances on anonymity cost.

5.3 Security Analysis

In our system, all of the anonymous process is completed in client-side and no information besides users' queries Q is sent to any party. By doing this, it is hard to gather users' data for adversary. Strong adversary can conduct inference attack to infer users' real location based on obtained information.

As we know, a scheme can prevent inference attack if the probability that any cell in dummy anonymity set C is the user's real location is equal from the adversary's point of view. p_i and p_j represent respectively the probabilities that

c_i and c_j are the user's real location, where c_i and c_j are any two locations in C. Our scheme can prevent inference attack if and only if $p_i = p_j$. Since a strong adversary knows location privacy protection mechanism and a lot of the user's visited locations data, he can compute the visit probability of each cell as the probability that the cell is the user's real location. However, we can know the probability is tremendously similar in anonymity set according to Eq. (12), namely $p_i = p_j$. Therefore, Our scheme can prevent inference attack.

6   Conclusion

In this paper, we proposed an online/offline privacy protection approach and utilized KDE to calculate personal visit probability. We firstly calculated the personal visit probability of all cells offline. Then, dummy location set was selected according to personal visit probability online by the dummy selection problem formulated as MCDM model, and a heuristic approach was used to solve this optimal problem. Experimental results showed the effectiveness of our approach.

Acknowledgment. This work was supported by NSFC grants (Nos. 61170085, 61472141, 61321064), Shanghai Knowledge Service Platform Project (No. ZF1213), Shanghai Agriculture Science Program (2015) Number 3-2 and Project of Shanghai Science and Technology Committee under Grant (No. 15110500700).

References

1. Yiu, M.L., Jensen, C.S., Moller, J., et al.: Design and analysis of a ranking approach to private location-based services[J]. ACM Trans. Database Syst. (TODS) (2011)
2. Machanavajjhala, A., Kifer, D., Gehrke, J., et al.: l-diversity: Privacy beyond k-anonymity. ACM Trans. Knowl. Disc. Data (TKDD) **5**, 1–47 (2007)
3. Pan, X., Xu, J., Meng, X.: Protecting location privacy against location-dependent attacks in mobile services. IEEE Trans. Knowl. Data Eng. **24**, 1506–1519 (2012)
4. Mokbel, M.F., Chow, C.Y., Aref, W.G.: The new Casper: Query processing for location services without compromising privacy. In: Proceedings of the 32nd International Conference on Very Large Data Bases. VLDB Endowment (2006)
5. Niu, B., Li, Q., Zhu, X., et al.: Achieving k-anonymity in privacy-aware location-based services. In: INFOCOM, 2014 Proceedings. IEEE (2014)
6. Zhang, C., Huang, Y.: Cloaking locations for anonymous location based services: A hybrid approach. Geoinformatica **13**, 159–182 (2009)
7. Gruteser, M., Grunwald, D.: Anonymous usage of location-based services through spatial and temporal cloaking. In: Proceedings of the First International Conference on Mobile Systems, Application, and Services. USENIX, San Francisco (2003)
8. Che, Y., Chiew, K., Hong, X., et al.: SALS: Semantics-aware location sharing based on cloaking zone in mobile social networks. In: Proceedings of the First SIGSPATIAL International Workshop on Mobile Geographic Information Systems (2012)
9. Bamba, B., Liu, L., Yigitoglu, E.: Road network-aware aonymization in mobile systems with reciprocity support. In: 2015 24th International Conference on Computer Communication and Networks (ICCCN). IEEE (2015)

10. Palanisamy, B., Liu, L.: Attack-resilient mix-zones over road networks: Architecture and algorithms. IEEE Trans. Mob. Comput. **14**, 495–508 (2015)
11. Kido, H., Yanagisawa, Y., Satoh, T.: An anonymous communication technique using dummies for location-based services. In: International Conference on Pervasive Services, 2005 ICPS 2005, Proceedings. IEEE (2005)
12. Lu, H., Jensen, C.S., Yiu, M.L.: PAD: Privacy-area aware, dummy-based location privacy in mobile services. In: Proceedings of the Seventh ACM International Workshop on Data Engineering for Wireless and Mobile Access. ACM (2008)
13. Jia, J., Zhang, F.: K-anonymity algorithm using encryption for location privacy protection. Int. J. Multimedia Ubiquit. Eng. **10**, 155–166 (2015)
14. Ghinita, G., Kalnis, P., Khoshgozaran, A., et al.: Private queries in location based services: Anonymizers are not necessary. In: Proceedings of the 2008 ACM SIGMOD International Conference on Management of Data. ACM (2008)
15. Lu, R., Lin, X., Shi, Z., et al.: PLAM: A privacy-preserving framework for local-area mobile social networks. In: 2014 Proceedings IEEE INFOCOM. IEEE (2014)
16. Andrs, M.E., Bordenabe, N.E., Chatzikokolakis, K., et al.: Geo-indistinguishability: Differential privacy for location-based systems. In: Proceedings of the 2013 ACM SIGSAC Conference on Computer Communications Security. ACM (2013)
17. Clifton, C., Tassa, T.: On syntactic anonymity and differential privacy. In: 2013 IEEE 29th International Conference on Data Engineering Workshops (ICDEW). IEEE (2013)
18. Zhang, J.D., Chow, C.Y.: iGSLR: Personalized geo-social location recommendation: a kernel density estimation approach. In: Proceedings of the 21st ACM SIGSPATIAL International Conference on Advances in Geographic Information Systems. ACM (2013)
19. Silverman, B.W.: Density Estimation Stat. Data Anal. CRC Press, Boca Raton (1986)
20. Cho, E., Myers, S.A., Leskovec, J.: Friendship and mobility: User movement in location-based social networks. In: Proceedings of the 17th ACM SIGKDD International Conference on Knowledge Discovery and Data Mining. ACM (2011)

Detecting Data-model-oriented Anomalies in Parallel Business Process

Ning Yin[1,2], Shanshan Wang[1,2], Hongyan Li[1,2(✉)], and Lilue Fan[1]

[1] Key Laboratory of Machine Perception, Ministry of Education, Peking University, Beijing 100871, China
[2] School of Electronics Engineering and Computer Science, Peking University, Beijing 100871, China
lihy@cis.pku.edu.cn

Abstract. Currently, most information systems are data intensive. The data models of such are posing notable influence on business processes. However, the predominance of existed process verification methods leave out the impact of data models on process models. Meanwhile, with parallel structures in business processes multiplying, business process structures are becoming increasingly intricate and large in size. A parallel structure engenders also uncertainty, and consequently increases the chances and decreases the detectability of anomalies occasioned by process and data model conflicts. In this paper, these anomalies are analyzed and classified. A data state matrix and data operation algebra is introduced to establish the relation between the parallel-process model and the data model. Then, an anomaly detection method under the divide-and-conquer framework is proposed to ensure efficiency in detecting anomalies in business processes. Both theoretical analysis and experimental results prove this method to be highly efficient and effective in detecting data model oriented anomalies.

Keywords: Parallel business process · Data model · Data-model-oriented anomalies · Anomalies detection · Semantic verification

1 Introduction

With the increasing scale of businesses and their process management, the business process model has grown unprecedentedly in complexity and size [8]. The prevalence of distributed computing and businesses' demand for resource utilization and work efficiency render it necessary to use process models with as many parallel structures as possible. A parallel business process requires an exponential increase in the process state space, which poses a challenge to existing means of verifying business processes in a short time period. Even worse, the parallel structure also engenders uncertainty, and consequently increases the probability of data-model-oriented anomalies in business processes.

This work was supported by Natural Science Foundation of China (No. 61170003).

B. Cui et al. (Eds.): WAIM 2016, Part II, LNCS 9659, pp. 65–77, 2016.
DOI: 10.1007/978-3-319-39958-4_6

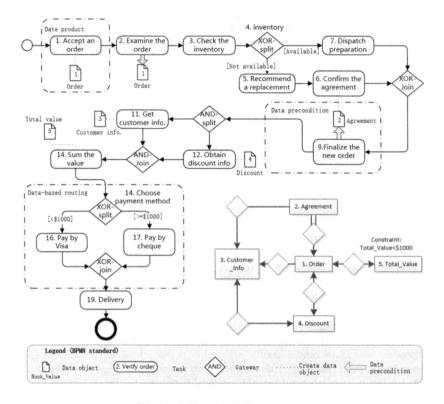

Fig. 1. Online bookshop process

In the past, there were too little anomalies caused by data to draw enough attention. But nowadays, most information systems are data intensive, leading to increasing data-model-oriented anomalies, which could no longer be ignored. The relations among data objects are also becoming more and more complicated. Previously, data is regarded as a part of the process, treated and managed as variables. For the size of data objects is so large and the relations of those are so complex that beyond the ability of variables means, also considering the concept of loose coupling, data objects are detached from process and primarily managed by data models independent of business process management. The separation of and interaction between business process model and data model have the potential for conflict.

For a more intuitive understanding of likely anomalies, please refer to Fig. 1, a flow chart displaying the process of an on-line book vendor, including its paralleled branches and a legion of data operations, as well as a corresponding data model. Both the data model and process model are flawless. However, three kinds of anomalies are still to be expected while executing the process, leading to an unexpected execution or resulting in deadlock. For example, a data object *Discount* is created for the task *Obtain discount info*, yet *Discount* is dependent on another data object *Customer_info*, which is created for the task *Get customer info*. In addition, *Get customer info* is performed in parallel with *Obtain discount info*. If *Customer_info* is yet to be created when creating *Discount*, an anomaly will inevitably occur. The scenario would be much more complicated in large-scale parallel business processes. Please refer to Sect. 3 for a more detailed data-model-oriented anomaly analysis.

The prime difficulties confronting the anomaly detection are as follows:

(1) Anomalies are triggered by the inconsistency between parallel-process models and data models. Even in a scenario where both models are flawless, anomalies are still to be expected.
(2) Uncertainty caused by paralleled branches renders it more difficult to detect anomalies. To make sure no anomaly shows up in actual operating, every possible state of the process model must be checked.

However, a mass of paralleled branches lead to a huge state space which extends exponentially and makes it unlikely for all anomalies to be detected within limited time frame. Simply searching the state space or traversing possible execution sequence [7,18] to detect potential anomalies will lead to a combination explosion. There are already $(5!/(2!3!)+6!/(2!4!))22=100$ possibilities in a case as simple as the business process segment shown in Fig. 1. It is next to impossible to complete a detection for large-scale parallel business processes within stipulated time frame. So efficiency of the anomaly detection method is of first priority.

This paper investigates the challenging problem of inconsistencies between parallel business processes and data models, and identifies three types of data-model-oriented anomalies. A divide-and-conquer-framed anomaly detection algorithm is proposed to ensure an efficient detection of anomalies in parallel business processes. Under this framework, the process is divided into blocks. For each block, data operations are deduced and checked by utilizing a newly invented data-operation algebra system, which offers an efficient way to merge and simplify data operation with algebraic method. Then, a data state matrix model is raised to establish the relation between process model and data model. Finally, an anomalies detection measure based on the data state matrix will be employed to conduct the detection work.

2   Related Work

The business process verification is stratified into three levels in this paper:

Syntax Checking. Syntax checking that verifies whether a business process model meets the grammar standard of certain modeling language [3] is one of the most fundamental segments in the business process verification.

Structure Verification. Structure verification is mainly used to examine whether there are structure conflicts that will lead to operational malfunctions. The majority parts of this research focus on control flow verification, able to verify whether control structures in verification process meet certain properties such as soundness or weak soundness. Some formalized methods are used in verifications, e.g. linear algebraic [4], finite state verification [18], coverability graph analysis [15]. Other verification methods include: Model Checking, Reachability Analysis, Graph Reduction Techniques, etc. However, the structure verification is restricted by the control structure of the business process itself. This on one hand simplifies and is conducive to formalizing the problems. Yet on the other, it cannot guarantee a zero anomalies process in practice.

Semantic Verification. Semantic verification is the highest-level verification and the most complete one. It examines semantic information in business process operations, making no anomaly show up in actual operating process. Data, as an operation object in the process, is of great importance. More emphases are now placed on the verification of data flow in business process models. They manages to detect the anomalies that business processes have in data manipulations, such as Missing data [2,5] Conflicting Data [2,5,12], Redundant Data [2]. Nevertheless, those methods are not really complete Semantic Verification, for data is regarded merely as process variables and isolated objects, and the relations between them are neglected. As shown in Experiment, anomalies stemmed from the conflicts between the process and data model are unpredictable.

To verify the consistency between process model and data model, a Process-Data Graph based verification method has been proposed [17]. However, it needs to explore the state space in an exhaustive manner. A business process with numerous paralleled branches is very likely to result in the combination explosion.

In conclusion, few of current business process verification methods take into consideration the impact of data model, whilst they are inefficient in verifying parallel processes. Hence, a new approach to detect anomalies in parallel business processes, especially on large-scale ones, is proposed here, which looks at both the data model and process model and guarantee no conflicts between them.

3   Problem Definitions

Definition 1 (Data Model). *A data model can be described by a tuple (D, D_{dp}, D_{cstr}), where D stands for the set of data objects, D_{dp} stands for the dependence relationship, $D_{dp} \subseteq D \times D$. $(d_1, d_2) \in D_{dp}$ means that d_1 is dependent on d_2. The data constraints set is denoted as D_{cstr}, in which all the constraints are represented by predicates of data objects.*

Definition 2 (Business Process). *The business process is denoted as a tuple $(T, G, Cond, CF)$, where T is a task set and G is a gateway set. A gateway is exactly one of $\{and\text{-}join, and\text{-}split, xor\text{-}join, xor\text{-}split, or\text{-}join, or\text{-}split\}$. -join gateway and its correspondent -split always emerge in pairs. $CF \subseteq (T \cup G) \times (T \cup G)$ defines the sequential relationship of the process.*

To guarantee the universality of the method, a more generalized definition is adopted. We assume a process can be converted into a directed acyclic graph [9,16], and most modeling languages can be transferred to such a definition easily.

Anomalies are triggered by the conflicts during the interaction of process model and data model. Three interaction modes now prevail in most of today's processes: (1) Data Products: Data products and mid-products of a business process are defined by the data model [10, 12–14]. (2) Tasks' Data Preconditions: The execution of a task in a business process usually depends on some specified data objects [10]. (3) Data-driven Process Control: The control flow of business processes can be influenced by data through many mechanisms [10]. For example, a gateway always uses a set of data items to decide the routing.

Here we give the definition of the three interaction modes. Data creating and deleting are deemed more important than reading and updating, for the former two operations decide the existence of data in a task.

Definition 3 (Interaction Modes). *Business process model and data model is connected in three ways* $(DataOpr, PreData, Cond)$. *Data Operation is denoted as a tuple* $DataOpr = (OprTyp, T, D)$, *where* $OprTyp \in \{create, delete, null\}$. *Precondition Data is denoted as a tuple* $PreData = (T, D)$. *If the execution of task* t_j *is dependent on data* d_i, *then the relation between the two is defined as* $(t_j, d_i) \in PreData$. $Cond = \{c_i\}$ *refers to the set of Control Conditions in the process, and each condition is represented by its data object's predicate.*

These interaction modes may be accompanied with conflicts, thus leading to mainly three kinds of anomalies.

(1) **Data Dependence Unsatisfied(DDU)**: In certain tasks, time-lags, even dead-locks, can be caused when the data object on which a new data object to be created depends, has not been created yet. In Fig. 1 task *Finalize the new order* depends on the data object *Agreement*; if *Agreement* has not been created yet, either task waiting or a dead-lock will ensue.

(2) **Precondition Data Undefined(PDU)**: When a task is set off by the control flow and one of the data objects on which it depends has not been created yet, the task will fail to go on its execution. In Fig. 1 when task *Obtain discount info* creates the object *Discount*, and *Discount* depends on object *Customer Info*, which has not been created yet.

(3) **Data Condition Contradiction(DCC)**: When the process control condition conflicts with data constraints, there will be a dead-lock or leading to inaccessibility of certain branches in the process. It happens in Fig. 1 since there is a conflict between the route condition $Total\_Value > 1000$ of the gateway *Choose payment method*, and the data constraint $Total\_Value < 1000$ in database.

The Data State Matrix is introduced to record the results of interactions and to detect potential conflicts.

Definition 4 (Data State Matrix). *When task* t_j *is completed, the state of the data object* d_i *is denoted as* ds_{ij}, $ds_{ij} \in \{lack, exist, uncertain\}$. *For all the possible* i *and* j, *all the fitting* ds_{ij} *can thus form Data State Matrix* $[ds_{ij}]$.

Before a business process starts, the initial states for data are all definite. Later on, if $\exists t_j, d_i, \text{s.t.}(create, t_j, d_i) \in DataOpr$, then $ds_{ij} = exist$. Similarly, if $\exists t_j, d_i$ s.t.$(delete, t_j, d_i) \in DataOpr$, then $ds_{ij} = lack$. $ds_{ij} = uncertain$ if there exists two sequences, one of which results in state $exist$ while the other $lack$.

Definition 5 (Data Model Oriented Anomaly). *Given the data state matrix* $[ds_{ij}]$. *If* $\exists t_j, d_a, d_b$, *s.t.* $(create, t_j, d_a) \in DataOpr$ & $(d_a, d_b) \in D_{dp}$ & $ds_{bj} \neq exist$, *then* t_j *would lead to anomaly DDU. If* $\exists t_j, d_i, \text{s.t.}(t_j, d_i) \in PreData$ & $ds_{ij} \neq exist$, *then* t_j *would lead to anomaly PDU. If* $\exists c_i, c_j$, *s.t.* $c_i \in D_{cstr}$ & $c_j \in Cond$ & c_i *conflicts with* c_j, *then* c_i, c_j *would set off anomaly DCC.*

Given that the definition of DDU and PDU relies heavily on the matrix $[ds_{ij}]$, the anomalies detection can be converted into the sub-problem: How to effectively obtain the data state matrix?

4   Divide and Conquer Framed Anomaly Detection

Obtaining data state matrix is a non-trivial task since the state space of a parallel business process can be too large for an exhaustive search [17]. Therefore, the divide-and-conquer strategy is adopted to obtain the matrix in polynomial time (less than square). First,the process is decomposed into a process tree. Then, data operations within each fragment are reduced using data operation algebra.

4.1   Process Fragments

Definition 6 (Process Graph). *Process graph is a directed acyclic graph* $G = (V, E, v_s, v_e)$, *where* V *is the node set,* $V = T \cup G$, E *is an edge set defined by* CF, $E = CF$. *The Start node and End node are denoted as* v_s, v_e *respectively.*

Definition 7 (Process Single-Entry-Single-Exit(SESE) Fragment). *Let* $G = (V, E, v_s, v_e)$ *be a process graph. A SESE fragment* $F = (V', E', v_{pre}, v_{post})$ *is a nonempty subgraph of* G, *i.e.* $V' \subseteq V$, *and* $E' = E \cap (V' \times V)$ *such that there exists edge* $e, e' \in E$ *with* $E \cap ((V \setminus V') \times V') = e$ *and* $E \cap (V' \times (V \setminus V')) = e'$. $e = (v_{pre}, v_i)$ *and* $e' = (v_j, v_{post})$ *stand for the entry and the exit edge of* F, *respectively.* v_{pre} *and* v_{post} *stand for the entry and the exit node of* F.

Fragments F and F' are thought to be in sequence if the exit edge of F is the entry edge of F' or reversely. The union of two fragments F and F' $F \cup F'$, which are in sequence, becomes a single fragment again. A fragment F is non-canonical if there are fragments X, Y, Z such that X and Y are in sequence, $F = X \cup Y$, F and Z are in sequence; otherwise F is said to be canonical.

Canonical fragments do not overlap. Two canonical fragments are either nested or disjoint [6]. Therefore, it is possible to organize the canonical fragments in a unique tree, similar to the Program Structure Tree shown in [6]. The tree is named the process structure tree of a process graph. It can be computed within linear time in size of the process graph [1,16]. Since we are only interested in canonical fragments here, we mean 'canonical fragment' whenever we say 'fragment' in the following. It can be proved with ease that to any fragment $F = (V', E', v_{pre}, v_{post}), v_{pre} \in \{and\_split, or\_split, xor\_slpit\}$, and $v_{post} \in \{and\_join, or\_join, xor\_join\}$. The parent of a fragment F is the smallest fragment F' that contains F. Then, F is called the child fragment of F'. The parent fragment of a node v_i is the smallest fragment that contains v_i. The level of a fragment $level(F)$ is the depth of F in the process structure tree and a node is of the same level with its parent fragment.

4.2   Data Operation Algebra

To estimate the influence that each fragment has on the data states more efficiently, data operation algebra is introduced.

Definition 8 (Data Operation Algebra). $(Dom, \odot, \oplus)$ *is an algebra system over domain* $Dom = \{\alpha, \delta, \varepsilon, \tau\}$. $\odot$ *is a closed binary operation:* $Dom \times Dom \rightarrow Dom$, *and the mathematical table is defined in Table 1.* $\oplus$ *is another closed binary operation:* $Dom \times Dom \rightarrow Dom$; *and the table is also in Table 1.*

Obviously, the operations are of certain properties: (1) $\odot$ is associative, idempotent and τ is the identity element. (2) $\oplus$ is associative, commutative, idempotent and τ is also the identity element. (3) $\odot$ is distributive over $\oplus$.

α, δ are respectively denoted as creating data and deleting data. For the sake of brevity, the operations discussed in the following act on single data object. The same rule applies to multiple objects once expanded into the vector form.

As to the sequence structures in the process model, the operation executed by the successive task will cover the previous one. $\odot$ is used to depict the sequence reduction rules. As to the parallel tasks whose data operations are different, since the execution sequence cannot be decided before execution, the result is consequently undetermined. Therefore, a third operation ε is introduced, and $\oplus$ is used to depict the parallel reduction rules. The operation of a task or fragment on a data is defined as void operation τ if it does not create or delete that data object. Meanwhile, τ is the identity element of operation $\odot$ and $\oplus$.

As to the fragments operations, we begin from the bottom of the tree. The data operation of a fragment F or task t is denoted as $op(F)$ and $op(t)$. All the operations on a specific data object by the tasks from a set T' is denoted as $\{op(t)|t \in T'\}$. Since a leaf fragment $F' = (V', E', v_{pre}, v_{post})$ only consists of

Table 1. Mathematical table of operation $\odot$ and $\oplus$

$\alpha \odot \alpha = \alpha$	$\delta \odot \alpha = \alpha$	$\varepsilon \odot \alpha = \alpha$	$\tau \odot \alpha = \alpha$	$\alpha \oplus \alpha = \alpha$	$\delta \oplus \alpha = \varepsilon$	$\varepsilon \oplus \alpha = \varepsilon$	$\tau \oplus \alpha = \alpha$
$\alpha \odot \delta = \delta$	$\delta \odot \delta = \delta$	$\varepsilon \odot \delta = \delta$	$\tau \odot \delta = \delta$	$\alpha \oplus \delta = \varepsilon$	$\delta \oplus \delta = \delta$	$\varepsilon \oplus \delta = \varepsilon$	$\tau \oplus \delta = \delta$
$\alpha \odot \varepsilon = \varepsilon$	$\delta \odot \varepsilon = \varepsilon$	$\varepsilon \odot \varepsilon = \varepsilon$	$\tau \odot \varepsilon = \varepsilon$	$\alpha \oplus \varepsilon = \varepsilon$	$\delta \oplus \varepsilon = \varepsilon$	$\varepsilon \oplus \varepsilon = \varepsilon$	$\tau \oplus \varepsilon = \varepsilon$
$\alpha \odot \tau = \alpha$	$\delta \odot \tau = \delta$	$\varepsilon \odot \tau = \varepsilon$	$\tau \odot \tau = \tau$	$\alpha \oplus \tau = \alpha$	$\delta \oplus \tau = \delta$	$\varepsilon \oplus \tau = \varepsilon$	$\tau \oplus \tau = \tau$

sequence tasks $T' \subseteq V'$, the operation of F is $op(F) = \odot\{op(t)|t \in T'\}$, where $\odot$ here is used as an iterated binary operation on the set $\{op(t)|t \in T'\}$.

For a non-leaf fragment $F = (V, E, v_{pre}, v_{post})$, whose child fragments are $X = (V_x, E_x, v_{pre}, v_{post}), Y = (V_y, E_y, v_{pre}, v_{post}), Z = (V_z, E_z, v_{pre}, v_{post})$, the children fragments all parallel with each other. And the remaining tasks, if there are any, are in series. The data operation of F is $op(F) = (\odot\{op(t)|t \in V_1\}) \oplus op(X) \oplus op(Y) \oplus op(Z) \oplus (\odot\{op(t)|t \in V_2\})$, where $V_1 \cup V_2 = V \setminus V_x \setminus V_y \setminus V_z$, and V_1 or V_2 is empty if $v_{pre} \neq v_s$ and $v_{post} \neq v_s$, thus $\odot\{op(t)|t \in V_1\} = \odot\{op(t)|t \in V_2\} = \tau$. The properties proved in Theorems 3–5 in the appendix can then be used to reduce the expression.

4.3   Data State Matrix

In this section we discuss how to calculate the matrix.

When a task with only one immediate predecessor executed, the data state will be influenced by 3 factors: (1) The task's operation on data if there is any. in this case, the state can be determined at once, either *exist* or *lack*. (2) If the task imposes a void operation, the state remains the same. (3) The state can also be affected by the operations of other tasks parallel with the current activity; these tasks are set off before the execution of the current task, but after that of the predecessor task. For the second factor, the immediate predecessor's state can be determined with depth-first traverse on the whole process, for all the states are definite initially. For the third factor, parallel sets of all nodes are required. The definitions and properties of the parallel sets are listed as follows:

Definition 9 (Parallelability and Parallel Sets). *Parallelability is a relation defined on the node set of process graph $V = T \cup G$. $v_a \sim v_b$ if there is no sequential relation between them, in other words, they can be conducted simultaneously and their execution sequence is not known. The parallel set of v_a is the set containing all nodes paralleling to v_a, $PS(v_a) = \{v_i|\forall v_i, v_i \sim v_a\}$.*

Theorem 1 (Parallelability). $v_a \sim v_b \iff$ *there exists no path starting from v_s and ending at v_e, on which v_a and v_b are located simultaneously, and the two do not under a mutual XOR-split gateway.*

Theorem 2 (Properties of Parallel Sets). *If v_c has only one immediate predecessor, v_a, then the condition is $PS(v_a) \subseteq PS(v_c)$. If v_a has a single immediate successor, node v_c, then the condition is $PS(v_c) \subseteq PS(v_a)$. If v_a and v_c are in sequential relation (v_a, v_c), that is, v_a is the only immediate predecessor of v_c, and v_c is the only immediate successor of v_a, then it comes $PS(v_c) = PS(v_a)$.*

When deciding the parallel set of a node v_c, the above-mentioned properties can be utilized. Let $v_a = pre(v_c)$. There are three situations: (1) $level(v_c) = level(v_a) + 1, \Rightarrow type(v_a) \in \{and\_split, or\_split, xor\_slpit\}, \Rightarrow v_a$ is the only immediate predecessor of $v_c, \Rightarrow PS(v_a) \subseteq PS(v_c)$ and the newly-augmented part $PS(v_c)$ as to $PS(v_a)$ is the other branches sharing a mutual gateway v_a with v_c. (2) $level(v_c) = level(v_a), \Rightarrow PS(v_a) = PS(v_c)$. (3) $level(v_c) = level(v_a) - 1$

$\Rightarrow type(v_c) \in \{and\_join, or\_join, xor\_join\}, \Rightarrow v_c$ is the only immediate pre-decessor of v_a, $\Rightarrow PS(v_c) \subseteq PS(v_a)$ and the newly-reduced part $PS(v_c)$ as to $PS(v_a)$ is exactly the other branches sharing a mutual gateway v_c with v_a.

Since the execution sequence of tasks in $PS(v_c)$ could not be decided. The influence of them should be described as $\oplus op(PS(v_c)) = \oplus\{op(t)|t \in PS(v_c)\}$.

Therefore, all the three factors affecting the data states of a single-immediate-predecessor node is fully decided. $op(v_c), ds(pre(v_c)), \oplus op(PS(v_c))$ is used to represent the task's operation, the predecessor's data state and the influence of the parallel set respectively. The data state of v_c can thus be decided in a function: $ds(v_c) = f(op(v_c), ds(pre(v_c)), \oplus op(PS(v_c)))$. The mapping rules of function $f()$ is designated by Table 2.

Table 2. $ds(v_c) = f(op(v_c), ds(pre(v_c)), \oplus op(PS(v_c)))$

$op(v_c)$	$ds(pre(v_c))$	$\oplus op(PS(v_c)))$	$ds(v_c)$
α	$exist/lack/uncertain$	$\alpha/\delta/\varepsilon/\tau$	$exist$
δ	$exist/lack/uncertain$	$\alpha/\delta/\varepsilon/\tau$	$lack$
τ	$exist/lack/uncertain$	ε	$uncertain$
τ	$exist$	α	$exist$
τ	$lack/uncertain$	α	$uncertain$
τ	$lack$	δ	$lack$
τ	$exist/uncertain$	δ	$uncertain$

When the data state of a -join gateway node v_c has to be determined, things get a little complicated. Three factors should also be considered: (1) The data state of the correspondent -split gateway $v_{c-split}$. (2) The data operations conducted by the fragment set $\{F' = (V', E', v_{pre}, v_{post})|v_{post} = v_c\}$, where v_c serves as an immediate successor. (3) The influence of parallel set $\oplus op(PS(v_c))$.

Similar to $PS(v_c)$, since the execution sequence of tasks in fragment set $\{F'|v_{post} = v_c\}$, which is exactly the set of branches among v_c and $v_{c-split}$, could not be decided, the influence of it should be described as $\oplus op(\{F'|v_{post} = v_c\}) = \oplus\{op(F)|F \in \{F'|v_{post} = v_c\}\}$.

Therefore, the data state of -join gateway v_c can thus be decided in a function: $ds(v_c) = f_{join}(\oplus op(\{F'|v_{post} = v_c\}), ds(v_{c-split}), \oplus op(PS(v_c)))$. The mapping rules of function f_{join} are designated by Table 3.

To get the whole Data State Matrix, a depth-first traverse on the process graph is needed. During the traverse a stack is skilfully introduced to calculate $\oplus op(PS(v_c))$, which is equal to $\oplus\{s|s \in Stack\}$. According to the properties of $\oplus op(PS(v_c))$, when the current node's level remains unchanged, so does the parallel set and no operations is done on the stack. When the node level decreases, the parallel set decreases and, the reduced part is on top of the stack

Table 3. $ds(v_c) = f_{join}(\oplus op(\{F'|v_{post} = v_c\}), ds(v_{c-split}), \oplus op(PS(v_c)))$

$\oplus op(\{F'|v_{post} = v_c\})$	$ds(v_{c-split})$	$\oplus op(PS(v_c))$	$ds(v_c)$
$\alpha/\delta/\varepsilon/\tau$	$exist/lack/uncertain$	ε	$uncertain$
ε	$exist/lack/uncertain$	$\alpha/\delta/\varepsilon/\tau$	$uncertain$
τ	$exist$	α	$exist$
τ	$lack/uncertain$	α	$uncertain$
α	$exist/lack/uncertain$	α	$exist$
δ/ε	$exist/lack/uncertain$	α	$uncertain$
τ	$lack$	δ	$lack$
τ	$exist/uncertain$	δ	$uncertain$
α/ε	$exist/lack/uncertain$	δ	$uncertain$
δ	$exist/lack/uncertain$	δ	$lack$
α	$exist/lack/uncertain$	τ	$exist$
δ	$exist/lack/uncertain$	τ	$lack$
τ	$exist/lack/uncertain$	τ	$ds(v_{c-split})$

and should be popped out. When the node level increases, so does the parallel set; the augmented part is $\{F' = (V', E', v_{pre}, v_{post})|v_{pre} = pre(v_c)\}$ and $\oplus op(\{F'|v_{pre} = pre(v_c)\})$ is pushed into the stack.

For each node v_c during the traverse, the corresponding function can be used and the factors are all known. Thus the Data State Matrix, can be achieved. Data model oriented anomalies can be detected by checking the data state matrix according to the Definition 5.

Lastly, we briefly summarize the algorithm and analyze its time complexity. Assume that the number of nodes in the graph is n, the number of data objects is m, and the maximum number of paralleled branches is p. In the preprocessing stage, the process graph is segmented, constructing the process structure tree in $O(n)$. Later on, the time complexity for traversing the process structure tree, calculating the data operations of each fragment is $O(m * n)$. A traverse of the whole process is needed to calculate the Data State Matrix. When calculating the data states of each node v_c, since the auxiliary structure, stack, is introduced, the time complexity of calculating $\oplus PS(v_c)$ is $O(p*m)$, and for $\oplus op(\{F'|v_{post} = v_c\}$ is $O(p * m)$. Therefore, the overall time complexity of the abovementioned step is $O(p * m * n)$. Finally, the time complexity for traverse the data state matrix and conducting the anomaly detection is $O(m * n)$. To sum up, the overall time complexity of the anomaly detection method proposed in the paper is $O(p*m*n)$. When the number of the process nodes is definite, the overall time complexity is linear with the maximum number of paralleled branches. Therefore, the method fit the process verification of massive paralleled branches quite well.

5   Experimental Validation

Firstly, the effectiveness of this paper is put to test. Artifact State Diagram (ASD) [5] and activity dependency analysis based data-flow verification [13], are chosen for comparison since they have relatively higher anomaly detection rates. Experiments was conducted over the synthetic database $D1$ generated randomly by a process simulator and a real world process in the CMCC(China Mobile Communications Corporation) assets management system. Please refers to PASE system [19] for more information about $D2$. Parameters of these two data set are shown in Table 4(a). The results are shown in Table 4(b). The number of anomalies detected by the method proposed in this essay far outweighs that by employing the other two methods, for it has given full consideration to the impact of the data model and all execution sequences possible in the parallel business process. The detection rate in $D2$ has not reached 100 %, for there are some advanced flowage structures and impromptu processes definition [11] in $D2$, still incompatible with the algorithm employed at present stage.

The second comparison experiment is conducted with PDGV algorithm, based on simulated processes. With its basic parameter identical to the data set $D1$ in Table 4(a) and other parameters fixed, the number of gateway nodes and the number of paralleled branches under each gateway were adjusted respectively, generating two groups of experiments. Each corresponding parameter has

Table 4. Data sets and results

(a) Properties of test sets

	D1	D2
Number of Process	10	1
Avg. Tasks	500	153
gatewayNum	5	34
branchNum	5	2-4
taskPerBranch	20	variable
Data Objects	20	124

(b) Numbers of anomalies detected

	Total	This Paper	[5]	[13]
D1-DDU	20	20	0	0
D1-PDU	20	20	20	20
D1-DCC	20	20	0	0
D2-DDU	17	15	0	0
D2-PDU	41	38	38	38
D2-DCC	37	37	0	0

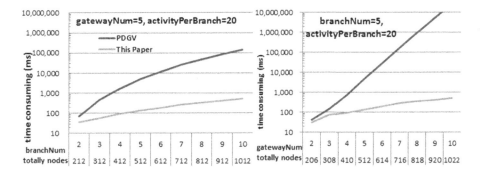

Fig. 2. Efficiency comparison as various branchNum / gatewayNum

generated 5 different structured process models, of whom the running times were averaged. The results are shown in Fig. 2.

As shown, PDGV algorithm is not adaptable to a system of high-level parallelism. An increase in gateway leads to an exponential growth of time needed. The method we have proposed, however, enables a linear growth for the time needed, proves itself to be highly efficient in detecting paralleled business process anomalies, and at the same time, substantiates the time complexity theory.

6   Conclusions

The influence that data models have on process models should not be neglected. Besides, a process model's increasing growth in size, complexity and parallelization level increases both the time complexity of the process verification, and the conflict chances between the process model and data model. To solve the problems in detecting the data-model-oriented anomalies in parallel business processes, an anomaly detection method under the divide-and-conquer framework that includes the process structure tree, the data-operation algebra system and the parallel set stack was then proposed, a data state matrix model was employed to establish the relation between the parallel-process model and data model. The method proposed in this paper has been successfully applied to system PASE [19]. Both the theoretical analysis and the experimental results prove this method to be highly efficient and effective in detecting anomalies.

References

1. Scott Ananian, C.: The static single information form. Massachusetts Institute of Technology (1999). phdananian1999static
2. Bhattacharya, K., Gerede, C.E., Hull, R., Liu, R., Su, J.: Towards formal analysis of artifact-centric business process models. In: Alonso, G., Dadam, P., Rosemann, M. (eds.) BPM 2007. LNCS, vol. 4714, pp. 288–304. Springer, Heidelberg (2007)
3. Chinosi, M., Trombetta, A.: Modeling and validating BPMN diagrams, pp. 353–360 (2009)
4. Crampton, J.: An algebraic approach to the analysis of constrained workflow systems, pp. 61–74 (2004)
5. Hsu, C.-L., Wang, F.-J.: Analysing inaccurate artifact usages in workflow specifications. IET Softw. 1(5), 188–205 (2007)
6. Johnson, R., Pearson, D., Pingali, K.: The program structure tree: Computing control regions in linear time. ACM SIGPLAN Not. 29, 171–185 (1994)
7. Leymann, F., et al.: Production Workflow: Concepts and Techniques. Prentice-Hall, Upper Saddle River (2000)
8. Li, M., Li, H., Tang, L., Qiu, B.: DOPA: A data-driven and ontology-based method for Ad Hoc process awareness in web information systems. In: Aberer, K., Peng, Z., Rundensteiner, E.A., Zhang, Y., Li, X. (eds.) WISE 2006. LNCS, vol. 4255, pp. 114–125. Springer, Heidelberg (2006)
9. Polyvyanyy, A., García-Bañuelos, L., Dumas, M.: Structuring acyclic process models. Inf. Syst. 37(6), 518–538 (2012)

10. Russell, N., ter Hofstede, A.H.M., Edmond, D., van der Aalst, W.M.P.: Workflow data patterns: identification, representation and tool support. In: Delcambre, L.M.L., Kop, C., Mayr, H.C., Mylopoulos, J., Pastor, Ó. (eds.) ER 2005. LNCS, vol. 3716, pp. 353–368. Springer, Heidelberg (2005)
11. Russell, N., et al.: Workflow controlflow patterns: A revised view (2006)
12. Sadiq, S., Orlowska, M., Sadiq, W., Foulger, C.: Data flow and validation in workflow modelling, pp. 207–214 (2004)
13. Sun, S.X., et al.: Formulating the data-flow perspective for business process management. Inf. Syst. Res. **17**(4), 374–391 (2006)
14. Van der Aalst, W.M.P., et al.: Case handling: a new paradigm for business process support. Data Knowl. Eng. **53**(2), 129–162 (2005)
15. van Dongen, B.F., Jansen-Vullers, M.H., Verbeek, H.M.W., van der Aalst, W.M.P.: Verification of the sap reference models using EPC reduction, state-space analysis, and invariants. Comput. Ind. **58**(6), 578–601 (2007)
16. Vanhatalo, J., Völzer, H., Koehler, J.: The refined process structure tree. Data Knowl. Eng. **68**(9), 793–818 (2009)
17. Wang, L., Li, H., Qu, Q., Zhang, H., Zhou, B.: Verifying the consistency between business process model and data model, pp. 171–174 (2009)
18. Wynn, M.T., et al.: Business process verification-finally a reality!. Bus. Process Manage. J. **15**(1), 74–92 (2009)
19. Zhou, B., et al.: PASE: A prototype for ad-hoc process-aware information system declaratively constructing environment, pp. 473–474 (2008)

Learning User Credibility on Aspects from Review Texts

Yifan Gao, Yuming Li, Yanhong Pan, Jiali Mao, and Rong Zhang[✉]

Shanghai Key Laboratory of Trustworthy Computing,
Institute for Data Science and Engineering, East China Normal University,
Shanghai, China
{yfgao,52151500018,51151500037,521415011}@ecnu.cn, rzhang@sei.ecnu.edu.cn

Abstract. Spammer detection has been popularly studied these years which aims at filtering unfair or incredible customers. Most users have different backgrounds or preferences so that they make distinct reviews/ratings, however they can not be treated as spammers. To date, the existing previous spammer detection technology has limited usability. In this paper, we propose a method to calculate user credibility on multi-dimensions by considering users difference related to their personalities e.g. background and preference. Firstly, we propose to evaluate customer credibilities on aspects with the consideration of different concerns given by different customers. A boot-strapping algorithm is applied to detect the intrinsic aspects of review text and the aspect ratings are assigned by mining semantic polarity. Then, an iteration algorithm is designed for estimating credibilities by considering the consistency between individual ratings and overall ratings on aspects. Finally, experiments on the real dataset demonstrate that our method outperforms baseline systems.

1 Introduction

Most e-commerce sites provide platforms for customers to share their experiences on products or services by giving ratings and writing reviews. These feedback information enables potential customers judge products and make decisions when doing online-shopping. Generally, there is an overall rating for products by averaging individual ratings, which plays an important role for making purchase decision. Since high overall ratings may attract more customers and orders, there are so called spammers to boost ratings intentionally. Spammer detection is designed to tackle with this kind of defraudation by labeling customers with "0" or "1" representing incredible or credible. However, "0" and "1" labeling has limited the usability. This is based on that different customers may have distinct backgrounds or preferences, which leads to various credibilities on product ratings and can not be simply divided into two categories. For example, a person came from Shanghai may provide a more accurate review comment for Shanghai-food than Sichuan-food. Hence, in a bid to evaluate users fairly, it is crucial to learn the credibility of reviewers instead of filtering spammers [16,20].

Corresponding author is Rong Zhang

© Springer International Publishing Switzerland 2016
B. Cui et al. (Eds.): WAIM 2016, Part II, LNCS 9659, pp. 78–91, 2016.
DOI: 10.1007/978-3-319-39958-4_7

Though a little work is devoted to evaluating customer by expertness [2,24], which commonly uses the overall evaluations for calculations and can not mimic customers their own personalities. Since customers care about different aspects even for the same services, such as tastes, environments or services, they should have distinct credibilities on different aspects. In work [10], it proposes a topic-biased model (TBM) to estimate user reputation on several topics. Though it admits users have different credibilities on latent topics, it uses the probabilistic topic model to discover topics which is difficult to explain and verify. We discuss user credibility on different aspects which are predefined.

Definition 1 (User Credibility on Aspects (UCA)). *It is defined as the evaluations to user ability to write reliable reviews and ratings on the aspects, represented as c_k.*

So in this paper, we explore User Credibility on Aspects (UCA) defined in Definition 1 by mining ratings and reviews. We propose an aspect-oriented credibility estimation method to calculate user credibilities on aspects. Firstly, a bootstrap algorithm is used to detect aspects from review texts. Then, the polarity of emotional words is taken into account when calculating the rating on each aspect. Finally, an iterative algorithm is designed to measure user credibilities based on the comparison between predicted ratings and the overall ratings. Compared with the work [10] which is the most relevant to our proposal, our method has the following advantages: (1) presenting explicit aspects instead of latent topics discovered by probabilistic method, which is explainable; (2) predicting the rating for each aspect instead of using the overall rating for each latent topic, which is more reasonable and accurate; (3) calculating user credibility on aspects to mimic customers expertness on different domains. Experiments on food dataset crawled from the biggest Chinese review site *Dianping*[1] demonstrate the preponderance of our proposal over state-of-the-art solutions.

The rest of this paper is organized as follows. Section 2 introduces the related work. Section 3 describes our method to show how to calculate the user credibility. Section 4 demonstrates the experimental results. Finally, in Sect. 5 we summarize the main points of our work.

2 Related Work

Spammer detection and expert finding are the most representative researches for labeling user credibility. Spammer detection prefers to find incredibile customers that user with zero credibility. Generally, it can take user linguistic features [10,13] and behavioral features [11,14,20] into consideration in detecting spammers who deliberately mislead readers by writing fake reviews and ratings. Many studies have been done on Web spam detection [3,16,19], Email spam detection [6], and social networks spammer detection [7,9]. Work [8] is a supervised-based

---

[1] Dianping: http://www.dianping.com.

work typically on spam analysis and comes up with the assumption that duplicate reviews should be fake. Work [12] builds an unsupervised bayesian framework and formulates spammer detection as a clustering problem. Work [17] considers the relationship among reviews, reviewers and stores to present a graph-based method. On the contrary, expert finding aims at mining persons with rich experience who are supposed to produce truthful and reliable evaluations for a specific domain. Most of studies are based on the person local information [2,15] and relationships [22,24]. [1] shows a principal approach which is concentrated on capturing experts from all candidates. Language model applies in the process of identifying experts in [2]. Work [24] attempt to find authoritative users with special knowledge for a relevant category in QA communities and they provide an extended category link graph and a topical link analysis approach.

However, spammer detection and expert finding are two kinds of extreme use cases for credibility detection. The most credible users are expects and most incredible ones are spammers. Between these two categories, there are users who may not be as strong as expects or as weak as spammers. However, non-spammers may be strong or credible on ratings for some specific aspects. After detecting the credibilities of users, we can adjust their ratings which affect product overall ratings by integrating their aspect-based credibilities finally. Most of the previous work [4,5,10,23] focus on evaluating the quality of reviewers according to reviews and ratings. A simple but standard work [23] designs a feedback mechanism to adjust customer credibility values. Work [4] combines reputation-based trust assessments with provenance information (i.e., how data has been produced) to determine trust values. The work most related to ours is [10]. It proposes a topic-biased model(TBM) to estimate user reputation and focuses on proving algorithm's convergence compared to most of previous work. However, the probabilistic topic model is used to discover topics, which is still difficult to explain users with the latent topics. Instead, in a specific application domain, aspects are easily found or listed and then the interpretability to users are more obvious. Additionally, their default topic distributions in review text are taken as the degree of which one item belongs to the topics. But topic distributions can't exactly represent user credibilities on topics.

3   Our Method

In this section, we propose a method to calculate User Credibility on Aspects (UCA) by analyzing review text. The major challenges are to detect aspects hidden in review text and to mine the credibilities using user ratings and item ratings. Here, we assume that an overall rating from a user is the weighted combination of ratings corresponding to different aspects in review text.

Our method includes two main modules: *Semantic-based Aspect Rating Predictor* and *Aspect-oriented Credibility Calculation*, as shown in Fig. 1. Semantic-based aspect rating predictor detects aspects and predicts ratings for the aspects according to semantic analysis of review text. In Aspect-oriented Credibility Calculation, the aspect ratings are used to calculate the scores of User Credibility on Aspects.

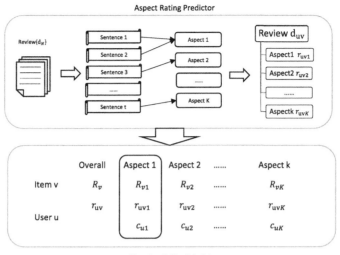

Fig. 1. Framework of our method

3.1 Aspect Rating Predictor

In this module, we assign an aspect to each sentence in a review comment, summarize the aspects in review-level, and predict ratings for aspects. There are two steps to obtain aspect ratings:

- **Aspect Extraction.** We first split reviews into sentences and assume that each sentence covers only one aspect. According to the correlation between words in a sentence and aspects, we assign each sentence an aspect.
- **Aspect Rating.** According to the polarity of emotional words for an aspect, we predicate the rating of the aspect for each sentence. Then, rating for a piece of review is computed from these aspect ratings.

The Aspect Rating Predictor in Fig. 1 aims at discovering the aspect ratings in each individual review. We first divide the review text into sentences and each sentence is treated as a semantic unit. Then, we use Ansj[2], a Natural Language Processing (NLP) tool, to process the sentences including word segmentation (if necessary), and part-of-speech (POS) tagging. Supposing that we have a review comment d from user u to item v, it is split into a set of h sentences $X_d = \{x_{d1}, x_{d2}, ..., x_{dh}\}$. Our target is to predicate the aspect ratings $r_{uv} = \{r_{uv1}, r_{uv2}, ..., r_{uvK}\}$ for review d with K as the number of aspects.

Aspect Extraction. Supposing each aspect can be described by a set of keywords and in each sentence there is at most one aspect discussed, we need to

---
[2] http://www.ansj.org/.

Algorithm 1. Aspect Keyword Learning Algorithm

Input:

    Then collection of N reviews,

    Aspect seed words $\{A_1, ..., A_K\}$, where A_k is the keyword set for the kth aspect a_k,

    Review word dictionary W, Correlation matrix $\mathcal{C}$ with $\mathcal{C}_{ik} = 0$ for w_i and a_k

    $(w_i \in W)$,

    Selection threshold t and correlation factor f;

Output:

    Lists of extended aspect keywords for K aspects

1: If $w_i \in A_k (1 \leq k \leq K$ and $w_i \in W)$, set $\mathcal{C}_{ik} = B$ (B as the maximum correlation value);

2: Split reviews into sentences $X = \{x_1, ..., x_m, ..., x_M\}$;

3: Match aspect keywords in each sentence of X;

4: Assign sentence x_m an aspect label a_k with the maximum correlation score by: $\arg\max_k \sum_{w_i \in x_m} \mathcal{C}_{ik}$ without any tie;

5: Calculate correlation score $\mathcal{C}_{jk}$ for $w_j \notin A_k$ and the aspect a_k by $\mathcal{X}^2$[21];

6: Rank the words with respect to $\mathcal{C}_{jk}$ and add the top t words for each aspect; remove word w_j from A_k, if $\mathcal{C}_{jk} \times f \leq \mathcal{C}_{jk'}$, where $1 \leq k' \leq K$ and $k' \neq k$;

7: Iterate from Step 3 to Step 6 until no change to any A_k $(1 \leq k \leq K)$;

---

generate keyword sets for each aspect. Inspired by the work of [18], we design a boot-strapping algorithm to learn representative keywords for each aspect as shown in Algorithm 1. At the beginning, we give a small set of seed words for each aspect. The learning algorithm finds words which frequently occur together with the seed words and these words are treated as the representatives for an aspect. The learning starts from line 3 to line 6 until no more changes to aspect words. In each iteration, we first mark the aspect keywords in each sentence using A_k $(1 \leq k \leq K)$ set as line 3. Then we assign each sentence an aspect a_k with the maximum correlation score $max_k \sum_{w_i \in x_m} \mathcal{C}_{ik}$ as in line 4. In line 5, we calculate the correlation for any pair of word w_j (w_j is not in A_k) and aspect a_k with the function $\mathcal{X}^2$ [21]. Then, we add the top related words for each aspect a_k in Step 6. The output of the algorithm is the lists of keywords for all the K aspects. Finally, based on the learned keywords, we are able to assign each sentence to an aspect as done in Line 4 in the algorithm.

Aspect Rating. Each sentence x_m is assigned an aspect label a_k by Algorithm 1. We calculate the rating of a sentence based on the polarity of each emotional word in the sentence. Different from [18], we treat the word emotional polarity as the most important factor that determines aspect ratings rather than the model parameters. We use an existed emotional polarity dictionary to compute the sentence rating r_m for sentence x_m,

$$r_m = \frac{1}{E} \sum_{e=1}^{E} p_e, \tag{1}$$

where E is the number of emotional words in x_m and p_e is the emotional polarity of word e. Since sentence x_m can be assigned only one aspect, e.g. a_k, we define the aspect rating r_{mk} as r_m and for other aspects the ratings are zero. For review comment d (associated with user u and item v), we compute the rating r_{dk} on the aspect a_k,

$$r_{dk} = r_{uvk} = \frac{1}{T} \sum_{i=1}^{T} r_{ik}, \tag{2}$$

where T is the number of sentences which have label a_k in d. Thus, for each review d, we have K aspect ratings: $r_{uv} = \{r_{uv1}, r_{uv2}, ..., r_{uvK}\}$.

3.2   Calculating User Credibility on Aspect

In most of previous work, the rating of an item is averaged with all users having equal weights. In our method, we add the credibility factor to reduce the influence of unreliable users. We design a credibility-based algorithm to estimate item ratings and user credibilities iteratively. We assume that if user u always gives ratings closed to overall ratings of item v, u will have high credibility, and vice versa. The overall item rating is derived from the weighted individual user ratings.

Given a set of user U and a set of item V, r_v is the rating of the item v, and r_{uv} ($u \in U, v \in V$) represents the rating user u gives to item v. c_u is defined as the credibility of user u. The general credibility-based algorithm works as follows:

$$r_v^{s+1} = \frac{1}{|U_v|} \sum_{u \in U_v} r_{uv} c_u^s \tag{3}$$

$$c_u^{s+1} = 1 - \frac{1}{|V_u|} \sum_{v \in V_u} |r_{uv} - r_v^{s+1}|, \tag{4}$$

where U_v is the set of users who have rated to item v, V_u is the set of items that u has rated and s is the round of iteration. The credibility values are initialized randomly. So far, we do not consider aspects in the Eq. 4 and we can extend the calculation as follows with respect to aspect a_k:

$$r_{vk}^{s+1} = \frac{\sum_u r_{uvk} c_{uk}^s}{\sum_u c_{uk}^s} \tag{5}$$

$$\mathcal{D}_{uk}^{s+1} = \frac{1}{|N_u|} \sum_{v \in N_u} |r_{uvk} - r_{vk}^{s+1}| \tag{6}$$

$$c_{uk}^{s+1} = 1 - \frac{\mathcal{D}_{uk}^{s+1} - \mathcal{D}_{min}}{\mathcal{D}_{max} - \mathcal{D}_{min}} \tag{7}$$

where r_{uvk} is the aspect rating of review d on aspect a_k, which is from user u to item v. In Eq. 5, item rating is considered as the average of weighted ratings by taking user credibilities as weight coefficients. In the calculation, the item ratings should be divided by the total of credibilities rather than the number of users as defined in Eq. 3. $\mathcal{D}_{uk}$ defined in Eq. 6 is a divergence measurement function to calculate the difference between user rating and overall item rating on aspect a_k. Normalization is done as Eq. 7. Our calculation will stop only when $\mathcal{D}_{uk}$ is unchanged.

3.3   Category-Based Method

Besides considering the fluctuation of credibilities among aspects, we also expect that the credibilities on different categories may help to produce better evaluation. For example, catering industry can be classified according to cuisine categories and hotels can be divided according to the stars. Usually, people give more accurate ratings and reviews on the domain that they are familiar. The actual effect is demonstrated in the experimental section.

4   Experiments

4.1   Data Set

We crawled the review data from Dianping which is the biggest restaurant review site in China. We remove the restaurants which have less than 20 reviews and obtain a data set including 1,074,931 reviews from 8,988 restaurants in Shanghai area. Each review contains review text, user ID, restaurant ID, an overall rating, and three aspect ratings including taste, surroundings, and service. All the ratings take values from 1 to 5. In Fig. 2, we report the log-log plot of the distributions in terms of the number of users with various amount of reviews. The distribution on users follows the power law. Since the reviews are written in Chinese, we pre-process those reviews by Ansj, an open source tool, which does word segmentation and Part-of-Speech tagging. Finally the reviews are split into 12,081,022 sentences according to the punctuation.

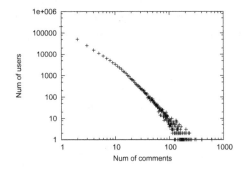

Fig. 2. The log-log plot on the number of comments and the number of users

4.2   Aspect Rating Prediction

In this evaluation, the task is to predict ratings of aspects for each review. In the Dianping reviews, the ratings of taste, surroundings, and service can be used as ground-truth data for this task.

In order to evaluate the predicated ratings for apects, we take the Aspect Mean Square Error (AMSE) and Mean Absolutely Error (MAE) as evaluation metrics, defined as follows,

$$AMSE = \frac{1}{K \times |D|} \sum_{n=1}^{|D|} \sum_{i=1}^{K} (\hat{r}_{ni} - r_{ni})^2, \tag{8}$$

$$MAE = \frac{1}{K \times |D|} \sum_{n=1}^{|D|} \sum_{i=1}^{K} |\hat{r}_{ni} - r_{ni}|, \tag{9}$$

where K is the number of aspects, $|D|$ is the amount of reviews, $\hat{r}_{ni}$ and r_{ni} are the predicted rating and the user assigned rating in review n on aspect a_i, respectively. Since in our dateset, we have three aspects with ratings, which are taste, surroundings, and service, and we use those aspects as ground-truth for this task.

Table 1. Aspect seed keywords for each aspect (Translated from Chinese)

Aspect	Seed keywords
taste	味道(taste), 美味(delicious), 甜(sweet), 辣(spicy), 酸(sour), 苦(bitter), 咸(salty)
surroundings	环境(surroundings), 吵闹(noisy), 装修(adornment), 景色(landscape), 画(painting), 情人(lovers), 情感(sentiment)
service	服务(service), 服务员(waiter), 小费(tip), 礼貌(polite), 怠慢(slight), 态度(attitude), 管理(manage)

We build a baseline (named as LARA) based on [18] which is a probabilistic rating regression model to analyze individual reviewer's latent opinion on aspects in reviews. Different from ours, they apply a Latent Rating Regression (LRR) model to learn aspect ratings and aspect weights. The implementation of LARA is available online[3]. Another baseline is Collaborative Filtering (CF), which is the most traditional but effective method to predict ratings according to users' historical records.

Results of Learning Aspect Keywords. We list the example seed keywords for each aspect as shown in Table 1. Those seed keywords are fed into the Aspect Keyword Learning Algorithm shown in Algorithm 1. The algorithm learns new keywords for each aspect and then we assign sentences with the aspect labels.

---

[3] http://sifaka.cs.uiuc.edu/~wang296/Codes/LARA.zip.

As mentioned in Algorithm 1, we select top $t \doteq 50$ words and correlation factor $f \doteq 100$ according to our dataset. Table 2 shows the most top 10 words. From Table 2, we can find that most of the extended keywords are highly related to the corresponding aspects. This indicates that our method is efficient to learn new aspect keywords.

Table 2. Top 10 aspect extended words for each aspect (Translated from Chinese)

Aspect	Extended keywords
taste	吃(eat), 菜(dish), 柔软(soft), 肉(meat), 鸡肉(chicken),汤(soup), 面条(noodle), 烤得(bake), 牛肉(beef), 巧克力(chocolate)
surroundings	家庭(family), 特别(special), 位置(location), 家具(fitment), 新(new), 房间(room), 舒服(comfortable), 杯子(cup), 桌子(table), 位置(seat)
service	花费(cost), 便宜(cheap), 饮料(drink), 推荐(recommend), 快速(fast), 热情(enthusiasm), 茶(tea), 优惠券(coupon), 老板(boss), 询问(ask)

Results of Aspect Rating Prediction. We modify NTUSD[4], an emotional polarity dictionary, based on the real data to help predict aspect ratings. NTUSD is provided by National Taiwan University which only covers two types of polarities with 1992 positive and 1553 negative words. In order to make our work more accurate, we use it together with DAD dictionary[5], which includes degree of expressive words. Subsequently, we put those emotional words into three sets according to their emotional degrees. Based on NTUSD and DAD, we revise Eq. 1 as,

$$r_x = v_m + v_b \times p_x \times d_x \tag{10}$$

where v_m refers to the average value, v_b refers to the bias value, p_x is the polarity of sentence x, and d_x is the emotional degree value of sentence x. In our experiments, v_m and v_b are 3 and 1 respectively, which are tuned on the real dataset introduced above. If the polarity of sentence is positive, then $p_x = 1$, otherwise $p_x = -1$. As for d_x, we set 0, 1, and 2 according to DAD respectively.

The results of aspect rating prediction are listed in Table 3, where OURS refers to our proposed method, and LARA and CF are baselines. From the table, we find that our system performs better than LARA on both AMSE and MAE. CF method who can be rival with our method has good robustness and performs well. But CF can not predict aspect ratings without users's historical aspect ratings while our method can mining aspect ratings from review text. This indicates that our proposed method can work well to predict aspect ratings that can be used as the input to calculate User Credibility on Aspect (UCA).

---

[4] http://www.datatang.com/data/44317/.
[5] http://pan.baidu.com/s/1sjoqp1z.

Table 3. Rating prediction

	AMSE	MAE
OURS	1.25	1.01
LARA	1.56	1.53
CF	1.26	0.97

Table 4. Rating prediction and repeat customers

Average of predict ratings	4	3
OURS	46.4 %	83.7 %
LARA	38.3 %	73.6 %
CF	42.8 %	79.2 %

Rating Prediction Vs Repeat Customers. In our dataset, there are 1347 restaurants with repeat customers who consume the same services more than once. Generally, if users satisfy with the restaurants, they may probably go there again. Based on such an assumption, restaurants with high ratings should have more repeat customers. Table 4 shows the relationship between predict ratings and repeat customer distributions. "4" means predicated score ≥ 4 (5 is the best one). Our method performs better than LARA and CF methods with the percentage of repeat customers more than those coming from the other two methods.

4.3   Calculation of User Credibility on Aspect

In this section, we evaluate the results of calculation of UCA. We first analyze the relationship between credibility of a user and the number of positive feedback the user receives. And then we evaluate the systems on the overall rating prediction.

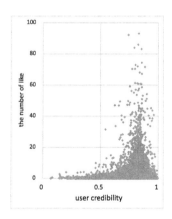

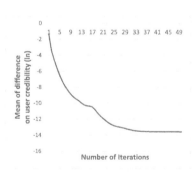

b. Algorithm Convergence Rate

a. The distribution of user credibility and "like"

Fig. 3. Experiment statistics

Credibility Vs Positive Feedback. In Dianping.com, if a user thinks one review written by another user is useful, he can click "like" to give a positive feedback. Though "like" mechanism is rough for it offers only a bool-value attributes instead of accurate quantization values, it is possible that a person with higher credibility receives more "like". Thus, we check the relationship between the credibility scores and the numbers of "like" that each user receives. As Fig. 3 (a) shows we find that users with higher credibilities deserve more recognition. This indicates that our method can profile user credibility accurately.

Results of Weighted Ratings. On the basis of user credibility affects rating, we generate a new overall rating by a weighted combination of aspect ratings and user credibility scores by:

$$\hat{r}_v = \frac{1}{K \times |M_v|} \sum_{i=1}^{K} \sum_{u \in M_v} r_{uvi} c_{ui} \tag{11}$$

where K is the number of aspects, M_v is the set of users who have commented on item v, and c_{ui} is the credibility score of user u on the ith aspect. We evaluate the performance by Root Mean Square Error(RMSE) and Mean Absolutely Error(MAE) as:

$$RMSE = \frac{1}{|V|} \sqrt{\sum_{v=1}^{|V|} (\hat{r}_v - R_v)^2} \tag{12}$$

$$MAE = \frac{1}{|V|} \sum_{v=1}^{|V|} |\hat{r}_v - R_v| \tag{13}$$

where R_v is the rating of item v provided by Dianping.com and V is item set.

We compare our work with TBM [10], which proposes six topic-biased model to estimate User Reputation(UR) in terms of different topics as well as item scores. We compare six algorithms described in [10] with our system according to the damping factor $\lambda = 0.03$.

We show the results in Table 5, where "TBM-" refers to the systems described in [10]. From the table, we find that our system (OURS) performs much better than all the TBM systems. This indicates that the user credibility on aspects are useful for generating overall ratings for items. And in Table 6, we list all the results for different categories with MAE and $RMSE$ values. Since "Sichuan Food" with the large set of reviews, it performs better than "Western Food" with the least data.

User Rating and Credibility. We assume that the user credibility is higher if its predicated rating is always closer to the restaurant's overall rating. Table 7 shows average user credibilities considering the difference between predicated user ratings and restaurant overall ratings. In Table 7, "±1" or "±2" means the rate difference. We can see that "TBM-L1-AVG" can not distinguish users by

Table 5. Rate predication on credibility **Table 6.** Category-based rate predication

	MAE	RMSE
OURS	0.59	0.85
TBM-L1-AVG	0.84	1.38
TBM-L1-MAX	0.86	1.34
TBM-L1-MIN	0.93	1.37
TBM-Square-AVG	0.88	1.37
TBM-Square-MAX	0.87	1.36
TBM-Square-MIN	1.02	1.36

Category	MAE	RMSE
Western Food	0.58	0.83
Cantonese Food	0.53	0.78
Shanghai Food	0.52	0.77
Japanese Food	0.55	0.80
Hot Pot	0.55	0.81
Sichuan Food	0.51	0.77

Table 7. Distribution of user ratings and credibilities

	± 1	± 2	$\pm 3, \pm 4, \pm 5$
OURS	0.88	0.76	0.64
TBM-L1-AVG	0.92	0.84	0.78

credibilities so much as ours with credibility difference (0.14=0.92-0.78) smaller than ours (0.24 =0.88-0.64).

Convergence. Besides of accuracy, we also check algorithm convergence shown in Fig. 3 (b). From the figure, we find that our algorithm can converge at 35th round.

4.4   Results of Category-Based Method

According to the categories of the restaurants, we split the data set into 20 categories. We apply the category-based method to Top 6 categories. The experimental results are listed in Table 6. From the table, we find that our systems perform similarly for each category. The scores of category-based method are lower than the scores of OURS in Table 5. This indicates that the detail information of categories is quite useful for our method.

5   Conclusion

In this paper, we present a novel method to calculate user credibility on aspects derived from review text. We first identify aspects for each review text and predict ratings on the aspects. Then we utilize these predicated scores to calculate ratings based on review text. User credibility are calculated by comparing individual ratings with the overall ratings. To identify aspects, we design a learning algorithm to find keywords for each aspect given a small set of seed keywords. As for calculating the user credibility, we design an iterative calculation method. The experimental results demonstrate that our proposed method works very well on real data set. For future work, there are several ways in which this research could be extended. First, we plan to utilize other data sets, such as Yelp and Amazon, to verify our proposed method. Second, we can apply the proposed method to other tasks, such as detecting spammer users. Finally, we can also involve use social network information into our method.

Acknowledgment. This work is partially supported by National Science Foundation of China under grant (No. 61103039 and NO. 61402180), and National Science Foundation of Shanghai (No. 14ZR1412600).

References

1. Balog, K., Azzopardi, L., De Rijke, M.: Formal models for expert finding in enterprise corpora. In: Proceedings of the 29th Annual International ACM SIGIR Conference on Research and Development in Information Retrieval, pp. 43–50. ACM (2006)
2. Balog, K., Azzopardi, L., de Rijke, M.: A language modeling framework for expert finding. Inf. Process. Manag. **45**(1), 1–19 (2009)
3. Castillo, C., Donato, D., Becchetti, L., Boldi, P., Leonardi, S., Santini, M., Vigna, S.: A reference collection for web spam. ACM Sigir. Forum. **40**, 11–24 (2006). ACM
4. Ceolin, D., Groth, P.T., Van Hage, W.R., Nottamkandath, A., Fokkink, W.: Trust evaluation through user reputation and provenance analysis. URSW **900**, 15–26 (2012)
5. Chen, B.C., Guo, J., Tseng, B., Yang, J.: User reputation in a comment rating environment. In: Proceedings of the 17th ACM SIGKDD International Conference on Knowledge Discovery and Data Mining, pp. 159–167. ACM (2011)
6. Chirita, P.A., Diederich, J., Nejdl, W.: Mailrank: using ranking for spam detection. In: Proceedings of the 14th ACM International Conference on Information and Knowledge Management, pp. 373–380. ACM (2005)
7. Ghosh, S., Viswanath, B., Kooti, F., Sharma, N.K., Korlam, G., Benevenuto, F., Ganguly, N., Gummadi, K.P.: Understanding and combating link farming in the twitter social network. In: Proceedings of the 21st International Conference on World Wide Web, pp. 61–70. ACM (2012)
8. Jindal, N., Liu, B.: Opinion spam and analysis. In: Proceedings of the 2008 International Conference on Web Search and Data Mining, pp. 219–230. ACM (2008)
9. Kolari, P., Java, A., Finin, T., Oates, T., Joshi, A.: Detecting spam blogs: A machine learning approach. In: Proceedings of the National Conference on Artificial Intelligence. vol. 21, p. 1351. AAAI Press, Menlo Park, CA; MIT Press, Cambridge, MA; London 1999 (2006)
10. Li, B., Li, R.H., King, I., Lyu, M.R., Yu, J.X.: A topic-biased user reputation model in rating systems. Knowl. Inf. Syst. **44**, 1–27 (2014)
11. Li, F., Hsieh, M.H.: An empirical study of clustering behavior of spammers and group-based anti-spam strategies. In: CEAS (2006)
12. Mukherjee, A., Kumar, A., Liu, B., Wang, J., Hsu, M., Castellanos, M., Ghosh, R.: Spotting opinion spammers using behavioral footprints. In: Proceedings of the 19th ACM SIGKDD International Conference on Knowledge Discovery and Data Mining, pp. 632–640. ACM (2013)

13. Ntoulas, A., Najork, M., Manasse, M., Fetterly, D.: Detecting spam web pages through content analysis. In: Proceedings of the 15th International Conference on World Wide Web, pp. 83–92. ACM (2006)
14. Ramachandran, A., Feamster, N.: Understanding the network-level behavior of spammers. ACM SIGCOMM Comput. Commun. Rev. **36**(4), 291–302 (2006)
15. Serdyukov, P., Hiemstra, D.: Modeling documents as mixtures of persons for expert finding. In: Macdonald, C., Ounis, I., Plachouras, V., Ruthven, I., White, R.W. (eds.) ECIR 2008. LNCS, vol. 4956, pp. 309–320. Springer, Heidelberg (2008)
16. Spirin, N., Han, J.: Survey on web spam detection: principles and algorithms. ACM SIGKDD Explor. Newsl. **13**(2), 50–64 (2012)
17. Wang, G., Xie, S., Liu, B., Yu, P.S.: Review graph based online store review spammer detection. In: 2011 IEEE 11th International Conference on Data mining (ICDM), pp. 1242–1247. IEEE (2011)
18. Wang, H., Lu, Y., Zhai, C.: Latent aspect rating analysis on review text data: a rating regression approach. In: Proceedings of the 16th ACM SIGKDD International Conference on Knowledge Discovery and Data Mining, pp. 783–792. ACM (2010)
19. Wu, B., Goel, V., Davison, B.D.: Topical trustrank: Using topicality to combat web spam. In: Proceedings of the 15th International Conference on World Wide Web, pp. 63–72. ACM (2006)
20. Wu, C.H.: Behavior-based spam detection using a hybrid method of rule-based techniques and neural networks. Expert Syst. Appl. **36**(3), 4321–4330 (2009)
21. Yang, Y., Pedersen, J.O.: A comparative study on feature selection in text categorization. ICML. **97**, 412–420 (1997)
22. Zhang, J., Ackerman, M.S., Adamic, L.: Expertise networks in online communities: structure and algorithms. In: Proceedings of the 16th International Conference on World Wide Web, pp. 221–230. ACM (2007)
23. Zhang, R., Gao, M., He, X., Zhou, A.: Learning user credibility for product ranking. Knowl. Inf. Syst. **46**, 679–705 (2014). Springer
24. Zhu, H., Chen, E., Xiong, H., Cao, H., Tian, J.: Ranking user authority with relevant knowledge categories for expert finding. World Wide Web **17**(5), 1081–1107 (2014)

Detecting Anomaly in Traffic Flow from Road Similarity Analysis

Xinran Liu[1], Xingwu Liu[2], Yuanhong Wang[1],
Juhua Pu[1,3(✉)], and Xiangliang Zhang[4]

[1] The State Key Laboratory of Software Development Environment,
Beihang University, Beijing, China
{liuxinran,lucienwang,pujh}@buaa.edu.cn
[2] Institute of Computing Technology Chinese Academy of Sciences, Beijing, China
liuxingwu@ict.ac.cn
[3] Research Institute of Beihang University in Shenzhen, Shenzhen, China
[4] King Abdullah University of Science and Technology, Jeddah, Saudi Arabia
xiangliang.zhang@kaust.edu.sa

Abstract. Taxies equipped with GPS devices are considered as 24-hour moving sensors widely distributed in urban road networks. Plenty of accurate and realtime trajectories of taxi are recorded by GPS devices and are commonly studied for understanding traffic dynamics. This paper focuses on anomaly detection in traffic volume, especially the *non-recurrent traffic anomaly* caused by unexpected or transient incidents, such as traffic accidents, celebrations and disasters. It is important to detect such sharp changes of traffic status for sensing abnormal events and planning their impact on the smooth volume of traffic. Unlike existing anomaly detection approaches that mainly monitor the derivation of current traffic status from history in the past, the proposed method in this paper evaluates the abnormal score of traffic on one road by comparing its current traffic volume with not only its historical data but also its neighbors. We define the neighbors as the roads that are close in sense of both geo-location and traffic patterns, which are extracted by matrix factorization. The evaluation results on trajectories data of 12,286 taxies over four weeks in Beijing show that our approach outperforms other baseline methods with higher precision and recall.

1 Introduction

With more and more taxies equipped with GPS device, plenty of accurate, real-time vehicle trajectory data are available. Taxi trajectory data are considered to be rich and reliable for studying traffic dynamics in a city, because a large amount of taxies are 24-hour moving sensors widely distributed in urban road networks and reflecting the traffic status.

In this paper, we analyze taxi GPS trajectories for detecting the *non-recurrent traffic anomaly* caused by unexpected or transient incidents, such as traffic accidents, control pretests, celebrations, disasters, emergent road maintenance. This type of traffic anomaly results in sharp changes of traffic status

© Springer International Publishing Switzerland 2016
B. Cui et al. (Eds.): WAIM 2016, Part II, LNCS 9659, pp. 92–104, 2016.
DOI: 10.1007/978-3-319-39958-4_8

(speed or volume) in the local area of urban. Unlike the traffic jams caused by the peak hours or holidays, traffic anomaly is difficult to anticipate. It is thus important to detect the anomalies for sensing abnormal events and planing for their impact on the smooth volume of traffic [1].

Generally, there are two different ways for detecting traffic anomaly. The first type of approaches follow an intuitive way of general anomaly detection in any field: model the normal behavior and report abnormal behaviors if they differ much from the normal ones. To detect traffic anomalies, a traffic volume forecasting system is built for modeling normal traffic patterns. Anomalies are reported when the system is not able to predict correctly, i.e., the observed traffic deviates much from what the system predicts as normal traffic [2]. The detection accuracy highly depends on the forecasting system used for normal traffic prediction. A poor prediction algorithm can result in false and missed alarms. In the second type, abnormal traffic patterns are identified based on comparing current observations with those in adjacent days [3] or in history up to this time [1]. An anomaly is reported if a large derivation or difference is found. For example, traffic volume patterns at the link connecting one region to another are compared with those at the same time in adjacent days [3]. The link is suspected to be abnormal if the traffic volume patterns behave differently from other days. To detect topological variation in traffic flow between two regions, [1] takes the distribution pattern of traffic volumes over all paths from one region to another at current time and compares it to the median of all distribution patterns up to this time. More discussion of approaches in this type will be presented in Sect. 5.

Our proposed approach in this paper is different from the existing ones on the following aspects:

(1) We detect anomalies in road level, rather than region or link level. It thus improves the spatial accuracy of the detected anomalies from region level to road segment level.

(2) We evaluate the abnormal score of traffic on one road by comparing its current traffic volume with both its historical data and its neighbors. The neighboring roads are selected according to their closeness in geo-location and similarity in traffic patterns. This new evaluation of anomaly reduces false alarms caused by only comparing to the historical data. Traffic often has fluctuations. The day-to-day or day-of-week variability due to temporal factors are easy to caught by monitoring periodically regular changes. However, irregular "surges" or "decreases" caused by circumstantial factors or special events introduce high variation to historical data. For example, the traffic volume sharply increased on the first workday of the Victory Day on September 3rd 2015 resulted from the no limits for driving on roads, while the traffic volume decreased obviously on December 8th 2015 due to the hazy weather in Beijing. By additionally considering the traffic status in neighboring roads, our approach is therefore more robust to irregular variation than the method in [4,5].

(3) We compare one road segment with those that are close in sense of both geo-location and traffic patterns, rather than comparing with only spatially

close roads or all road segments [6] in two reasons. First, comparing with all segments is not computationally efficient, as the number of road segments is large. Second, comparing with only spatially close roads potentially hides anomalies that affect several spatial adjacent roads in a region [6], e.g., a major traffic accident. We factorize the traffic volume matrix by Nonnegative Matrix Factorization (NMF) and extract a pattern matrix and a coefficient matrix, which represent the types of traffic patterns in all road segments and the distribution of these types in each road segment, respectively. Similarity of two roads is measured by what types of traffic running on the roads and how geographically close they are. Bringing in roads similar in traffic patterns for comparison helps on filtering out anomalies that causes traffic variation jointly in spatial adjacent roads.

We evaluate our proposed approach on trajectories data of 12,286 taxis over 4 weeks in Beijing, and compared with baseline methods in [4–6]. The experimental results show that our approach outperforms baseline methods with higher precision and recall, and improves the F1-measure by at least 10 %.

The rest of the paper is organized as follows. Section 2 introduces the traffic pattern extraction. Section 3 presents the proposed traffic anomaly detection method. Section 4 reports the experimental evaluation results. Section 5 discusses the related work. Finally Sect. 6 concludes our paper.

2   Traffic Pattern Extraction

2.1   Basic Traffic Pattern Extraction Based on NMF

Definition 1 (Traffic Volume Matrix): The traffic volume of each road on day d is represented by traffic volume matrix V^d, where each element v_{ij}^d is the traffic volume of i-th road on day d at j-th hour. For example, $V^d \in \mathbb{R}_+^{m \times 24}$ is a traffic volume matrix containing 24 h traffic volumes of m roads.

$$V^d = \begin{bmatrix} 151 & 108 & \cdots & 73 \\ 45 & 29 & \cdots & 50 \\ \cdots & \cdots & \cdots & \cdots \\ 22 & 32 & \cdots & 20 \end{bmatrix}$$

To avoid the disturbance from the noise in the historical volume, traffic volume matrix is not used directly in our study. Inspired by [7] where Nonnegative Matrix Factorization (NMF) was employed to find human mobility patterns and land uses of urban area, we use NMF to extract basic traffic patterns of roads, because it produces two nonnegative matrices with physical meanings that match our analysis requirement [8]. Meanwhile, NMF reduces data dimensionality and thus makes the anomaly detection more efficient.

NMF decomposes a nonnegative matrix into two:

$$V \approx CP \tag{1}$$

where $C \in \mathbb{R}_+^{m \times r}$, $P \in \mathbb{R}_+^{r \times h}$, m is the number of road segments, h is the number of time slots, 24 h of one day in our case, and r is the number of traffic patterns (usually specified according to application need).

To minimize the loss of factorization, an objective function is defined as

$$J(C,P) = \frac{1}{2} \sum_{i,j} [V_{ij} - (CP)_{i,j}]^2 \tag{2}$$

Given r, matrix C and P can be iteratively updated until convergence by

$$C_{ik} \leftarrow C_{ik} \cdot \frac{(VP^T)_{ik}}{(CPP^T)_{ik}}, \quad P_{kj} \leftarrow P_{kj} \cdot \frac{(C^TV)_{kj}}{(C^TCP)_{kj}} \tag{3}$$

The matrix C and P in our application have their own physical meanings.

Definition 2 (Pattern Matrix): P, pattern matrix, has rows each of which corresponds to a basic traffic pattern, representing a stable mode of human collective mobility.

Definition 3 (Coefficient Matrix): C, coefficient matrix, includes the coefficients of each road segment with respect to the corresponding basic patterns. In other words, one row of C indicates the proportion and scale of each basic traffic patterns on a road, and thus can be employed to study road similarity in next section.

The factorization formula $V \approx CP$ indicates that the traffic volume of one road (e.g., i-th row of V) is a linear combination of traffic patterns (P), with coefficient in the i-th row of C. Next, we will demonstrate and discuss the P pattern matrix and C coefficient matrix obtained from our traffic volume data.

2.2   Discussion of Pattern Matrix

Our data set includes traffic of 28 days. We first construct the Traffic Volume Matrix V^d for each day, $d = 1, ..., 28$. The factorization rank r in NMF is a critical parameter. In order to obtain stable and meaningful decomposition results, we tried r to be any integer from 2 to 20. For each setting of r, V^d ($d = 1, ..., 28$) is factorized into C^d and P^d.

Conforming with the finding in [7], when $r = 3$, the obtained matrix P is most stable over time, namely, the row-wise deviation of $\{P^d\}_{d=1}^{28}$ is minimum.

Figure 1(a) shows the basic patterns on workdays, P_1, P_2 and P_3 denoting the three row of P. Solid lines represent the mean value, while dashed lines represent the positive and negative deviations averaged on different days. The basic pattern P_1 is in accord with working or education mobility pattern. The peak hour of P_1 is at 7:00–8:00 am and 5:00–6:00 pm, when people go to work and get off work, or students go to school and leave school. The basic pattern P_2 reveals people moving in daytime, about from 10:00 am to 4:00 pm, which may be the traveling between two workplaces for business. The peak hour of P_3 is at around 8:00 pm, which is related with entertainment and other activities.

2.3 Discussion of Coefficient Matrix

Each row of C contains r coefficients, describing how much different basic patterns contribute to the traffic volume on each road. In our case of $r = 3$ for example, a large value in the first coefficient means the traffic volume of this road is relatively high in 7:00–8:00 am and 5:00–6:00 pm. Figure 1(b) visualizes the coefficients of road segments in C (the three columns corresponding to P_1, P_2 and P_3 are represented in red, blue and yellow, respectively). The major areas in red are Xizhimen and Zhongguancun, which are the vital communication lines in Beijing and have high traffic volume during peak hours. The blue areas include two famous business zones: Sanlitun and Xidan. Areas in yellow are widely distributed over the whole city, as well as roads heading to Beijing Capital Airport. Therefore, we can see roads with similar coefficients have the same property, even though they are not adjacent. Considering the similarity in traffic patterns can compensate for the deficiency of finding similar roads only by closeness in geo-location.

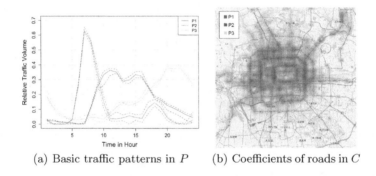

(a) Basic traffic patterns in P (b) Coefficients of roads in C

Fig. 1. The results of NMF (Color figure online)

3   Traffic Anomaly Detection

After applying NMF, we develop our proposed traffic anomaly detection approach. Generally, the approach consists of two steps: first, select similar road segments for comparison; second, calculate the anomaly score.

3.1   Finding Similar Roads by Clusters

As mentioned in the Sect. 2, the traffic volume of a road can be considered as a linear combination of different traffic patterns. The combination coefficients indicate how the traffic on this road relevant to different patterns. In other words, one row of C matrix characterizes each road segment as a new representation vector. Therefore, two road segments i and j hosting similar types of traffic pattern will have a small distance calculated from the i-th and j-th row of C.

Note, road segments are compared based on C, rather than the original volume matrix V, due to 1) efficiency (C has a smaller size than V); and 2) robustness (C is road's features mapped into stable space P). The neighboring roads are selected according to their closeness in geo-location and similarity in traffic patterns. We thus define the distance between road i and road j as follow:

$$D(i,j) = \alpha D_t^T(i,j) + (1-\alpha)\frac{D_g(i,j)}{\tau} \tag{4}$$

Distance $D(i,j)$ is composed of two parts: D_t^T is the distance measured based on traffic patterns; D_g is the physical distance between two roads. τ is a scaling constant for making D_t^T and D_g comparable. α is a parameter adjusting the weight of two distances, whose setting will be studied in Sect. 4.4. We set $T = 21$ (the number of training days) and $\tau = 10000$ (scaling down the geo-distance measured in meters) in this paper.

While D_g can be computed by the coordinates of two roads, D_t^T is defined as:

$$D_t^T(i,j) = \sqrt{\sum_{d=1}^{T}\sum_{k=1}^{3}(C_{ik}^d - C_{jk}^d)^2} \tag{5}$$

where C_{ik}^d is the k-th coefficient of i-th road on day d in coefficient matrix C^d. $D_t^T(i,j)$ actually measures the difference of coefficients of two roads on T continuous days, for enhancing the stability.

The neighboring roads are formed based on clustering. Given the pairwise distance from Eq. (4), roads can be clustered by exemplar-based method, which takes actual roads as cluster centroids. Affinity Propagation (AP) [9] is the most appropriate algorithm for our problem. It a clustering algorithm proposed to find out the most representative actual items, called *exemplars*, in a data set. It provides optimality guarantee about minimizing the clustering distortion, compared to k-medoids. It can be applied with or without a specified number of clusters [10]. Since our problem has no determined number of clusters, we assign each road the same preference to be an exemplar and let AP method form the optimal number of clusters. To ensure a road has enough neighbors, a cluster with a small number of members will be merged with another cluster to which it has the nearest center.

3.2   Anomaly Scoring

We evaluate the abnormal score of traffic on one road by comparing its current traffic volume with not only its historical data but also its neighbors. Normally, traffic volume at current moment is similar to that at the same moment of previous days. However, irregular factors, such as the atrocious weather and the particular event, cause variations in traffic volume. We define the abnormal score by estimating the existing probability of traffic volume. The abnormal score of road i on day d at time slot t is:

$$S_{it}^d = \beta P_a^T(v_{it}^d) + (1-\beta)P_r(v_{it}^d) \tag{6}$$

where P_a^T is the existing probability among history data (till T days), while P_r is the probability among the neighbors' volume. We use Kernel Density Estimation (KDE) [11] algorithm to evaluate the probability. β is the weight for balancing the two factors, whose sensitivity will be discussed in Sect. 4.4. The P_a^T and P_r are defined as follow.

$$P_a^T(v_{it}^d) = \frac{1}{Th_1} \sum_{j=1}^{T} \phi(\frac{v_{it}^d - v_{it}^j}{h_1}) \tag{7}$$

$$P_r(v_{it}^d) = \frac{1}{Nh_2} \sum_{j=1}^{N} \phi(\frac{v_{it}^d - v_{jt}^d}{h_2}) \tag{8}$$

where v_{it}^d is the volume of road i on day d at time slot t. N is the number of neighbors in the cluster of road i. The KDE kernel ϕ is set to be Gaussian. Bandwidth $h_1 = \frac{5\sigma}{T}$ and $h_2 = \frac{5\sigma}{N}$, and σ is the standard deviation of the history volume or neighboring volume.

4   Experiments and Analysis

4.1   Settings

Trajectory Data: Real world taxi GPS trajectories are used to evaluate the effectiveness and efficiency of the proposed algorithm, with statistics shown in Table 1. As about 20 % of traffic on road surfaces in Beijing is generated by taxicabs, the taxi trajectories represent a significant portion of traffic volume on the network [1]. We used ST-Matching algorithm to process the low-sampling-rate GPS trajectories [12], and counted the traffic volume for each road segment every hour. The first 3 weeks trajectory data are used as training set, while the last 1 week data are used as testing set.

Road Network: The used road networks of Beijing are shown in Table 1.

Table 1. Statistics of dataset

Trajectories	data duration	2–29 Nov. 2011
	No. of taxies	12,286
	No. of effective days	28
	No. of Avg. sampling intervals(s)	60
Roads	No. of road segments	180,350
	No. of road nodes	132,273
Weibo Reports	data duration	23–29 Nov. 2011
	Avg. reports of day	19

Traffic Anomaly Reports: WeiBo is a Twitter-like social site in China. We extract the traffic anomaly reports from the news posted at WeiBo and use them for evaluating the detected anomalies. The statistics is shown in Table 1.

Baseline: We choose three different methods for comparison. ICDE2009 [6] detects road anomaly based on a comparison with all other roads. ICDM2012 [4] takes all roads and history data into consideration. SPIE2015 [5] applies KDE to investigate the probability value of a road traffic status given history data.

Measurement: The reported traffic accidents at WeiBo are taken as the ground truth. Three metrics are used:

1. **Recall** is the percentage of the number of the correctly detected anomalies over the number of all reported anomalies.
2. **Precision** is the percentage of the number of correctly detected anomalies over the number of all detected anomalies.
3. $F1$-**measure** is the equally-weighted mean of precision and recall:

$$F1 - measure = \frac{2 * recall * precision}{recall + precision} \tag{9}$$

4.2 Experiments Using Real-World Data

The detection performance of our proposed method is compared with that of baseline methods on the last seven days data. As shown in Fig. 2(a), the recall of the proposed method is 92.76 %, which is about 20 % higher than the highest recall of baseline. The precision of the proposed method is 74.55 % while the highest precision of baseline is 62.11 %. The $F1$-measure of proposed method is the highest.

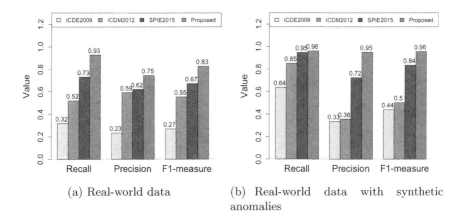

(a) Real-world data   (b) Real-world data with synthetic anomalies

Fig. 2. Detection performance comparing to baseline methods (Color figure online)

According to [6], the data in California is considerably stable with an average deviation of less than 10 %. While the data of Beijing is relatively unstable and irregular. Thus, our method performs the best detecting anomalies in traffic with high variations in Beijing, though ICDE2009 performed well with the stable traffic data in California. The performance of ICDM2012 is consistent with [1]. SPIE2015 is a special case of our method when $\beta = 1$.

The reported *Precision* in Fig. 2(a) is not as high as *Recall*, mainly because the ground truth we used from Weibo is just a subset of the true ground truth. The facts are: not all accidents were reported at Weibo; not all anomalies are caused by accidents; a reported anomaly event may have impact on several road segments and last for a long time but only have been recorded at one road and at one time stamp.

We have two case-studies on this unsatisfactory precision issue due to imperfect ground truth. In the first case, an anomaly was detected by our method at Workers' Stadium North Road around 10:00 pm on 11/26/2011. However, it was not reported in the Weibo record. After looking back the historical news, we find this anomaly was caused by a concert. In the evening of 11/26/2011, a vocal concert of Sodagreen (a popular idol groups in Taiwan, China) was hold in Beijing Workers' Sports Complex, which was nearby the road. In the Fig. 3(a), the red mark is the Beijing Workers' Sports Complex, and the blue mark is the Workers' Stadium North Road. The road is in the north of the stadium. Figure 3(c) shows the traffic volume patterns of the road and its neighbors. The traffic patterns of roads are accordant most of the time except the 10:00 pm of 11/26/2011, when the volume of the anomalous road is relatively larger than that of its neighbors. The increasing traffic volume may be caused by the leaving of the people when the concert was over.

In the second case, there was an anomaly on Xueyuan Road at 8:04 am on 11/23/2011 reported by Weibo, while two road segments are identified as anomalies by the proposed method. As illustrated in Fig. 3(b), the red mark is the reported anomaly and the two blue marks are the detected anomalies. Actually, both of the two road segments are the parts of Xueyuan Road. Due to the lack of the perfect ground truth, the proposed method cannot be evaluated

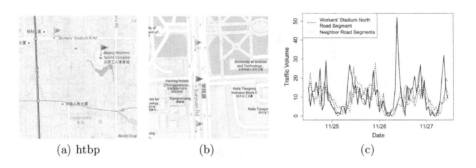

(a) htbp   (b)   (c)

Fig. 3. Two study cases (Color figure online)

Table 2. Results with different parameter α and β tested on real-world data

β \ α	0			0.5			1		
	Recall	Precision	F1	Recall	Precision	F1	Recall	Precision	F1
0	65.38%	43.07%	51.93%	78.86%	63.84%	70.56%	81.22%	48.87%	61.03%
0.5	73.43%	53.10%	61.63%	**92.76%**	**74.56%**	**82.67%**	89.86%	69.40%	78.32%
1	73.20%	62.12%	67.21%	73.20%	62.12%	67.21%	73.20%	62.12%	67.21%

Table 3. Results with different parameter α and β tested on real-world data with synthetic anomalies

β \ α	0			0.5			1		
	Recall	Precision	F1	Recall	Precision	F1	Recall	Precision	F1
0	65.53%	41.20%	50.60%	72.53%	82.29%	77.11%	80.83%	40.86%	54.30%
0.5	81.70%	68.74%	74.66%	**96.12%**	**95.04%**	**95.58%**	93.70%	81.70%	87.29%
1	94.77%	72.44%	82.11%	94.77%	72.44%	82.11%	94.77%	72.44%	82.11%

properly. Therefore, we design the semi-synthetic experiments, which will be discussed in the next section.

4.3   Experiments Using Semi-synthetic Data

To better evaluate our method, the experiment of semi-synthetic data still uses the last week data as testing set but with added synthetic anomalies.

Since the traffic volume on different days at the same time slot demonstrates normal distribution, anomalies are generated with volumes larger than $\mu + 4\sigma$, where μ is the average and σ is the standard deviation of traffic volumes. The results of the all methods are shown in Fig. 2(b). Our proposed method performs the best, while the precision is also improved.

4.4   Parameter Discussion

In this section, we take experiments on both real-world and semi-synthetic data to explore how parameters α and β have an influence on anomaly detection, as well as the clustering results.

The α and β. α is the weight to balance the distance of geo-location and of traffic patterns in Eq. (4). β is the weight to balance the history score and neighbor score in Eq. (6). Tables 2 and 3 show the results of different α and β tested on real-world and real-world data with synthetic anomalies, respectively.

From the results, we can see $\alpha = \beta = 0.5$ has the best performance. That is to say, considering both closeness in geo-location and traffic patterns finds better neighboring roads for comparison, and combining the comparison with history traffic data (time domain) and neighboring road traffic status (space domain) results in better abnormal score.

To further study the impact of these two parameters, we fix $\beta = 0.5$ and change α. The F1-measure with different setting of α is shown in Fig. 4(a).

Table 4. Sensitivity analysis of parameter p in AP clustering

Setting of p	0.5*m	m	2*m
No. of clusters	638	376	220
Recall	90.42 %	**96.12 %**	92.84 %
Precision	91.83 %	**95.04 %**	89.99 %
F1-measure	91.12 %	**95.58 %**	91.39 %

Figure 4(b) presents the results when changing β with $\alpha = 0.5$. As shown in the figures, when $\beta = 0.5$ and $\alpha = 0.5$, the performance is the best.

When $\alpha = 0$, the distance measure in Eq. (4) takes only the road closeness in geo-location, while when $\alpha = 1$ the distance relies only on the closeness in traffic patterns. Neither of these two settings can select a proper neighboring road cluster as a reference for anomaly detection. Bringing roads with similar traffic patterns together with spatial adjacent roads, anomalies affecting regions from a single road or a large district can be detected.

The β parameter controls how much the anomaly score relying on neighboring data (solely, when $\beta = 0$) and historical data (solely, when $\beta = 1$). Relying on only historical data causes false alarms or miss anomalies when traffic volume varies with irregular "surges" or "decrease" caused by circumstantial factors or special events. Neighboring (close in space and in road characteristics) traffic status helps the detection system resist such variations.

The Parameter of Clustering. We adjust the preference p in AP clustering method, and obtain different clustering results. Table 4 shows the results of anomaly detection with different parameter p. When $p = m$, we can obtain the best result, where m is the median of all pairwise distance values and is popularly

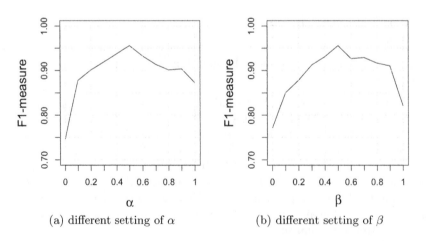

(a) different setting of α (b) different setting of β

Fig. 4. Sensitivity analysis of parameter α and β

suggested as the best setting of p in many applications [9]. When p changes, the number of obtained clusters varies a lot. However, the detection performance only slightly affected. Generally, the proposed method is not sensitive to the clustering results.

5   Related Work

The study of anomaly detection on trajectory data can be classified into two categories: **trajectory** anomaly detection and **traffic** anomaly detection. [13,14] focus on the problem of detecting anomalous trajectories by comparing with the historically "normal" routes. [1] identifies traffic anomalies by investigating drivers' routing behaviors that significantly differ from their original patterns. [15] detects traffic anomalies using microscopic traffic variables, such as relative speed, inter vehicle time gap, and lane changing. [3] proposes the STOTree algorithm based on both spatial and temporal properties of detected outliers. [16] adapts likelihood ratio tests to describe traffic patterns. Since these two methods detect anomalies on the level of regions, they may only find the very large scale events and miss the road segment-level traffic anomalies, such as traffic accident. Kuang et al. propose an anomaly detection method based on wavelet transform and PCA [17]. Even though [17] are efficient, its recall value is just 74.62 %. The work in this paper focuses on traffic anomaly detection on road segment level. The proposed detection method identifies anomalies by taking into account historical data in time and neighboring traffic status in space and in road characteristics.

6   Conclusion

In this paper we have studied the problem of detecting anomalies. We evaluate the abnormal score of traffic on one road by comparing its current traffic volume with not only its historical data but also its neighbors. The neighboring roads are selected according to their closeness in geo-location and high similarity in traffic patterns. We evaluated our method with GPS trajectories generated by 12,286 taxis in Beijing. The recall, precision and $F1$-measure of our method with real word data are 92.76 %, 74.56 % and 82.26 %, which is better than the baselines. In the next, we will focus on how the anomalies propagate.

Acknowledgment. This work was supported by the Chinese 863 Programs(2013 AA01A601), Science Foundation of Shenzhen in China(JCYJ20140509150917445), the State Key Laboratory of Software Development Environment(SKLSDE-2015ZX-25) and Fundamental Research Funds for the Central Universities.

References

1. Pan, B., Zheng, Y., Wilkie, D., Shahabi, C.: Crowd sensing of traffic anomalies based on human mobility and social media. In: ACM SIGSPATIAL GIS (2013)

2. Guo, J., Huang, W., Williams, B.M.: Real time traffic flow outlier detection using short-term traffic conditional variance prediction. Transp. Res. Part C: Emerg. Technol. **50**, 160–172 (2015)

3. Liu, W., Zheng, Y., Chawla, S., Yuan, J., Xing, X.: Discovering spatio-temporal causal interactions in traffic data streams. In: SIGKDD, pp. 1010–1018 (2011)

4. Chawla, S., Zheng, Y., Hu, J.: Inferring the root cause in road traffic anomalies. In: ICDM, pp. 141–150 (2012)

5. Ngan, H.Y., Yung, N.H., Yeh, A.G.: A comparative study of outlier detection for large-scale traffic data by one-class SVM and kernel density estimation. In: IS&T/SPIE Electronic Imaging. International Society for Optics and Photonics, pp. 94050I–94050I-10 (2015)

6. Li, X., Li, Z., Han, J., Lee, J.-G.: Temporal outlier detection in vehicle traffic data. In: ICDE, pp. 1319–1322. IEEE (2009)

7. Peng, C., Jin, X., Wong, K.-C., Shi, M., Liò, P.: Collective human mobility pattern from taxi trips in urban area. PloS One **7**(4), e34487 (2012)

8. Lee, D.D., Seung, H.S.: Learning the parts of objects by non-negative matrix factorization. Nature **401**(6755), 788–791 (1999)

9. Frey, B.J., Dueck, D.: Clustering by passing messages between data points. Science **315**(5814), 972–976 (2007)

10. Zhang, X., Wang, W., Nørvåg, K., Sebag, M.: K-AP: generating specified K clusters by efficient affinity propagation. In: ICDM, pp. 1187–1192 (2010)

11. Parzen, E.: On estimation of a probability density function and mode. Ann. Math. Stat. **33**, 1065–1076 (1962)

12. Lou, Y., Zhang, C., Zheng, Y., Xie, X., Wang, W., Huang, Y.: Map-matching for low-sampling-rate GPS trajectories. In: ACM SIGSPATIAL (2009)

13. Chen, C., Zhang, D., Samuel Castro, P., Li, N., Sun, L., Li, S.: Real-time detection of anomalous taxi trajectories from GPS traces. In: Puiatti, A., Gu, T. (eds.) MobiQuitous 2011. LNICST, vol. 104, pp. 63–74. Springer, Heidelberg (2012)

14. Chen, C., Zhang, D., Castro, P.S., Li, N., Sun, L., Li, S., Wang, Z.: iboat: Isolation-based online anomalous trajectory detection. IEEE Trans. Intell. Transp. Syst. **14**(2), 806–818 (2013)

15. Barria, J.A., Thajchayapong, S.: Detection and classification of traffic anomalies using microscopic traffic variables. IEEE Trans. Intell. Transp. Syst. **12**(3), 695–704 (2011)

16. Pang, L.X., Chawla, S., Liu, W., Zheng, Y.: On mining anomalous patterns in road traffic streams. In: Advanced Data Mining and Applications, pp. 237–251 (2011)

17. Kuang, W., An, S., Jiang, H.: Detecting traffic anomalies in urban areas using taxi GPS data. Math. Probl. Eng. **501**, 809582 (2015)

Query Processing and Optimization

PACOKS: Progressive Ant-Colony-Optimization-Based Keyword Search over Relational Databases

Ziyu Lin[1](✉), Qian Xue[1], and Yongxuan Lai[2]

[1] Department of Computer Science, Xiamen University, Xiamen, China
{ziyulin,xueqian2015}@xmu.edu.cn
[2] School of Software, Xiamen University, Xiamen, China
laiyx@xmu.edu.cn

Abstract. Keyword search over relational databases makes it easier to retrieve information from structural data. One solution is to first represent the relational data as a graph, and then find the minimum Steiner tree containing all the keywords by traversing the graph. However, the existing work involves substantial costs even for those based on heuristic algorithms, as the minimum Steiner tree problem is proved to be an NP-hard problem. In order to reduce the response time for a single search to a low level, a progressive ant-colony-optimization-based algorithm, called PACOKS, is proposed here, which achieves the best answer in a step-by-step manner, through the cooperation of large amounts of searches over time, instead of in an one-step manner by a single search. Through this way, the high costs for finding the best answer, are shared among large amounts of searches, so that low cost and fast response time for a single search is achieved. Extensive experimental results based on our prototype show that our method can achieve better performance than those state-of-the-art methods.

1 Introduction

Keyword search over relational databases and XML documents makes it an easier task to retrieve information from structural and semi-structural data, as users, especially casual users, do not need to learn query languages such as SQL or to know anything about data schemas. Keyword search over these data has been recently thoroughly studied (e.g., [2,3,8,13,16]), and the work can be typically classified into two categories, namely data graph based (e.g., BANKS [2] and EASE [9]) and schema graph based (e.g., DISCOVER [6] and DBXplorer [1]). Schema-graph-based approach enumerates results by directly running SQL statements against DBMS, which means that such approach can only be used in relational data, while the data-graph-based approach is to find out the minimum

Supported by the Natural Science Foundation of China (61303004), the National Key Technology Support Program (2015BAH16F00/F01) and the Key Technology Program of Xiamen City (3502Z20151016).

B. Cui et al. (Eds.): WAIM 2016, Part II, LNCS 9659, pp. 107–119, 2016.
DOI: 10.1007/978-3-319-39958-4_9

Steiner tree by traversing through a data graph, which makes it appropriate for all the data that can be represented as graph.

However, the existing work encountered a challenging issue, i.e., high cost for a single search. As the minimum Steiner tree problem is proved to be an NP-hard problem, the previous data-graph-based solutions, most based on heuristics, usually involve substantial cost for a single search. It is also the case for those schema-graph-based solutions, since there are large amounts of possible expressions to run over the DBMS.

To address the above problem, we first propose an ant-colony-optimization-based algorithm, called ACOKS, to find the minimum Steiner tree from the data graph. Furthermore, we propose a novel progressive ant-colony-optimization-based algorithm, called PACOKS, an abbreviation for Progressive Ant Colony Optimization based Keyword Search, which achieves the best answer in a step-by-step manner through the cooperation of large amounts of searches over time, instead of in an one-step manner by a single search. In other words, the later search result is further optimized based on the earlier one, and the global optimal solution, i.e., the minimum Steiner tree, is gradually achieved through many searches. This way, the high costs for finding the best answer are shared among large amounts of searches, so that low cost and fast response time for a single search is achieved.

To sum up, the main contributions of this paper include:

- An ant-colony-optimization-based algorithm for approximating the minimum or top-k Steiner trees problem.
- A progressive ant-colony-optimization-based algorithm to share the high cost of Steiner tree problem among large amounts of searches so as to achieve very low cost for a single search.
- A prototype to carry out extensive experiments, which confirm the superior performance of our approach over the state-of-the-art methods.

The remainder of this paper is organized as follows. We formalize the minimum Steiner tree problem in Sect. 2. Section 3 discusses the ACOKS algorithm. We discuss in detail the PACOKS algorithm in Sect. 4. Experimental studies are given in Sect. 5, followed by discussions over related work in Sect. 6. Finally, we conclude in Sect. 7.

2   Data-Graph-Based Approach

2.1   Data Graph

The principal of data-graph-based solutions to keyword search over relational databases, is to enumerate the minimum cost Steiner trees by traversing through the data graph constructed from a relational database.

A relational database D can be considered as a directed graph $G(V, E)$, called *data graph*, where V represents the set of nodes, and E the set of edges. A node $u \in V$ represents a tuple in D. A tuple may be inserted into or deleted from a relation in D, resulting in the change of the data graph. Given two nodes $u \in V$ and $v \in V$, there exists a forward edge $e(u, v)$ in the data graph, from u to v, denoted $u \rightarrow v$, and a backward edge $e(v, u)$, from v to u, denoted $v \rightarrow u$, if there is primary-foreign-key relationship between u and v, where the foreign key is defined on u referencing to the primary key defined on v. Both edges and nodes may have weights, which are identified by the functions of $w(e)$ and $w(u)$ respectively. The weight of a data graph, denoted $w(G)$, is the sum of the weights of all edges, i.e., $w(G) = \sum_{e \in E} w(e)$. Keyword search algorithms score the candidate Steiner trees based on node weights and edge weights, and then rank these trees in the order of decreasing score. Since the focus of this paper is not how to define the weight function, we simply take the weight function proposed in BANKS [2]. In other words, the weight of a node u, denoted $w(u)$, is a function of in-degree. The weight of an edge $e(u, v)$ depends on the its type, i.e., $w(u, v)=1$ for a forward edge representing a primary-foreign-key relationship, and $w(v, u) = w(u, v) * log_2(1 + D_{in}(v))$ for a backward edge, where $D_{in}(v)$ represents the in-degree of a node v.

Example 1. Figure 1 shows a simple example on the publication database DBLP. It consists of four relation schemas (see Fig. 1(a)), i.e., *Authors*, *Papers*, *Cites* and *Writes*. *Authors* has two attributes, *AID* and *Name*, and the primary key is defined on *AID*. *Papers* has two attributes (*PID* and *Title*) with *PID* as the primary key. *Cites* has two foreign keys, *Citing* and *Cited*, both referring to the primary key defined on *Papers*. *Writes* has two foreign keys, *AID* (refer to the primary key defined on *Authors*) and *PID* (refer to the primary key defined on *Papers*). Figure 1(b) and (c) show the database conforming to the relation schemas above and its corresponding data graph respectively.   □

2.2   Steiner Tree Problem

In a relational database, the task of keyword search is to find those Steiner trees containing the keywords. These nodes containing the keywords are interconnected by sequences of primary/foreign key relationships among tuples.

Example 2. Figure 1(d) shows two Steiner trees. The left one may be one of the answers for the 4-keyword search $\{database(k_1), XML(k_2), Jim(k_3), Steiner(k_4)\}$, and the right one for the 3-keyword search $\{Jim(k_5), Steiner(k_6), Kate(k_7)\}$, over the publication database in Fig. 1(b).   □

The best answer for a keyword search is a minimal Steiner tree.

Definition 1. *[Minimum Steiner tree] Given a data graph $G(V, E)$ and a node set $V' \subseteq V$, a tree T is called a Steiner tree over V', if T contains all the nodes in V'. Let $c(T) = \sum_{e \in E} w(e)$ be the cost of T, where $w(e)$ is the weight of an edge e. We say that T is the minimum Steiner tree, if $c(T)$ is the minimum among all the Steiner trees over V' in G.*   □

An extension of minimum Steiner tree is minimum group Steiner tree. Keyword search over a data graph can also be seen as a minimum group Steiner tree problem.

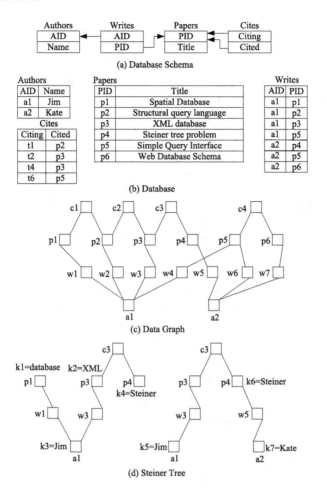

Fig. 1. A simple publication database.

Definition 2. *[Minimum group Steiner tree] Given a data graph $G(V, E)$ and groups $V_1, V_2, ..., V_n \subseteq V$, we say that T is the minimum group Steiner tree over $V_1, V_2, ..., V_n$ in G, if T is a minimum Steiner tree and contains at least one node from each group $V_i (1 \leq i \leq n)$.* ☐

3   The ACOKS Algorithm

Large amounts of experiments in recent years proved that, ant colony optimization (or ACO in abbreviation) is able to achieve high efficiency in dealing with various kinds of NP-hard problem. Therefore we here propose an ant-colony-optimization-based algorithm, called ACOKS, an abbreviation of ACO-based Keyword Search, to deal with the Steiner tree problem. The basic idea of ACOKS is to find Steiner trees containing the keywords in the data graph through the cooperation of many ants. When searching for the optimal solution, it aims to find the minimum Steiner tree, while for the top-k answers, it outputs the top-k Steiner trees with minimum costs. As far as the solution for the Steiner tree problem is concerned, it usually uses the method of spanning and cleanup. In this method, it starts from any node of a group V_i and spans to its neighboring nodes, until covering at least one node from each group. Finally, those redundant nodes are deleted from the result tree. The main algorithm of ACOKS is outlined in Algorithm 1. The function of ONE_ANT_MOV(ant) in Algorithm 1 is given in Algorithm 2. The transition rule and pheromone updating for ant colony optimization can be found in many other related research work, so they are not discussed here.

Algorithm 1. ACOKS

Input : keyword search $K = \{k_1, k_2, ..., k_m\}$;
        data graph G;
Output: top-k Steiner trees;

begin
    foreach $k_i \in K$ **do**
        get the content set C_i of k_i;

    $iteration\_time \leftarrow 0$;
    initialize the pheromone matrix M;
    $result\_heap \leftarrow \Phi$;
    $t \leftarrow 0$;
    while $iteration\_time < max\_iterlation\_time$ **do**
        select one node from each C_i and put p ants on it to form ant set S_{ants};
        $ant\_num \leftarrow m * p$;
        foreach $ant \in S_{ants}$ **do**
            $step\_num \leftarrow 0$;
            $S_{visited}(ant) \leftarrow \Phi$;
            $S_{spanned}(ant) \leftarrow \{k_i\}$;

        while $ant\_num > 0$ **do**
            foreach $ant \in S_{ants}$ **do**
                ONE_ANT_MOV(ant)

        $iteration\_time \leftarrow iteration\_time + 1$;
    output the result trees in $result\_heap$;

---

4 The PACOKS Algorithm

4.1 The Defect of ACOKS

The ACO algorithm has characteristics that do not exist for ants in the natural world. For example, the results can be optimized little by little with the accumulation of pheromone on the paths. By optimization, it means that the

---

Algorithm 2. ONE_ANT_MOV(ant)

---

begin
    if $ant.step\_num > max\_step\_num$ **then**
        | destroy ant;
    else
        $ant.step\_num \leftarrow ant.step\_num + 1$;
        $t \leftarrow t + 1$;
        move to next node v satisfying that $v \notin S_{visited}(ant)$;
        if $there\ is\ no\ other\ ant\ arriving\ at\ v\ at\ time\ t$ **then**
            | $S_{visited}(ant) \leftarrow S_{visited}(ant) \cup v$
        else
            $S_{visited}(ant) \leftarrow S_{visited}(ant\_other) \cup S_{visited}(ant)$;
            $S_{spanned}(ant) \leftarrow S_{spanned}(ant\_other) \cup S_{spanned}(ant)$;
            destroy ant_other;
            if $S_{spanned}(ant) = K$ **then**
                output the Steiner tree s_tree composed of nodes in
                $S_{visited}(ant)$;
                remove redundant nodes from s_tree;
                if $score(s\_tree) > score(s\_tree\_min)$ **then**
                    | remove s_tree_min from $result\_heap$;
                    | put s_tree into $result\_heap$;
                else
                  | discard s_tree;
            destroy ant;
    update the pheromone matrix M at time t;

---

later solution is closer to the global optimal solution (or GOS in abbreviation), compared with the previous one. In other words, for ACOKS, the solution of each iteration is a successive approximation to the GOS. Also the ACO algorithm usually can achieve good convergence in most real applications, though the theoretical proof of its convergence is still not available for some cases. Gutjahr et al. [4] discussed the convergence of a graph-based ACO, and proved that the GOS can always be found if the amount of ants is large enough. Yang et al. [15] proved that the ACO algorithm for the Steiner tree problem is able to converge to the GOS, with the probability of 1-ϵ, when the the number of iterations is large enough, where ϵ is an arbitrarily small value. Therefore, for the above ACOKS algorithm, we can get a conclusion as follows:

– it can not be assured to find the GOS if the number of ants is limited; and
– there must be enough ants if the GOS is to be achieved.

However, the contradiction is that it will take too long and be unacceptable for user, if it involves a large amount of ants for a single user search, though the GOS can be finally achieved through this way. On the other hand, if we want to limit the response time for a single search to a low level, the number of ants must be limited to a small amount. However, this may influence the result quality and lead to low search effectiveness.

4.2   The PACOKS Algorithm

To resolve the issue above, we here propose the progressive-ACO-based algorithm, or PACOKS in abbreviation, which aims to achieve the GOS through a step-by-step approach rather than a one-step approach. In other words, the result of the current search is a further optimization upon that of the previous one, so that the result of every search is a successive approximation of the GOS. In this way, the expensive costs of finding the GOS are shared among a large amounts of searches, and the cost for a single search is reduced to a very low level, resulting in fast response times for it.

Now we consider how to find the best answer for a search by PACOKS, which can be easily extended to support top-k search.

Algorithm 3. PACOKS

> **Input**   : keyword search $K = \{k_1, k_2, ..., k_m\}$
> **Output**: the best answer $result\_tree$
>
> **begin**
> > construct complete graph G_K with $k_i(\in K)$ as vertex;
> > **foreach** $e(k_i, k_j) \in G_k$ **do**
> > > $s\_tree(k_i, k_j) \leftarrow$ ACOKS(k_i, k_j);
> > > $weight(k_i, k_j) \leftarrow$ COST$(s\_tree(k_i, k_j))$;
> >
> > $min\_tree \leftarrow$ MINTREE(G_K);
> > **foreach** $e(k_i, k_j) \in min\_tree$ **do**
> > > $min\_tree \leftarrow$ REPLACE$(min\_tree, e(k_i, k_j), s\_tree(k_i, k_j))$;
> >
> > $result\_tree \leftarrow min\_tree$;

Algorithm 3 shows the main steps of PACOKS, in which the functions are as follows:

- COST(s_tree): compute the cost(or weight) of the Steiner tree s_tree;
- ACOKS(k_i, k_j): the ant-colony-optimization based algorithm, as is shown in Algorithm 1, for keyword pair$\{k_i, k_j\}$;
- MINTREE(G_K): compute the minimum spanning tree of G_K;
- REPLACE($min\_tree, e(k_i, k_j), s\_tree(k_i, k_j)$): replace the edge $e(k_i, k_j)$ in min_tree with $s\_tree(k_i, k_j)$.

Since the three algorithms, COST, MINTREE, and REPLACE, are so simple, they are not given in detail here.

Example 3. Figure 2 is an example explaining the process of PACOKS, which includes the following steps:

- Step 1: Given a keyword search K, including four keywords k_1, k_2, k_3 and k_4. Construct a complete graph G_K, with k_1, k_2, k_3 and k_4 as vertexes;

- Step 2: Call the algorithm ACOKS(k_1, k_2) for the edge $e(k_1, k_2)$, so as to get the minimum Steiner tree $s\_tree(k_1, k_2)$ containing the keywords k_1 and k_2; Similarly, we can get $s\_tree(k_1, k_3)$, $s\_tree(k_1, k_4)$, $s\_tree(k_2, k_3)$, $s\_tree(k_2, k_4)$ and $s\_tree(k_3, k_4)$;
- Step 3: Compute the cost of $s\_tree(k_1, k_2)$, denoted g_1, as the weight of the edge $e(k_1, k_2)$ in G_K. Similarly, we can get g_2, g_3, g_4, g_5 and g_6 as the weights of the other edges in G_K ;
- Step 4: Compute the minimum spanning tree min_tree from G_K, which includes three edges, say, $e(k_1, k_2)$, $e(k_1, k_4)$ and $e(k_3, k_4)$;
- Step 5: For each edge $e(k_i, k_j)$ in min_tree, replace $e(k_i, k_j)$ with $s\_tree(k_i, k_j)$. Therefore, $e(k_1, k_2)$, $e(k_1, k_4)$ and $e(k_3, k_4)$ are replaced with $s\_tree(k_1, k_2)$, $s\_tree(k_1, k_4)$ and $s\_tree(k_3, k_4)$ respectively. Figure 2(b) shows $s\_tree(k_1, k_2)$, $s\_tree(k_1, k_4)$ and $s\_tree(k_3, k_4)$, where $u(k_i)$ represents a node containing the keyword k_i. There exist common nodes, i.e., u_1 and $u(k_1)$, between $s\_tree(k_1, k_2)$ and $s\_tree(k_1, k_4)$, and $u_3, u(k_4)$ between $s\_tree(k_1, k_4)$ and $s\_tree(k_3, k_4)$. Therefore the three Steiner trees can be combined into one result tree based on their common nodes.   □

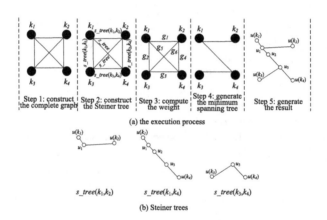

(a) the execution process

(b) Steiner trees

Fig. 2. The execution process of PACOKS.

Note: as Fig. 2 shows, in PACOKS, it only contains two keywords, k_i and k_j, whenever it calls the algorithm ACOKS. If a search K contains more than two keywords, it will call the algorithm ACOKS for each pair of keywords in K. Fortunately, a user will not start a search containing too many keywords.

5   Experimental Study

Here we report the performance evaluation of our method. The algorithms are implemented in JAVA. All the experiments were conducted on Intel i7-2600 3.40 GHz CPU, 16.0 GB memory DELL OptiPlex990 PC running Windows Server 2003 and Oracle 11g. As with most other work, we downloaded the

DBLP data (http://dblp.uni-trier.de/xml/) as the testing datasets. The DBLP database involves about 800MB data in the form of XML documents, which are uploaded to the Oracle database through a simple program developed by us. In this way, relational database called DBLP in Oracle database can be populated. Then we write a program to generate data graphs from the relational database DBLP, and put them in the memory. For DBLP, the data graph includes 7,270,404 nodes and 9,047,382 edges. The whole processes of generating the data graphs from DBLP take 89 s. Another program is written to automatically extract terms from DBLP, and a total of 534,124 terms are found. Then an inverted index is built for all the terms, which takes 105 seconds. We compare our methods with BANKS [2] and BLINKS [5].

5.1   Algorithm Training

Algorithm Convergency. We select many keyword pairs to run against ACOKS in the experiments, but only part of the experimental results will be presented here since they all share similar features. The results of two keyword pairs are reported here, i.e., $K_1 = \{Database, Design\}$, $K_2 = \{Zhang, Ullman\}$. Figure 3 shows the relationship between the score of the optimal Steiner tree and the number of iterations of ACOKS, in which the scoring method of BANKS is adopted. We can see from Fig. 3 that the result quality is improved step by step during the algorithm training process, and it can achieve high result quality after the algorithm converges.

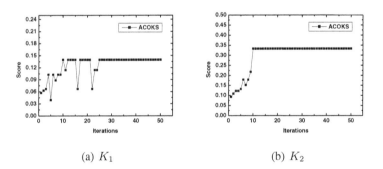

(a) K_1 (b) K_2

Fig. 3. The convergence curve of the algorithm training process

Compression Ratio of Pheromone Matrix. During the process of algorithm training, we analyzed the compression ratio of pheromone matrix for many paired keywords, so as to show that our method of constructing pheromone matrix based on paired keywords spanning graph is able to greatly reduce the space costs of the pheromone matrix. Here the compression ratio of the pheromone matrix, denoted f, is defined as N_{pksg}/N_{dg}, where N_{pksg} means the number of elements in the pheromone matrix based on a paired keywords spanning graph, and N_{dg} means the number of elements in the pheromone matrix based on a data graph. For the pheromone matrix based on a data graph, when it is not optimized, the number of elements in the matrix is the square of the number of nodes in the data graph. In theory, the number of elements in the pheromone matrix based on paired keywords spanning graph is the square of the number of nodes in such graph. However, for a directed graph, an edge may exist from u to v, but there is no edge from v to u, so we will not store any pheromone information for the latter. Furthermore, in our experiments, the maxium step number an ant may move forward, is limited to 50, which is large enough to satisfy any requirements in real applications. Based on the above two aspects, the pheromone matrix in our method can be further simplified. We here select 8 groups of paired keywords, and Table 1 shows the pheromone matrix compression ratios for them. From Table 1, we can see that our method of constructing a pheromone matrix based on paired keywords spanning graph is able to achieve a very high compression ratio. Through our method, all the pheromone matrices for various paired keywords are able to be memory-resident so as to greatly enhance the performance of ACOKS.

Table 1. Pheromone matrix compression ratio

K	N_{ant}	N_{dg}	N_{kp}	$f(\times 10^{-10})$
adriano,gianluca	3029	7270404^2	90735	17.2
alfred,ullman	35176	7270404^2	698974	132
design,database	4095	7270404^2	120231	22.7
information,retrieval	3506	7270404^2	44981	8.51
manfred,joachim	5091	7270404^2	9608	1.82
nikolay,aphrodite	47171	7270404^2	894337	169
pagerank,algorithm	3747	7270404^2	132846	25.1
query,processing	3300	7270404^2	25169	4.76
relational,database	2199	7270404^2	2446	0.463
robert,stephan	7542	7270404^2	213302	40.4
search,keyword	3012	7270404^2	1815	0.343
sunita,soumen	26938	7270404^2	727019	138
vassilis,grigoris	2477	7270404^2	154577	29.2

5.2 Search Efficiency

We now evaluate search efficiency of BANKS, BLINKS and PACOKS. We design 50 keyword searches, and the number of keywords, m, alternates between 2 and 6. After the response times for every search are acquired, they are averaged to be the evaluation index. Figure 4(a) shows the performance comparison results for BANKS, BLINKS [5] and PACOKS. We can make the following observations:

- PACOKS can outperform both BANKS and BLINKS for various keyword number. When $m = 2$, it takes more than 6,000 ms for BANKS to return the answers, and approximately 3000 ms for BLINKS on the DBLP dataset. While PACOKS only runs for 410 ms to get the results. It is because PACOKS is an ant-colony-optimization-based algorithm, which is able to take full advantage of parallel execution among various ants. Furthermore, when PACOKS provides online service, it has gone through the algorithm training period, and the later searches can make full use of the previously accumulated pheromone information, thus being able to achieve high efficiency during its calling ACOKS for each paired keywords $\{k_i, k_j\}$. BANKS and BLINKS, however, do not have such feature as sharing high costs among various searches.

– It performs best for PACOKS when the keyword number is 2, because there is no result subtree merging process. When the keyword number is 3, 4, 5 or 6, the search in PACOKS is split into several keyword pairs and there exists result subtree merging process, which leads to the increase of search response time.

– PACOKS is with good scalability. When m changes from 2 to 6, the response time for PACOKS only increases from 410 ms to 1,040 ms. The increase in time is mainly due to the subtree merging process.

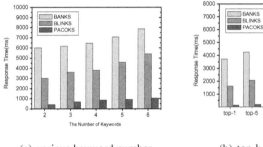

(a) various keyword number   (b) top-k search, $m = 3$

Fig. 4. Response time comparison

Furthermore, we compare the three algorithms on the aspect of response time for top-k answers in Fig. 4(b). One can see that our method is able to achieve much better performance than BANKS and BLINKS. For example, PACOKS only runs 190 ms to return the top-5 answers on the DBLP dataset, while BANKS and BLINKS take more than 4,000 ms and 2,000 ms respectively. This is also attributed to the "cost sharing mechanism" adopted by our method.

6 Related Work

Keyword search over structural data (e.g., relational database) and semi-structural data (e.g., XML and HTML documents), has received much attention in the database community in recent years. Existing approaches to support keyword search over relational databases can be typically classified into two categories, namely those based on data graph (e.g., BANKS [2] and EASE [9]) and those based on schema graph (e.g., DISCOVER [6] and DBXplorer [1]).

BANKS [2] is a representative of data-graph-based method, which uses a back expanding search algorithm to traverse the graph. However, back expanding search algorithm deteriorates a lot when it meets a node with large in-degrees. BANKS-II [7] improves the performance of BANKS through bidirectional search. In BLINKS [5], He et al. they proposed a partition-based method and a strategy called cost-balanced expansion, and at the same time, used an additional bi-level index to speed up the traversing process, which achieves better search efficiency than BANKS-II. Li et al. [9] proposed the concept of "r-radius Steiner tree", and keyword search problem is converted into a r-radius Steiner tree problem, which is able to identify some complicated and meaningful structures from database. Simple structures, such as tuple joining trees, are the focus of the literature in the initial research stage. Then research interest is extended to search for more sophisticated structures, such as r-radius Steiner tree [9], community, frequent co-occurring term [14].

DISCOVER [6] and DBXplorer [1] run SQL statements directly against database, while other work, such as [12], take middleware to execute SQL statements. DISCOVER and DBXplorer only focus on the search efficiency instead of result effectiveness, so Liu et al. [10] proposed a new weighted ranking mechanism and carried out extensive experiments to improve effectiveness of keyword search. Luo et al. [11] discussed how to support efficient top-k keyword search over relational database.

7   Conclusion

In this paper, we have addressed the problem of keyword search over relational databases. We proposed ant-colony-optimization-based algorithm, called ACOKS, to deal with minimum group Steiner tree problem. To limit the cost of a single search to a very low level, we further proposed progressive-ant-colony-optimization-based algorithm, called PACOKS, which aims to achieve the final global optimal solution, through the cooperation of searches continuously arriving at the system along the time. Thus the huge costs for finding the global optimal solution, are shared among large amounts of searches, and a single search can achieve very fast response time. The experimental results show that our proposed scheme can achieve both high efficiency and effectiveness.

References

1. Agrawal, S., Chaudhuri, S., Das, G.: DBXplorer: A system for keyword-based search over relational databases. In: Proceedings of ICDE, pp. 5–16 (2002)
2. Bhalotia, G., Hulgeri, A., Nakhe, C., Chakrabarti, S., Sudarshan, S.: Keyword searching and browsing in databases using banks. In: Proceedings of ICDE, pp. 431–440 (2002)
3. Djebali, S., Raimbault, T.: SimplePARQL: a new approach using keywords over SPARQL to query the web of data. In: Proceedings of the 11th International Conference on Semantic Systems, SEMANTICS 2015, Vienna, Austria, 15–17 September 2015, pp. 188–191 (2015)

4. Gutjahr, W.J.: A graph-based ant system and its convergence. Future Gener. Comput. Syst. **16**(1), 873–888 (2000)
5. He, H., Wang, H., Yang, J., Yu, P.S.: BLINKS: ranked keyword searches on graphs. In: Proceedings of SIGMOD, pp. 305–316 (2007)
6. Hristidis, V., Papakonstantinou, Y.: DISCOVER: Keyword search in relational-databases. In: Proceedings of VLDB, pp. 670–681 (2002)
7. Kacholia, V., Pandit, S., Chakrabarti, S., Sudarshan, S., Desai, R., Karambelkar, H.: Bidirectional expansion for keyword search on graph databases. In: Proceedings of VLDB, pp. 505–516 (2005)
8. Kim, I.-J., Whang, K.-Y., Kwon, H.-Y.: SRT-rank: Ranking keyword query results in relational databases using the strongly related tree. IEICE Trans. Inf. Syst. **97**(D(9)), 2398–2414 (2014)
9. Li, G., Ooi, B.C., Feng, J., Wang, J., Zhou, L.: EASE: an effective 3-in-1 keyword search method for unstructured, semi-structured and structured data. In: Proceedings of SIGMOD, pp. 903–914 (2008)
10. Liu, F., Yu, C., Meng, W., Chowdhury, A.: Effective keyword search in relational databases. In: Proceedings of SIGMOD, pp. 563–574 (2006)
11. Luo, Y., Wang, W., Lin, X., Zhou, X., Wang, J., Li, K.: Spark2: Top-k keyword query in relational databases. TKDE **23**(12), 1763–1780 (2011)
12. Markowetz, A., Yang, Y., Papadias, D.: Keyword search on relational data streams. In: Proceedings of SIGMOD, pp. 605–616 (2007)
13. Park, C.-S., Lim, S.: Effective keyword query processing with an extended answer structure in large graph databases. IJWIS **10**(1), 65–84 (2014)
14. Tao, Y., Yu, J.X.: Finding frequent co-occurring terms in relational keyword search. In: Proceedings of EDBT, pp. 839–850 (2009)
15. Yang, W., Guo, T.: An ant colony optimization algorithm for the minimum steiner tree problem and its convergence proof. Acta Math. Appl. Sinica **29**(2), 352–361 (2006)
16. Zhou, J., Liu, Y., Yu, Z.: Improving the effectiveness of keyword search in databases using query logs. In: Li, J., Sun, Y., Yu, X., Sun, Y., Dong, X.L., Dong, X.L. (eds.) WAIM 2015. LNCS, vol. 9098, pp. 193–206. Springer, Heidelberg (2015). doi:10.1007/978-3-319-21042-1_16

Enhanced Query Classification with Millions of Fine-Grained Topics

Qi Ye$^{(\boxtimes)}$, Feng Wang, Bo Li, and Zhimin Liu

Sogou Inc., Beijing, China
{yeqi,wangfeng,libo202442,liuzhimin}@sogou-inc.com

Abstract. Query classification is a crucial task to understand user search intents. Although this problem has been well studied in the past decades, it is still a big challenge in real-world applications due to the sparse, noisy and ambiguous nature of queries. In this paper, we present another important issue called "the pomegranate phenomenon". This phenomenon is named for the gap between manually manageable small taxonomy and massive coherent topics in each category. Furthermore, the fine-grained topics in the same category of the taxonomy may be textually more relevant to the topics in other categories. This phenomenon will hurt the performances of most traditional classification methods. To overcome this problem, we present a practical approach to enhance the performances of traditional query classifiers. First, we detect millions of fine-grained query topics from two years of click logs which can represent different query intents and give them category labels. Second, for a given query, we calculate the K most relevant topics and select the label by majority voting, then try to use this label to improve the results of classical query classification methods. Empirical evaluation confirms that our topic based classification algorithms can significantly enhance the performances of traditional classifiers in read-world query classification tasks.

Keywords: Multi-class query classification · Large-scale classification · Search log mining · Query clustering

1 Introduction

In real-word search query classification tasks, to analyze query intents, domain experts prefer to maintain a small taxonomy which can be easily managed manually. The taxonomy of web query classification tasks may contain from tens of to thousands of categories, and some of them are organized in hierarchical structure [3–5,10]. Traditional classification algorithms such as the Naive Bayesian method and linear support vector machines have been widely used in short text classification [4,10,13,18]. These methods suppose that the feature space is separable. Recent studies have also shown that there are tremendous fine-grained intent topics in daily search traffic [9]. However, there is a big gap between predefined small size taxonomy and the massive coherent topics in each category.

© Springer International Publishing Switzerland 2016
B. Cui et al. (Eds.): WAIM 2016, Part II, LNCS 9659, pp. 120–131, 2016.
DOI: 10.1007/978-3-319-39958-4_10

The gap between predefined taxonomy and fine-grained topics shows up not only in quantity but also in semantics which leads us to find an interesting phenomenon in query classification which we call "the pomegranate". The phenomenon indicates that there are massive fine-grained topics in each category of the taxonomy, and these relatively independent fine-grained topics in the categories are just like seeds in pomegranates. Furthermore, the fine-grained topics in the same category of the taxonomy may be textually more relevant to the topics in other categories. As the existence of this phenomenon, the hyperplanes between different categories in the feature space are hard to find. The traditional query classification approach based on the query textual features may be confused by the lack of intent features. Furthermore, in sponsored search, we have to identify queries with commercial intents from others which are not suitable for displaying Ads in web search tasks. If queries are misclassified, it might cause relevance issues in search applications and hurt user experience.

To have a deep insight of "the pomegranate phenomenon", let us show some examples from real-world search engine. "Canon printer" seems more relevant to "the driver of Canon printer" than to "Gree air conditioner", if we only consider textual features. However, we prefer to categorize "Canon printer" and "Gree air-conditioner" into the same category as "electronics", as users who search for these queries usually tend to buy these electronic products. While we tend to classify "the driver of Canon printer' as "software" category, as users are more likely to have already bought the printer and just want to download the software from web-sites. There are more examples. The textual relation of "gynecology doctors online viewing" seems more textually relevant to "gynecological examining" than to "free high-definition movies". To be precise, "gynecology doctors" is the name of a TV play and "gynecology doctors online viewing" belongs to the category of "entertainment", as users want to find the TV play from online websites. Meanwhile, most users who search for "gynecological examining" intend to find information of medical service. So is the case of "gaming PC" and "PC game". The former belongs to the "electronics" category, and the latter should be classified as "computer game".

To overcome the problem caused by "the pomegranate phenomenon", we propose a general method to enhance the performance of real-world query classification, and our contributions are as follows:

- We find an interesting issue called "the pomegranate phenomenon" in real-world query classification, which indicates the big gap between manually predefined small taxonomy and massive coherent intent topics.
- We present a methodology to extract millions of topics from click logs and propose the K-Nearest Topic classification algorithm to boost the performances of online classifiers. Furthermore, our approach is very practical which can be easily applied in various real-world query classification tasks.

The rest of the paper is organized as follows. In the next section, we first review related work. Section 3 describes our methods in detail. In Sect. 4, we show the experimental results in real-world search traffic. Section 5 gives the conclusion and discussion.

2   Related Work

Query classification is difficult due to the short, sparse and noisy nature of search queries [10,12,13]. There are many proposed algorithms for short text and search query classification. Phan et al. [8] present a framework to build classifiers for short and sparse text by making use of the hidden topics discovered by the LDA algorithm from huge web text data collections. Wang and Manning [13] show that the Naive Bayes classifier actually does better than SVMs in short text classification tasks while for longer documents the opposite result holds. Yu et al. [17] find some interesting differences between short title classification and general text classification. They find that stemming and stop-word removal are harmful, and bi-grams are very effective in title classification. Shen et al. [10] describe a hierarchical product classification system at eBay. They use the K-NN algorithm at the first level and use the SVM classifiers for the fine-grained levels. Bekkerman and Gavish [2] find that in short text classification a document's class label is mostly concentrated in a few n-grams. Yuan et al. [18] give a comprehensive survey on recent development of large-scale linear classification. They find that linear classifiers yield comparable accuracy to nonlinear ones in all their text classification tasks.

Another approach to enhance the precision of query classifier is to obtain extra data. To classify rare queries, Broder et al. [3] use a blind feedback technique to classify web queries by using the web search results. Their method can give each query one of 6000 class labels with high accuracy. However, their method must send the query to a search engine, and this manner makes it computationally infeasible in many online real-time tasks. Wang et al. [12] propose a framework for short text classification applications based on "bag of concepts", and the concepts are sets of entities learned from the Probase[1] knowledge base. However, there are many queries do not contain entities and the coverage of this method is still limited for daily web search queries. As we mainly focus on Chinese search query classification, some widely used extra resources such as FreeBase[2] and ProBase cannot be easily applied to our tasks.

3   Methodology

In this section, we propose a new approach to enhance the performances of query classifiers by making use of millions of fine-grained topics. After extracting millions of fine-grained topics with consistent labels, for a given query, we first give its initial label by a traditional classifier, then try to revise the label and output the final result by employing our proposed K-Nearest Topic (KNT) classifier.

3.1   Fine-Grained Cluster Detection

We first extract the query co-click relations from click logs and form a query co-click graph. The basic assumption is that if users click the same web search

---

[1] http://research.microsoft.com/en-us/projects/probase/.
[2] https://www.freebase.com/.

results, the intents of queries they used should probably be similar. Without loss of generality, we use a community detection algorithm to discover clusters of queries with similar intents in the co-click graph. We implement a nearly linear time complexity community detection algorithm named MMO [15], as it is very fast and performs well in avoiding formation of very large communities and captures various fine-grained query intents easily. More details can be found in this work [16].

3.2   Cluster Classification & Topic Formation

After cluster detection, we propose a straight-forward algorithm to assign labels to clusters based on the labels of their queries by majority voting.

Query Classification & Cluster Label Prediction. Given a labeled query training dataset, our goal is to infer the class labels of the unlabeled queries in the clusters. We consider the textual features (i.e., n-grams and n = 1, 2, 3) in the training dataset to train a multi-class classifier, then we classify all the unlabeled queries in the clusters. Based on the classified queries, we assign the most likely class label l^* to each fine-grained cluster $\mathbf{c}$ by majority voting using the following purity function $f(l)$, i.e.,

$$l^* = \operatorname*{argmax}_{l \in L} f(l) = \operatorname*{argmax}_{l \in L} \sum_{q \in \mathbf{c}} \frac{\mathbb{I}(l_q = l)}{|\mathbf{c}|}, \qquad (1)$$

where $\mathbb{I}(\cdot)$ is an indicator function and L is the set of class labels in the dataset. To make the class label of each cluster more trustable, we filter the clusters whose purities are less than an empirical threshold λ. The basic assumption of the cluster labeling method is that although the traditional query classifier might do not perform well in individual query classification, the cluster classification method can obtain much better performance using the results of majority voting provided by all queries in the clusters. Finally, we call these fine-grained query clusters with consistent labels topics.

3.3   K-Nearest Topic Classification Algorithm

To classify a new query q, we propose the K-Nearest Topic classifier (KNT) to assign the class label to the query based on the K most relevant topics.

Topic Features. As the topics are clusters of queries with consistent labels, we can form a longer document for each topic which contains more meaningful textual features than individual queries in it. The features of each topic are composed by n-grams of all queries in the topic. The vector $\mathbf{v}_q$ is the feature vector of the query q which contains n weighted n-grams, that is $\mathbf{v}_q = \langle \text{TF-IDF}(x_1), \cdots, \text{TF-IDF}(x_n) \rangle$, where $\text{TF-IDF}(x_i)$ is the TF-IDF weight of the n-gram x_i. We use $\mathbf{v}_q(x_i)$ to denote the weighted feature of x_i in $\mathbf{v}_q$.

For the sake of precision and efficiency, we only use 2-gram and 3-gram here as features of each query in our algorithm. A topic $\mathbf{c}$ with k n-gram features can also be represented in the same way as a TF-IDF weighted vector $\mathbf{v_c} = \langle \text{TF-IDF}(x_1), \cdots, \text{TF-IDF}(x_k) \rangle$.

Topic Likelihood Ranking. Given the feature vector $\mathbf{v_q}$ of a query q, the likelihood score between a topic $\mathbf{c}$ and the given query q can be inferred by the following equation using the Naive Bayes assumption:

$$y(\mathbf{c}|\mathbf{v_q}) \propto p(\mathbf{v_q}|\mathbf{c}) \times p(\mathbf{c}) = \prod_{i=1}^{n} p(x_i|\mathbf{c}) \times p(\mathbf{c}) \propto \sum_{i=1}^{n} \log p(x_i|\mathbf{c}) + \log p(\mathbf{c}), \quad (2)$$

where x_i is the ith n-gram of query q. The inference process of Eq. 2 can be greatly sped-up by building an inverted index of n-grams, as the n-gram features of each topic are very sparse.

K-Nearest Topic Classification. We describe the details of our query classification in Algorithm 1. To speed up the algorithm, for queries in the topics, we directly return their labels as shown in the first three lines in Algorithm 1. After the topic likelihood ranking, we get a list of candidate topics $C_N = \{\mathbf{c}_i | 1 \le i \le N\}$ sorted by their likelihood scores in Eq. 2 in descending order. To make the assigning process more trustable, motivated by the rejection algorithm proposed by Ye et al. [16] we also introduce two similarity scores, i.e., $s_q(\mathbf{v_q}, \mathbf{c})$ and $s_c(\mathbf{v_q}, \mathbf{c})$ as metrics. Finally, the given query q is labeled as the most common class among its K nearest neighbors in C_N by majority voting.

The score $s_q(\mathbf{v_q}, \mathbf{c})$ is defined as query side similarity score:

$$s_q(\mathbf{v_q}, \mathbf{c}) = \frac{\sum_{x_i \in \mathbf{x_q} \cap \mathbf{x_c}} \mathbf{v_q}(x_i)}{\sum_{x_i \in \mathbf{x_q}} \mathbf{v_q}(x_i)}, \quad (3)$$

which shows the ratio of common feature weights in query q. To keep a reliable precision, we set an empirical threshold λ_q and will give more details about the selection of λ_q in the experiment section.

The score $s_c(\mathbf{v_q}, \mathbf{c})$ of topic side similarity is defined in similar manner:

$$s_c(\mathbf{v_q}, \mathbf{c}) = \frac{\sum_{x_i \in \mathbf{x_q} \cap \mathbf{x_c}} \mathbf{v_c}(x_i)}{\sum_{x_i \in \mathbf{x_c}} \mathbf{v_c}(x_i)} \propto \sum_{x_i \in \mathbf{x_q} \cap \mathbf{x_c}} \mathbf{v_c}(x_i). \quad (4)$$

As it is hard to set a common threshold for all the topics, we use the following method to calculate a local threshold λ_c for each topic $\mathbf{c}$:

$$\lambda_c = \frac{\sum_{q \in \mathbf{c}} s_c(\mathbf{v_q}, \mathbf{c})}{|\mathbf{c}|} = \frac{\sum_{q \in \mathbf{c}} \sum_{x_i \in \mathbf{x_q} \cap \mathbf{x_c}} \mathbf{v_c}(x_i)}{|\mathbf{c}| \times \sum_{x_i \in \mathbf{x_c}} \mathbf{v_c}(x_i)} \propto \frac{\sum_{q \in \mathbf{c}} \sum_{x_i \in \mathbf{x_q} \cap \mathbf{x_c}} \mathbf{v_c}(x_i)}{|\mathbf{c}|}.$$
$$(5)$$

It indicates the average query similarity of topic $\mathbf{c}$ and can be calculated offline. To simplify the trustable cluster selection process in practice, we omit the identical denominator $\sum_{x_i \in \mathbf{x_c}} \mathbf{v_c}(x_i)$ in Eqs. 4 and 5.

Algorithm 1. The K-Nearest Topic classifier algorithm.

Require:
    The given query q, the set of all topics C, the parameter K, the maximal length N of the candidate topic set C_N.
1: **if** q in a topic $\mathbf{c} \in C$ **then**
2:     **return** the class label of $\mathbf{c}$.
3: **end if**
4: Get a set of candidate topics $C_N \subset C$ using Eq. 2.
5: $C_{cand} = \varnothing$
6: **for** $c \in C_N$ **do**
7:     Get the query side similarity score $s_q(\mathbf{v}_q, \mathbf{c})$ in Eq. 3.
8:     Get the topic side similarity score $s_c(\mathbf{v}_q, \mathbf{c})$ in Eq. 4.
9:     **if** $s_q(\mathbf{v}_q, \mathbf{c}) > \lambda_q$ and $s_\mathbf{c}(\mathbf{v}_q, \mathbf{c}) > \lambda_\mathbf{c}$ **then**
10:       $C_{cand} = C_{cand} \bigcup \{\mathbf{c}\}$
11:     **end if**
12: **end for**
13: Sort C_{cand} by their likelihood scores in Eq. 2 in descending order.
14: Get a topic set C_K by selection the top K topics in C_{cand}.
15: Get class label l^* from the labels in C_K by majority voting.
16: **return** The class label l^*.

4 Experiments

In this section, we first introduce the dataset and the evaluation metrics. After that, we conduct several experiments to verify the effectiveness of our algorithm.

4.1 Training Dataset

Our training dataset is extracted from real-world query logs. We maintain a pre-defined taxonomy with nearly 300 mutually exclusive categories which primarily focuses on classifying search queries. As most hierarchial classification algorithms focus on handling a coarse-level classification task [10,17], we only consider 24 first-level categories of the taxonomy in this paper. We sample about 17 million unique queries from the query logs, and ask 6 teams each of which contains 3 to 10 persons to annotate these queries. Each query should be exactly in only one class. To make this annotating process more quickly and easily, the annotators may use some tricks such as manual classification, semi-supervised learning and hand-craft rules. The labeled dataset still contains some noisy whose average precision is about 95.5 %. As the existence of noisy data is unavoidable in real-world tasks, we do not attempt to correct the wrong labels in our training dataset.

Figure 1(a) shows the histogram of the number of labeled queries in the training dataset. Clearly, the numbers of queries in different classes are highly imbalanced. The smallest class only contributes 0.21 % of the whole training dataset, while the largest one contributes about 17.72 %. The search traffic of each class follows similar distribution of the training data, and the classes are significantly imbalanced.

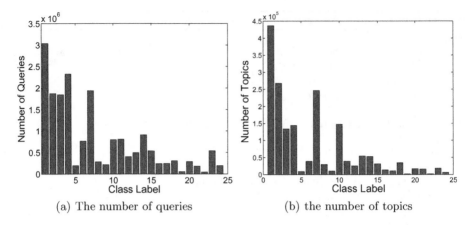

(a) The number of queries   (b) the number of topics

Fig. 1. The histogram of the number of queries and topics in training dataset in each category.

4.2 Metrics for Classification

The quality of a classification algorithm is usually measured in terms of precision and recall. However, in this paper, we use coverage instead of recall because coverage is more practical and meaningful in real-world query classification applications [11,14]. There are at least two advantages of using coverage rather than recall in real-world evolving dataset. First, coverage is much easier to estimate than recall. We would need 24 labeling tasks to know the ground truth label of the query if we were to estimate recall. Second, coverage can reveal the performance of classifier in evolving search queries rather than recall which is obtained in closed test set.

To evaluate the overall performance of classifiers in imbalanced multi-class classification task, we use the following two metrics, i.e., the average precision (AvgPre) and the geometric mean of precision (GMPre) [1,7]:

$$AvgPre = \frac{1}{|L|} \sum_{i=1}^{|L|} Pre_i, \tag{6}$$

and

$$GMPre = \sqrt[|L|]{\prod_{i=1}^{|L|} Pre_i}, \tag{7}$$

where $|L|$ is the number of classes and Pre_i is the precision of class i.

4.3 Cluster Classification & Topic Formation

We first extract web search query-URL pairs from more than 2 years of anonymized click logs of the Sogou search engine, and we remove the noisy

Table 1. Examples of classified queries in topics.

Class Label	Sampled Queries
Electronics	双卡三星手机(dual-card samsung mobile phones), 三星双卡(samsung dual-card), 三星双卡手机报价(price of samsung dual-card mobile phones), 三星双卡双待手机(Samsung dual-card-two-standby mobile phones)
Software	网页版msn(web version of msn), msn中文网页版(Chinese web version of msn), msn+web, 在线msn(online msn)
Entertainment	盗梦空间(Inception), 国语盗梦空间(Inception in Chinese), 盗墓空间电影高清版(high-resolution version for Inception)
Health	瘦腿方法小妙招(tips on the methods of slimming legs), 瘦腿小妙招(tips on slimming legs)
Hobbies	纯种金毛犬图片(pictures of purebred golden retrievers), 金毛犬多少钱一只(price for a golden retriever), 金毛多少钱(prices for golden retrievers)

edges with the least click counts. After that, we create a query co-click graph by connecting two queries with the same URL in the bipartite graph. Then we apply the MMO algorithm [15] to find 2.03 million query clusters in the co-click graph which contains about 10 million nodes and 191 million edges.

We use the training dataset mentioned above to train a Multinomial Naive Bayes classifier to predict the class labels of unlabeled queries in the clusters. For the sake of generating more features to provide more useful information as suggested by Yu et al. [17], we use n-gram ($n = 1, 2, 3$) here as features for traditional classifiers. There are about 35 million features. After all the queries in the topics have been classified, we get the label l^* of each topic using Eq. 1 satisfying that $f(l^*)$ is larger than an empirical threshold $\lambda = 0.5$. Finally, we get about 1.78 million label consistent topics. The histogram of the number of topics in different classes are shown in Fig. 1(b). The largest class contains 436303 topics, while the smallest one only contains 1724 topics. The result shows that the fine-grained topics are also highly imbalanced. This phenomenon may indicate that if we use the under-sampling technique to get training dataset, the data points in some small fine-grained topics may be missed. Furthermore, we also evaluate the average accuracy of the classified queries in these purified topics by human annotation. For each topic, we randomly select some query pairs for labeling. We conduct a manual annotating on 5000 query pairs randomly selected from the topics. For each query pair, annotators are asked to determine its correctness. Each query pair is assigned to 3 annotators and finally be labeled by majority voting. The average precision is about 96.96 %. Table 1 shows some examples of the topics including their class labels and sampled queries, respectively. As most queries in the topics are in Chinese, we also give their English explanations in Table 1.

4.4 The Performance of the KNT Classifier

To improve the precision of the KNT algorithm, we need to choose a threshold of λ_q. We set the number of the nearest neighbors $K = 3$ as the default setting in the following experiments. We randomly sample 20000 queries from daily search

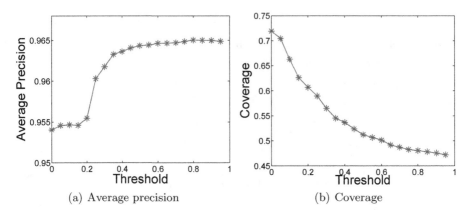

Fig. 2. The values of average precision and coverage got by the K-Nearest Topic algorithm with different λ_q threshold in daily real-world traffic.

traffic, and ask 3 annotators to select the class labels for each query. These labeled queries are used to evaluate the performances of the KNT classifier with different values of λ_q as shown in Fig. 2(a) by considering the query counts in real-world search traffic. The KNT classifier is very precise. We choose a critical point of the parameter, i.e., $\lambda_q = 0.25$, and the average precision of the KNT classifier remains 96.02 %.

Different coverage values with corresponding thresholds are shown in Fig. 2(b). Coverage drops dramatically as the increasing of the thresholds of λ_q. When the value of λ_q equals to 0.25, the coverage of the KNT classifier is about 58.95 %. As both the precision and the coverage of the KNT classifier are remarkable high with the parameter $\lambda_q = 0.25$, we use this setting in the following experiments.

4.5   Performances of Classifiers

In this part, we study the performances of classifiers by combining the KNT classifier with the Multinomial Naive Bayes (MNB) classifier and with the support vector machine (SVM) classifier with a linear kernel [6], respectively. We also use the Multinomial Naive Bayes classifier with a uniform prior distribution (MNB_U) to overcome the imbalanced class problem in priors as a comparison. The linear SVM classifier with default parameters is trained by the liblinear library [6]. We evaluate their performances by real-world search queries by considering search counts. For each task, we randomly select 2000 queries from daily search traffic, and ask three annotators to judge the correctness. These are about 0.2 million "stop words" of n-grams which are selected by both automatical methods and manual labeling strategies. Short queries which only contain "stop words" will not be covered in classification. As how to select the "stop words" is not the focus of this paper, we do not give the details here.

Table 2. Performances of different classification algorithms in real-world datasets.

Model	AvgPre	GMPre	Coverage
SVM	85.10 %	84.39 %	95.13 %
MNB	86.09 %	85.75 %	95.13 %
MNB_U	86.97 %	86.77 %	95.13 %
KNT + SVM	89.75 %	89.50 %	99.35 %
KNT + MNB	90.31 %	90.06 %	99.35 %
KNT + MNB_U	**91.16%**	**90.90%**	99.35 %

Table 2 shows that both the MNB and the MNB_U classifiers work better than the SVM classifier. This result is similar with the conclusion given by Wang and Manning [13] in short text classification. We also find that MNB_U classifier works slightly better than the MNB classifier in real-world query classification. As mentioned above, the KNT classifier achieves the highest average precision of 96.02 % with the coverage of 58.95 %. By combining the KNT classifier with traditional classifiers, both the precision and the coverage values of the traditional classifiers are greatly improved. The last line in Table 2 shows that the KNT + MNB_U classifier for imbalanced data works best considering the average precision metric (AvgPre) and the geometric mean of precision metric (GMPre). The KNT + MNB_U classifier achieves the average precision of 91.16 % with nearly 100 % coverage. While the precision got by the single SVM classifier is only about 85.10 % covering about 95.13 % daily traffic, there is 7.12 % relatively improvement in the average precision by using our best method, i.e., the KNT + MNB_U classifier to replace the SVM classifier. By combining the KNT algorithm, the traditional classifiers get about 5.46 %, 4.90 % and 4.82 % relatively improvement in the average precision, respectively. The coverage values of the traditional classifiers are greatly improved by introducing our KNT classifier as shown in Table 2.

4.6 Efficiency

Suppose the given query q contains n n-grams. If the query q has already existed in the topics, it only takes $O(1)$ time complexity to get its class label. It takes $O(n \times L')$ time to find the candidate topic set C_N where L' is the largest retrieved topic size of each n-grams. The size of C_N is $O(n \times L')$ in Algorithm 1. To make the following process faster, In real applications, we restrict C_N contains at most N topics. In the loop from line 6 to line 11, the time complexity is $O(N \times 2n)$, as it takes $O(n)$ time to get the similarity scores of $s_q(\mathbf{v}_q, \mathbf{c})$ and $s_\mathbf{c}(\mathbf{v}_q, \mathbf{c})$, respectively. In the rest of the algorithm, the time complexity is about $O(N \times \log(N) + 2K)$. As L' is much less than N in our practical setting, our algorithm can be very fast with nearly linear time complexity, i.e., $O(N \times n)$. We implement the KNT classification algorithm in C++ and deploy it on a server with 128 GB memory, Intel Xeon 3.2 GHz CPU with 24 cores, and Redhat OS system. By processing the web queries randomly selected from daily web search traffic, the whole classification takes only approximately 0.2 ms for handling every incoming search query in a single thread on average. So it fast enough in real-world query classification task in multi-thread environment.

5   Conclusion and Discussion

In this paper, we find the interesting "pomegranate phenomenon" which indicates the big gap between the predefined small size taxonomy and the massive coherent topics in categories. This phenomenon will hurt the performances of most common-used query classification algorithms. To overcome this problem, we introduce a practical query classification algorithm called the K-Nearest topic classifier to enhance real-world online query classification by using millions of fine-grained topics. First, we divide queries into fine-grained clusters from a query co-click graph extracted from two years click logs. Appropriate class labels are assigned to these fine-grained query clusters, and 1.78 million topics have been found finally. With the help of the KNT algorithm, the traditional classifiers can get significantly improvement by considering the average precision metrics. Our evaluations show that by combining the KNT algorithm with the MNB_U classifier we obtain the average precision of 91.16 % and covers most of the daily web search traffic. Moreover, our method is quite practical and efficient since it only requires millions of fine-grained query topics which can easily be detected in search logs of any real-world search engines.

Acknowledgments. The authors would like to thank all the members in ADRS (ADvertisement Research for Sponsered search) group in Sogou Inc. especially Ruining Wang for the help with parts of the data processing and experiments.

References

1. Barandela, R., Sánchez, J.S., et al.: Strategies for learning in class imbalance problems. Pattern Recogn. **36**(3), 849–851 (2003)
2. Bekkerman, R., Gavish, M.: High-precision phrase-based document classification on a modern scale. In: Proceedings of the 17th ACM SIGKDD International Conference on Knowledge Discovery and Data Mining, KDD 2011, pp. 231–239. ACM, New York (2011)
3. Broder, A.: A taxonomy of web search. SIGIR Forum **36**(2), 3–10 (2002)
4. Broder, A., Fontoura, M., et al.: A semantic approach to contextual advertising. In: Proceedings of the 30th Annual International ACM SIGIR Conference on Research and Development in Information Retrieval, SIGIR 2007, pp. 559–566. ACM, New York (2007)
5. Broder, A.Z., Fontoura, M., et al.: Robust classification of rare queries using web knowledge. In: Proceedings of the 30th Annual International ACM SIGIR, pp. 231–238 (2007)
6. Fan, R.-E., Chang, K.-W., Hsieh, C.-J., Wang, X.-R., Lin, C.-J.: LIBLINEAR: A library for large linear classification. J. Mach. Learn. Res. **9**, 1871–1874 (2008)

7. Galar, M., Fernández, A., et al.: Empowering difficult classes with a similarity-based aggregation in multi-class classification problems. Inf. Sci. **264**, 135–157 (2014)
8. Phan, X.-H., Nguyen, L.-M., Horiguchi, S.: Learning to classify short and sparse text & web with hidden topics from large-scale data collections. In: Proceedings of the 17th International Conference on World Wide Web, WWW 2008, pp. 91–100. ACM, New York (2008)
9. Radlinski, F., Szummer, M., Craswell, N.: Inferring query intent from reformulations and clicks. In: Proceedings of the 19th International Conference on World Wide Web, WWW 2010, pp. 1171–1172. ACM, New York (2010)
10. Shen, D., Ruvini, J.-D., Sarwar, B.: Large-scale item categorization for e-commerce. In: Proceedings of the 21st ACM International Conference on Information and Knowledge Management, CIKM 2012, pp. 595–604, ACM, New York (2012)
11. Sun, C., Rampalli, N., Yang, F., Doan, A.: Chimera: Large-scale classification using machine learning, rules, and crowdsourcing. Proc. VLDB Endowment **7**(13), 1529–1540 (2014)
12. Wang, F., Wang, Z., et al.: Concept-based short text classification and ranking. In: Proceedings of the 23rd ACM International Conference on Conference on Information and Knowledge Management, pp. 1069–1078. Shanghai, 3–7 November 2014
13. Wang, S.I., Manning, C.D.: Baselines and bigrams: Simple, good sentiment and topic classification. In: Proceedings of the ACL, pp. 90–94 (2012)
14. Yang, S., Kolcz, A., Schlaikjer, A., Gupta, P.: Large-scale high-precision topic modeling on twitter. In: Proceedings of the 20th ACM SIGKDD International Conference on Knowledge Discovery and Data Mining, pp. 1907–1916. ACM New York (2014)
15. Ye, Q., Bin, W., Bai, W.: The influence of technology on social network analysis and mining. In: Özyer, T., Rokne, J., Wagner, G., Reuser, A.H.P. (eds.) Detecting Communities in Massive Networks Efficiently with Flexible Resolution, pp. 373–392. Springer, Heidelberg (2013)
16. Ye, Q., Wang, F., Li, B.: Starrysky: A practical system to track millions of high-precision query intents. In: 8th International Workshop on Web Intelligence & Communities, April 2016 (to appear)
17. Yu, H.-F., Hoy, C.-H., et al.: Product title classification versus text classification. Technical report, Department of Computer Science, The University of Texas, Austin (2012). http://www.csie.ntu.edu.tw/~cjlin/papers/title.pdf
18. Yuan, G.-X., Ho, C.-H., Lin, C.-J.: Recent advances of large-scale linear classification. Proc. IEEE **100**(9), 2584–2603 (2012)

A Hybrid Machine-Crowdsourcing Approach for Web Table Matching and Cleaning

Chunhua Li[1], Pengpeng Zhao[1(✉)], Victor S. Sheng[2], Zhixu Li[1], Guanfeng Liu[1], Jian Wu[1], and Zhiming Cui[1]

[1] School of Computer Science and Technology, Soochow University,
Suzhou, China
ppzhao@suda.edu.cn
[2] Department of Computer Science, University of Central Arkansas,
Conway, USA

Abstract. Table matching and data cleaning are two crucial activities in integrating data from different web tables, which have traditionally been considered as separate activities. We show that data cleaning can effectively help us discover table matches, and vice versa. In this paper, we study a hybrid machine-crowdsourcing approach to handle the two activities together with a well-developed knowledge base. Understanding the semantics of tables is fundamental to both matching and cleaning. We select the most valuable columns to crowdsourcing validation and infer others by consolidating crowdsourcing results and machine-generated results. When resolving inconsistency between data and semantics, relative trust is taken into account to validate data or semantics via crowd. Our experimental results show the effectiveness of the proposed approach for matching and cleaning web tables using real-life datasets.

Keywords: Crowdsourcing · Table matching · Data cleaning

1 Introduction

The web contains a vast amount of structured data in the form of tables. These tables can be extracted from the web, which provides an opportunity to build a knowledge repository by integrating these tables [2,3]. However, these data suffer from many data quality issues. Head rows exist in few cases and column names can sometimes be meaningless or unreliable [15]. It is also well known that web tables often contain errors and inconsistencies.

Table matching and data cleaning are two crucial activities in consolidating or integrating data from different web tables [1,13,14]. So far, they have been traditionally studied in isolation. Even though numerous solutions [1,13] have been proposed in the past for solving a schema matching problem, previous work does not explicitly assume the presence of dirty data. On the other side, most of the recent data cleaning algorithms concentrate on a single inconsistent table [4,5]. However, simply pipelining existing schema matching and data cleaning algorithms may not always work well for web table integration.

© Springer International Publishing Switzerland 2016
B. Cui et al. (Eds.): WAIM 2016, Part II, LNCS 9659, pp. 132–144, 2016.
DOI: 10.1007/978-3-319-39958-4_11

For example, there are two tables shown in Fig. 1. The first table contains the capital information of states, and the capital of Connecticut "Bridgeport" in the table is incorrect. The second table contains the largest city of states. Notice that the value of capital also refers to a city. An instance-based conventional schema matching techniques may create a correspondence between the two columns (capital and largest city). Then, Sacramento may be considered as incorrect data and changed to Los Angeles in a data cleaning phase. In fact, Bridgeport should be changed to Hartford.

tid	state	capital
t_1	Arizona	Phoenix
t_2	Arkansas	Little Rock
t_3	California	Sacramento
t_4	Connecticut	Bridgeport

tid	state	largest city
t_1	Arizona	Phoenix
t_2	Arkansas	Little Rock
t_3	California	Los Angeles
t_4	Connecticut	Bridgeport

Fig. 1. Two example tables

Intuitively, if we can identify errors in data during the process of table matching, we will improve the results of table matching. On the other side, a more accurate understanding of tables will enhance the ability of detecting errors. Therefore, to integrate data from different web tables, putting the two phases together will be beneficial for both table matching and cleaning.

While pure machine-based methods may not be able to achieve satisfactory results, some recent works look into leveraging human intelligence to help processing [4,7]. For table matching and cleaning, it is difficult for machines to discern whether the data or the intended semantics of tables are incorrect when inconsistency occurs between data and semantics, while such tasks are actually quite effortless for human beings. In addition, since experts may be limited and expensive, crowdsourcing has been proven to be a viable and cost-effective alternative solution. For the problem of table matching and cleaning, dealing with the two tasks together will also effectively reduce the cost of crowdsourcing.

In this paper, we study a hybrid machine-crowdsourcing approach to handle the two activities together with well-developed knowledge bases, such as Yago [11]. We leverage human intelligence to discern the types of columns or relationships between two columns for those that machines consider "difficult" and to validate the correctness of data not covered by the knowledge base. To the best of our knowledge, we are the first to exploit both knowledge bases and crowdsourcing to deal with table matching and data cleaning together. We summarize our contributions below.

- We propose a unified solution for table matching and cleaning, which generate table matches based on table semantics discovered from data itself and identify incorrect data, jointly using knowledge bases and the crowd.
- We present a hybrid machine-crowdsourcing framework for discovering table semantics, including column types and relationships.
- We conduct extensive experiments to demonstrate the effectiveness of the proposed approach using real-world datasets and knowledge bases.

2 Preliminaries and Approach Overview

2.1 Preliminaries

Given a dirty corpus of web tables Γ and a knowledge base K, our goal is to build a table match M and clean tables Γ' simultaneously.

Knowledge Base. We consider a knowledge base (KB) as RDF-based data consisting of *resources*, whose schema is defined using the Resource Description Framework Schema (RDFS). A resource is a unique identifier for a real-word entity. A *property* (a.k.a. *relationship*) is a binary predicate that represents a relationship between two resources or between a resource and a literal. We denote the property between resource x and resource (or literal) y by $P(x, y)$. The *type* relationship associates an instance to a class e.g., *type*(Italy, Country). A more specific class C_1 can be specified as a subclass of a more general class C_2 by using the statement $subClassOf(C_1, C_2)$. Similarly, a property P_1 can be a sub-property of a property P_2 by the statement $subPropertyOf(P_1, P_2)$.

Table Semantics. There are two basic kinds of semantics on a table, i.e., a type of each column and a relationship between a column pair. Table semantics is a set of column types and column relationships. We say that a tuple t of the table T matches table semantics Σ containing m column types $\{C_1, ..., C_m\}$ and n column relationships $\{R_1, ..., R_n\}$ among m columns w.r.t a KB K, denoted by $t \models \Sigma$, if there exist m distinct columns $\{A_1, ..., A_m\}$ in T and m resources $\{X_1, ..., X_m\}$ in K such that:

1. There is one-to-one mapping from $t[A_i]$ to X_i for $i \in [1, m]$;
2. Either $type(X_i, C_i)$ or $type(X_i, C')$ and $subClassOf(C', C_i)$;
3. For each relationship R between two columns A_i and A_j, there exists a relationship R' for the corresponding resources X_i and X_j in K such that $R' = R$, or $subPropertyOf(R', R)$.

Table Matching. Table matching is to find correspondences between columns from different tables. We say that there is a semantic correspondence between two columns across two tables if they are semantically related. We create a semantics correspondence between two columns that are assigned to the same type in KB. A table match is a set of column correspondences.

Data Cleaning. Data cleaning deals with detecting and removing errors and inconsistencies from data in order to improve the quality of data [14]. Given the semantics of a table, we annotate each tuple t as correct or incorrect through validating whether t matches the semantics jointly by KB and crowd.

2.2 Approach Overview

Table matches and clean data are computed through the following three phases.

Phase I: Candidate Type/Relationship Generation. We first generate candidate types and relationships for all columns and column pairs, and compute the likelihood of matching a table column or column pair to a candidate type or relationship in the knowledge base. A Table Semantics Graph is derived to maintain such associations (Sect. 3.1).

Phase II: Semantics and Data Annotation. In this phase, we leverage both the machines and the crowd to determine the types/relationships that best model the data in tables and to annotate each tuple as correct or incorrect incrementally. We present a utility-based method to select most beneficial questions for semantics validation and refine the semantics leveraging the results aggregated so far from machines and crowdsourcing (Sect. 3.2). When inconsistency occurs between the data and the semantics, we introduce relative trust to guide our algorithm to validate semantics or data correctness (Sect. 3.3).

Phase III: Table Match and Data Repair Generation. We create a semantic correspondence between two columns from distinct tables that are assigned to the same type. For the incorrect data identified in Phase II, we query the knowledge base to generate possible repairs. This phase is straightforward. Thus we shall focus on Phase I and Phase II hereafter.

3 Table Matching and Cleaning via Hybrid Machine-Crowdsourcing

3.1 Candidate Type/Relationship Generation

For table-KB mapping, we use an instance based approach that does not require the availability of meaningful column labels. For each column A_i in table T, we map each value $t[A_i]$ of a tuple t to possible resources in the knowledge base K and then retrieve all types and super-types of corresponding entities as candidate types of column A_i.

Similarly, the candidate relationships between two values $t[A_i]$ and $t[A_j]$ can be retrieved by extracting all properties hold between the corresponding entities in the KB. In addition, for two values $t[A_i]$ and $t[A_j]$, we consider them as an ordered pair, so there are two different relationships between them.

Matching Likelihood. We use the normalized version of tf-idf (term frequency-inverse document frequency) [4] to measure the likelihood of matching a column of a table to a type in the knowledge base.

$$w\left(m_{A_{ij}}\right) = \sum_{t \in T} tf - idf\left(t\left[A_i\right], T_j\right) = \sum_{t \in T} tf\left(t\left[A_i\right], T_j\right) \cdot idf\left(t\left[A_i\right], T_j\right) \quad (1)$$

where the term frequency $tf\left(t\left[A_i\right], T_j\right) = \frac{1}{\log(\text{Number of Entities of Type } T_j)}$ and inverse document frequency $idf\left(t\left[A_i\right], T_j\right) = \log \frac{1}{\text{Number of Types of } t[A_i]}$.

The tf-idf scores of all candidate types for the same column are normalized to $[0, 1]$ by dividing them by the largest tf-idf score. The score of tf-idf $(t\left[R_i\right], P_j)$ of a candidate relationship match can be defined similarly, where R_i is an ordered column pair $< A_{i1}, A_{i2} >$, P_j is a property (i.e., relationship) in the KB.

Table Semantics Graph. We present the matches between table columns (resp. column pairs) and their candidate types (resp. relationships) as a labelled graph $G(V_A \cup V_R \cup V_C \cup V_P, \varepsilon_T \cup \varepsilon_K \cup \varepsilon_M)$, where V_A and V_R are the set of columns and column pairs in table corpus Γ, V_C and V_P are the set of types and relationships in the knowledge base K, ε_T is the set of edges between columns and column pairs, ε_K is the set of edges between types and relationships in K, ε_M is the set of edges of matches between columns or column pairs in Γ and types or relationships in K. The weight $\omega(e_{T_{ij}})$ of an edge in ε_T between A_i and R_j is equal to 1 if A_i is the subject or the object of R_j, and 0 otherwise. The weight $\omega(e_{K_{ij}})$ of an edge in ε_K between C_i and P_j is equal to 1 if C_i is the domain or the range of P_j, and 0 otherwise. A match $m_{A_{ij}} = \langle A_i, C_j \rangle$ (resp. $m_{R_{ij}} = \langle R_i, P_j \rangle$) is presented as an edge between A_i and C_j (resp. R_i and P_j) with the corresponding matching likelihood as the weight. Figure 2 shows an example of table semantics graph.

3.2   Question Generation for Semantics Validation

We would like to exploiting the power of crowdsourcing to improve the matching results. We deliver to the crowd to discern the semantics only for the columns that machines consider "difficult". At the meanwhile, we prefer the columns, if verified by the crowd, whose results would have greater influence on inferring the semantics of other variables, especially on those whose current semantics are most likely to be incorrect. We select a small batch of variables each time.

Utility-based Column Selection. We use a utility function to capture the usefulness of variables (refer to columns or column pairs) by combining following three measures.

Difficulty. We model the difficulty of a column (or column pair) as the amount of entropy in the distribution of weights associated with the candidate types or relationships of that column. For $V \in \{V_A \cup V_R\}$, the difficulty of the variable V, denoted as $D(V)$, is defined as follows:

$$D(V) = - \sum_{m \in \varepsilon_m(V)} \frac{\omega(m)}{Z} \cdot log \frac{\omega(m)}{Z} \tag{2}$$

where $Z = \sum_{m \in \varepsilon_m(V)} \omega(m)$ is used for normalization.

Inconsistency. Data violations indicate deviations from intended data semantics and the probability of being fault semantics. We compute inconsistency between data and semantics as follows.

$$\text{Inc}(V) = 1 - \frac{|\{t[V] | t[V] \models \Sigma\}|}{|V|} \tag{3}$$

Influence. The knowledge of the column-to-type associations of some columns can help us infer the types of other columns [7]. We prefer to select the variables, if verified by the crowd, whose results would have greater influence on inferring

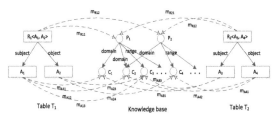

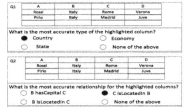

Fig. 2. An example of a table semantics graph

Fig. 3. Sample questions for type and relationship validation

the types or relationships of other variables. We consider two types of influence: *intra-table* and *inter-table*. For $V \in \{V_A\}$, the intra-table influence is defined as

$$P\left(m_{A_{ij}} \mid m_A^q\right) = \frac{|\langle C^q, C_j \rangle \cap \langle A^q, A_i \rangle|}{\sum_{C_k \in C(A_i)} |\langle C^q, C_k \rangle \cap \langle A^q, A_i \rangle|} \qquad (4)$$

where $\langle C^q, C_j \rangle$ is the set of instance pairs that participate in the relationship between C^q and C_j in the KB, and $\langle A^q, A_i \rangle$ is the set of value pairs in column A^q and A_i. We denote all the candidate types of column A_i as $C(A_i)$. For $V \in \{V_R\}$, there exists influence between two different column pairs only if there is a common column in the two column pairs. We defined the relationship influence $P(m_{R_{ij}} \mid m_R^q)$ with the set of instance group that participates in two relationships and the set of value group in two column pairs similarly.

For variables from different tables, we use the cosine similarity of their candidate vectors modeling the inter-table influence $P\left(m_{A_{iq}} \mid m_A^q\right)$ and $P\left(m_{R_{iq}} \mid m_R^q\right)$. Finally, the influence of a crowdsourced column A^q on another column A_i is computed as follow.

$$\text{Inf}\left(A_i \mid A^q\right) = \sum_{m^q \in \varepsilon_m(A^q)} \left(1 - \prod_{m_{ij} \in \varepsilon_m(A_i)} 1 - P\left(m_{ij} \mid m^q\right)\right) \cdot P\left(m^q\right) \qquad (5)$$

$\text{Inf}\left(R_i \mid R^q\right)$ is computed similarly. Combining the above three factors, we measure the utility of variables as

$$U\left(V^q\right) = \sum_{V_i \in \{V_A \cup V_R\}} D\left(V_i\right) \cdot \text{Inc}\left(V_i\right) \cdot \text{Inf}\left(V_i \mid V^q\right) \qquad (6)$$

We use a greedy algorithm to select a k-element subset of columns so that the sum of utility is maximized, by iteratively selecting the most useful variable given the ones selected so far.

Question Generation. For selected columns, we will create a series of questions accordingly for crowdsourcing. There are two types of tasks for table semantics validation: (1) column type validation; and (2) binary relationship validation, i.e., to validate the relationship between two columns. Crowd workers are prone to mistakes when values in tables are ambiguous. We generate five questions for each variable, each question containing five tuples randomly selected from table. We assign each question to three different workers and the majority answer is taken. We choose the result with the highest support from the workers. Example questions are shown in Fig. 3, where the current type or relationship of variable is default selected and some contextual columns are also presented to help workers better understand the question. Candidate types and relationships are shown as corresponding labels instead of URIselves.

Refining Semantics with Crowdsourced Answers. After crowdsourcing results obtained, we can proceed to determine which candidate mapping is the best for each column and column pair. To compute the probability $P(m_{ij})$, we consider two sources of evidences, i.e., machines and crowdsourcing. Based on the influence of the crowdsourcing results M^q, the confidence that m_{ij} is inferred to be correct is computed as follows:

$$P\left(m_{ij}|M^q\right) = 1 - \prod_{m_A^q \in M_A^q} \left(1 - P\left(m_{ij}|m^q\right)\right) \tag{7}$$

Then, we combine the two evidences as follows.

$$P\left(m_{ij}\right) = \alpha \cdot \omega\left(m_{ij}\right) + (1 - \alpha) \cdot P\left(m_{ij}|M^q\right) \tag{8}$$

3.3   Unified Semantics and Data Annotation

Data Annotation. After table semantics generated, we can annotate data in the table as correct or incorrect according to whether the data matches the semantics. We first validate each tuple using the knowledge base and annotate it as correct tuple if it is fully covered by the knowledge base, i.e., $t \models \Sigma$. For each semantic mapping missing from the knowledge base, we generate a boolean question to ask the crowd whether the semantics holds for the corresponding value. If the crowd says yes for all missing semantics matches of a tuple, we annotate it as a correct tuple. Otherwise, there exist errors in this tuple.

Algorithm. It is unclear whether the data or the intended semantics are incorrect when data violate the semantics. To guide our algorithm, two thresholds on inconsistency are introduced, τ_1 and τ_2 ($\tau_1 < \tau_2$). For tables whose inconsistency is lower than τ_2, we argue that their semantics get well supported by the data and validate the correctness of data using these semantics. If inconsistency is lower than τ_1 after data validated, we take the semantics as correct. The semantics with inconsistency greater than τ_2 are deemed incorrect.

Algorithm 1 describes the overall procedure for unified semantics and data annotation. It first initializes a variable set V as $V_A \cup V_R$ whose semantics are untrusted and initializes a table semantics Σ with candidates with the maximum matching likelihood of each variable (line 2–3). Next, it iteratively refines the semantics and data annotation (line 4–16). At each iteration: (1) it chooses m best variables for crowdsourcing validating the semantics based on the utility of variables (line 5); (2) it refines untrusted table semantics by consolidating the matching likelihood generated by machines and the influence of crowdsourcing results (line 7); (3) it computes KB-based inconsistencies between semantics and data using the refined semantics (line 8); (4) it annotates the data jointly by the KB and the crowd and refreshes the inconsistencies of tables using the annotations (line 9–14); (5) finally, it updates variables with untrusted semantics (line 15). Algorithm terminates when all table semantics are trusted or the maximum number of iterations (corresponding to the budget) is reached.

Algorithm 1. SemanticsAndDataAnnotation

Input: table corpus Γ, table semantics graph G, KB K, maximum iterations k, batch size m

Output: table semantics Σ, data annotation $\mathcal{A}$

1 **begin**
2     $V \leftarrow V_A \cup V_R$; $M^q \leftarrow \emptyset$; $\mathcal{A} \leftarrow \emptyset$
3     $\Sigma \leftarrow$ candidates with the maximum matching likelihood of each variable
4   **for** $i = 0$ **to** k **do**
5        $V^q \leftarrow$ select a subset with m variables from V based on *Utility*
6        $M^q \leftarrow M^q \cup \{$crowdsourced answers for $V^q\}$
7        $\Sigma \leftarrow$ compute type and relationship matches with M^q using Eq. 8
8        $I \leftarrow$ compute KB-based inconsistency for V using Σ
9   **for** *each* $v \in V$ **do**
10   **if** $v \in T \wedge Inc(T) < \tau_2$ **then**
11              $\mathcal{A}_v \leftarrow$ annotate data jointly by the KB and the crowd with Σ
12              $I \leftarrow$ refresh inconsistency with $\mathcal{A}_v$
13   **if** $Inc(T) < \tau_1$ **then**
14                 $\mathcal{A} \leftarrow \mathcal{A} \cup \mathcal{A}_v$

15        $V \leftarrow V \backslash \{V^q \cup V_{v \in T \wedge Inc(T) < \tau_1}\}$
16   halt when $V = \emptyset$

---

4 Experiments

4.1 Experimental Setup

Knowledge Base. We used Yago [11] as the knowledge base. Yago is a huge semantic knowledge base derived from Wikipedia, WordNet and GeoNames, with a confirmed accuracy of 95 %. We examine the coverage of Yago over the values in a table column. As shown in Fig. 4(a), 82 % (61 %) of the columns of WebTables (WikiTables) contain more than 50 % values. This illustrates that Yago has a large coverage on the column values, and thus it can be employed for table matching and cleaning. There is still a fraction of values not covered which requires human intelligence involved.

Datasets. We used two real-world web table datasets, WikiTables and WebTables [4]. WikiTables contains 28 tables from Wikipedia pages with average 32 tuples per table. WebTables contains 30 tables from Web pages with average 67 tuples per table.

All the tables were manually annotated using types and relationships in Yago for columns and column pairs, which we considered as *ground truths* of table semantics. On both datasets, we derive correct column matches based on the ground truths. Then, we measure the value overlap between each pair of matched columns. As shown in Fig. 4(b), the value overlap of more than 90 % of the column pairs is smaller than 0.1 in each dataset. All the tuples not fully covered by the KB with the ground truth semantics were manually annotated as correct or incorrect, which we considered as *ground truths* of data annotation. The data error rates of WikiTables and WebTables are 2.4 % and 6.8 %. The distribution of error rates among tables is shown in Fig. 4(c).

We use a crowd with 10 students for semantics and data validation. All algorithms were implemented in JAVA. Experiments were conducted on Windows 7 with an Intel i5 CPU@3.10 Ghz and 4 GB memory.

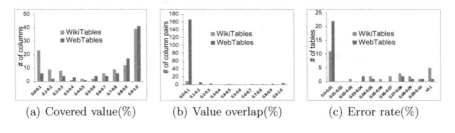

(a) Covered value(%)   (b) Value overlap(%)   (c) Error rate(%)

Fig. 4. Datasets and KB characteristics

4.2   Experimental Results

Evaluation of Hybrid Machine-Crowdsourcing Method. We first examine our hybrid machine-crowdsourcing method for generating table semantics.

For hybrid machine-crowdsourcing method, we set the two threshold τ_1 and τ_2 to 0.2 and 0.5 according to the value coverages in Yago and the priori knowledge of data error rates, set the batch size $m = 5$. We vary α from 0.2 to 0.8 and report the results for different settings to examine the effect of α. The algorithm terminates at the 17*th* iteration and achieves 95.3 % on F-measure, totally 83 variables selected for crowdsourcing validation. For comparisons, we also evaluate the pure machine-based method and the crowdsourcing method. The pure machine-based method computes the matching likelihood for each candidate match and selects the ones with the maximum matching likelihood.

Our experimental results are shown in Fig. 5(a). The vertical axis presents the value of F-measure. The horizontal axis presents the number of iterations. From Fig. 5(a), we can see that both our hybrid machine-crowdsourcing method and the crowdsourcing method improve significantly with the increment of the number of iterations. The pure machine-based method achieves only 44.7 %, 56.7 % and 50 % on precision, recall and F-measure respectively. However, our hybrid machine-crowdsourcing method outperforms significantly better than both the pure machine-based and the pure crowdsourcing-based method. We observe that the hybrid method achieves the best F-measure for $\alpha = 0.8$. Thus, we set $\alpha = 0.8$ in the rest of experiments.

We further study the effectiveness of the variable selection model. We compare our *Utility* model with considering only *Difficulty*, *Inconsistency*, *Influence* respectively. As illustrated in Fig. 5(b), our model *Utility* outperforms the other three settings at each iteration.

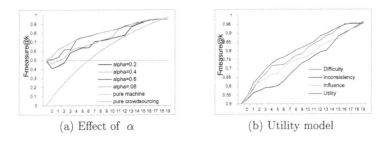

(a) Effect of α (b) Utility model

Fig. 5. Evaluation on the table semantics (WebTables)

Evaluation on Table Matching. Next, we evaluate the performance of our unified (matching and cleaning) approach against the standalone versioin on discovering column correspondences. Our experimental results are shown in Fig. 6. Note that for the table matching standalone approach, we select the same number of variables to validate by the crowd at once, and then determine the best table semantics without data validation. From Fig. 6, we can see that although many matched columns have a few values in common, our approach successfully discovered column correspondences. Our unified approach performs better than the approach considering table matching alone, especially at the beginning of

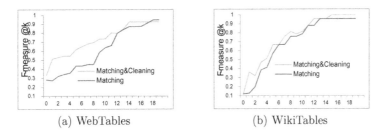

(a) WebTables   (b) WikiTables

Fig. 6. Evaluation on discovering column correspondences

the iterations. There is a fewer margin for WikiTables due to a small quantity of column matches in ground truths. The results confirm our claim that considering table matching and cleaning together improves the matching results and reduce the crowdsourcing cost.

Evaluation on Data Annotation. Values are annotated w.r.t the obtained table semantics jointly by the KB and the crowd. We annotate the data in three categories: when a value is covered by the KB, we annotate the value as true validated by the KB; when a value is not covered by the KB, we annotate the value as true validated by the crowd if crowdsourcing results say yes, and as erroneous otherwise. Table 1 shows the percentage of values in each category.

Table 1. Data annotation by KB and crowdsourcing

	type			relationship		
	KB	crowd	error	KB	crowd	error
Webtables	0.77	0.21	0.02	0.63	0.33	0.04
Wikitables	0.73	0.25	0.02	0.60	0.36	0.04

5 Related Work

Geerts et al. [9] is the most closely related work. It investigated the problem of bring transforming data using schema mappings and data repairing together. Such techniques are appropriate for the scenario where the data is noisy but the schemas of tables are known and fixed. Unfortunately, the schema information of a web table may not always be available. Another critical difference from [9] is that we exploit the crowd to help improve the accuracy of table semantics annotation and data annotation.

Numerous studies have attempted to improve the quality of data using integrity constraints [5,8,10]. Different from constraint based approaches, we leverage KBs as reference data. There are previous works on understanding web tables using knowledge bases [4,6,7,12,15]. Most of these approaches are pure machine-based. Fan et al. [7] leveraged crowdsourcing to improve the quality of column concepts determination and studied the influence between table columns to help select valuable columns to conduct crowdsourcing validation. We adapt a similar idea to select questions for semantics (i.e., types and relationships) validation. Chu et al. [4] also derives both types of columns and relationships between column pairs. However, [4] considers a single table and cannot utilize the influence between columns from different tables.

Some recent works have concentrated on employing crowdsourcing for different sub tasks of data cleaning such as deduplication [16]. Our study is complementary to these works. There are also many works on studying improving the quality of crowdsourcing answers. These techniques can be easily adopted and may further improve the accuracy of our approach.

6   Conclusions

In this paper, we studied the problem of dealing with table matching and cleaning together. We proposed a unified approach for table matching and cleaning via hybrid machine-crowdsourcing, which first annotates tables with types and relationships in a knowledge base and then derives the table matches and data repairs. We refined the annotations incrementally by selecting a small batch questions for semantics and data validation in turn. Our experiments on two real-world web table datasets showed the effectiveness of our proposed approach.

Acknowledgments. This work was partially supported by Chinese NSFC (61170020, 61402311, 61440053), Jiangsu Province Colleges and Universities Natural Science Research project (13KJB520021), Jiangsu Province Postgraduate Cultivation and Innovation project (CXZZ13_0813), and the US National Science Foundation (IIS-1115417).

References

1. Bernstein, P.A., Madhavan, J., Rahm, E.: Generic schema matching, ten years later. Proc. VLDB Endowment **4**(11), 695–701 (2011)
2. Cafarella, M.J., Halevy, A., Khoussainova, N.: Data integration for the relational web. Proc. VLDB Endowment **2**(1), 1090–1101 (2009)
3. Cafarella, M.J., Halevy, A., Wang, D.Z., Wu, E., Zhang, Y.: Webtables: exploring the power of tables on the web. Proc. VLDB Endowment **1**(1), 538–549 (2008)
4. Chu, X., Morcos, J., Ilyas, I.F., Ouzzani, M., Papotti, P., Tang, N., Ye, Y.: Katara: A data cleaning system powered by knowledge bases and crowdsourcing. In: Proceedings of the 2015 ACM SIGMOD International Conference on Management of Data, pp. 1247–1261. ACM (2015)
5. Cong, G., Fan, W., Geerts, F., Jia, X., Ma, S.: Improving data quality: Consistency and accuracy. In: Proceedings of the 33rd International Conference on Very Large Data Bases, pp. 315–326. VLDB Endowment (2007)
6. Deng, D., Jiang, Y., Li, G., Li, J., Yu, C.: Scalable column concept determination for web tables using large knowledge bases. Proc. VLDB Endowment **6**(13), 1606–1617 (2013)
7. Fan, J., Lu, M., Ooi, B.C., Tan, W.C., Zhang, M.: A hybrid machine-crowdsourcing system for matching web tables. In: 2014 IEEE 30th International Conference on Data Engineering (ICDE), pp. 976–987. IEEE (2014)
8. Fan, W., Ma, S., Tang, N., Yu, W.: Interaction between record matching and data repairing. J. Data Inf. Qual. (JDIQ) **4**(4), 16 (2014)
9. Geerts, F., Mecca, G., Papotti, P., Santoro, D.: Mapping and cleaning. In: 2014 IEEE 30th International Conference on Data Engineering (ICDE), pp. 232–243. IEEE (2014)
10. Geerts, F., Mecca, G., Papotti, P., Santoro, D.: The llunatic data-cleaning framework. Proc. VLDB Endowment **6**(9), 625–636 (2013)
11. Hoffart, J., Suchanek, F.M., Berberich, K., Weikum, G.: Yago2: A spatially and temporally enhanced knowledge base from wikipedia. In: Proceedings of the Twenty-Third International Joint Conference on Artificial Intelligence, pp. 3161–3165. AAAI Press (2013)

12. Limaye, G., Sarawagi, S., Chakrabarti, S.: Annotating and searching web tables using entities, types and relationships. Proc. VLDB Endowment **3**(1–2), 1338–1347 (2010)
13. Rahm, E., Bernstein, P.A.: A survey of approaches to automatic schema matching. VLDB J. **10**(4), 334–350 (2001)
14. Rahm, E., Do, H.H.: Data cleaning: Problems and current approaches. IEEE Data Eng. Bull. **23**(4), 3–13 (2000)
15. Venetis, P., Halevy, A., Madhavan, J., Paşca, M., Shen, W., Wu, F., Miao, G., Wu, C.: Recovering semantics of tables on the web. Proc. VLDB Endowment **4**(9), 528–538 (2011)
16. Wang, S., Xiao, X., Lee, C.H.: Crowd-based deduplication: An adaptive approach. In: Proceedings of the 2015 ACM SIGMOD International Conference on Management of Data, pp. 1263–1277. ACM (2015)

An Update Method for Shortest Path Caching with Burst Paths Based on Sliding Windows

Xiaohua Li[1(✉)], Ning Wang[2], Kanggui Peng[1], Xiaochun Yang[1], and Ge Yu[1]

[1] School of Computer Science and Engineering,
Northeastern University, Shenyang, China
lixiaohua@ise.neu.edu.cn, {yangxc,yuge}@mail.neu.edu.cn
[2] Department of Information Management, Shanghai University, Shanghai, China
ningwang@shu.edu.cn

Abstract. Caching shortest paths is an important problem, and it is widely used in traffic networks, social networks, logistic networks, communication networks and other fields. By far, few existing methods has considered burst paths, which are common in real life. For instance, queries of a certain path can sharply increase due to a promotion or vacations. Such burst queries usually last for a relatively short period, and their frequencies are too low to be loaded into caches according to conventional methods. In this paper, we propose two methods: the basic and incremental cache update methods. Both methods are based on sliding windows. Specially, the incremental update method quantifies burst paths and updates the cache incrementally. Comprehensive experiments show that our methods surpass current best algorithm SPC by more than 30 % in terms of hit ratio on real road network data with burst paths.

Keywords: Burst path · Sliding window · Incremental update

1 Introduction

In recent years, with the development of information technology, a large number of shortest path queries [1–3] have been produced in road networks. How to respond to users' queries quickly is a problem. Caching, to a large extent, resolves the quick response problem. But how to load the cache is complex, because it is hard to capture the pattern of history queries, queries have high diversity and continuously arrived new queries become the history.

In a network, if queries for a path rarely appear initially, then queries occur in rapid succession, after that queries are quiet again, we call such irregular shortest path queries as **burst queries**, and corresponding path as a **burst path** [4–6]. The duration of a burst path is generally not long; it may last several hours or a few days. The concept of burst is not strange; on the contrary, irregular

The work is partially supported by the NSF of China (Nos. 61272178, 61572122), the NSF of China for Outstanding Young Scholars (No. 61322208), and the NSF of China for Key Program (No. 61532021).

© Springer International Publishing Switzerland 2016
B. Cui et al. (Eds.): WAIM 2016, Part II, LNCS 9659, pp. 145–158, 2016.
DOI: 10.1007/978-3-319-39958-4_12

burst paths are widespread in our life. For example, when a shopping mall has a promotion, a large number of customers will query the path to the mall. When the promotion ends, the query amount will recover normal. In this situation, the path to the mall is a burst path. Another example is queries in golden weeks. Many families choose to go on holiday or a picnic outside the city. In this case, queries in golden weeks will be totally different from queries on working days. Paths for golden-week queries are bursts.

The features of burst paths include: (1) from the time horizon, the query frequency for a burst path is high in the middle, and low in both ends; (2) durations and burst latitudes of different paths vary a lot; and (3) the duration of a burst path is generally short. Due to these features, the challenges of caching burst paths are: (1) burst path detection. Burst paths do not have a uniform pattern, thus it is hard to quantify and detect them; (2) burst paths approach and leave quickly. It needs to detect and load burst paths into cache in a timely manner; otherwise, the hit ratio is affected; (3) balance the loading of conventional hot paths with burst paths.

Though many academic works have talked about caching shortest paths to improve the response time, no one considers burst paths. As the features of burst paths have not been carefully considered in existing methods [7–9], they cannot resolve burst paths very well. In this paper, we address how to detect burst paths timely. Two cache update methods are proposed. One is the basic cache update method (BCU), and another is the incremental cache update method (ICU). Both methods are based on sliding windows and balance conventional hot paths and burst paths. Specially, the incremental update method quantifies burst paths and updates the cache incrementally. The hit ratio and update time of ICU are considerable. Comprehensive experiments show that our methods surpass current best algorithm SPC [10] by more than 30 % on real road network data with burst paths.

The main contributions of this paper lie in four aspects:

- Sliding windows are considered to resolve the shortest path caching problem with burst paths. The shortest path caching is a dynamic update process, and the characteristic of queries vary along the time. Sliding windows can filter queries far away and pay more attention to new queries.
- We design a mechanism to detect burst paths. The challenge is how to describe burst paths quantitatively.
- An incremental cache update algorithm is proposed. The algorithm guarantees the timeliness and usefulness of the paths in the cache, so that the cache can keep a high hit ratio.
- Experiment results show that our algorithms can adapt to the changes of users' queries and are better than existing best algorithm SPC by more than 30 % in terms of hit ratio on real road benchmark data sets.

The rest of this paper is organized as follows. Section 2 gives a review of related works, and Sect. 3 explains the basic cache update method. Section 4 quantifies burst paths and designs an incremental cache update method.

Section 5 represents the comparison with SPC and the effects of different parameters. Finally, Sect. 6 closes the paper.

2   Related Work

The existing methods for caching the shortest paths include HQF, LRU [13], and SPC [10].

HQF (Highest-Query-Frequency) is a typical static caching method [13]. In the off-line phase, shortest paths with highest frequencies based on the query log are loaded into the cache, and the content of the cache remains unchanged during the cache works. Its disadvantage is that high frequency queried paths may have many queries in earlier periods but are not queried in later periods. Hence, even if the cache is full, the hit ratio is not necessarily high.

Another static method is SPC. In [10], SPC method chooses to load paths which can answer most queries, including queries of sub-paths. By far, SPC has the best performance.

HQF and SPC are static methods; once the cache is loaded according to the query log, the contents in the cache remain unchanged. Actually, the road network changes, as well as coming queries. [14] presents an incremental update method to handle changes of networks. As edge weights change, the shortest paths in the cache are no more shortest. The authors design an algorithm to detect affected shortest paths due to network changes, and four strategies are proposed to update the cache.

LRU (Least-Recently-Used) is a typical dynamic caching method to handle query changes [11–13]. At the beginning, the cache is empty. When a new query comes, the path for this query is inserted into the cache. When the cache is full, least-recently-used paths are replaced by recently-requested paths. The shortcoming of LRU is that if the cache is small, the cache will be updated frequently, which reduces the hit ratio.

In this work, the query changes will be considered in a batch (a sliding window), which is different from LRU. LRU is short-sighted as it changes the cache once upon a new query arrives. The main focus of this paper is burst paths; as far as we know, no work has talked about burst paths.

3   A Basic Cache Update Method

In this section, the basic cache update method (BCU) is introduced. The method uses sliding windows to cope with new coming queries. Sliding windows can filter out old queries and focus on new queries. As shown in Fig. 1, the size of the ith sliding window is $w_i = w$ queries and the window moves by $\Delta w_i = \Delta w(w = k\Delta w, k$ is an integer) queries every time. w and Δw are parameters and the experiments on their size will be explained in Sect. 5. A tuple of three elements $< v_s, v_t, f^w_{Q_{a,b}} >$ is used to represent a path $P_{a,b}$ with nodes v_s and v_t, and its query frequency $f^w_{Q_{a,b}}$ in sliding window w.

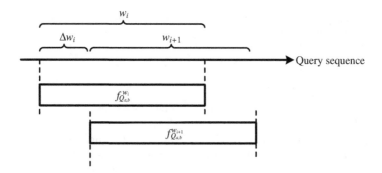

Fig. 1. A sliding window

In a sliding window w_i, the benefit of loading a path $P_{a,b}$ depends on the query frequency $f_{Q_{a,b}}^{w_i}$ and the computational cost saved $C_{a,b}$, which is:

$$B_i(P_{a,b}) = f_{Q_{a,b}}^{w_i} \cdot C_{a,b} \tag{1}$$

This paper regards $C_{a,b}$ as a constant for any path, because even if we do not introduce the caching system, the path is stored in a local system. Therefore, the benefit is proportional to the query frequency. With the assistance of sliding window, the appearance of burst paths is reflected by the increase of its benefit value in certain windows.

The basic cache loading method is to perform the process of cache loading in every window. In each window, the cache Ψ is firstly cleared. Window w_i is scanned to count $f_{Q_{a,b}}^{w_i}$ for each $P_{a,b}$ appeared in w_i. The paths with the highest benefit values are loaded into the cache until the cache is full. The sliding window slides once the caching system receives another Δw queries. And the loading process is triggered again. It is clear that if the total size of non-duplicated queries in w are smaller than that of Ψ, Ψ is not fully used.

Now we analyze the complexity of the above algorithm. First, the queries in a window are scanned and the corresponding query frequencies are recorded. The time complexity of such an action is $\mathcal{O}(w)$. The complexity of calculating benefits of paths is also $\mathcal{O}(w)$. The time complexity of heap sorting is $\mathcal{O}(w \log w)$, and the time complexity of loading paths is $\mathcal{O}(w)$. In summary, the complexity of the overall algorithm is $\mathcal{O}(w \log w)$.

4   Incremental Cache Update Method

The section demonstrates an improved cache update method, i.e., incremental cache update method (ICU). BCU separates windows and performs cache loading process for every window. The link between windows is lost. ICU proposed in this section is more effective and takes less time.

4.1 Burst Path Detection

To load burst paths into the cache, one critical problem is to determine which are burst paths. From the time horizon, burst paths usually have fewer number of queries than other hot paths in the whole life, but they boom in a certain short period of time. The spread of query frequencies in two consecutive windows is the focus. Hence, we define query frequency - inverse query frequency (QF-IQF) score $\text{QF-IQF}_i(Q_{a,b})$ of each path $P_{a,b}$ in a certain window w_i, which is:

$$\text{QF-IQF}_i(Q_{a,b}) = \frac{f_{Q_{a,b}}^{w_i} - f_{Q_{a,b}}^{w_{i-1}}}{\Delta w_i} \cdot \ln \frac{w_i}{f_{Q_{a,b}}^{w_i}} \tag{2}$$

QF-IQF value in this paper is designed based on document frequency - inverse document frequency (DF-IDF) of [15,16]. In [15,16], online documents are used to detect burst events. To achieve this goal, events are first described by several features (keywords and demographic information). Detection of burst events is converted to detection of burst features. If features of an event are burst, then the event is a burst. To detect burst features, the authors define:

$$\text{DF-IDF}_f(t) = \frac{DF_f(t)}{N(t)} \log \frac{N}{DF_f} \tag{3}$$

$\text{DF-IDF}_f(t)$ is the DF-IDF score of feature f in time t. $DF_f(t)$ and $N(t)$ are the number of documents which contain feature f and the total number of documents at time t, respectively. Likewise, DF_f and N are the number of documents which contain feature f and the total number of documents over the whole observing period. If $\text{DF-IDF}_f(t)$ score of feature f is higher than a threshold in time t, f is determined to be a burst.

QF-IQF score imitates the idea of DF-IDF score and is modified a little. N and DF_f in DF-IDF require that all documents are at hand, instead of coming one by one. However, in our case, queries arrive continuously, therefore N and DF_f cannot be computed. Hence, we define our QF-IQF in the form of Eq. 2. Window w corresponds to the overall documents in DF-IDF, and sliding step Δw corresponds to the documents in time t.

Basically, QF-IQF formula captures the features of burst paths. The QF-IQF formula amplifies the difference of query frequencies between two consecutive windows $(f_{Q_{a,b}}^{w_i} - f_{Q_{a,b}}^{w_{i-1}})$ while it discounts the query frequency in the new window $f_{Q_{a,b}}^{w_i}$. Burst paths have higher QF-IQF scores while conventional paths have ordinary QF-IQF scores. When a path has a QF-IQF score higher than a threshold, we regard it as a burst path; otherwise, it is a conventional path. The threshold is a user-defined parameter.

4.2 Incremental Benefit Function

This section introduces how to update benefit values of paths in the cache incrementally. In the basic update method, the benefit value of each path is calculated in each window. As the changes in each window are queries which slide out of and newly arrive at the window. We consider incrementally update the window and associate calculations. Although incremental update does not have any effect on the calculation results, it will save calculation time. According to Eq. 1, we have:

$$
\begin{aligned}
B_{i+1}(P_{a,b}) &= f_{Q_{a,b}}^{w_{i+1}} \cdot C_{a,b} \\
&= (f_{Q_{a,b}}^{w_i} + f_{Q_{a,b}}^{w_{i+1}} - f_{Q_{a,b}}^{w_i}) \cdot C_{a,b} \\
&= B_i(P_{a,b}) + (f_{Q_{a,b}}^{w_{i+1}} - f_{Q_{a,b}}^{w_i}) \cdot C_{a,b} \\
&= B_i(P_{a,b}) + (f_{Qa,b}^{\Delta w_{i+k}} - f_{Q_{a,b}}^{\Delta w_i}) \cdot C_{a,b}
\end{aligned}
\tag{4}
$$

See Fig. 2 for reference, every time the window slides, $f_{Q_{a,b}}^{w_{i+1}} - f_{Q_{a,b}}^{w_i}$ is calculated incrementally. Each window $w_i = w$ is divided into k times of sliding step $\Delta w_i = \Delta w$. The query frequency of each path is recorded based on the unit of Δw. When the window slides, $f_{Q_{a,b}}^{w_{i+1}} - f_{Q_{a,b}}^{w_i} = f_{Qa,b}^{\Delta w_{i+k}} - f_{Q_{a,b}}^{\Delta w_i}$. The time to scan the overlapping part of two windows is saved, and only the newly emerging small window Δw needs scanned.

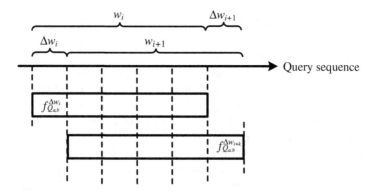

Fig. 2. Incremental update

4.3 Incremental Cache Update Method

When the cache is empty, the queries in the first sliding window are used to load the cache. In the following, an incremental update method is deployed to dynamically update the cache.

Algorithm 1 gives the pseudo code of the incremental cache update method. Only one round of update is illustrated; other rounds are the same.

The input of the Algorithm 1 is $G(V, E)$, a series of queries Q, and the cache space Ψ. In Line 4, the benefit values of paths in Ψ are updated by Eq. 4 incrementally. Line 5 sorts the cache by a heap. Line 6 determines the threshold for admitting paths outside Ψ. In this paper, the threshold is $\alpha \cdot \min_{P_{a,b} \in \Psi} B_i(P_{a,b})$.

It can be adjusted based on the query log information. Line 8 justifies whether a path outside Ψ should be put into Ψ. A tuple $(B_i(Q_{a,b}), \text{QF-IQF}_i(Q_{a,b}))$ is used. The first element is to test whether $P_{a,b}$ is of high benefit value and the second element is to test whether $P_{a,b}$ is a burst path. Either of the two cases will bring $P_{a,b}$ into Ψ. Line 8 to 9 insert all qualified-for-cache paths in a batch. This action is superior to inserting these paths into Ψ one by one. Because each insertion is followed by sorting the cache (sorting a heap from the implementation perspective). We sacrifice the storage space (S) a little to save the running time.

Algorithm 1. Incremental cache update algorithm

Input: $G(V, E)$, a series of queries Q, a full cache Ψ;
Output: Updated cache Ψ;

1 Scan the newly emerging small window Δw_{i+k-1}, count the query frequencies of paths in it, and store these frequencies;
2 **for** *each path $P_{a,b}$ in new sliding window w_i* **do**
3 $\quad$ Calculate $f_{Q_{a,b}}^{w_i}$ and $B_i(P_{a,b})$;

4 Update the benefit values of paths in Ψ by Eq. 4;
5 Sort Ψ according to $B_i(P_{a,b})$;
6 $b = \alpha \min_{P_{a,b} \in \Psi} B_i(P_{a,b})$;
7 **for** *any $P_{a,b} \in w_i$* **do**
8 $\quad$ **if** $P_{a,b}$ *satisfies* $\max(B_i(Q_{a,b}), QF\text{-}IQF_i(Q_{a,b})) > b$ *and is out of Ψ* **then**
9 $\quad\quad$ Insert $P_{a,b}$ to a temporary storage S;

10 Delete paths with the minimum $B_i(P_{a,b})$ in Ψ, so that paths in S can be inserted into Ψ;
11 Insert paths in S;
12 **return** Ψ

Once more, we analyze the complexity of the algorithm. The query frequencies in the newly emerging small window are scanned, so the time complexity of this part is $\mathcal{O}(\Delta w)$. The time of calculating benefit values of paths in new window w is $\mathcal{O}(w)$. The time complexity of sorting the cache is $\mathcal{O}(\Psi log \Psi)$. The complexity of justifying whether a path in w is qualified for Ψ is $\mathcal{O}(w)$. The total complexity of Algorithm 1 is $\mathcal{O}(w + \Psi log \Psi)$. Recall that the time complexity of the basic cache update method is $\mathcal{O}(w + w \log w)$. If $w < \Psi$, the basic method has lower time complexity than the incremental cache update method, but the cache is not fully used in basic method. If $w \geq \Psi$, the basic method has higher time complexity than the incremental method.

5   Experimental Study

In this paper, a series of experiments have been conducted to evaluate the performance of the proposed algorithms. [10] designs a similar benefit model to evaluate paths; we use this method as a benchmark method. Moreover, this paper also compares the basic cache update algorithm (BCU) with the incremental cache update algorithm (ICU).

5.1   Experimental Setup

All experiments were done on a PC. The operating system is Ubuntu 14.04 64 bit with an Intel Core processor clocked at Q8400 2.66 GHz. The memory is 4 GB. Our algorithms were implemented in C++. The test data sets were generated based on Aalborg[1] and Beijing[2]. Aalborg and Beijing are two commonly used data sets in the cache loading problem. Each of the two data sets has a road network and historical trajectories accumulated by GPS.

Table 1 shows the relevant information of the two data sets, including the number of nodes and the number of edges.

Table 1. Information of data set

Data set	Number of nodes	Number of edges	Description
Aalborg	3.9 Million	4.6 Million	Road network
Beijing	2.4 Million	2.7 Million	Road network

As this paper talks about the dynamic update of caches, there is no concept of query log. We just generate queries, named query sequence. The query sequence emphasizes burst paths. The data generation based on Aalborg is explained in this section, and the data set based on Beijing is similar.

First, 1000 paths were selected from the trajectories of Aalborg. The selected 1000 paths are not duplicated and act as ordinary paths. For each ordinary path, its appearance number in the query sequence were generated randomly. Each path was inserted into the query sequence randomly and the insertion repeated its appearance number of times.

When generating burst paths, we consider the need of SPC, since we will compare our algorithms with SPC in Sect. 5.2. In SPC, it needs a training set and a test set. We marked the queries in the first part and last part in the query sequence as training and test sets, respectively. For our algorithms, we do not distinguished training and test sets. The burst paths were generated according to the fact whether a burst path which appears in the test set has appeared in the training set. This feature can be used to detect the efficiency of loading burst paths. As shown in Table 2, Q_A (Q_C) includes 5 burst paths in the training data while only 3 of 5 burst paths are randomly queried in the test data. Q_B (Q_D) includes 3 burst paths in the training set, and all are queried in the test set. In addition, another 3 burst paths not in the training set are queried in the test set.

---

[1] http://www.dbxr.org/experimental-guidelines/.
[2] http://arxiv.org/abs/cs/04100001.

Table 2. Information of query sequence

ID	Data set	Traing set size	Test set size	No. of bursts in training	No. of bursts in test
Q_A	Aalborg	10000	15000	5	3
Q_B	Aalborg	10000	15000	3	6
Q_C	Beijing	20000	35000	5	3
Q_D	Beijing	20000	35000	3	6

Burst paths were generated and inserted into the query sequence manually. The durations and the peak frequencies of burst paths are different from each other. There is a total of 25,000 queries in query sequence for data set based on Aalborg data. For data set based on Beijing data, 55,000 duplicated queries were generated based on 1000 non-duplicated queries.

5.2 Performance Comparison with SPC

In this section, we compare the performance of our algorithms and SPC. Training data were used to load cache in both our algorithms and SPC. For the test set, queries arrived one by one. The cache for SPC was static and the hit ratio was calculated; the cache for BCU and ICU were updated while the queries arrived.

Parameters of our algorithms are as follows: $w = 300$, $\Delta w = 100$, the size of the cache is 4 KB. It is noteworthy that 4 KB can store 15 paths on average.

Table 3 shows the hit ratio of SPC, BCU and ICU, in which the first column indicates the ID of each data set. And the second column demonstrates the performance of SPC. In the following, column three and five show the performance of BCU and ICU, respectively. 'Impr.pct' denotes the improvement percentage of our methods (BCU and ICU) compared to SPC, which are displayed in column four and six.

Table 3. Hit ratios of SPC vs. BCU and ICU

ID	SPC	BCU		ICU	
	Hit ratio(%)	Hit ratio(%)	Impr. pct	Hit ratio(%)	Impr. pct
Q_A	17.3	25.2	45.7 %	24.5	41.6 %
Q_B	17.3	32.0	85.0 %	31.4	81.5 %
Q_C	21.1	30.0	42.2 %	28.7	36.0 %
Q_D	21.1	38.3	81.5 %	36.9	74.9 %

From Table 3 we can see that the hit ratios of BCU and ICU are apparently higher than that of SPC: the improvement of hit ratios in any data set is higher than 30 %. Specially, our algorithms have higher improvement compared to SPC in Q_B and Q_D than in Q_A and Q_C. Note that Q_B and Q_D are more difficult as some burst paths arrive without appearance in training set. The result proves the efficiency of our algorithms in detecting burst paths and update the query sequence dynamically.

5.3   Effect of Window Size

In this section, we talk about the effect of window size on the hit ratio and cache update time, respectively. The data sets tested were Q_B and Q_D. As BCU and ICU are dynamic methods, no concept of training and test data is used in the following. The queries in the query sequence Q_B and Q_D arrive sequently. The queries in the first window were used to initialize the cache. From then on, the cache was updated with every slide. Here, $\alpha = 40(60)$ for Aalborg (Beijing), $\Delta w = 100$ and the cache size was 4 KB. Figures 3 and 4 show the results on hit ratio and update time, respectively, in which the horizon axes represent window size, and the vertical axes represent hit ratio (%) and update time (ms), respectively.

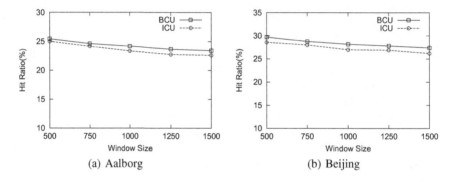

(a) Aalborg   (b) Beijing

Fig. 3. Hit ratio vs. window size

From Fig. 3(a) and (b) we can see that the hit ratio of BCU is higher than that of ICU in whatever case. This is because BCU reloads the cache in every window while ICU is only a dynamically incremental algorithm. With the increase of the window size, the hit ratio slowly decreases. The decrease of the hit ratio is due to that in the case of large window size, the cache is updated according to more old queries in the window, so the hit ratio decreases. The speed of decrease is slow, because the query distribution is stable from the time horizon. The cache contents in different windows differ a little. Therefore, the decrease of the hit ratio is slight.

Next, we explore the effect of window size on the update time. The update time refers to the total update time for the query sequence, not including the cache initiation time in the first window. The update time is related to the update frequency and the time of single update. As the algorithm ceases when the last w queries are in a window, small window size $w1$ will have $(w_2 - w_1)/\Delta w$ more times of update than large window size w_1 when Δw remains unchanged.

Figure 4(a) and (b) show that as the window becomes large, the update time of BCU becomes longer apparently. This is because every time the cache is updated in BCU, all queries in a window have to be processed. A larger window size will inevitably lead to the time of a single update increase. Even though the update becomes less frequent when the window gets larger, the total update time still expands significantly.

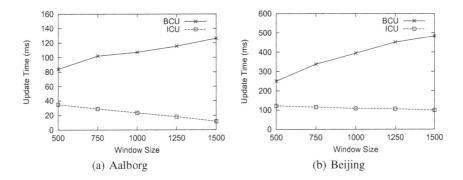

(a) Aalborg (b) Beijing

Fig. 4. Update time vs. window size

For the effect of window size on ICU's performance, the update time for a window is almost the same for all window sizes as the sliding step unchanged. But a large window has lower update frequency; thus, the line for ICU declines a little. For a certain window size, the update time of ICU is far less than that of BCU, because ICU is incrementally updated and the update time of a single update is proportional to the sliding step while the time of a single update of BCU is proportional to the window size.

5.4 Effect of Sliding Step

This section discusses the effect of sliding step on the hit ratio and update time. The data sets were again Q_B and Q_D. $\alpha = 40(60)$ for Aalborg (Beijing). The window size was set 1000, and the cache size was 4 KB. The experimental results are shown in Figs. 5 and 6, in which the horizontal axes represent sliding step and the vertical axes represent hit ratio (%) and update time (ms), respectively.

As Fig. 5(a) and (b) show, the hit ratio does not display an obvious trend along with the sliding step. Whether a window covers burst paths will highly affect the hit ratio. If queries for a burst path are just across two consecutive sliding windows, then the burst path cannot be detected and loaded into the cache. Generally speaking, small sliding steps will lead to high hit ratio as the cache update is more frequent.

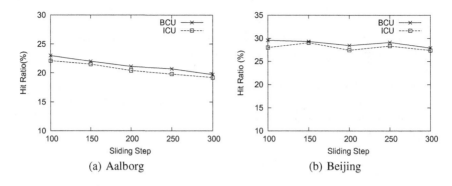

Fig. 5. Hit ratio vs. sliding step

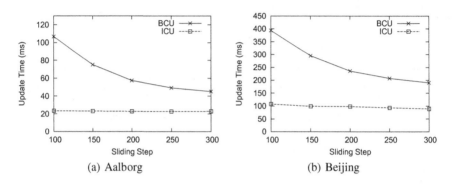

Fig. 6. Update time vs. sliding step

Figure 6(a) and (b) show the trend of update time with the sliding step. The update time of BCU decreases with increasing sliding step, because the cache update time of BCU is related to sliding steps. When the window size does not change, the increase of sliding step brings the decrease of update frequency and update time. For ICU, the increasing of sliding step makes the update time of each sliding longer, but the update frequency decreases at the same time. The update time of ICU has only slight fluctuations.

From Figs. 3 and 5, we can see that the difference between hit ratios of ICU and BCU are tiny. From Figs. 4 and 6, we can see that the difference between update time of ICU and BCU is notable. Moreover, the update time of BCU is more sensitive to the change of window size and sliding step; in contrast, the update time of ICU is stable across different window sizes and sliding steps. ICU is superior to BCU.

6   Conclusion

Existing works about shortest path caching focus on the query log and the ability to deal with dynamic queries is weak. This paper addresses the dynamic update of cache so as to satisfy continuously arrived queries. Burst paths, which are common seen in real life, are the main selling point of this work. By far, no work has considered burst paths and the extant methods are not effective in dealing with burst paths.

Burst paths usually arrive unpredictably; the total number of appearance of burst paths are fewer than other hot paths while the burst paths can boom in a relatively short period. Based on the features of burst paths, we define QF-IQF value in this paper to detect burst paths, and we propose basic and incremental cache update algorithms to update cache on a basis of sliding windows. It shows that our algorithms can achieve higher hit ratio than SPC algorithm on well-known data sets.

In the future, the features of queries can be better described, not subject to burst paths. Statistical variables, such as mean, deviation, minimum and maximum, can be deployed to describe the features of queries. Season and cycle factors can also be investigated.

References

1. Wu, L., Xiao, X., Deng, D., et al.: Shortest path and distance queries on road networks: an experimental evaluation. Proc. VLDB Endow. **5**(5), 406–417 (2012)
2. Altingovde, I.S., Ozcan, R., Ulusoy, Ö.: A cost-aware strategy for query result caching in web search engines. In: Boughanem, M., Berrut, C., Mothe, J., Soule-Dupuy, C. (eds.) ECIR 2009. LNCS, vol. 5478, pp. 628–636. Springer, Heidelberg (2009)
3. Baeza-Yates, R., Gionis, A., Junqueira, F., et al.: The impact of caching on search engines. In: Proceedings of the 30th Annual International ACM SIGIR Conference on Research and Development in Information Retrieval, pp. 183–190. ACM (2007)
4. Vlachos, M., Wu, K.L., Chen, S.K., et al.: Correlating burst events on streaming stock market data. Data Min. Knowl. Discov. **16**(1), 109–133 (2008)
5. Parikh, N., Sundaresan, N.: Scalable and near real-time burst detection from eCommerce queries. In: Proceedings of the 14th ACM SIGKDD International Conference on Knowledge Discovery and Data Mining, pp. 972–980. ACM (2008)
6. Subašić, I., Castillo, C.: The effects of query bursts on web search. In: Web Intelligence and Intelligent Agent Technology, pp. 374–381. International Conference on IEEE (2010)
7. Lee, K., Lee, W.C., Zheng, B., Xu, J.: Caching complementary space for location based services. In: EDBT, pp. 1020–1038 (2006)
8. Wei, F.: TEDI: efficient shortest path query answering on graphs. In: Proceedings of the 2010 International Conference on Management of Data, pp. 99–110. ACM (2010)
9. Cheng, J., Ke, Y., Chu, S., et al.: Efficient processing of distance queries in large graphs: a vertex cover approach. In: Proceedings of the 2012 International Conference on Management of Data, pp. 457–468. ACM (2012)
10. Thomsen, J.R., Yiu, M.L., Jensen, C.S.: Effective caching of shortest paths for location-based services. In: Proceedings of the 2012 International Conference on Management of Data, pp. 313–324. ACM (2012)
11. Gan, Q., Suel, T.: Improved techniques for result cache in web search engines. In: Proceedings of the 18th International Conference on World Wide Web, pp. 431–440. WWW (2009)

12. Long, X., Suel, T.: Three-level caching for efficient query processing in large web search engines. World Wide Web-Internet Web Inf. Syst. **9**(4), 369–395 (2006)
13. Markatos, E.P.: On caching search engine query results. Comput. Commun. **24**(2), 137–143 (2001)
14. Li, X., Qiu, T., Yang, X., Wang, B., Yu, G.: Refreshment strategies for the shortest path caching problem with changing edge weight. In: Jia, Y., Sellis, T., Liu, G., Chen, L. (eds.) APWeb 2014. LNCS, vol. 8709, pp. 331–342. Springer, Heidelberg (2014)
15. He, Q., Chang, K., Lim, E.P.: Analyzing feature trajectories for event detection. In: Proceedings of the 30th Annual International ACM SIGIR Conference on Research and Development in Information Retrieval, pp. 207–214. ACM (2007)
16. Chen, W., Chen, C., Zhang, L., Wang, C., Bu, J.: Online detection of bursty events and their evolution in news streams. J. Zhejiang Univ. **11**(5), 340–355 (2010)

Low Overhead Log Replication for Main Memory Database System

Jinwei Guo, Chendong Zhang, Peng Cai$^{(\boxtimes)}$, Minqi Zhou, and Aoying Zhou

School of Computer Science and Software Engineering, ECNU,
Shanghai 200062, People's Republic of China
guojinwei@stu.ecnu.edu.cn, zhangcd_encu@ecnu.cn,
{pcai,mqzhou,ayzhou}@sei.ecnu.edu.cn

Abstract. Log replication is the key component of high available database system. To guarantee data consistency and reliability, modern database systems often use Paxos protocol to replicate log in multiple database instance sites. Since the replicated logs need to contain some metadata such as committed log sequence number (LSN), this increases the overhead of storage and network. It has significantly negative impact on the throughput in the update intensive work load. In this paper, we present an implementation of log replication and database recovery, which adopts the idea of piggybacking, i.e. committed LSN is embedded in the commit logs. This practice not only retains virtues of Paxos replication, but also reduces disk and network IO effectively, which enhances performance and decreases recovery time. We implemented and evaluated our approach in a main memory database system (Oceanbase), and found that our method can offer 1.3x higher throughput than traditional log replication with synchronization mechanism.

Keywords: Log replication · Database recovery · Paxos · OceanBase

1 Introduction

Through the smart phone we can submit transaction processing requests to the database at any time, and in the scenario of Internet application highly concurrent requests have overwhelmed the traditional database system. For example, in Chinese "Single Day", i.e. Double 11 shopping carnival, the total transactions may hit the level of hundreds of millions in the first minute. To resolve this challenge, many NoSQL and NewSQL systems were designed and implemented [1]. NoSQL refers to the data storage system which is non-relational, distributed and not guaranteed to follow the ACID properties. Compared to the relational database system, NoSQL systems have some excellent characteristics such as without

This work is partially supported by National High-tech R&D Program (863 Program) under grant number 2015AA015307, and National Science Foundation of China under grant number 61332006.

B. Cui et al. (Eds.): WAIM 2016, Part II, LNCS 9659, pp. 159–170, 2016.
DOI: 10.1007/978-3-319-39958-4_13

needing to predefine the data schema, high scalability, share nothing architecture and asynchronous replication. These features provide strong support for the Internet application in the web 2.0. In the other hand, both the industrial and academic community hope to use NoSQLs unique features to solve the massive data processing problems. NoSQL systems have got extensive attentions, and main industry players including Google, Amazon and Facebook have developed their NoSQL database products which play a key role in their services.

NoSQL systems have some limitation when used in the mission critical applications which require high data consistency. For example, asynchronous replication and the final consistency mechanism provided by NoSQL is not applicable for the bank system. If the delay of inconsistency window is too long, during the delay the primary has the risks of shutdown. Then, the update information may be lost because the committed update-transactions have not been synchronized to the secondary. In this procedure, it is possible that a customer performs a withdraw operation, but the final balance of the account is not reduced accordingly.

Log replication based on Paxos [3] can achieve the strong data consistency. The Paxos algorithm is proposed by Leslie Lamport in 1990 which is a consistency algorithm based on the message passing model. The algorithm solve the problem of reaching agreement among majority processes or threads under the distributed environment. Recently, there have may systems adopt Paxos algorithm to the log replication [17–19,23]. As long as the log records have been replicated in the majority of servers, the primary node can submit the transaction. This method can guarantee the strong consistency between primary and secondary nodes. When the primary node failed, the majority of the system nodes can select at least one and only one new primary to achieve a seamless takeover of the old primary.

The remainder of the paper is organized as follows: Preliminary work is presented in Sect. 2. Traditional log replication procedure is presented in Sect. 3. We introduce related work of Paxos replication in Sect. 4. Sections 5 and 6 introduce the log replication and recovery mechanism for OceanBase, which aim to reduce overhead. Section 5 presents experimental results. We conclude the paper in Sect. 6.

2   Preliminary

OceanBase [4] is a scalable relational database management system developed by Alibaba. It supports cross-table and cross-row transactions over billions of records with hundreds of terabyte data.

OceanBase can be divided into four modules: the master server (RootServer), update server (UpdateServer), baseline data server (ChunkServer) and data merge server (MergeServer).

- **RootServer:** It manages all servers meta information in an OceanBase cluster, as well as data storage location.

- **UpdateServer:** It stores updated data in OceanBase. UpdateServer is the only node responsible to execute any update requests such as Delete or Update SQL statements. Thus, there is no distributed transaction in OceanBase because any update operations are processed in a single node.
- **ChunkServer:** It stores OceanBase baseline data, which is also called static data.
- **MergeServer:** It receives and parses SQL requests, and forwards them to the corresponding ChunkServers or UpdateServer after lexical analysis, syntax analysis, query optimization and a series of operations.

UpdateServer is a key component in OceanBase, and it has some characteristics, which we utilize to implement our Paxos replication, as follows:

- UpdateServer can be seen as a main memory database, which store updated data in memory table.
- One transaction only corresponds to one commit log, which is generated until the transaction is finished.
- Log records are stored on disk continuously. Therefore, there are no holes in log files.

OceanBase can configure multiple clusters, e.g. one master cluster and one slave cluster. Only master cluster can receive write transactions. When master cluster breaks down, the whole system is not available for clients. For this reason, we should implement Paxos replication in OceanBase.

2.1 Log Replication Model

Using Paxos to replicate log records has two phases: leader election and log replication. The servers participating in these phases are called election members. To briefly describe, we use *member* to refer to the election member. Each member processes one of two election roles: Leader and Follower.

During the leader election phase, there may be no Leader in the system. Therefore, every member should take part in the leader election. Each member reports its local last log sequence number (LSN) to the election service. The local last LSN may be comprised of log id, generated log timestamp or epoch number. The election service elects a Leader from the reporting members in consideration of their LSN. When the majority of members acquire the election result and succeed to register to the Leader, the election phase finished successfully. It is noteworthy that the Leader should renew its election lease at each interval. If the majority of members note that the Leader's lease is expired, the system will enter into the leader election phase again.

During the log replication phase, as is shown in Fig. 1, the Leader receives write requests from clients, generates commit log, and replicates the log record to each follower. When the Followers receive the log message, they can do different actions according to reliability strategy:

- **Durability:** The Followers responses the Leader until they force the log records to disk.
- **Non-durability:** The Followers responses the Leader immediately when they receives the log message.

The flow of point 1 in Fig. 1 shows the situation of non-durability. We find that the delay of write request is smaller than the flow of point 2, which means the situation of durability.

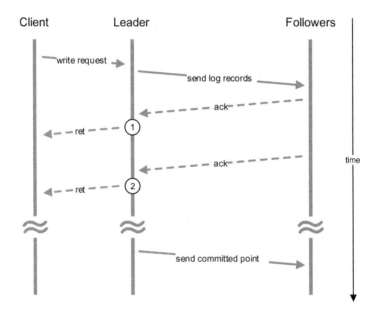

Fig. 1. Log replication

When the Leader gets majority of acks, it can update local committed LSN, which should be flushed to disk and sent to each follower. Committed LSN is a important metadata in log replication. It enables the Follower to provide timeline consistency and simplifies recovery from failure. However, we note that the traditional synchronization mechanism of Committed LSN increases the storage and network overhead. Therefore, we design and implement low overhead log replication adopting durability strategy for OceanBase.

3   Related Work

ARIES [5] has been the actual standard for transaction logging. It gives a reference for log model and recovery mechanism. Databases adopting ARIES can provide a reliable single server, which is a basis for log replication in distributed database system.

NoSQL provides us with scalable and high available datastore technology. Dynamo [6] is Amazons high available Key-value storage system. Its replication resorts to NWR strategy, which permits clients to decide to balance availability against consistency. Cassandra [7,8] was initially developed at Facebook. Its architecture is similar to Dynamo. At present, it has became an open source distributed database management system in Apache. Yahoo's PNUTS [9] is a scalable datastore, which is focused on cross-datacenter replication. These systems use eventual consistency to provide high availability.

Paxos is a consensus protocol for solving consistency problem in distributed system. It is described basically by Lamport in [3]. Multi-Paxos introduced in [10] is an important protocal for Paxos replication. And more variants are introduced by him in [11].

Using Paxos for replication is a common choice for implementing scalable, consistent, and highly available datastore. Chubby [15] is google's service aiming at providing a coarse grained lock service for loosely-coupled distributed systems. Zookeeper [16] is its open source implementation. Google has developed MegaStore [17], Spanner [18] and F1 [19], and these database systems have used Paxos for log replication. Megastore is a storage system providing strong consistent. Spanner is a scalable, multi-version, global distributed, synchronous replication database. F1 provides the functionality of the SQL database. Raft [12,13] is a consensus algorithm for RAMCloud [14]. It is designed to be easy to understand and equivalent to Paxos. [20] introduces relatively complete technology solution to build a datastore using Paxos. [21] analyzes and discusses the replication based on Paxos.

In recent years, with the rapid development of new hardware, e.g. NVM (Non-volatile memory) and RDMA (Remote direct memory access), new log replication and recovery mechanism emerge. [22] introduces one of implementations.

4   Log Replication

This section describes the log replication protocol. To simplify the discussion, we adopt 3-way replication. In order to give a simple explanation, we treat MergeServer and ChunkServer as the clients which issue write requests in the replication protocols, UpdateServer as the election members in log replication model as mentioned before, which is responsible for generating and keeping the log replication among all the UpdateServer.

During the log replication phase, only the Leader can accept the write request. When the client issues write requests, it would acquire which UpdateServer is the Leader and send write requests to the Leader.

Once the Leader accepts the updating requests from clients, transaction logs will be generated. In addition to LSN and the eventual value being contained in each transaction log, the committed LSN also should be in it too. The format of the log record is shown as follows:

$$< LSN, CommittedLSN, LogData >$$

Server:	Leader	Follower 1	Follower 2
Durability :	120	110	100

Fig. 2. Last LSN of each Follower stored at the memory of Leader

After the Leader has generated the operation log, it sends the log record to all the Followers by an asynchronized network function. This manner do not block the Leader's commit thread, it is able to flush the commit log to local disk without waiting for the responses of the Followers.

When the Followers receive the log message from the Leader, it would extract the committed LSN from the log record and compare it with the local cached committed LSN. If the local committed LSN is less than the new committed LSN, it should be updated with the new LSN. Then the Followers append the commit log to the end of the log file. Once the appending operation has finished or overrun by a certain period of time, the Follower would get the maximal flushed LSN whose corresponding log record is stored into disk and respond to the Leader through the RPC message containing the flushed LSN.

The committed LSN would be checked frequently by the replay threads in Followers. When its value is changed, the replay threads will fetch the logs which has not been replayed, namely the log before the committed LSN from the log file, and replay and apply them in the memory table. If the Followers cannot find the corresponding logs in log file, it will fetch them from the Leader by themselves.

The Leader has stored the flushed LSNs of all the Followers. When receiving the responding message from Follower, the Leader would extract the flushed LSN in the message and compare it with the local cached flushed LSN of this Follower. If the new flushed LSN is greater than the local one, the Leader would replace the local value. Based on all the local cached Followers' flushed LSN and the Quorum Policy Protocol, the Leader calculates a new committed LSN. If the new committed LSN is greater than the previous one cached in local memory, the local committed LSN would be updated with the new one. Finally, the Leader would commit the transactions at their last stage, based on the local cached committed LSN, and responds to the clients.

Figure 2 shows an example of two followers' flushed LSN stored in Leader. We find that the log records whose LSN is not greater than 110 are durable in majority of all servers. Therefore, the Leader can commit and end the transactions whose commit log's LSN is not greater than 110.

The above procedure is the main process of the log replication protocol. To improve the throughput of the database system, we adopt the group commit policy. When the Leader generates an operation log, it caches the log in the log buffer. Once the log buffer reaches the max length or at regular intervals, the

Leader packages the logs in buffer and sends the package to the Followers, and flushes all of them into local disk.

5 Recovery

This section describes how a election member recovers from failure. Failure is a common phenomenon in distributed systems, e.g. power failure, administrator mistakes, software or hardware errors and so on.

There are two kinds of system states in Paxos replication systems, i.e. DURING_ELECTION and AFTER_ELECTION, which indicate whether there exists a Leader in the system. If the system is in AFTER_ELECTION state, it shows that the system in the log replication phase. When a member is restarting, its election role is definitely determined. Therefore, it can take predetermined actions in accordance with the role. If the system is in DURING_ELECTION state, it means that the system in the leader election phase. The restarting member need to take part in the election. It is only when new Leader is elected that the restarting member can continue to recover.

If election role of the recovering member is Leader, it is not until the Leader replicates its local log records to a majority of the servers that it can service requests of the clients. Firstly, it scans local log files to get local last LSN and max committed LSN, which are cached in local variables. Then the Leader starts up threads to replay whole local logs from checkpoint. By this time, the Leader can not service for clients. Thirdly, it appends a special commit log which only contains max committed LSN, and replicates the log record to other members. Lastly, the master receives responses of other Followers and updates corresponding information of the servers. When the master detects that the committed LSN is not less than the previous cached local last LSN, it can provide service for clients.

Leader adopting above recovery protocol can provide strong consistency. Its main steps are summarized in Procedure 1. In some cases, we would like high availability rather than strong consistency. Therefore, when the Master starts

---

Procedure 1. Leader Recovery

1: $local\_last\_LSN$ = the last LSN from local log files;
2: $committed\_LSN$ = the max committed LSN from local log files;
3: start up threads to replay local log;
4: push a NOP task to commit queue;
5: **while** $local\_last\_LSN > committed\_LSN$ **do**
6:    $temp\_committed\_LSN$ = get committed LSN from Follower responses;
7:    **if** $committed\_LSN \leq temp\_committed\_LSN$ **then**
8:      $committed\_LSN = temp\_committed\_LSN$;
9:    **end if**
10:    sleep for a while;
11: **end while**
12: $state$ = ACTIVE; // can provide service

Procedure 2. Follower Recovery

1: *local_last_LSN* = the last LSN from local log files;
2: *committed_LSN* = the max committed LSN from local log files;
3: *local_committed_LSN* = *committed_LSN*;
4: start up threads to replay local log to committed LSN;
5: register and report local committed LSN to the Leader;
6: wait for receiving log from master;
7: *master_LSN* = the max LSN from master's log;
8: **while** *last_local_LSN* $\leq$ *committed_LSN* **do**
9:     discards log records after local committed log;
10:    appends new log to log files;
11:    sleep for a while;
12: **end while**
13: *state* = ACTIVE; // can response to Leader

the replaying task, it can service for clients as long as replaying local logs to the committed LSN. The remainder of the local log is replayed along with the committed LSN.

The recovery of the Follower is different from the Leader's. As the Follower can not judge whether the log records after the committed LSN should be replayed and committed, it must get necessary information from the Leader. In order to reduce the network overhead, we implement recovery mechanism as below. To begin with, the Follower scans local log files to update local variables, e.g. local last LSN, committed LSN. As described before, the committed LSN is the max committed LSN stored in the log files. Then, it starts to replay local logs to the committed LSN, and it discards the remaining log records. At the same time, the Follower reports its committed LSN to the Leader. When the Leader receives this message, it sends log records after that LSN to the Follower. Finally, the Follower receives new log records and refresh the committed LSN, which triggers itself to replay the log continuously.

If the role of Master is frequently switched in different members, the log records of committed transactions will be lost. To prevent this, it is not until the Slave receives Leader's LSN which is greater than the local last LSN that Follower discards log records after the committed LSN. The main steps are described in Procedure 2.

6   Experiments

This section evaluates the performance of several different implementations, i.e. piggybacking method, synchronization method, and asynchronous method, which are implemented in OceanBase 0.4.2:

- **Piggybacking method:** This method is our implementation in this work. We encode the committed LSN into the log records.

- **Synchronization method:** This method is different from piggybacking method. When the Leader detects that the committed LSN has be changed, it would call Linux interface fsync() to flush the committed LSN to the disk and call RPC to send the LSN to the Followers.
- **Asynchronization method:** This method is different from the above two methods. The Leader starts an additional thread, which is responsible for flushing and sending the committed LSN.

6.1 Experimental Setup

This subsection describes the cluster configuration and benchmarks.

Cluster Configuration: We deploy three clusters, and each cluster consists of one RootServer, one UpdateServer, four Chunkservers and four Mergeservers. Each server's configuration is shown in Table 1:

Table 1. Experimental setup

Software and Hardware Setup	
CPU	E5606@2.13G * 2
CPU cores	8 (Hyper-threading disabled)
Memory	16 GB PC3L-12800R * 6
Disk	100 GB SSD * 1
Network	Gigabit Ethernet
Operating System	CentOS 6.5

Benchmarks: We adopt a micro-benchmark to evaluate our implementations. Firstly, each Client gets a connection with MergeServer. Then it executes *replace* auto-commit transaction repeatedly, which updates about fifty bytes of data. We observe the statistics of database and operating system during the execution.

6.2 Log Replication Performance

We measure the performance by TPS and wrequests/s. The TPS refers to the number of transactions performed by system per second, and the wrequests/s refers to the write requests issued to disk per second. The experimental results are presented in Figs. 3 and 4. Each client connection commits 6000 transactions. The total transactions count is determined by the client connections count. For example, if the client connection number in each MergeServer is 100, the total transaction count is 100 * 6000 * 4. The wrequests/s is used to indicate the balance between the disk performance and the TPS. While the clients issue more connections, the speed of the wrequests/s would decrease. In the contrary, the TPS would increase as the client connection number grows. This illustrates

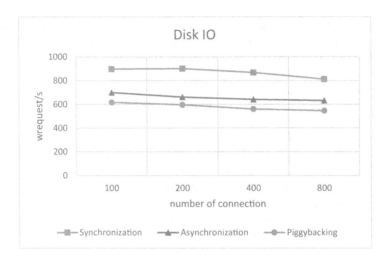

Fig. 3. Average write requests over the disk

the positive effect of the group commit method which is used to improve the efficiency of committing logs by reducing the frequency of write requests to disk. The more parallel transactions are processed, the greater improvement it will achieve.

Through the comparison of the three methods in Fig. 3, we also could find that the Piggybacking has the lowest wrequests/s, the middle is the Asynchronization method, and the highest is the Synchronization. The Piggypacking method has the lowest wrequests/s because it doesn't have to store committed LSN to local disk, which may increase additional IO. The Synchronization method has the highest wrequests/s because committed LSN is written to local disk when the Leader refreshes its value. So when the TPS increases the frequency of writing committed LSN will grow simultaneously and the wrequests/s would increase as well. The Asynchronization method makes use of an additional thread to write committed LSN to local disk every 10 ms, which is asynchronous with the committing log thread. In this method, its wrequests/s doesn't have direct relationship to TPS.

In Fig. 4, the Piggybacking method has the highest TPS, as it has no burden of writing committed LSN. But by the comparison of the Asynchronization method, the Synchronization method has a much more frequency writing committed LSN, which will inevitably reduce the performance of the transaction processing. But the effect is not obvious because the write latency of committed LSN is less than 2 ms, while a common transaction processing time is about 30–40 ms at high load, which is about 20 times longer than the former. The experiment results significantly demonstrate the features of the committed LSN.

From the above experiment results, we could draw a conclusion that the committed LSN could improve the system availability, but it also leads to much more write requests to disk and thus decreases the maximum IO bandwidth used

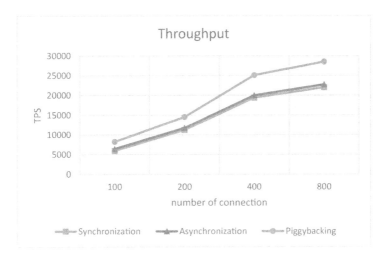

Fig. 4. Average throughput of transactions

to store transaction log. Moreover, the Synchronization method could provide more availability than Asynchronization method without significantly negative impact on performance.

7   Conclusion

Log replication based on Paxos can provide database with scalability, consistency and highly availability. This paper described an implementation mechanism of Paxos replication for OceanBase, which is scalable and has a memory transaction engine. Unlike traditional implementation, our method takes into account the overhead of storage and network, which have a significant impact on performance.

We find that the synchronization of committed LSN used for timeline consistency may improve the overhead of the system. Therefore, we make use of piggybacking technology to implement log replication and database recovery. Compared to the synchronization mechanism, our method improves throughput of update operations by 1.3x.

References

1. Cattell, R.: Scalable SQL and noSQL data stores. SIGMOD Rec. **39**(4), 12–27 (2010)
2. Stonebraker, M., Cetintemel, U.: "One size fits all": an idea whose time has come and gone. In: Proceedings of ICDE, pp. 2–11 (2005)
3. Lamport, L.: The part-time parliament. TOCS **16**(2), 133–169 (1998)
4. OceanBase website. https://github.com/alibaba/oceanbase

5. Mohan, C., et al.: ARIES: a transaction recovery method supporting fine-granularity locking and partial roll backs using write-ahead logging. TODS **17**(1), 94–162 (1992)
6. DeCandia G., Hastorun D., Jampani M., Kakulapati, G., Lakshman, A., Pilchin, A., Sivasubramanian, S., Vosshall, P., Vogels, W.: Dynamo: amazons highlyavailable key-value store. In: Proceedings of SOSP, pp. 205–220 (2007)
7. Lakshman, A., Malik, P.: Cassandra: a decentralized structured storage system. SIGOPS **44**(2), 35–40 (2010)
8. Cassandra website. http://cassandra.apache.org/
9. Cooper, B.F., Ramakrishnan, R., Srivastava, U., et al.: PUNTS: Yahoo!'s hosted data serving platform. In: Proceedings of VLDB, pp. 1277–1288 (2008)
10. Lamport, L.: Paxos made simple. SIGACT **32**(4), 18–25 (2001)
11. Lamport, L.: Fast paxos. Distrib. Comput. **19**(2), 79–103 (2006)
12. Ongard, D., Ousterhout, J.: In search of an understandable consensus algorithm. In: Proceedings of ATC (2014)
13. Raft consensus algorithm website. https://raft.github.io
14. Ousterhout, J., Agrawal, P., Erikson, D., et al.: The case for RAMCloud. CACM **54**, 121–130 (2011)
15. Burrows, M.: The chubby lock service for loosely coupled distributed systems. In: Proceedings of OSDI, pp. 335–350 (2006)
16. Zookeeper website. https://zookeeper.apache.org/
17. Baker, J., Bond, C., Corbett, J.C., Megastore, et al.: Providing scalable, highly available storage for interactive services. In: Proceedings of CIDR, pp. 223–234 (2011)
18. Corbett J.C., Dean J., Epstein, M.: Spanner: Googles globally distributed database. In: Proceedings of OSDI (2012)
19. Shute, J., Vingralek, R., Bart, S., et al.: F1: a distributed SQL database that scales. In: Proceedings of VLDB, pp. 1068–1079 (2013)
20. Rao, J., Shekita, E.J., Tata, S.: Using paxos to bulid a scalable, consistent, highly available datastore. In: Proceedings of VLDB, pp. 243–254 (2011)
21. Patterson, S., et al.: Serializability, not serial: concurrency control and availability in multi-datacenter datastores. Proc. VLDB Endow. **5**(11), 1459–1470 (2012)
22. Dragojevic, A., Narayanan, D., et al.: No compromises: distributed transactions with consistency, availability, and performance. In: Proceedings of SOSP, pp. 54–70 (2015)
23. Thomson, A., Diamond, T., et al.: Calvin: fast distributed transactions for partitioned database systems. In: Proceedings of SIGMOD, pp. 1–12 (2012)

Diversification of Keyword Query Result Patterns

Cem Aksoy[1], Ananya Dass[1], Dimitri Theodoratos[1(✉)], and Xiaoying Wu[2]

[1] New Jersey Institute of Technology, Newark, NJ 07103, USA
{ca64,dth}@njit.edu
[2] Wuhan University, Wuhan, China

Abstract. Keyword search allows the users to search for information on tree data without making use of a complex query language and without knowing the schema of the data sources. However, keyword queries are usually ambiguous in expressing the user intent. Most of the current keyword search approaches either filter or use a scoring function to rank the candidate result set. These techniques do not differentiate the results and might return to the user a result set which is not the intended. To address this problem, we introduce in this paper an original approach for diversification of keyword search results on tree data which aims at returning a subset of the candidate result set trading off relevance for diversity. We formally define the problem of diversification of patterns of keyword search results on tree data as an optimization problem. We introduce relevance and diversity measures on result pattern sets. We design a greedy heuristic algorithm that chooses top-k most relevant and diverse result patterns for a given keyword query. Our experimental results show that the introduced relevance and diversity measures can be used effectively and that our algorithm can efficiently compute a set of result patterns for keyword queries which is both relevant and diverse.

1 Introduction

Keyword search is a popular technique for retrieving information from tree data which has a widespread use with different forms (XML, JSON, clustering and categorization hierarchies, ontologies etc.) in current data-oriented applications. The reason of the popularity of keyword search on tree-structured data is twofold: (a) the user does not need to master a complex query language (e.g., XQuery), and (b) the user can issue queries against the data without having full or even partial knowledge of the structure (schema) of the data sources. Further, the same query can be issued against differently structured data sources on the web.

Although convenient for the users, there is a major drawback of keyword search systems: keyword queries are usually ambiguous in expressing the user intent.

X. Wu—Supported by the National Natural Science Foundation of China under Grant No. 61202035.

B. Cui et al. (Eds.): WAIM 2016, Part II, LNCS 9659, pp. 171–183, 2016.
DOI: 10.1007/978-3-319-39958-4_14

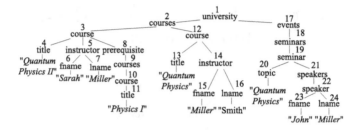

Fig. 1. A university database involving courses and seminars

As a consequence, they usually return a large number of results. The same keyword query might represent multiple intents and it is possible that none of these intents can satisfy all the users [11,12]. Hence, the answer of keyword queries may contain different types of results which are all meaningful with respect to the query even though the user is interested in only few of them. Consider the following example.

Example 1. Let $Q = \{Quantum, Physics, Miller\}$ be a query on the data tree of Fig. 1, which represents a university database recording courses and seminars. There are courses titled "Quantum Physics" and taught by an instructor whose last name is "Miller", and a seminar whose topic is "Quantum Physics" given by a speaker whose last name is "Miller". The user might be interested in both courses and seminars or she can be interested only in one of them.

Since inferring the user intent from the queries is difficult, systems have to provide results which cover a diverse set of interpretations of the query. This can also be seen as a way of minimizing the risk of user dissatisfaction. That is, if you return a diverse set of results, even if it is impossible to guess what a user is looking for, the user will find their intended results in the query answer.

Another motivation of diversification is provided by the exploratory queries where the users are interested in an overall look of the result set for the query [12]. These might be queries issued for conducting research on a specific topic. Ideally, the answer of exploratory queries should contain results which cover different aspects of the entire result space and the user finds the desired results by further exploring the result set.

Search result diversification attracted a lot of attention in both Information Retrieval and Recommendation Systems [12,14]. With the help of a diversified result set, the systems can address the problems of keyword query ambiguity and of the multitude of the results, and can support exploratory search. When the user intent cannot be inferred from a keyword query, result diversification can minimize the user dissatisfaction. It can also address the over-specialization problem [20] which might be caused by relevance-based ranking of the results which is usually based on statistical information (frequency or popularity of the keywords). For example, consider the query *chelsea*. With this query, a user might be referring to the Chelsea district in London, the Chelsea Football Club in

UK or the Chelsea neighborhood in Manhattan, New York. A ranked list solely based on relevancy may exhaustively return results of one of the types above while ranking the results of the other types far behind. A satisfactory result set should include results of different aspects by keeping a balance between relevancy and diversity.

In this paper, we introduce an original approach for diversification of keyword search results on tree data which aims at returning a subset of the candidate result set trading off relevance for diversity. In order to reduce the result space to a manageable size, we apply our technique to patterns of results which cluster together results of the same type. These result patterns represent alternative interpretations of the keyword query. We formally define the problem of diversification of result patterns on tree data as an optimization problem. We introduce relevance and diversity measures for sets of result patterns. We design a greedy heuristic algorithm that chooses top-k most relevant and diverse result patterns for a given keyword query. Finally, we run experiments to show that the introduced relevance and diversity measures can be used effectively and that our algorithm can efficiently compute a set of result patterns for keyword queries which is both relevant and diverse.

2 Data Model, Query Semantics and Result Patterns

Data Model. We assume that our data is represented as node labeled trees. These trees have two types of nodes: *element nodes* and *value nodes*. Value nodes are leaf nodes. All the nodes are labeled. Value nodes are labeled by text (a set of keywords). Element nodes are labeled by a single keyword. We allow keywords to match *both* the labels of element nodes and value nodes. We say that a node n *contains* keyword k if its label contains k. If n contains keyword k then, node n is an *instance* of k.

Query Semantics. A *(keyword) query* Q is a set of keywords $\{k_1, k_2, \ldots, k_n\}$. Keyword queries are embedded to data trees.

An *embedding* of a query Q to a data tree T is a function e from Q to the nodes of T that maps every keyword k in Q to an instance of k in T.

Definition 1 (Instance Tree of a Query). *Let Q be a query, T be a data tree and e be an embedding of Q to T. The* instance tree (IT) *of Q on T for e is the minimum subtree S of T such that: (a) S is rooted at the root of T and comprises all the images of the keywords under e and (b) a node of S which is the image of some keywords in Q under e is labeled only by these keywords which are explicitly marked and they are called annotating keywords of this node.*

Consider the tree of Fig. 1 and the keyword query $Q = \{Quantum, Physics, Miller\}$. Figure 2(a) shows the IT for the query instance of the embedding $\{(Quantum, 4), (Physics, 4), (Miller, 7)\}$ of Q to T. In the figures, the annotating keywords are shown between square brackets by the nodes.

In our approach, we study keyword search on data trees by assuming that the results of keyword queries are ITs. An IT is a rich construct in terms of the

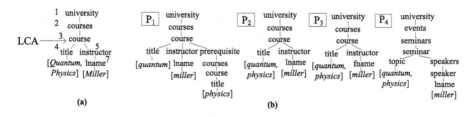

Fig. 2. (a) an instance tree (b) four patterns for $Q = \{Quantum, Physics, Miller\}$ on the tree of Fig. 1

information it provides since it shows: (a) how the keyword images are linked under their LCA, and (b) how the LCA is linked to the root of the data tree.

Definition 2. *The answer of a query Q on a data tree T is the set of ITs of Q on T.*

Result Patterns. A query may have many results. In order to cope with the large number of results, we introduce *result patterns*.

Definition 3 (Result Pattern). *A pattern P of a query Q on a data tree T is a tree which is isomorphic (including the annotations) to an IT of Q on T.*

A pattern has all the information of an IT except the physical location of that one in the data tree. As an example, Fig. 2(b) shows four patterns (out of 32 in total) of the keyword query $Q = \{Quantum, Physics, Miller\}$ on the data tree T of Fig. 1. Pattern P_2 is the pattern of the IT of Fig. 2(a). Patterns represent alternative interpretations of a keyword query.

3   Diversification of Result Patterns

3.1   Formal Problem Definition

Our diversification goal aims at providing the users with a result set that contains relevant and diverse results. We define the problem of diversification of keyword search result patterns as an optimization problem. Given a data tree T and a query Q, let S be the set of result patterns of Q on T and k be a positive integer. The goal of diversification is to choose a subset R of S of size k whose elements are both relevant and diverse. Therefore, R is defined as:

$$R \in \underset{R' \subseteq S, |R'| = k}{\arg\max} \left(\lambda \, relevance(R', Q) + (1 - \lambda) \, diversity(R') \right)$$

where λ is a parameter in the [0,1] range which tunes the importance of relevance and diversity. This tuning factor allows us to give more importance to the relevance or diversity. If $\lambda = 1$, the result set would be formed solely based on the relevance of the patterns, whereas if $\lambda = 0$, the result set will contain the most diverse set of patterns without considering the relevance of the patterns. We call the function $\lambda \, relevance(R', Q) + (1 - \lambda) \, diversity(R')$ objective function.

Here, we assume that the relevance of a result is independent of the relevance of the other results in R'. Therefore, the relevance of R' with respect to Q is defined as:

$$relevance(R',Q) = \frac{1}{k} \sum_{P \in R'} rel(P,Q)$$

where $rel(P,Q)$ stands for the relevance of result pattern P with respect to Q. We define later $rel(P,Q)$ as a value in the interval $[0,1]$. Since the sum is divided by the number k of results in R', $relevance(R',Q)$ ranges between 0 and 1.

A set of patterns is diverse when the pairwise dissimilarity of its patterns is maximized. We define the pairwise dissimilarity of two patterns as the negative of their similarity. Therefore, the diversity of R' is given by the following formula:

$$diversity(R') = -\frac{1}{k(k-1)} \sum_{P,P' \in R', P' \neq P} sim(P,P')$$

where $sim(P,P')$ denotes the similarity between the patterns P and P', and is defined later as a value in the range $[0,1]$. The sum is divided by the number of pairs of patterns in R' to guarantee that $diversity(R') \in [0,1]$.

One can see from the formula above that we need to quantify the relevance of a given pattern and the similarity of two given patterns. In the next sections, we describe how the relevance of a pattern and the similarity between two patterns can be measured. Our definition of the diversification problem does not involve ranking of the patterns. We only select a set of patterns. However, the patterns returned to the user are ranked by our algorithm.

3.2   Relevance of Patterns

In this section, we introduce our relevance scoring scheme. We adapt a statistical approach that incorporates both semantic and structural information of the patterns. Our approach utilizes the TF-IDF measure customized to tree data. TF-IDF [21] has been widely used in information retrieval for assigning weights to characterize the relevance of a term in relation to a document. It combines the term frequency of a term in a document and its inverse document frequency. Inverse document frequency is computed as the reciprocal of the fraction of documents that contain a term. TF-IDF has also been adapted by some keyword search approaches for identifying relevant result types in tree structured data [6,7,17]. Our approach adapts the one presented in [17] so that it can be applied to the ranking of result patterns.

We define the weight of a keyword instance node n_k annotated by keyword k in a pattern as follows: $weight(n_k) = ln(ief(k))$ where ief is the *inverse element frequency* of keyword k in the data tree. The value of ief of keyword k is calculated as the ratio of the number of elements in the data tree over the number of elements in the data tree that contain k in the subtrees rooted at them. This weighting scheme gives nodes that contain the same keyword the same weight value. Consider, for instance, the data tree in Fig. 1. The ief of keyword *physics* is $ln(24/14) = 0.54$ and ief of the keyword *quantum* is $ln(24/10) = 0.88$. These numbers also indicate the importance of the nodes that contain them in a result pattern and therefore, we use them to represent the nodes' contribution to the total relevance of the pattern. However, using merely the node weights is not enough for designing an accurate relevance score.

In order to take into account the structure of the result patterns, we reduce the weight of each keyword instance node accordingly by its distance from the LCA node, denoted as $dist(n_k, LCA(P))$. This reduced score is then aggregated to compute the relevance scores of patterns. The reason behind this reduction is that intuitively, the nodes that appear farther from the LCA node contribute to the relevance of the patterns to a lesser extent. Given a result pattern P for a query Q that consists of n keywords,

$$rel(P, Q) = \sum_{k=1}^{n} \frac{weight(n_k)}{dist(n_k, LCA(P))}$$

where $dist(n_k, LCA(P))$ is the distance of node n_k from the LCA node of pattern P increased by 1.

Consider the patterns shown in Fig. 2 for the keyword query $Q = \{Quantum, Physics, Miller\}$ on the tree of Fig. 1. Patterns P_1, P_2 and P_3 represent courses that contain all the keywords under their subtrees. Pattern P_1 contains the keyword *physics* under the title of a prerequisite course, whereas P_2 and P_3 contain all the information under a single course (a course titled "Quantum Physics" and taught by an instructor whose first name or last name is "Miller"). Pattern P_4 represents a seminar that contains all the query keywords. According to the relevance formula given above, $rel(P_1, Q) = 0.73$, $rel(P_2, Q) = 0.89$, $rel(P_3, Q) = 0.89$ and $rel(P_4, Q) = 0.84$. One can see that patterns that have keywords closer to their LCA node usually represent a more meaningful relationship between the keywords (as a counter example, pattern P_1 represents a loose relationship between the keywords). The relevance scores are normalized into the $[0,1]$ range before being combined with the similarity scores in the objective function.

3.3   Similarity of Patterns

We define the similarity of patterns based on the similarity of their corresponding paths. The similarity of the paths, in turn, is defined based on the similarity of the edges they contain. This similarity scoring takes advantage of both structural and semantic information of the patterns.

Given an edge $e = (n_1, n_2)$ in a pattern, let $c_k(e)$ be the set of annotating keywords which occur in the subtree rooted at n_2. Let $e = (n_1, n_2)$ and $e' = (n_1', n_2')$ be two edges in two patterns P and P', respectively, such that $label(n_1) = label(n_1')$ and $label(n_2) = label(n_2')$. The similarity of e and e', $sim(e, e') = |c_k(e) \cap c_k(e')|$. Intuitively, the number of shared annotating keywords under two edges indicates the shared context under the subtrees below

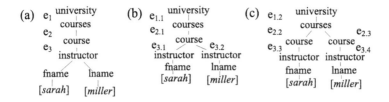

Fig. 3. Three patterns which match the keyword query $\{sarah, miller\}$

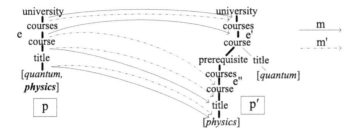

Fig. 4. Two paths p and p' shown with bold edges and possible mappings between their edges depicted with different dashed arrows

these edges. For instance, consider the patterns depicted in Fig. 3. These patterns represent keyword query results for the query $\{sarah, miller\}$ on a university database. The corresponding edges in the paths of the patterns are depicted with the same subscript followed by a dot and an index. The similarity of edges e_1 and $e_{1.1}$ and the similarity of e_1 and $e_{1.2}$ are both 2, since these edges have the same set of annotating keywords. All three patterns contain the edge (university, courses) in all their root-to-annotated-node paths so this edge indicates information which is strongly shared among the patterns. For the courses, course edge, $sim(e_2, e_{2.1}) = 2$ whereas $sim(e_2, e_{2.2}) = 1$. This shows that e_2 and $e_{2.1}$ are more similar than e_2 and $e_{2.2}$ because they contain two common keyword instances in their subtrees while e_2 and $e_{2.2}$ contain only one common keyword instance in their subtrees.

Given two patterns P and P' for the same keyword query Q and two root-to-annotated-node paths p and p', respectively, with the same annotated leaf node, a similarity mapping between p and p' is a one-to-one function m from edges of p to edges of p' satisfying the following properties: (a) the mapped edges have the same labels (that is, if $e = (n_1, n_2)$, $e' = (n_1', n_2')$ and $m(e) = e'$, then $label(n_1) = label(n_2)$ and $label(n_1') = label(n_2')$), and (b) the domain of m, $dom(m)$, is maximal in the sense that m cannot be extended to map any additional edges from p. Clearly, a similarity mapping might not map all the edges of p to edges of p'. Further, there can be many similarity mappings between p and p' since there can be more than one edges in p (or in p') with the same pair of labels. Consider, for instance the two paths in Fig. 4. There are two possible mappings, say m and m', from the the edges of p to p' which are depicted with different dashed arrows. The difference between m and m' is the mapping of the (courses, course) edge, e in p to either e' or e'' in p'. According to our edge scoring scheme, mapping e to e' gives a better score because these edges share more annotated keywords under their subtrees.

Let now $M(p, p')$ be the set of all similarity mappings between two paths p and p'. The similarity of p and p' is

$$sim(p, p') = max\{s \mid s = \sum_{e \in dom(m)} sim(e, e') \text{ where } e' = m(e) \text{ and } m \in M(p, p')\}$$

On the paths of Fig. 4, $sim(p, p') = 6$ which is obtained by the mapping m.

We aggregate the similarity of the corresponding root-to-annotated-node paths in patterns to compute the similarity of the patterns. Let P and P' be two patterns for the same keyword query Q. We first define the path mapping similarity between P and P' as $pathmap(P, P') = \sum_{p \in P, p' \in P'} sim(p, p')$. The similarity of patterns P and P' is defined as follows:

$$sim(P, P') = \frac{pathmap(P, P')}{\frac{1}{2}(pathmap(P, P) + pathmap(P', P'))}$$

The denominator of the formula guarantees that the similarity value lies in the $[0, 1]$ range. For example, consider the four patterns in Fig. 2(b). $sim(P_1, P_2) = 0.91$ and $sim(P_1, P_4) = 0.03$. One can see that patterns that represent similar concepts (that is, their LCAs are nodes of the same type) will get higher similarity values. This is very useful in choosing a diverse set of results. For instance, suppose that the pattern P_2 has been selected to appear in the result set. Even though the pattern P_3 has a higher relevance value than P_4, the diversification scheme might choose P_4 over P_3 since it is less similar to P_2 than P_3 and therefore, it might contribute more to maximizing the objective function.

4   Algorithm

As described in Sect. 3.1, selecting a diverse and relevant set of result patterns from the set of all candidate result patterns is an optimization problem. Given a query Q on a data tree T and the set S of all patterns that match Q on T, the brute force solution to this optimization problem would be to generate all the k-sized subsets of S and select the ones that maximize the objective function. Clearly, this solution is not computationally feasible as the number of possible subsets is $\binom{|S|}{k}$. Different variations of the diversification problem have been proven to be NP-complete [1,9,12,13]. Therefore, in this section, we introduce a heuristic algorithm, Algorithm 1, for computing a set of keyword search result patterns. This algorithm incrementally builds the pattern set by making greedy choices for the inclusion of every new pattern in the pattern set. The resulting pattern set is an approximation of the optimal set.

The algorithm takes as input the set of patterns of a query Q on a data tree and a parameter k which is the desired size for the diverse pattern set.

Algorithm 1. *Diversify* algorithm

```
 1  Diversify(Q = {k₁,...,kₙ}: query, S: set of patterns, k: size of pattern set)
 2      R = {L[0]} /* list initialized with pattern L[0] */
 3      L = sortByRelevance(S)
 4      i = 1
 5      while i < k do
 6          j = i; maxScore = 0; maxIndex = −1
 7          while j < |S| do
 8              if λrel(L[j], Q) < maxScore then
 9                  break
10              sumSimilarity = Σ|R|_{l=1} sim(L[j], R[l])
11              curScore = λrel(L[j], Q) − (1 − λ)(sumSimilarity/|R|)
12              if curScore > maxScore then
13                  maxScore = curScore
14                  maxIndex = j
15              j = j + 1
16          R.add(L[maxIndex])
17          tmp = L.remove(maxIndex)
18          L.insert(i, tmp)
19          i = i + 1
20      return R
```

The patterns of the query results are generated efficiently using the PatternStack algorithm [3,4]. We sort the input patterns with respect to their relevance to the query in descending order. The first pattern, the most relevant pattern, is included in the pattern set by default (line 2). Indeed, if $k = 1$ this would be a trivial choice. Then, we iterate over the remaining list of patterns in L to find the next pattern to be included in the pattern set R (lines 5–19). The iteration terminates when all patterns in L have been examined or the score of a pattern cannot be greater than the current maximum score (line 8). The algorithm stops when the size of R is equal to k.

5 Experimental Evaluation

We performed experiments to evaluate the effectiveness of our relevance and diversity measures and the efficiency of our diversification algorithm.

We use the Mondial[1] dataset for the experiments which were conducted on a 2.9 GHz Intel Core i7 machine with 3 GB memory running Ubuntu.

Evaluation Metrics. In order to evaluate the effectiveness of the relevance measure, we use the popular *precision@N* metric. Precision is computed as the ratio of the relevant results to the total number of results retrieved while precision@N of a ranking retrieval system is the precision at rank N.

---

[1] http://www.cs.washington.edu/research/xmldatasets.

Table 1. Queries used in the experiments.

Query ID	Keywords
Q1	*government, democracy, muslim*
Q2	*organization, name, members*
Q3	*country, ethnicgroups, german*
Q4	*france, territory*
Q5	*singapore, country*
Q6	*country, government, republic*
Q7	*province, houston, dallas*
Q8	*city, washington, province*
Q9	*jewish, percentage*
Q10	*new, york, population*

We use *normalized discounted cumulative gain (NDCG)* for measuring the success of our similarity measure. The NDCG of a ranked list of results is computed as the discounted cumulative gain of the ranked list of results over the ideal discounted cumulative gain for the set of results included in that list. The discounted cumulative gain for a ranked list of results is computed as

$$DCG(list) = \sum_{i=1}^{|list|} \frac{2^{rel_i} - 1}{log_2 i + 1}$$

where rel_i is the graded relevance of the result at rank i. For our experiments, we use $NDCG@3$ which is the NDCG value for the highest three ranks of the ranked lists.

For both metrics, we report two values, *best* and *worst*. This is because of the fact that some patterns might display the same value for a measure, resulting in an equivalence class of patterns. Therefore, we consider the best and the worst possible rankings of all the equivalence classes in a ranking and report on the measured respective best and worst values of the metric. The ground truth was determined by expert users.

Effectiveness of Relevance Measure. We performed retrieval experiments using our relevance measure by running the queries list in Table 1 on the Mondial dataset. Figure 5 shows the precision@3 values measured for each query. According to these values, our relevance measure can retrieve almost always retrieve the correct results in the top ranks. For query $Q3$ even though the worst precision@3 score is around 0.5, the best version is still 1.0. This shows that our relevance measure can be effectively used as part of a diversification scheme.

Effectiveness of Similarity Measure. In order to assess the effectiveness of our similarity measure, we choose the most relevant pattern for each query listed in Table 1 on the Mondial dataset and retrieved the most similar three patterns

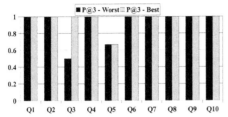

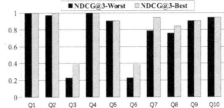

Fig. 5. Precision@3 scores achieved by our relevance measure

Fig. 6. NDCG@3 scores achieved by our similarity measure

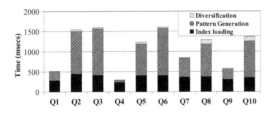

Fig. 7. Execution times of the diversification algorithm $(k = 3)$ over 10 queries on the Mondial dataset.

to it with respect to our similarity measure. Figure 6 shows the NDCG@3 values measured for each case. These values show that the similarity measure can accurately find the most similar patterns of a given pattern. In all cases, the similarity measure was able to retrieve at least two of the most similar patterns in the top three ranks. When scores slightly lower than 1.0 are obtained, the top three similar patterns are still selected by our system but in an order different than the ideal one. Overall, the NDCG@3 values show that the similarity measure can effectively retrieve the similar patterns of a given pattern.

Efficiency. We present the computation times of our algorithm $(k = 3)$ for each query of Table 1 in Fig. 7. As one can see, most of the execution times are dominated by the pattern generation process while the time necessary for diversification is not significant since our technique of dealing with patterns allows us to reduce the result space and consequently the diversification time.

6   Related Work

Search result diversification attracted a lot of attention in both Information Retrieval and Recommendation Systems. The need for diversification raised from the imprecision of keyword queries in expressing the user intent. In general, the diversification problem has been defined as selecting a subset of the result set such that the diversity among these results within the subset is maximized [12,14]. In one of the earliest works, Carbonell and Goldstein [8] described the concept of maximal marginal relevance (MMR), a trade-off between relevance and novelty. Gollapudi and Sharma [14] gave an axiomatic approach for result diversification. Aksoy et al. [2] and Liu et al. [18] addressed the problem of query disambiguation by clustering the search results. Drosou and Pitoura [12] review different definitions of diversity, and examine the algorithms and evaluation metrics. They categorize diversity definitions as content-based [23], novelty-based [10,24] and coverage-based [1]. Diversification is also an attempt to solve the over-specialization problem [12,22,23] where a highly homogenous set of results is returned to the user due to relevance-based ranking and/or personalization. Several algorithms have been proposed for the problem of result diversification [5,12].

There is a limited amount of works on diversification of search results on databases. Demidova et al. [11] proposed a technique to diversify keyword search results on structured databases. Li et al. [16] addressed the ambiguity of XML keyword queries by expanding the keyword queries with additional keywords to narrow down their result set. Liu et al. [19] introduced an approach to differentiate search results which can be used as a basis for diversification. Hasan et al. [15] introduced an extension of tree edit distance for diversification of XML search results for structured queries. In this work, we directly diversify the results of keyword search on tree databases by specifying diversification as a trade off between relevance and diversity.

7   Conclusion

We have addressed the problem of balancing the relevance and diversity of the results in the answer of keyword queries on tree data. A diverse result set helps the users to view different aspects of the results. To cope with the possibly large number of results, we formally defined this problem as an optimization problem for sets of result patterns. Result patterns cluster together results of the same type. We introduced measures for the relevance and the diversity of result sets which take into account both structural and semantic information of the result patterns. Since the diversification problem is hard, we presented a heuristic algorithm which incrementally builds a relevant and diverse result set in a greedy manner. We conducted experiments to measure the quality of the proposed diversification scheme and the efficiency of the heuristic algorithm.

References

1. Agrawal, R., Gollapudi, S., Halverson, A., Ieong, S.: Diversifying search results. In: WSDM, pp. 5–14 (2009)
2. Aksoy, C., Dass, A., Theodoratos, D., Wu, X.: Clustering query results to support keyword search on tree data. In: Li, F., Li, G., Hwang, S., Yao, B., Zhang, Z. (eds.) WAIM 2014. LNCS, vol. 8485, pp. 213–224. Springer, Heidelberg (2014)
3. Aksoy, C., Dimitriou, A., Theodoratos, D.: Reasoning with patterns to effectively answer XML keyword queries. VLDB J. **24**(3), 441–465 (2015)

4. Aksoy, C., Dimitriou, A., Theodoratos, D., Wu, X.: XReason: A semantic approach that reasons with patterns to answer XML keyword queries. In: DASFAA, pp. 299–314 (2013)
5. Angel, A., Koudas, N.: Efficient diversity-aware search. In: SIGMOD, pp. 781–792 (2011)
6. Bao, Z., Ling, T.W., Chen, B., Lu, J.: Effective XML keyword search with relevance oriented ranking. In: ICDE, pp. 517–528 (2009)
7. Bao, Z., Lu, J., Ling, T.W., Chen, B.: Towards an effective XML keyword search. IEEE Trans. Knowl. Data Eng. **22**(8), 1077–1092 (2010)
8. Carbonell, J., Goldstein, J.: The use of MMR, diversity-based reranking for reordering documents and producing summaries. In: SIGIR, pp. 335–336 (1998)
9. Carterette, B.: An analysis of NP-completeness in novelty and diversity ranking. Inf. Retr. **14**(1), 89–106 (2011)
10. Clarke, C.L., Kolla, M., Cormack, G.V., Vechtomova, O., Ashkan, A., Büttcher, S., MacKinnon, I.: Novelty and diversity in information retrieval evaluation. In: SIGIR, pp. 659–666 (2008)
11. Demidova, E., Fankhauser, P., Zhou, X., Nejdl, W.: DivQ: Diversification for keyword search over structured databases. In: SIGIR, pp. 331–338 (2010)
12. Drosou, M., Pitoura, E.: Search result diversification. SIGMOD Rec. **39**(1), 41–47 (2010)
13. Erkut, E., Ulkusal, Y., Yenicerioglu, O.: A comparison of p-dispersion heuristics. Comput. Oper. Res. **21**(10), 1103–1113 (1994)
14. Gollapudi, S., Sharma, A.: An axiomatic approach for result diversification. In: WWW, pp. 381–390 (2009)
15. Hasan, M., Mueen, A., Tsotras, V., Keogh, E.: Diversifying query results on semi-structured data. In: CIKM, pp. 2099–2103 (2012)
16. Li, J., Liu, C., Yu, J.: Context-based diversification for keyword queries over XML data. IEEE Trans. Knowl. Data Eng. **27**(3), 660–672 (2015)
17. Li, J., Liu, C., Zhou, R., Wang, W.: Suggestion of promising result types for XML keyword search. In: EDBT, pp. 561–572 (2010)
18. Liu, Z., Natarajan, S., Chen, Y.: Query expansion based on clustered results. Proc. VLDB Endow. **4**(6), 350–361 (2011)
19. Liu, Z., Sun, P., Chen, Y.: Structured search result differentiation. PVLDB **2**(1), 313–324 (2009)
20. Radlinski, F., Dumais, S.: Improving personalized web search using result diversification. In: SIGIR, pp. 691–692 (2006)
21. Salton, G., Buckley, C.: Term-weighting approaches in automatic text retrieval. Inf. Process. Manage. **24**(5), 513–523 (1988)
22. Yu, C., Lakshmanan, L., Amer-Yahia, S.: Recommendation diversification using explanations. In: ICDE, pp. 1299–1302 (2009)
23. Zhang, M., Hurley, N.: Avoiding monotony: Improving the diversity of recommendation lists. In: RecSys, pp. 123–130 (2008)
24. Zhang, Y., Callan, J., Minka, T.: Novelty and redundancy detection in adaptive filtering. In: SIGIR, pp. 81–88 (2002)

Efficient Approximate Substring Matching in Compressed String

Yutong Han$^{(\boxtimes)}$, Bin Wang, and Xiaochun Yang

School of Computer Science and Engineering, Northeastern University, Shenyang
110819, Liaoning, China
han_shuangxue@163.com, {binwang,yangxc}@mail.neu.edu.cn

Abstract. The idea of LZ77 self-index has been proposed for repetitive
text in compressed forms. Existing methods of approximate string matching based on LZ77 focus on space efficiency. We focus on how to efficiently
search similar strings in text without decompressing the whole text. We
propose RS-search algorithm to merge all the occurrences of substring
efficiently to narrow down the potential region and design novel filterings
to reduce the scale of candidates. The experiments show that our algorithm achieves outstanding performance and an interesting time-space
trade-off in approximate matching for compressed string.

Keywords: LZ77 · Self-index · Approximate string matching · Edit
distance

1 Introduction

To deal with string matching problem for massive strings, LZ text compression
is a popular approach to index original text, which only stores the differences to
previous string and can refer to any content of the original text. Self-index is an
index structure used to solve matching problems on compressed text instead of
original text.

Since approximate string matching is a fascinating problem in many applications and requires high performance in practice, existing works based on LZ
compressed representation concentrate on approximate pattern matching problems, and most of them focus on space consumption or theoretical analysis.
There are a large amount of works following pigeon-hole principle [1] that hunt
for signatures in the uncompressed text. However, for compressed text original
characters are invisible because of the representation transformation. Displaying
all the areas around signatures is not realistic in practice. Moreover, a large size
of signatures will lead to a large number of candidates to be verified.

The work is partially supported by the NSF of China for Outstanding Young Scholars
(No. 61322208), the NSF of China (Nos. 61272178, 61572122), and the NSF of China
for Key Program (No. 61532021).

B. Cui et al. (Eds.): WAIM 2016, Part II, LNCS 9659, pp. 184–197, 2016.
DOI: 10.1007/978-3-319-39958-4_15

In this paper, we adapt pigeon-hole principle to LZ compressed text and propose a novel approximate substring matching method. The main contributions of this paper are as follows. (1) We provide Reverse Suffix LZ77 (RSLZ77) with self-index, and propose an efficient algorithm to acquire signatures in the pattern. (2) We propose a merging algorithm to integrate the signatures into similar strings. (3) Combining the property of LZ77 parsing with the composition of RSLZ77 self-index, we design novel filterings to reduce the scale of candidates. The ultimate goal of our algorithm is to balance the cost between exact substring matching and verification of potential region to avoid decompressing the whole string.

Paper Outline: Sect. 2 reviews works related to approximate string matching with compression representation. Section 3 briefly introduces self-index based on LZ77 and exact matching strategy. Section 4 gives a variant index structure RSLZ77 self-index and improves exact pattern matching. Section 5 presents our efficient algorithm and discusses novel filterings. Section 6 shows the experiments in details.

2   Related Work

Approximate string matching methods can be grouped into many categories depending on the string representation.

The simple and direct way is processing original text. There are many methods in which index structures are designed for long strings. The general idea of this method is based on an extension of pigeon-hole principle [2,3]. When the pattern is split into j pieces with the threshold k, at least k/j errors would occur in more than one pieces. Hence, to balance the cost between searching suffix tree and verifying candidates, the method searches every piece of the pattern through suffix array and integrate the occurrences according to a hybrid index scheme. For collections of sequences, [4] proposed DivideSkip that improve merging process of inverted lists which describe corresponding q-grams and integrate proposed filterings. For compressing sequences with references, Sebastian Wandelt and Ulf Leser proposed index structure RCSI [5] and a variant MRCSI [6] using multiple references to compress genomes. [7] described a novel index structures to access compressed sequences. After appending a large amount of bits to q-gram inverted list, they use several techniques to reduce the size of inverted index. Self-index based on compression method is another direction. Navarro present the first self-index based on LZ77 or LZ-End, where LZ-End is a variant LZ compression scheme that can efficiently access text at random in [8]. Such as [9–13], these works studied on approximate matching. [10] proposed a hybrid compressed indexing method on LZ77 self-index, which could be considered as extending for approximate q-sample index. They also described a hierarchical verification that can support pattern matching bidirectionally.

3 Preliminary

3.1 Definition

Let T be a long sequence and each character $T[i]$ belongs to Σ, where Σ is a finite alphabet set. $T[i,j]$ indicates the substring of T from the i-th character to the j-th character. Specifically, $T[1,j]$ is a prefix of T whose end is the j-th character.

Problem. In this paper we pay more attention to approximate string matching based on LZ77 compressed representation. If there is a long string T of length n, a pattern P of length m $(m \ll n)$, which is much shorter than T, and threshold k, what we aim to do is to locate all the substrings whose edit distances [14] compared to pattern P are no larger than threshold k.

The LZ77 parsing of text $T[1,n]$ is a sequence $Z[1,n']$ of phrases such that $T = Z[1]Z[2]...Z[n']$, built as in [8]. Assume we have already processed $T[1,i-1]$ producing the sequence $Z[1,p-1]$. Then, we find the longest prefix $T[i,i'-1]$ of $T[i,n]$ which occurs in $T[1,i-1]$, set $Z[p] = T[i,i']$ and continue with $i = i'+1$. The occurrence in $T[i,i'-1]$ of the prefix $T[i,i']$ is called the source of the phrase $Z[p]$.

Example 1. Given a sequence $T = $ abdacadbedabbedacbdacad. Figure 1 shows an example of LZ77 parsing. We give an identifier on the top of every phrase. There is no prefix of $T[3]$ in $T[1,2]$, so we parse $Z[3] = T[3]$. $T[13,16]$ is the longest prefix generated from $T[8,11]$ so that the 8th phrase is $Z[8] = T[13,17]$.

3.2 LZ77 Self-Index

LZ77 self-index structure was build up based on LZ77 parsing. Figure 2 shows an example for LZ77 self-index. The index structure consists of two tries and a range structure. A suffix trie on the top of Fig. 2 indexes all the suffixes starting from phrases. On the left of Fig. 2 is the reverse trie. Each leaf node of both tries stores the identifier of a phrase. The range structure connects the point between adjacent phrases in the grid.

According to the definition of LZ77 parsing, we split a text T of length n into n' phrases such that $T = Z[1]Z[2]\dots Z[n']$. Given a pattern P, there are three types of exact occurrences in the LZ77 parsing.

Example 2. In Fig. 1 ab spanning the first two phrases is a *primary occurrence*. ab appearing as the suffix of the seventh phrase is a *special primary occurrence*. The substring dac beginning at position 19 in the last phrase is regarded as a *second occurrence*.

1	2	3	4	5	6	7	8	9
a	b	d	ac	ad	be	dab	bedac	bdacad

Fig. 1. Example for LZ77 parsing

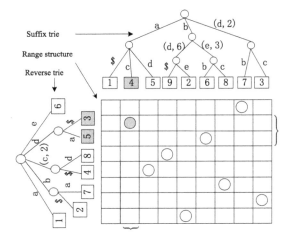

Fig. 2. LZ77 self-index for string $T =$ abdacadbedabbedacbdacad

Navarro's algorithm emphasized on how to locate the precise position of the pattern in the string. Firstly, the algorithm divides the pattern $P[1, m]$ into two pieces for every i from 1 to m for $P = P_l$ and P_r. Secondly, the algorithm searches each part of substring from the suffix trie and reverse trie separately. A range structure is built to concatenate id (the identifier of the phrases) to $revid$ (the identifier of the reversed phrases) pertained to their respective region.

Example 3. For an exact match example, we match pattern $P =$ dac. P is divided into two parts $P_l =$ d and $P_r =$ ac. On the top of Fig. 2, the gray leaf node in the suffix trie implies the phrase begins with right part of the pattern $P_r =$ ac. In the left there is a reverse trie, where the gray leaf node implies the phrase ends with left part $P_l =$ d. We could get all the results in crossing area between interval $[2, 3]$ and $[2, 2]$, such that $id = revid + 1 = 2 + 1 = 3$, and there exists a primary occurrence started in phrase 2. Then we process the other partitions in the same way to cover all the primary occurrences.

Before looking for similar substrings in the long sequence, we need drop anchors depending on exact matching. In LZ77 self-index, time complexity of exact matching depends not only on pattern length and text scale, but also the number of occurrences. As a matter of fact, the experiments show that exact matching mainly depends on snowballed localization of the second occurrences. It is a choke point in approximate matching.

4   Exact Substring Matching

In this section, we propose an optimizing index RSLZ77 and demonstrate an efficient algorithm to solve exact substring matching. Due to the properties of our index, the locating time of second occurrences can be significantly reduced.

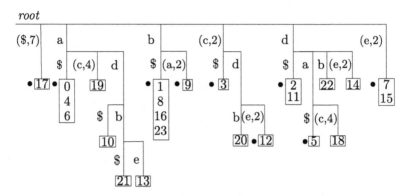

Fig. 3. Reverse suffix trie of RSLZ77 self-index for string T

4.1 RSLZ77 Index

We propose Reverse Suffix LZ77 Index (RSLZ77), which integrates all the suffixes of the reverse phrases instead of reverse trie in LZ77 self-index. Intuitively, we compute all matches for second occurrences going through the *reverse suffix trie* that has stored all the suffixes of one phrase. Such as in Fig. 3, the number of the leaf nodes *rsid* in the rectangle represents the identifiers of all the suffixes of the reverse phrases. The black dot called *phrase node* marks the *rsid* started from end position of every phrase. We can get a permutation of *rsid* after a depth-first traversal of reverse suffix trie. We build a bitmap R_S associated with the permutation, in which symbol 1 and symbol 0 correspond to phrase node and leaf node respectively.

4.2 Locating Occurrences

As the suffix trie and range structure, we take use of Navarro's pattern match strategy mentioned above for primary occurrences. We append substring of phrases into reverse suffix trie. We match the pattern from right to left and traverse the reverse suffix trie from the root irrespective of many skip characters on the path. We can extract an interval, in which all the results of second occurrences and special primary occurrences are involved. It is necessary to check one of the occurrences in the region compared to pattern P to make sure that all of the other results are the valid. RSLZ77 index can weaken the influence of occurrence frequency and speed up locating time. There is a relationship between *rsid* and *revid*.

- Range connection: Suppose the i-th element in the permutation is denoted as $rsid_i$, and it is commensurate with $revid_j$ which is the j-th element in the array of $revids$. Then $j = rank_1 (R_S,\ i)$[1].
- Phrase $rsid$ to text position: Suppose the i-th element in the permutation is denoted as $rsid_i$. If the value at position $rsid_i$ in the bitmap B is 0, $revid_i$ can be obtained via Formula (1). Then an occurrence terminates at text position $pos = 2 * mid - rsid_i$, and $mid = end - \lfloor length/2 \rfloor$ is the median of $rsid_i$ from one phrase, where end and $length$ stand for end position and length of $rsid_i$ respectively.

$$revid_i = \begin{cases} rank_1 (B,\ rsid_i) & B[rsid_i] = 0, \\ rank_1 (B,\ rsid_i) - 1 & B[rsid_i] = 1. \end{cases} \tag{1}$$

Example 4. Reexamine Example 3, we match pattern $P = $ dac. An interval contains leaf node 20 and 12. Since the value at position 20 in bitmap B is 0, $revid = rank_1 (B,\ 20) = 8$. As the median mid of the 8th phrase is equal to 20, the end position corresponding to leaf node is $pos = 2 * mid - rsid = 20$. After checking the substring ended up at position 20 is a match, the same procedure is carried out for phrase node 12 ended at position 16.

5   Approximate Substring Matching

In this section, we describe how to do approximate substring matching RS-search without decompressing all the context around anchors. We also provide novel filterings based on RSLZ77 properties to reduce the candidate size.

Lemma 1. *Let A and O be strings such that $ed(A, B) \leqslant k$. Let $A = A_0 A_1 \ldots A_j$, for $j = k + s$, $s > 1$. Then there is at least s substrings O' of O and $ed(A_i, O') = 0$.*

Example 5. Given a threshold $k = 1$ and $s = 2$, pattern $A = $ abcdef is divided into $j = k + s = 3$ pieces, $A_0 = $ab, $A_1 = $bc and $A_2 = $ef. If there is an occurrence $O = $abcaef with $ed(A, O) = 1$, we can conclude that at least two substrings in occurrence O satisfy $ed(A_i, O') = 0$, $ed(A_0, ab) = 0$ and $ed(A_2, ef) = 0$.

5.1   Basic Algorithm

The main idea in our method depends on Lemma 1. We divide the pattern into $j = k + s$ non-overlapping pieces to locate candidates with high accuracy. RSLZ77 can achieve an occurrence list for each piece. According to combinational logic, we choose s positions from $j = k + s$ different lists of substring for universally composing occurrences randomly.

Figure 4 shows an example. As we match pattern $P = $ bdacad with threshold $k=1$ and $s=2$, the pattern is split into three pieces bd, ac, and ad. All the

---

[1] $rank_b(B,\ i)$ is the number of occurrences of bit b in $B_{1,\ i}$.

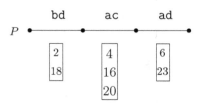

Fig. 4. Example of basic algorithm

occurrences occ correspond to pattern pieces in three lists. A candidate set consists of $CS_s^k (T, P) = \{(2, 4), (2, 16), (2, 20), \ldots, (20, 23)\}$, in which a potential region $\Delta_C = (2, 4)$, where 2 and 4 denote start position of start and end phrases respectively. There are at most $\binom{k+s}{s} d^s$ candidates to be verified, where d is length of the longest list. However, the permutation requires so much time consumption. Therefore, we develop RS-search algorithm to merge exact substring occurrences.

5.2 An Algorithm RS-Search for Merging Positions

We maintain a heap to integrate lists of pieces by their occ into one sorted alignment denoted as L. Each element in $L = l_1 l_2 \ldots$ is represented by a binary group $l_i = \{occ, id\}$, where id indicates the label of a piece in the pattern P. We inspect the alignment L to hunt for potential regions.

Property 1. Let A be a sequence and B be a pattern. For $B = B_0 B_1, \ldots, B_{k+s-1}$ in Fig. 5, assume $A[a, b]$ is one of the occurrences of piece B_i in sequence A, denoted as $B_i = A[a, b]$. If $B_{i+1} = A[c, d]$ and $c < a$, $d < a$, one possibility is that either $A[a, b]$ or $A[c, d]$ would be inserted compared to pattern B, and the other is $A[a, b]$ or $A[c, d]$ could not be included in the same potential region.

According to Property. 1, we decide whether put l_j into Δ_C. Let id_i denote the id of l_i, occ_i denote the occ of l_i, $id_i.length$ denote the length of i-th piece in the pattern, and l_j the next elements for $j = i + 1$. As in Algorithm 1, there are three cases:

Case 1. If id of element $id_j = id_i + 1$, and $occ_i + id_i.length = occ_j$, we recognize l_j as part of C_s^k and immediately insert l_j into the potential region (Line 6).

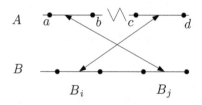

Fig. 5. Position relationship

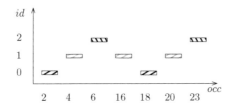

Fig. 6. OccBacktrack algorithm

Case 2. If *id* of element $id_j < id_i + 1$ and l_{j+1} is the element behind l_j, we ignore element l_{j+1} and concentrate on l_{j+1} (Line 13).

Case 3. If *id* of element $id_j > id_i + 1$, we get two branches:

If l_j is part of Δ_C, we check whether the next element l_{j+1} and following elements can satisfy bound s in the same way. There is a potential region Δ_C (Line 9).

Otherwise, we check the next element l_{j+1}. In the next period, we concentrate on the relationship between id_{j+1} and $id_i + 1$, and then restart (Line 12).

Example 6. Reexamine Example 4, we maintain all the occurrences of pieces in the alignment in ascending order for $L = l_1\, l_2\, \ldots\, l_7$, where $l_1 = (2,\, 0)$, $l_2 = (4,\, 1)$ and so on. In Fig. 6, *occ* is displayed on the horizontal axis and $id = 0$ in short segment on the vertical. We take $l_1 = (2,\, 0)$ into account and $id_2 = sid_1 + 1 = 0 + 1 = 1$, $occ_2 = occ_1 + id_1.length = 2 + 2 = 4$, $pnum = 2$. Since $pnum = s = 2$, there is a potential region starting from l_1. We can skip l_2 and l_3 based on continuous *id*. Then we consider remaining elements in L and achieve $CS_s^k = \{(2,\, 6), (18,\, 23)\}$.

5.3   Fliterings in RS-search

During *OccBacktrack* procedure, a large number of potential regions are generated. Do all of them need to be verified? Reminding of LZ77 compress scheme, occurrences link each other. Therefore we discover approaches to prune false positives. We provide two novel filterings (Location filtering and Duplication filtering) combined with length filtering [4].

Length Filtering. Suppose string A and pattern $B = B_0\, B_1\, \ldots\, B_{k+s-1}$ with $ed(A,\, B) \leqslant k$. There are two adjacent exact matchings $B_i = A\,[a, b]$ and $B_j = A\,[c, d]$ in L, and the offset between B_i and B_j is denoted as δ_i. If $|(c-b) - \delta_i| > k$, $A[a, b]$ and $A[c, d]$ could not belong to the same potential region.

Theorem 1. *Let B be a pattern, O be an occurrence, for $B = B_0 B_1,\ \ldots,$ B_{k+s-1} and $ed(B, O) \leqslant k$. Suppose O' is the first segment matched with B_j, which implies that no exact substring matching come from O and B_q for $q < j$. The edit distance between B and O must meet $ed(B, O) \leqslant k - j$.*

Proof. In an occurrence O of pattern $B = B_0B_1, \ldots, B_{k+s-1}$, O' is the first segment for $ed(B_j, O') = 0$. We can not concentrate on a substring in $B_0, B_1, \ldots, B_j$ precisely. Therefore for every piece in $B_0, B_1, \ldots, B_j$, at least one edit operation take place in the current occurrence. There are at least j mismatches in O. The upper bound of edit distance declines to $ed(B, O) \leqslant k - j$.

Algorithm 1. OccBacktrack

Input: The alignment of list sorted by occurrence position L; the threshold of edit distance k; the lower bound of exact substring s; position in L is to verify *start*, current position in L *current*, position in L is added into s at present *eff*, piece *id* is added into s at present *currentId*, the counter of current exact substring matched *pnum*

Output: The candidate set $CS_s^k(T, P)$;

1 **if** $pnum = s$ **then**
2   Verify substring start from $l_{start}.occ - offset$;

3 **while** *next* in L && *dif* $\|$ $l_{next}.id < currentId$ **do**
4     $next = next + 1$;
5     $dif = l_{next}.occ - l_{eff}.occ$;

6 **if** $l_{next}.id = currentId + 1$ **then**
7   **return** $OccBackTrace(L, mk, s, start, next, next, currentId + 1, pnum + 1)$;

8 **if** $l_{next}.id > currentId + 1$ **then**
9   **if** $OccBackTrace(L, mk, s, start, next, eff, currentId, pnum)$ **then**
10   Add Δ_C started from L into $CS_s^k(T, P)$;
11   **return** $CS_s^k(T, P)$;

12   **return** $OccBackTrace(L, mk, s, start, next, next, l_{next}.id, pnum + 1)$;

13 **return** $CS_s^k(T, P)$;

Location Filtering. According to Theorem 1, if $id > k$, $l_i = (occ, id)$ could not be regarded as the start position of a potential region. For instance, assume $l_1 = (8, 3)$ and edit distance threshold $k = 2$. Since $id > k$, there are at least one character mismatch compared to each pieces B_0, B_1, and B_2. Therefore it is not necessary to consider l_1 as the beginning of exact matching (Fig. 7).

Duplication Filtering. In RSLZ77, the concept of second occurrences inspires that the primary occurrence contains fractional substring information of the duplicated second occurrence. In the process of occurrence integration, there are instinct relationships among these occurrences. We supplement two dimensions $(socc, sid)$ into alignment L, denoted as SL with elements $l' = (occ, id, socc, sid)$. If the current element is a primary or special occurrence, $socc = occ$. sid represents the phrase target of current occurrence. In order to locate second occurrences corresponding to the primary in the sorted sequence, we construct an inverted list to store the absolute locations of second occurrences in SL.

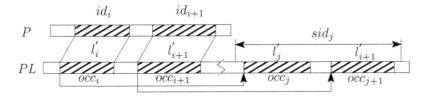

Fig. 7. Example of duplication filtering

SL	0	1	2	3	4	5	6
id	0	1	2	1	0	1	2
occ	2	4	6	16	18	20	22
$socc$	2	4	6	16	2	4	6
sid	2	4	6	8	9	9	9

Inverted list

$2 \longrightarrow 4$

$4 \longrightarrow 5$

$6 \longrightarrow 6$

Fig. 8. Example of SL structure

Example 7. Figure 8 describes SL structure with $P = $ bdacad, threshold $k = 1$, and $s = 2$. The inverted list on the right indexes all locations of second occurrences generated from the source element. As OccBacktrack algorithm, we can make sure there is a potential region starting from SL_0 and passing through verification. From the inverted list, SL_4, SL_5, and SL_6 are the second occurrences generated from SL_0, SL_1, and SL_2 respectively. Moreover, SL_4, SL_5, and SL_6 come from the same phrase as $sid = 9$. Therefore, we regard SL_4 as the start of a potential region and verify immediately.

As introduced above, length filtering and location filtering are applied to the stage of next element processing. Duplication filtering is useful for the skipping in the list of all occurrences when we have already found a potential region, vice versa.

6   Experiment

We evaluated our algorithm on different real datasets DNA, Chrome, Einstein, Kernel and Leaders. All of these datasets were downloaded from Pizza &Chili Corpus[2] which offers publicly available implementations of compressed indexes.

Our experiments were implemented in Intel Core CPU running at 3.40 GHz with 1TB main memory. The compiler was GCC version 4.8.2 in Ubuntu operating system. We evaluated our index on multiple datasets. Table 1 shows the ratio between index size and original text. LZ77 and LZEnd are self-index proposed by Kreft in [8]. The RS-B is built up by binary search instead of storing the reverse suffix tire. For DNA sequences, our index size almost needs 3 times space than LZ77 self-index, because we appended reverse suffixes of phrase into

---

[2] http://pizzachili.dcc.uchile.cl/index.html.

Table 1. Ratio between index and original text

Dataset	LZ77	LZEnd	RSLZ77	RS-B	RLCSA	GM
DNA	1.632	2.178	10.761	3.932	0.01562	3.632
Chrome	1.442	1.963	10.561	3.781	0.01562	3.473
Kernel	1.705	2.353	9.939	3.966	0.964	3.959
Leader	0.7604	1.058	10.291	3.232	0.997	1.726

our index. RLCSA and GenomeMapper (GM) are another compression schemes mentioned in [7,15].

6.1   Performance of Exact Substring Matching

Figure 9 shows the performance of proposed self-index in exact substring matching over Kernel, Leaders, and Chrome. Although we made use of binary search over reverse suffix tree in RS-B, the performance in query time of RSLZ77 and RS-B were close in practice. The function of frequency in our index was weakened. The result for Kernel in Fig. 9(a) provided an interesting insight in terms of heterogeneity. As is shown that the length of pattern at $m = 6$ reached a peak, LZ77 spent 150ms on locating 19885 occurrences, and RSLZ77 took 25ms on average. Our index needs much shorter time for substring with high frequency as shown in Fig. 9(b) and (c).

We evaluated our algorithm on different sizes of datasets in Fig. 10. With the increasing size of Einstein from 2MB to 10MB, the query time was linear to size of dataset in Fig. 10(a). Moreover the curve of pattern length $m = 4$ is higher than $m = 2$ and $m = 3$ in Fig. 10(c). This is associated with the frequency of occurrences.

6.2   Performance of Approximate Pattern Matching

Parameter Selection. We split pattern into $k + s$ pieces where s is the number of exact substring matching. According to our observation, the choices of parameter s had an effect on the searching performance. Figure 11(a) shows the

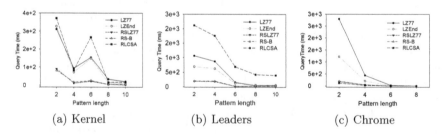

(a) Kernel   (b) Leaders   (c) Chrome

Fig. 9. Performance of exact substring matching in self-index

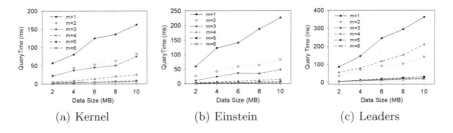

Fig. 10. Exact substring matching in different sizes of dataset

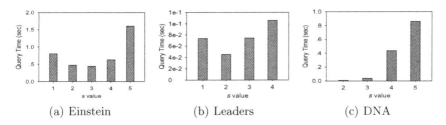

Fig. 11. s for different datasets

query time of our method in Einstein with threshold $k = 1$ and $m = 15$ for different s. The performance of $s = 3$ was much better than the others. It is because that a small s involved a larger candidates to be verified and a large s led to high time cost in exact substring match. Therefore we need to choose an appropriate parameter s. We tuned the parameter through experiments for our algorithm. s is bound up with m and k. As shown in Fig. 11(b) and (c), we have chosen $s = 2$ and $s = 1$ for dataset Leaders and DNA respectively, when the error thresholds were $k = 1$.

Pruning Power. Pruning power can be ascertained by the percentage of false positive candidates which have been overlooked in query processing. The results were shown in Fig. 12. Figure 12(a) tuned the error threshold k from 5 to 10. The performance of our filterings achieve more than 81 % in Einstein. Once we found an occurrence, Duplication filtering can skip the candidates snowballed. Figure 12(b) shown that 90 % false positives can be bypassed in Leaders sequences. In Fig. 12(c), our filterings are scalable with the increase of data size.

Running Time. We evaluated the performance of RS-search compared to state-of-the-art GM and the basic method LZ-search which located substring based on LZ77 self-index without Duplication filtering. Figure 13 shown the query time of these two approaches. When the query length increased from 12 to 20, RS-search needed half time of LZ-search to locate occurrences in the Leaders in Fig. 13(a). In Fig. 13(b), the GM and LZ-search got even worse because of homogeneous distribution of DNA sequences.

(a) Einstein   (b) Leaders   (c) Size of Dataset

Fig. 12. Pruning power

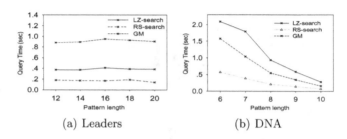

(a) Leaders   (b) DNA

Fig. 13. Running time

7   Conclusion

In conclusion, we focus on approximate pattern matching based on compressed representation. First, we developed a variant self-index RSLZ77 to support exact substring matching efficiently. Second, we proposed a new strategy RS-search for approximate pattern matching and presented novel filterings to optimize time-space efficiency. We evaluated our algorithm on five real datasets and the experiments show the good performance of our algorithm.

References

1. Qin, J., Wang, W., Xiao, C., Lu, Y., Lin, X., Wang, H.: Asymmetric signature schemes for efficient exact edit similarity query processing. ACM Trans. Database Syst. (TODS) **38**(3), 16 (2013)
2. Navarro, G., Baeza-Yates, R.: A hybrid indexing method for approximate string matching. J. Discrete Algorithms **1**(1), 205–239 (2000)
3. Deng, D., Li, G., Feng, J.: A pivotal prefix based filtering algorithm for string similarity search. In: ACM Sigmod International Conference on Management of Data, pp. 673–684 (2014)
4. Li, C., Lu, J., Lu, Y.: Efficient merging and filtering algorithms for approximate string searches. In: IEEE 24th International Conference on Data Engineering, ICDE 2008, pp. 257–266 (2008)
5. Wandelt, S., Starlinger, J., Bux, M., Leser, U.: RCSI: scalable similarity search in thousand(s) of genomes. Proc. VLDB Endow. **6**(13), 1534–1545 (2013)

6. Wandelt, S., Leser, U.: MRCSI: compressing and searching string collections with multiple references. PVLDB **8**(5), 461–472 (2015)

7. Yang, X., Wang, B., Li, C., Wang, J.: Efficient direct search on compressed genomic data. In: 2013 IEEE 29th International Conference on Data Engineering (ICDE), pp. 961–972 (2013)

8. Kreft, S., Navarro, G.: Self-indexing based on LZ77. In: Giancarlo, R., Manzini, G. (eds.) CPM 2011. LNCS, vol. 6661, pp. 41–54. Springer, Heidelberg (2011)

9. Gagie, T., Gawrychowski, P., Puglisi, S.J.: Approximate pattern matching in LZ77-compressed texts. J. Discrete Algorithms **32**, 64–68 (2014)

10. Russo, L.M.S., Navarro, G., Oliveira, A.L., Morales, P.: Approximate string matching with compressed indexes. Algorithms **2**(3), 1105–1136 (2009)

11. Bille, P., Fagerberg, R., Li Gørtz, I.: Improved approximate string matching and regular expression matching on Ziv-Lempel compressed texts. In: Ma, B., Zhang, K. (eds.) CPM 2007. LNCS, vol. 4580, pp. 52–62. Springer, Heidelberg (2007)

12. Navarro, G., Baeza-Yates, R., Sutinen, E., Tarhio, J.: Indexing methods for approximate string matching. IEEE Data Eng. Bull. **24**(85), 19–27 (2001)

13. Russo, L.M.S., Navarro, G., Oliveira, A.L.: Approximate string matching with Lempel-Ziv compressed indexes. In: Ziviani, N., Baeza-Yates, R. (eds.) SPIRE 2007. LNCS, vol. 4726, pp. 264–275. Springer, Heidelberg (2007)

14. Levenstein, V.: Binary codes capable of correcting spurious insertions and deletions of ones. Probl. Inf. Transm. **1**(1), 8–17 (1965)

15. Schneeberger, K., Hagmann, J., Ossowski, S., Warthmann, N., Gesing, S., Kohlbacher, O., Weigel, D.: Simultaneous alignment of short reads against multiple genomes. Genome Biol. **10**(9), R98 (2009)

Top-K Similarity Search for Query-By-Humming

Peipei Wang$^{(\boxtimes)}$, Bin Wang, and Shiying Luo

School of Computer Science and Engineering,
Northeastern University, Shenyang, China
wangpeipei@research.neu.edu.cn, binwang@mail.neu.edu.cn,
neulsy@hotmail.com

Abstract. As an important way of music retrieval, Query-By-Humming has gained wide attention because of its effectiveness and convenience. This paper proposes a novel Top-K similarity search technique, which provides fast retrieval for Query-By-Humming. We propose a distance function $MDTW$ for multi-dimensional sequence matching as well as a subsequence matching method $MDTW_{sub}$. We show that the proposed method is highly applicable to music retrieval. In our paper, music pieces are represented by 2-dimensional time series, where each dimension holds information about the pitch or duration of each note, respectively. In order to improve the efficiency, we utilize inverted lists and q-gram technique to process music database, and utilize q-chunk technique to process hummed piece. Then, we calculate the $MDTW$ distances between hummed q-chunks and music q-grams, and we can get the candidate music and their sensitive data areas. We proposes Top$K$-Brute and Top$K$-LB Algorithm to search the Top-K songs. The experimental results demonstrate both the efficiency and effectiveness of our approach.

Keywords: Query-by-humming · Sequence matching · DTW · Top-K

1 Introduction

With the rapid expansion of the Internet and the growth of digital music market, there is an increasing demand for convenient music search anywhere and anytime. Suppose you hear a piece of music but you don't konw its name. One solution is to hum a short part of the music and search the melody in a large music repository to find the music you are looking for or even music with similar melody [1]. In this paper, we focus on subsequence matching and approach the problem from the music retrieval perspective. The main work of a Query-By-Humming (QBH) [2] is the following: given a hummed query music, search a music database for the K most similar music. This can be formalized as a subsequence matching problem as the hummed query is typically a very small part of the target sequence.

The work is partially supported by the NSF of China (Nos. 61572122, 61272178), the NSF of China for Outstanding Young Scholars (No. 61322208), and the NSF of China for Key Program (No. 61572122).

B. Cui et al. (Eds.): WAIM 2016, Part II, LNCS 9659, pp. 198–210, 2016.
DOI: 10.1007/978-3-319-39958-4_16

Given the fact that the melodies hummed by users in QBH are monophonic, we consider only monophonic music here. Every piece of music is a sequence of notes characterized by a key, which defines the standard pattern of allowed intervals that the sequence of notes should conform with, and a tempo, which regulates the speed of the piece. Each note consists of two parts: the pitch (frequency) and the duration. A pitch interval is the distance between two pitches. Pitch and duration are two distinctive features for a music piece, and they should both be used for efficient music representation [3]. We could have two or more songs that share similar pitch values, but still have noticeably different melodies due to different individual pitch durations. In this work, we take into account both pitch and duration. Hence, melodies are defined as 2-dimensional time series of notes, where one dimension represents pitch and the other duration. Data conversion work is not the focus of this paper, and we use the existing data set.

Given a set of music sequences D and a query sequence Q, a basic method of Top-K approximate subsequence matching examines every sequence X in D and compute the Multi-dimensional Subsequence Dynamic Time Warping one by one. We refer to the brute-force algorithm as basic method. Since calculating the edit distance of subsequence is computationally expensive, we further optimize the efficiency. The experimental results validate the efficiency and effectiveness of our method. This paper makes the following contributions:

(1) We propose a distance function $MDTW$ for multi-dimensional sequence matching as well as a subsequence matching method $MDTW_{sub}$. This approach applies well to QBH for music retrieval method on 2-dimensional data collection.

(2) We use inverted lists and q-gram [4] to process music database, and use q-chunks [5] to index hummed piece. Then, we calculate the $MDTW$ distances between hummed q-chunks and music q-grams. Our algorithm returns the candidate music and their corresponding data areas. Based on $MDTW_{sub}$, we propose the basic Top-K similarity search method. In order to further improve the efficiency, we propose Top$K$-LB Algorithm by calculating the lower bound of distance between hummed piece and the sensitive data areas of candidate music set.

(3) The proposed approach has been implemented. Experimental results and performance studies verify the effectiveness of our approach.

2   Related Work

In the past ten years, many organizations have studied QBH, which promoted the rapid development in this field. The framework of QBH is shown in Fig. 1. The research of QBH are focus on finding a stable feature extraction algorithm and an accurate and fast matching algorithm. The research focus of this paper is the retrieval part, as for the feature extraction part, we mainly use the previous method [3]. There are a lot of algorithms about QBH, such as edit distance [2,6], linear scaling [7], dynamic time warping [8,9], hidden Markov model [10,11], local sensitive hash [12], subsequence matching method [3,13], and so on. In 2006, the

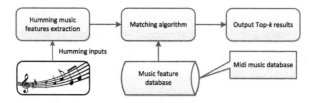

Fig. 1. The framework of QBH

international music information retrieval evaluation game was carried out for the first time, and attracted a wide range of organizations ever since. The game has played an important role in the development of music retrieval technique.

Although the research on QBH is gradually deepening, it still has a lot of room for improvement. In this paper, several factors are considered for QBH. Firstly, in a hummed song, errors may occur due to instant key or tempo loss. Thus, the matching method should be error-tolerant. Secondly, the number of consecutive mismatching position in both query and target sequences should be bounded, in order to provide a setting that controls the expansion of matched subsequences. This paper presents the technique to improve the effectiveness and efficiency of music retrieval.

3   Problem Definition

3.1   Representing Music Pieces

In this paper, we use both pitch and duration to describe a music piece, and this results in a 2-dimensional time series representation. We consider the encoding scheme (pitch interval, duration). Thus, we deal with note transitions, saving much computational time as we do not have to check for possible transpositions of a melody, nor do we have to scale in time when comparing melodies. Apart from this, this representation leads to highest accuracies for several synthetic and hummed query sets compared to all other possible 2-dimensional encoding

Id	Pitch sequence	Duration sequence
X_1	3, 5, 7, 4, 5	1, 2, 1, 1, 2
X_2	2, 4, 5, 7, 9, 2	3, 2, 4, 1, 2, 1
X_3	4, 6, 2, 3	2, 1, 2, 3

(a) Music database D

Humming	Pitch sequence	Duration sequence
Q	2, 5, 5, 6	2, 2, 4, 1

(b) Query

Fig. 2. An example of music sequence

schemes. Example for music sequence is shown in Fig. 2. X_1, X_2, X_3 are music files in database D, and Q is hummed query. They have 2-dimensional data as pitch sequence and duration sequence.

3.2   Definitions and Objective Function

In this subsection, we present relevant notations and definitions. Now, Let us give a more general problem definition. Consider the music sequence $X = (x_1, ..., x_n)$ to be a multi-dimensional time series, where n denotes the length of X. We use x_i^d to denote the d-th dimension of x_i, where $i = 1, ..., n$. In our case, where X represents a music piece, $X = (X^1, X^2)$, each $x_i = (x_i^1, x_i^2)$ is a pair of real values, where x_i^1 and x_i^2 correspond to pitch and duration respectively, as represented using the encoding of Sect. 3.1. A music database D is a set of time series: $D = \{X_1, ..., X_N\}$, where N is the number of music pieces in D. A subsequence of X, denoted as $X[s : e] = \{x_s, ..., x_e\}$. Let $Q = (q_1, ..., q_m)$ be another multi-dimensional time series as segments hummed by user with the same representation as X.

Definition 1 Dynamic Time Warping of Multi-dimensional Time Series ($MDTW$): Given two multi-dimensional time series Q and X, d represents dimensions of multi-dimensional time series, and $a_{[1...d]}$ represents their corresponding weights. Their $MDTW$ distance is defined as:

$$MDTW(Q^d, X^d) = F(m, n)$$
$$F(0,0) = 0$$
$$F(i,0) = F(0,j) = \infty$$
$$F(i,j) =$$

$$a_1 * dist^2(x_i^1, q_j^1) + ... + a_d * dist^2(x_i^d, q_j^d) + \min \begin{cases} F(i, j-1) \\ F(i-1, j) \\ F(i-1, j-1) \end{cases} \tag{1}$$

$$(i = 1, ..., m; j = 1, ..., n)$$

In Formula 1, the distance between x_i^d and q_j^d is denoted as $dist^2(x_i^d, q_j^d)$, and $dist(x_i^d, q_j^d) = |x_i^d, q_j^d|$. The time complexity for calculating the $MDTW$ distance is $O(mn)$, where m and n denote the length of two sequences, respectively. From the Formula 1, we can find that the $MDTW$ distance of 1-dimensional sequences is DTW distance.

In this paper, the music sequence Q and X are the 2-dimensional sequences. The weights of pitch sequence and duration sequence for a and b are given by user.

Definition 2 Best-match Query: Given multi-dimensional sequences X of length n and Q of length m, find the subsequence $X[s : e]$ whose $MDTW$ distance from Q is the smallest among all possible subsequences, i.e. $MDTW(X[s : e], Q) \leq MDTW(X[i : j], Q)$ for any pair of $i = 1, \ldots, n$ and $j = i, \ldots, n$.

Definition 3 Multi-dimensional Subsequence Matching: Given multi-dimensional sequences X of length n and Q of length m, Best-match query of X and Q denoted as $MDTW_{sub}(X, Q)$ in Formula 2. It contains the starting position in MSP, As well as the distance $F(i, j)$. MSP is the abbreviations of Multi-dimensional subsequence matching Starting Position. The calculation of $F(i, j)$ is the same as the Formula 1, but the initial value is different.

$$MDTW_{sub}(X, Q) = MDTW(X[s:e], Q) = \min(F(i, n))$$
$$F(i, 0) = 0; F(0, j) = \infty$$

$$MSP(i, j) = \begin{cases} MSP(i, j-1), if \ F(i, j-1) \ is \ the \ minimum \\ MSP(i-1, j), if \ F(i-1, j) \ is \ the \ minimum \\ MSP(i-1, j-1), if \ F(i-1, j-1) \ is \ the \ minimum \end{cases} \qquad (2)$$
$$MSP(i, 0) = 0; MSP(0, j) = j; (i = 1, ..., m; j = 1, ..., n)$$

Definition of Multi-dimensional subsequence matching is based on the DTW algorithm. It improves standard DTW matrix calculation by inserting a variable into each matrix element. We retain the starting position in each element of MSP. The optimal warping path is obtained using the distance computation. The starting position of the best subsequence is propagated through the matrix on the optimal warping path. The definition is different from SPRING which is aimed at 1-dimensional time series in paper [14], our work in this paper focuses on multi-dimensional subsequence matching.

Definition 4 Top-K Multi-dimensional   Approximate   Subsequence Matching: Given a set of multi-dimensional sequences D and a query sequence Q, the Top-K approximate subsequence matching problem is to find the k approximate sequences $Top_k(Q) = \{X_1, X_2, ..., X_k\}$ in D satisfying the following conditions:

(1) $MDTW_{sub}(X_1, Q) \leq MDTW_{sub}(X_2, Q) \leq ... \leq MDTW_{sub}(X_k, Q)$ holds.
(2) For every $X_j \in D$ such that $X_j \notin Top_k(Q)$, we have $MDTW_{sub}(X_k, Q) \leq MDTW_{sub}(X_j, Q)$.

4   Top-K Similarity Search with Filter and Refined Method

In this section, we propose a novel framework to implement Top-K Similarity Search for QBH. Firstly, we construct the musical fragment candidate set. This step is based on the idea of q-gram. We divide the music sequences into a number of grams, and establish a double elements inverted index. Secondly, a music candidate set is constructed to compute music candidate set. Compute the $MDTW$ distance of music sequence of alternative music grams and query chunks. Then we match the hummed segments in the database D according to the matching threshold β. If correct results are left out due to a too

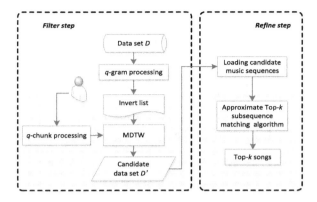

Fig. 3. A filter and refined framework

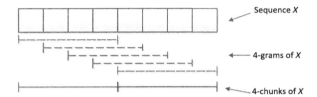

Fig. 4. The q-grams and q-chunks of sequence X, $q{=}4$

high threshold, we call it false negative. In contrast, false positive refers to the situation that wrong answers are returned because the threshold is too low. Therefore a reasonable threshold value is very important. We set threshold β to $(a * \sum Q_q^1 + b * \sum Q_q^2) * 0.7$, in which 0.7 is a reasonable value derived from the experiment. The retrieval results for music sequences are the grams whose $MDTW$ distance is no more than threshold β. By the inverted index information from the first step, it is possible to determine the position of the approximate sequence in the musical sequence, and find candidate music sets as well as the sensitive data areas of candidate music set. We use Top-K approximate subsequence matching algorithm to find out Top-K results. The framework is shown in Fig. 3.

4.1   Indexing and Query Processing

We develop an efficient filtering strategy by utilizing q-grams [4] and q-chunks [5] in the music and query sequences. In addition available inverted q-gram indexes are used to find approximate sequence matches.

   We use inverted lists and q-gram to index music database, and use q-chunk to process hummed piece. For a given music sequence X, we can slide sliding window with a length of q on the X to generate q-gram. In our case, the q-gram is recorded in the form of $(X[i,\ i+q\text{-}1],\ i)$, where $X[i,\ i+q\text{-}1]$ represents the subsequence with the length q of sequence X, where i represents the position of the subsequence $X[i,\ i+q\text{-}1]$ appearing in music sequence X.

Definition 5. The q-grams of a sequence X are $X[i,\ i+q\text{-}1]$s with all $1 \le i \le (|X| - q + 1)$. The positional q-grams of a string X are all pairs of q-gram in X and its corresponding position.

Example 1. Given a musical pitch sequence $X^1 = (2, 4, 5, 7, 9, 2, 4)$, and duration sequence $X^2 = (3, 2, 4, 1, 2, 1, 3)$, when $q = 4$, 4-grams of X are generated as:

$X = [(2, 4, 5, 7), (3, 2, 4, 1), 1];\ [(4, 5, 7, 9), (2, 4, 1, 2), 2];\ [(5, 7, 9, 2), (4, 1, 2, 1), 3];\ [(7, 9, 2, 4), (1, 2, 1, 3), 4]$

We use q-chunk to process hummed piece. The definition of q-chunk is similar to q-gram, but there is no overlapping part when dividing the fragment, as shown in Fig. 4.

Then we establish inverted index of music data in music database, which includes not only the *ID* number of the musical sequences *ID*, but also the offset of a certain gram in its music sequences. Therefore, using the inverted index of the two elements not only can determine the music sequences of similar sequence, but also find the similar sequence positions in the music sequences. Given a q-gram and a sequence containing the q-gram, a pair of its sequence *ID* and the position where the q-gram occurs in the sequence is called a posting. With a set of sequences D, the posting list of a q-gram is the list of postings for every occurrence of the q-gram in all sequences in D. Note that every posting list is primarily sorted in ascending order of sequence *ID*s and then sorted in ascending order of positions. With an inverted q-gram index of D, we can access the posting lists with q-grams as keys.

For the music database D and a given query sequence Q, the allowed *MDTW* error threshold of q-grams and q-chunks is β.

Although the music sequences in D are 2-dimensional, we make index for only one dimension instead of both. Because the index size would be too large if both pitch and duration are indexed, and efficiency might be reduced as a result. In the matching weights given to the user, we denote the pitch and duration as a and b respectively. When we use pitch sequence as indexing keywords, if $DTW(X^1, Q^1) \ge \beta/a$, the gram can be filtered out directly, and we do not need matching duration part. Because this formula is satisfied that one-dimensional data distance is greater than or equal to the threshold. If added another dimension of data, it will be more than the threshold. Otherwise, it's necessary to calculate the *MDTW* distance of duration and pitch gram, to determine whether the gram in the corresponding document meets the threshold. For the selected music files, according to the order of their selection, we arrange them in descending order.

Firstly, we calculate q-chunks of Q. According to these q-chunks, we will search the corresponding inverted list, and find out the candidate music X, where q-grams match q-chunks in Q with the threshold β. Then, the music files X need for further matching. The other music X does not satisfy the matching conditions, and thus does not need further matching. For the candidate music, we need to calculate the subsequence matching distance $MDTW_{sub}(Q, X)$ between hummed music Q and each candidate music X in the set of candidate music sequences D'.

The superiority of the q-gram and inverted index is that, when the music sequence is matched we don't need scanning all the sequences in D from beginning to end. We can find a result from the candidate set D' using the index, and verify the candidate music sequence one by one. To a large extent, the time overhead is reduced.

4.2   A Basic Method for Top-K Approximate Subsequence Matching

We first introduce the brute-force algorithm Top$K$-Brute which blindly examines every sequence X in a set of candidate sequences set D' and computes subsequence $MDTW$ distances $MDTW_{sub}(X, Q)$ one by one. We present its pseudo code in Algorithm 1.

We maintain a max-heap H storing the k sequences X' with the smallest $MDTW_{sub}(X', Q)$s which are used as the keys in the max-heap H. If the size of H is less than k, we just insert the sequence X to H. Otherwise, we check whether $MDTW_{sub}(X, Q)$ is smaller than $MDTW_{sub}(X_R, Q)$ of the sequence X_R at the root of H (i.e., whether $MDTW_{sub}(X, Q)$ is smaller than the k-th smallest subsequence $MDTW$ distance so far). If it is satisfied, we delete the sequence at the root X_R of H and insert the sequence X to H. If not, we move to the next sequence and repeat the above step until we encounter the last sequence in D. We refer the brute-force algorithm as basic method. The following task is to further optimize efficiency.

Algorithm 1. A Basic Method for Top-K Similarity Search (Top$K$-Brute)

  Input: the set of candidate music sequences D'; query sequence Q; the number
        of target songs k;
  Output: k sequences in the set of D';

1 H= an empty max-heap storing $< dist, str >$;
2 **for** *each sequence X in D'* **do**
3 | $f = MDTW_{sub}(X, Q)$;
4 | **if** $|H| < k$ **then**
5 | | $H.insert(< f, X >)$;
6 | **else if** $H.getMax().dist > f$ **then**
7 | | $H.insert(< f, X >)$;
8 | | $H.deleteMax()$;

9 Return H;

4.3 The Optimized Method with Lower Bound of Subsequence Matching

4.3.1 A Lower Bound of Subsequence Matching

In this section, we propose a lower bound function of multi-dimensional subsequence. We denote the lower bound of the $MDTW(X, Q)$ between a sequence X and a query sequence Q by $LBM(X, Q)$. LBM is computed by the distance between one sequence and the boundary of the maximum or the minimum of he other sequence, as the $MDTW$ lower bound function. As shown in Fig. 5. The LBM lower bound of distance is the difference between the part of Q beyond the minimum or maximum values of the sequence X and the minimum value or the maximum value of the sequence X. As the shadow part is shown in the Fig. 5. Then we give the definition of lower bound function of multi-dimensional subsequence as Definition 6.

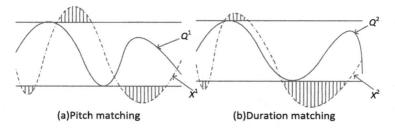

(a)Pitch matching   (b)Duration matching

Fig. 5. LBM lower bound for music sequence

Definition 6. *LBM* Lower Bound Function: given a multi-dimensional sequence $X \; =< x_1, x_2, ..., x_m >$ and $Q \; =< q_1, q_2, ..., q_n >$, and their DTW distance to the LBM lower bound of distance is shown in Formula 3:

$$LBM(X^d, Q^d) = \sqrt{\sum_{i=1}^{n} \begin{cases} (q_i{}^d - \min_{x_i \in X}(x_i{}^d))^2 & q_i{}^d < \min_{x_i \in X}(x_i{}^d) \\ (q_i{}^d - \max_{x_i \in X}(x_i{}^d))^2 & q_i{}^d > \max_{x_i \in X}(x_i{}^d) \\ 0 & others \end{cases}}$$

$$LBM(X, Q) = a_1 * LBM(X^1, Q^1) + ... + a_d * LBM(X^d, Q^d)$$

$\qquad (3)$

$LBM(X^d, Q^d)$ is the distance lower bound in one dimension, and $a_1, ..., a_d$ is the weight of each dimension. $LBM(X, Q)$ is the lower bound of X and Q.

4.3.2 Top-K Approximate Search Optimized Method

By calculating the lower bound of distance between hummed piece and the sensitive data areas of candidate music set, we quickly complete the approximate

---

Algorithm 2. Top$K$-LB Algorithm Using the *LBM* lower bound

Input: the set of candidate music sequences D'; query sequence Q; the number
  of target songs k;

Output: k sequences in the set of D';

1 H= an empty max-heap storing $< dist, str >$;
2 **for** *each sequence X in D'* **do**
3 | find X_{sub} in X with $X_{sub}.length = |Q|$;
4 | $LB = LBM(X_{sub}, Q)$;
5 | **if** $|H| < k$ *or* $LB < H.getMax().dist$ **then**
6 | | $f = MDTW_{sub}(X, Q)$;
7 | | **if** $|H| < k$ **then**
8 | | | $H.insert(< f, X >)$;
9 | | **else if** $H.getMax().dist > f$ **then**
10 | | | $H.insert(< f, X >)$;
11 | | | $H.deleteMax()$;

12 Return H;

---

retrieval of the target music. The following is a general description of the approximate matching query technique. We call the algorithm Top$K$-LB and present
its pseudo code in Algorithm 2.

We scan all sequences in D'. We can compute the lower bound *LBM*. If the
lowerbound is not smaller than the k-th smallest subsequence *MDTW* distance
so far, we do not need to calculate the actual value of $MDTWsub(X, Q)$ since X
cannot be a sequence of the Top-K approximate subsequence matches.

5 Experiments

In this section, we report our experimental results. All the algorithms were implemented with Java 1.7. Eclipse 4.2 was used as our Java IDE tool. The experiments were run on a PC with an Intel 3.10GHz Quad Core CPU i5 and 8GB
memory with a 1TB disk, running a Windows 7 operating system.

5.1 Data Sets

Database: Our music database consists of 40,891 2-dimensional time series
sequences of variable length and a total of 13,455,603 2-dimensional tuples. The
database was created based on a set of 5,643 MIDI files, freely available on the
web.

Hummed Queries: We also experimented with 100 hummed queries of lengths
in [14, 76]. Users were asked to sing to a microphone and avoid lyrics. All-
channels extraction was applied to the queries to obtain the representation of D.
The query set covered several genres, such as Classical, Blues, Jazz, and so on.
The database and queries are available online[1].

---

[1] http://vlm1.uta.edu/ akotsif/ismbgt/.

Synthetic Queries: Six synthetic query sets (100 queries per set) of lengths in [13, 137] were generated: Q_0, $Q_{.10}$, $Q_{.20}$, $Q_{.30}$, $Q_{.40}$, and $Q_{.50}$. Q_0 contained 100 exact segments of D, while $Q_{.10}$ - $Q_{.50}$ were generated by adding noise to each query in Q_0. For all queries of Q_0 we randomly modified 10 %, 20 %, 30 %, 40 %, and 50 % of their corresponding time series in both dimensions; pitch interval by $\pm z \in [3, 8]$ (simulates error performed when singing by memory and intrinsic noise that may be added by audio processing tools), duration by $\pm z \in [2, 4]$(simulates reasonable variations of duration and is outside the bounds of Eq. 4, avoiding bias in favor of SMBGT). In all query sets we allowed at most 3 consecutive elements to be modified (no insertions or deletions); in QBH we do not expect to have too many consecutive matching errors.

The weights of pitch sequence and duration sequence for a and b should be given by user. In this paper, we set default values of a and b are 0.5.

5.2   Performance Results

We evaluated the performance of Top$K$-Brute and Top$K$-LB Algorithm for the five noisy synthetic query sets and the hummed query set in terms of recall, runtime, and efficiency of filter and refined. Recall is the percent of queries for which the correct answer is in the Top-K results, while efficiency is a particularly useful measure for evaluating Top$K$-Brute and Top$K$-LB Algorithm that influences the retrieval runtime. Efficiency is the runtime evaluated during the filter and refined step by Top$K$-Brute and Top$K$-LB Algorithm.

We first evaluated Top$K$-Brute and Top$K$-LB in terms of accuracy. As illustrated in Fig. 6 we present the performance of Top$K$-Brute and Top$K$-LB in terms of recall when K values were 5, 10, 20, 50, 100, and returned results of the hummed queries and the synthetic queries, respectively. In Fig. 6(a) and (b), we divide sequences into q-grams and q-chunks, where $q = 3$. The situation for $q = 4$ is presented in Fig. 6(c) and (d). The figures illustrate what recall can be achieved when varying the number of the K. It is clear that the results are Similar for the two data sets, and high recall values increased for higner K. When $q = 4$, the accuracy is better. Top$K$-Brute is slightly better than Top$K$-LB algorithm.

In Fig. 7, the average runtime of Top$K$-Brute and Top$K$-LB Algorithm are shown for the synthetic and hummed queries. We assume $K=20$. Since query lengths are in [13, 137], we divide the set of query into five buckets. 13, 21, 33,

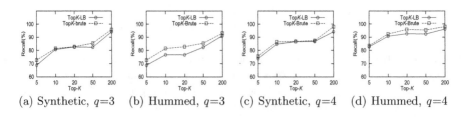

(a) Synthetic, $q=3$ (b) Hummed, $q=3$ (c) Synthetic, $q=4$ (d) Hummed, $q=4$

Fig. 6. Recall for Top$K$-Brute and Top$K$-LB

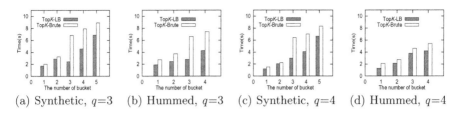

(a) Synthetic, $q=3$ (b) Hummed, $q=3$ (c) Synthetic, $q=4$ (d) Hummed, $q=4$

Fig. 7. Runtime for Top$K$-Brute and Top$K$-LB

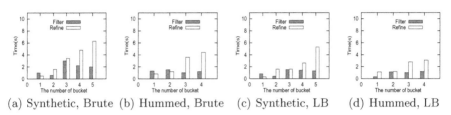

(a) Synthetic, Brute (b) Hummed, Brute (c) Synthetic, LB (d) Hummed, LB

Fig. 8. Efficiency for filter and refined step

53, and 86 is the minimum length of query in each bucket. In Fig. 7(a) and (b), we divide sequences into q-grams and q-chunks, where $q=3$, and $q=4$ in Fig. 7(c) and (d). There is no hummed query in 5th bucket, so Fig. 7(b) and (d) have less a value. With the growth of the length of the sequence, the query time is also in the corresponding growth. It is clear that the average runtime is much greater for the Top$K$-Brute than the Top$K$-LB for every bucket.

Lastly, we present the statistics for the efficiency evaluation measure Top$K$-Brute and Top$K$-LB for all queries. We assume $K = 20$ and $q = 4$. These statistics evaluated efficiency of the filter and refined step, as illustrated in Fig. 8. From the experimental results we can find that the time spent in filter is stable, and refined step essentially determines the total runtime of Top$K$-Brute and Top$K$-LB.

In summary, We have carried out experiments on Top$K$-Brute and Top$K$-LB algorithm proposed in this paper. The recall rate of the two algorithms is comparable. But for the running time, Top$K$-LB is significantly faster. Further, we have shown that refined step essentially determines the total runtime. We have tied the values of weights to data set characteristics so that they can be determined easily. So in future work, we can do more work in further optimizing the refining. In addition, because of the different coefficient settings, this paper does not compare with other people's methods. But it can be seen from the experimental results, our approach has achieved the best results in terms of recall and running time control.

6   Conclusion

Motivated by QBH we proposed an Top-K approximate subsequence matching indexing approach. In our paper, music pieces represented by 2-dimensional time series, so we propose a distance function for multi-dimensional sequence matching as well as a subsequence matching method. We use inverted lists and q-gram technique to process music database, and use Top$K$-Brute and Top$K$-LB Algorithm to search Top-K songs. The filter-and-refine method show excellent performance on the Top-K search for QBH.

References

1. Kageyama, T.: Melody retrieval with humming. In: Proceedings of the ICMC 1993, pp. 349–351 (1993)
2. Ghias, A., Logan, J., Chamberlin, D., Smith, B.C.: Query by humming: musical information retrieval in an audio database. In: Proceedings of the Third ACM International Conference on Multimedia, pp. 231–236. ACM (1995)
3. Kotsifakos, A., Karlsson, I., Papapetrou, P., Athitsos, V., Gunopulos, D.: Embedding-based subsequence matching with gaps-range-tolerances: a query-by-humming application. VLDB J. **24**(4), 1–18 (2015)
4. Sutinen, E., Tarhio, J.: On using q-gram locations in approximate string matching. In: Spirakis, P.G. (ed.) ESA 1995. LNCS, vol. 979, pp. 327–340. Springer, Heidelberg (1995)
5. Qin, J., Wang, W., Lu, Y., Xiao, C., Lin, X.: Efficient exact edit similarity query processing with the asymmetric signature scheme. In: Proceedings of the 2011 ACM SIGMOD International Conference on Management of data, pp. 1033–1044. ACM (2011)
6. Lemström, K., Ukkonen, E.: Including interval encoding into edit distance based music comparison and retrieval. In: Proceedings of the AISB, pp. 53–60 (2000)
7. Jang, J.S.R., Lee, H.R., Kao, M.Y.: Content-based music retrieval using linear scaling and branch-and-bound tree search. In: Null, p. 74. IEEE (2001)
8. Zhu, Y., Shasha, D.: Warping indexes with envelope transforms for query by humming. In: Proceedings of the 2003 ACM SIGMOD International Conference on Management of Data, pp. 181–192. ACM (2003)
9. Adams, N.H., Bartsch, M.A., Shifrin, J.B., Wakefield, G.H.: Time series alignment for music information retrieval. Ann Arbor **1001**, 48109-2110 (2004)
10. Unal, E., Chew, E., Georgiou, P.G., Narayanan, S.S.: Challenging uncertainty in query by humming systems: a fingerprinting approach. IEEE Trans. Audio Speech Lang. Processing **16**(2), 359–371 (2008)
11. Shifrin, J., Pardo, B., Meek, C., Birmingham, W.: Hmm-based musical query retrieval. In: Proceedings of the 2nd ACM/IEEE-CS Joint Conference on Digital Libraries, pp. 295–300. ACM (2002)
12. Ryynänen, M., Klapuri, A.: Query by humming of midi and audio using locality sensitive hashing. In: 2008 IEEE International Conference on Acoustics, Speech and Signal Processing, ICASSP 2008, pp. 2249–2252. IEEE (2008)
13. Kotsifakos, A., Papapetrou, P., Hollmén, J., Gunopulos, D.: A subsequence matching with gaps-range-tolerances framework: a query-by-humming application. Proc. VLDB Endowment **4**(11), 761–771 (2011)
14. Sakurai, Y., Faloutsos, C., Yamamuro, M.: Stream monitoring under the time warping distance. In: 2007 IEEE 23rd International Conference on Data Engineering, ICDE 2007, pp. 1046–1055. IEEE (2007)

Social Media

Restricted Boltzmann Machines for Retweeting Behaviours Prediction

Xiang Li[1,2](✉), Lijuan Xie[3], Yong Tan[1,2], and Qiuli Tong[4]

[1] Tsinghua National Laboratory for Information Science and Technology,
Tsinghua University, Beijing, China
{l-x14,ty15}@mails.tsinghua.edu.cn
[2] Department of Computer Science and Technology,
Tsinghua University, Beijing, China
[3] Department of Electronics Engineering and Computer Science,
Peking University, Beijing, China
xielijuan@pku.edu.cn
[4] Computer and Information Management Center,
Tsinghua University, Beijing, China
tql@tsinghua.edu.cn

Abstract. With the information explosion on social network, personalized recommendation is eagerly required to assist users to obtain interesting news or tweets within limited time. Since retweeting history reveals users personal preferences in some degree, retweeting behaviors predicting system could feed users with messages according to their probability of being retweeted. In this paper, based on the neural network model called Restricted Bolzmann Machine(RBM), we propose retweeting behaviours prediction methods adapting two scenarios: with or without detailed information of users and microblogs. When the dataset misses the detailed information, the predicting problem is treated as a collaborative filtering task and RBM plays the role of an independent classifier. The other is that RBM performs as a feature selector to detect the hidden similarity between users for a content-based model, logistic regression model(LR). Furthermore, users are clustered into different communities by our previously proposed community detection algorithm and community property is integrated into RBM to improve its performance. Experiment results indicate that features extracted by RBM can help get promotion of performance by 3.79 % in terms of F1-Score comparing with basic LR model and the community property ulteriorly improves the effectiveness of RBM.

1 Introduction

Social network, such as Sina Weibo and Twitter, is becoming an indispensable part of our daily life and plays an important role in the diffusion of information. At the same time, there comes significant quantity of junk information. Thus, mining massive valuable data contained in social network is eagerly required to assist users to gain effective information. Information transmission is based on

© Springer International Publishing Switzerland 2016
B. Cui et al. (Eds.): WAIM 2016, Part II, LNCS 9659, pp. 213–224, 2016.
DOI: 10.1007/978-3-319-39958-4_17

the relationship among users called follower-followee relation. The relationship is unidirectional in most social networks, which means that user A follows user B, user B doesn't follow back, the information posted by B is delivered to A and B is not aware of the information of A. Users post or retweet microblogs in a social network platform and their corresponding followers can browse those information or retweet them selectively.

Retweeting prediction is widely studied over the past decade and most work focus on two aspects, macro and micro level. Specially, macro level is aiming at predicting the scale and duration of microblogs' diffusion, while micro level is to predict whether the specified users will retweet a given microblog. Additionally, a quantity of work have been carried out to investigate the mechanism of the information diffusion including topic influence [1], community influence [2], pairwise influence [3] and influence locality [4].

In this paper, we mine out the hidden structural relations between users and integrate these features into retweeting behaviors model in micro level. In particular, RBM is employed to extract the hidden similarity between users and community features are further used to improve the effectiveness of RBM.

Considering different scenarios, we solve the predicting problem in two ways. Firstly, the prediction is solved as collaborative filtering task when detailed information of users or microblogs is not accessed. We apply collaborative filtering RBM(CFRBM) as an independent classifier and further improve the performance of CFRBM by adding community features. Secondly, CFRBM is applied to detect hidden features of similarity between users, and we append the extracted hidden features to the LR model with other content-related features. Our contributions are summarized as follows:

- We novelly employ CFRBM model to solve retweeting prediction which is regarded as a collaborative filtering problem.
- Experiments reveal that features of similarity between users extracted by CFRBM drastically improve the performance of logistic regression model with other content-related features.
- Community features detected by our proposed algorithm further improve the performance of CFRBM as they can represent distribution more accurately.

The paper is organized as follows. Section 2 reviews related work in retweeting behaviours prediction and RBM. Section 3 introduces the basic RBM method. Section 4 describes Collaborative Filtering RBM model and our previous proposed community detection algorithm. Section 5 extracts basic features of LR model and uses CFRBM as a feature selector. Section 6 analyzes the experimental results and Sect. 7 concludes our work.

2   Related Work

Retweeting Prediction. An increasing number of methods of retweeting prediction have been developed. Hong et al. [5] predicted the popularity of messages as measured by the number of future retweets and sheds some light on what kinds of factors influence information propagation in Twitter. Suh et al. [6] studied various features of content or personal properties. Boyd et al. [7] analyzed the retweeting behavior of conversation. Jing [4] investigated the social influence locality and predicted retweeting behaviors by using logistic regression model. Yang [8] developed RAIN model considering the distribution of users' roles. Chen et al. [9] proposed a recommender system which can recommend topics for users according to their real-time interests.

Restricted Boltzmann Machines. RBM model has been extensively studied. Hinton [10] employed RBM as a feature selector to reduce the dimensionality of data. Teh [11] applied RBM to deal with face recognition. Larochelle [12] introduced a discriminative component to RBM training and improved their performance as classifiers. Smolensky [13] firstly improved RBM to do collaborative filtering. Georgiev K [14] proposed a non-IID framework for using RBM to do collaborative filtering. Liu [15] solved the cold start problem of RBM by adding a preprocess of Naive Bayes classifier (Fig. 1).

3   Restricted Bolzmann Machine

The RBM model is represented as an undirected graph $G = \{V, H, E\}$, where V is the set of n nodes in visible layer, H is the set of m nodes in hidden layer, and E is the set of connections between two layers. $\boldsymbol{v} = \{v_1, v_2, ..., v_n\}$ and $\boldsymbol{h} = \{h_1, h_2, ..., h_m\}$ is the status (0 or 1) of visible nodes and hidden nodes, respectively. The energy function of RBM is defined as:

$$E(\boldsymbol{v}, \boldsymbol{h}|\boldsymbol{\theta}) = -\sum_n vbias_i v_i - \sum_m hbias_j h_j - \sum_n \sum_m v_i W_{ij} h_j \tag{1}$$

where $\boldsymbol{\theta} = \{\boldsymbol{W}, \boldsymbol{vbias}, \boldsymbol{hbias}\}$ are the parameters of RBM, $\boldsymbol{W} = \{W_{ij}, i \in n, j \in m\}$ and W_{ij} is the weight of edge between visible node v_i and hidden node v_j, $\boldsymbol{vbias} = \{vbias_1, ..., vbias_n\}$ and $vbias_i$ is the bias of visible node v_i, $\boldsymbol{hbias} = \{hbias_1, ..., hbias_m\}$ and $hbias_j$ is the bias of hidden node h_j. The marginal distribution of visible node is defined as:

$$P(\boldsymbol{v}) = \sum_{\boldsymbol{h}} \frac{exp(-E(\boldsymbol{v}, \boldsymbol{h}|\boldsymbol{\theta}))}{\sum_{\hat{\boldsymbol{v}}, \hat{\boldsymbol{h}}} exp(-E(\hat{\boldsymbol{v}}, \hat{\boldsymbol{h}}|\boldsymbol{\theta}))} \tag{2}$$

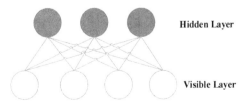

Fig. 1. Structure of RBM

Additionally, the activating condition of each node is independent when the status of the other layer is fixed. Formally,

$$P(h_j = 1|\boldsymbol{v}, \boldsymbol{\theta}) = \sigma(hbias_j + \sum_n v_i W_{ij})$$

$$P(v_i = 1|\boldsymbol{h}, \boldsymbol{\theta}) = \sigma(vbias_i + \sum_m W_{ij} h_j) \tag{3}$$

where $\sigma = 1/(1 + e^{-x})$ is the sigmoid function.

The training target is to obtain the $\boldsymbol{\theta}$ and the objective function is defined as:

$$\boldsymbol{\theta}^* = \underset{\boldsymbol{\theta}}{\text{argmax}} \log P(\boldsymbol{v}) \tag{4}$$

The parameters in $\boldsymbol{\theta}$ are updated as the gradient ascent of the log-likelihood objective function:

$$\Delta W_{ij} = \epsilon(< v_i h_j >_{data} - < v_i h_j >_{model})$$

$$\Delta vbias_i = \epsilon(< v_i >_{data} - < v_i >_{model}) \tag{5}$$

$$\Delta hbias_j = \epsilon(< h_j >_{data} - < h_j >_{model})$$

where ϵ is the learning rate, $< \cdot >_{data}$ and $< \cdot >_{model}$ is the expected distribution of training data and the model, respectively.

Gibbs sampling is employed to sample the distribution of RBM which includes several iterative processes resulting in the loss of speed. Hinton [16] proposed a fast training algorithm called Contrastive Divergence(CD). He proved that only K steps of Gibbs sampling can achieve sufficient performance. Specifically, K is often set as 1. The status of visible nodes is initialized as the training data. Equation 3 is employed to compute the status of hidden nodes according to the initial visible nodes, and visible nodes are reconstructed by the status of hidden nodes. The parameters are updated by the status of the training data and reconstructed data. Formally,

$$\Delta W_{ij} = \epsilon(< v_i h_j >_{data} - < v_i h_j >_{reconstructed}),$$

$$\Delta vbias_i = \epsilon(< v_i >_{data} - < v_i >_{reconstructed}), \tag{6}$$

$$\Delta hbias_j = \epsilon(< h_j >_{data} - < h_j >_{reconstructed})$$

where $< \cdot >_{reconstructed}$ is the status of each node in reconstructed data.

4   RBM for Collaborative Filtering

4.1   RBM Classifier on Social Network

In the collaborative filtering, according to Salakhudinov [17], we convert the structure of Restricted Boltzmann Machine into CFRBM. In the problem of retweeting prediction, there exists binary status, 0 and 1, to represent the retweeting action. Users in social network browse lots of microblogs, but only

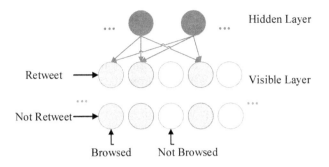

Fig. 2. Structure of CFRBM

retweet a fraction of them which they are interested in. Then, the predicting problem is defined as that there are N users, M microblogs and the target is to predict whether user u will retweet the microblogs v. Each user is an independent CFRBM. M groups of visible nodes are regarded as M microblogs and each group owns two nodes representing retweeting and not retweeting, respectively.

Different from basic RBM, if microblog m_i is invisible for user u, then group m_i doesn't connect to hidden layers in CFRBM. Assuming that invisible group m_i connects to the hidden layer, status of both nodes should be unactivated. But the unactivated status of one node indicates that the other node is active. It is obviously ambivalent. Thus, only the edge of visible microblogs is connected to hidden layer as shown in Fig. 2. Note that, N CFRBMs share the parameters $\boldsymbol{\theta}$ in the training process.

In each CFRBM, v_i^r is regarded as the status of node r in group i. We define the status of connection of group i as $\phi(i) = 1$ when group i connects to hidden layer and $\phi(i) = 0$ when group i doesn't connect to hidden layer. The active condition of each node should be modified as:

$$P(h_j = 1|\boldsymbol{v}, \boldsymbol{\theta}) = \sigma(hbias_j + \sum_r \sum_{i=1}^{n} v_i^r W_{ij}^r \phi(i)) \tag{7}$$

$$P(v_i^r = 1|\boldsymbol{h}, \boldsymbol{\theta}) = \sigma(vbias_i^r + \sum_r \sum_{i=1}^{m} W_{ij}^r h_j \phi(i)) \tag{8}$$

The updating functions are:

$$\begin{aligned}
\Delta W_{ij}^r &= \epsilon(< v_i^r h_j >_{data} - < v_i^r h_j >_{reconstructed}), \\
\Delta vbias_i^r &= \epsilon(< v_i^r >_{data} - < v_i^r >_{reconstructed}), \\
\Delta hbias_j &= \epsilon(< h_j >_{data} - < h_j >_{reconstructed})
\end{aligned} \tag{9}$$

Unretweeting data refers to the microblogs posted by a followee which not retweeted by his corresponding followers. In the training process of each CFRBM, the status of visible layer nodes is initialized by user u's data. Equation 7 is employed to compute the status of hidden layer. Then, Eq. 8 reconstructs the status of visible layer. Furthermore, reconstructed visible layer is applied to compute the status of hidden layer $< h_j >_{reconstructed}$ again. Finally, Eq. 9 updates the parameters $\boldsymbol{\theta}$.

In the predicting process, we initialize the status of visible layer with the user's data, and gain the status of hidden layer according to Eq. 7. Then, $\phi(i)$ of all microblogs required predicting is set to 1. Similarly, the status of visible layer is reconstructed by Eq. 8. Finally, we obtain the status of nodes in each visible group which need to be predicted. If the value of retweeting node is larger than the value of unretweeting node in visible group i, we conclude that user u will repost the microblog i.

4.2   RBM with Community on Social Network

With the precondition of sufficient hidden nodes, CFRBM can fit any distribution of training data but leading to obvious loss in speed. We intuitively speculate that applying the same scale of hidden layer in Sect. 4.1 to represent less patterns would improve the accuracy.

In social network, some users with the same hobbies form a community including tight inner connections and sparse external connections. Obviously, the retweeting behaviour among users in same community is more likely to be similar and users coming from diverse communities act far away from each other. Thus, we integrate the community features into CFRBM to represent less hidden patterns of retweeting behaviours.

The training process of CFRBM with community features(CFRBM-C) is similar to CFRBM. Firstly, we cluster the users into different communities. Then N RBMs only share their parameters $\boldsymbol{\theta}$ with those in the same community. Finally, we gain c groups of $\boldsymbol{\theta}$, where c is the number of communities. In classifying process, each CFRBM utilizes its own parameters $\boldsymbol{\theta}$ to predict the results of retweeting.

We previously proposed a community detection algorithm based on relationship density (CDRD)(paper under double-blind review). Here, the second process of CDRD algorithm is modified. Influential users, such as entertainers, politicians and other celebrities, often have tremendous impact on their followers and play an important role in information diffusion. Meanwhile, ordinary users have less influence in the social network. Influential user is called core user and ordinary user is called auxiliary user. Additionally, community formed by core users is called kernel community.

The proposed algorithm is formed by two pivotal parts which cluster core vertices and auxiliary vertices, respectively. More specifically: (1) Cluster kernel communities for core users in decreasing order of relationship density. (2) Cluster auxiliary users by assigning them into the existing kernel communities with the closest connection.

Core users are selected from users who have plenty of followers, formally:

$$Core(v) \Leftrightarrow |I(v)| \geq \varepsilon_{core} \tag{10}$$

where ε_{core} is a tunable parameter. Relationship density is defined as:

$$\rho(v) = \omega_{tr}|Tr(v)| + \omega_{ne}|Ne(v)| \tag{11}$$

where $|Tr(v)|$ is the number of $v's$ relation triangle circles which means that v follows u, u follows w and w follows v. $|Ne(v)|$ is the number of $v's$ mutual followers and ω_{tr} and ω_{ne} are the weights of these values.

Core Clustering. The pseudo-code is shown in Algorithm 1. In real social network, communities vary a lot in scale. Hence, the relationship density of marginal vertices in bigger community is sometimes close to the density of central vertices in smaller community. Our algorithm prevents the small kernel community from merging into big kernel community by setting a limit of maximum kernel community size max_core. $NC(v)$ is the set of $v's$ neighboring communities in which the fraction of users connecting to v in total users is greater than a given threshold μ_v. $Merge(C_i, C_j)$ is true if the number of edges between C_i and C_j exceed the ratio μ_C of product of two communities' scale. Considering the noisy data, if the size of a kernel community C is less than min_size, a tunable parameter, the vertices in this kernel community C are set as unclassified vertices in this stage.

Auxiliary Clustering. In a network, if user A follows user B who is a famous expert in specified area, we can infer that user A is the same as user B to a certain extent. Therefore, the community of user B can be regarded as the community of user A. After getting the kernel communities of most core vertices, these unclassified vertices are clustered by relationship with classified core vertices. Breadth-First-Search algorithm is employed to add the vertex which is connected to the classified vertex to existing communities iteratively. However, a bigger kernel community grows extremely quickly than a smaller kernel community in a dense network, so we grow each community to the same size before searching. Finally, those unvisited vertices are considered as isolated vertices and don't belong to any existing community.

5 CFRBM for Feature Selection

However, collaborative filtering models suffer from cold start problem [18]. In the retweeting prediction, it is hard to predict retweeting behaviour accurately when a microblog is retweeted just a few times. Additionally, CFRBM ignores the available detailed characteristic of users or microblogs. Fortunately, content-based models, such as logistic regression model(LR), overcome the cold start problem and take the content into consideration.

5.1 Logistic Regression with Basic Feature(LRB)

According to [4], we select basic features including users' features and the textual features extracted by Latent Dirichlet Allocation(LDA) method [19].

Algorithm 1. Core Users Clustering Algorithm

Input: $G = (V, E)$ and $\varepsilon_{core}, \omega_{tr}, \omega_{ne}, \mu_v, \mu_C, max\_core, min\_core$

Output: Kernel communities $\mathbf{KC} = \{C_1, C_2, ..., C_x\}$

Select core users and compute the relation density;

Sort all core users in decreasing order and assign each user as unclassified;

Let $\mathbf{KC} \leftarrow \oslash$ be empty;

for *each core user v in decreasing order of density* **do**

    if $|NC(v)| = 0$ **then**

        assign v as a new community C ;

        $\mathbf{KC} \leftarrow \{\mathbf{KC}, C\}$;

    end

    else if $|NC(v)| = 1$ *and* $|C| < max\_core$ **then**

        Let C be the existing neighbor community in $NC(v)$ and add v to C;

    end

    else

        Sort all $C \in NC(v)$ in decreasing $|C|$;

        Add v into to the first C which $|C| < max\_core$;

        for *each pair of communities (C_i, C_j)* **do**

            if $Merge(C_i, C_j)$ *and* $|C_i| + |C_j| \leq max\_core$ **then**

                merge C_i to C_j and $\mathbf{KC} \leftarrow \{\mathbf{KC} - C_i\}$;

            end

        end

    end

end

Set users in which community's size is smaller than min_core as unclassifed

---

Users' Feature. We select gender(0 indicates male and 1 indicates female), verified status(0 represents not verified and 1 represents verified), the number of followers, the number of followees, the number of friends and the number of his retweeting microblogs. Besides, we append the number of followees who retweet the microblog and $|Tri(v)|$ mentioned before as features.

Latent Dirichlet Allocation Feature. LDA topic model is applied to represent the topic distribution of each microblog. Given the number of topics, LDA generates the probability distribution on the topics for each microblog. We use the distribution as the feature of microblog and compute the user's distribution to the topics by averaging all the distribution of his historical retweeted microblogs. Further, Jensen-Shannon divergence [20] is employed to compute the distance between two distributions.

5.2 Features from CFRBM and CFRBM-C

For each training sample, given the status of visible layer, CFRBM is able to figure out the status of hidden layer which can be regarded as the distributions of the users or the features of the users. Firstly we compute distributions for

---

Algorithm 2. Auxiliary Users Clustering Algorithm

Input: Network $G = (V, E)$ and Kernel communities **KC**
Output: Communities $\mathbf{C} = \{C_1, C_2, ..., C_x\}$ and set of outliers Out $=$
$\quad\quad \{v_1, v_2, ..., v_o\}$
Grow each kernel community to the same size;
Set queue Q as empty;
for *each classified* v **do**
| Push v into Q;
end
while $Q \neq \oslash$ **do**
| let v be the front of Q;
| let C_v be the community of vertex v;
| **for** *each* $u \in I(v)$ **do**
| | **if** u *is not classified* **then**
| | | Add u to community C_v;
| | | Push u into Q;
| | **end**
| **end**
| Remove v from Q;
end
Set all of unvisited vertices as outliers.

---

every user using RBMs. Then the distributions of microblogs are computed by averaging the distributions of users who retweet those microbogs.

We apply the Jensen-Shannon divergence [20] to compute the distance between user and microblog according to their distributions. Additionally, sum of the numerical difference is also selected as a feature. (LRB-CFRBM and LRB-CFRBM-C).

6 Experimental Results

6.1 Social Network Data Set

Weibo social network dataset come from Jing [4], which is crawled from Weibo.com. It contains 1,776,950 users and 308,489,739 following relationships. 300,000 microblogs are retweeted 23,755,810 times on the network. The randomly selected evaluation data includes 10000 user who retweet more than 40 microblogs, and top 10000 microblogs retweeted by those chosen users (Table 1).

If user u retweets a microblog m_p, we consider it as a positive instance. If a followee v of user u retweets microblog m_n, but user u never retweeted the

Table 1. Statistics of weibo social network

	total users	following relations	orignal microblogs	total retweets
Weibo	1,776,950	308,489,739	300,000	23,755,810

m_n in our network, we consider u to m_n as a negative instance. However, the numbers of positive and negative instances are extremely unbalanced, so we sample balanced positive and negative instances from the selected data set.

6.2   Experimental Setup

Firstly, CDRD method is employed to cluster users into different communities. The parameters of CDRD are set as follows. We set ε_{core} to be 4000 to tag the core users, and set $\omega_{tr} = 1.0$, $\omega_{ne} = 0.4$, the threshold value of μ_v and μ_C are set to be 0.08 and 0.03, respectively. Set $max\_core = 100$ and $min\_core = 20$. Finally, users are clustered into two communities.

Each CFRBM has 50 hidden nodes, and 10,000 visible groups. Each visible group owns two nodes, representing positive instance and negative instance, respectively. The learning rate of CFRBM is set to be 0.01. We initialize **vbias** and **hbias** as 0, and set **W** to be a real number in range [-0.012, 0.012].

Total 954,110 instances are selected as the training and testing data. They are randomly divided into 5 parts and we conduct 5-fold cross-validation.

Effectiveness of experiments is evaluated in terms of precision, recall, f1-score. Precision is the number of true positive instances in all instances that are classified to be positive. Recall is the number of true positive instances that have been classified to be positive as some are classified to be negative. For example, $Precision_{positive} = \frac{\#true\ postive\ instances}{\#classified\ as\ positive}$, and $Recall_{positive} = \frac{\#true\ postive\ instances}{\#total\ positive\ instances}$.

6.3   Experimental Results and Analysis

CFRBM achieves 74.97 % of F1-Score without detailed feature of user and the corresponding microblogs. The CFRBM-C improves 2.67 % in term of F1-Score, 2.60 % and 2.98 % in terms of Recall and Precision. The result indicates that the CFRBM-C extracts hidden similarity between users better than CFRBM. At the same time, our assumption that users in the same community are more similar to each other obtains strong support. CFRBM-C with the fixed number of hidden nodes could represent the distribution of hidden relations more accurately (Table 2).

Table 2. Experimental results on Weibo social network

	Recall(%)	Precision%	F1-Score(%)
CFRBM	74.75	75.60	74.97
CFRBM-C	**77.35**	**78.58**	**77.64**
LRB	86.85	86.54	86.63
LRB-CFRBM	91.23	90.96	90.42
LRB-CFRBM-C	**91.27**	**91.03**	**90.45**

LRB achieves 86.63 % of F1-Score and performs better than both of CFRBM and CFRBM-C. The result reveals the importance of the content to the classification methods. LRB-CFRBM performs 3.79 % better than LRB in terms of F1-Score by using additional features which are represented by hidden nodes of CFRBM. LRB-CFRBM-C performs a little better than LRB-CFRBM because CFRBM-C represents the hidden relations more accurately than CFRBM. LRB only uses the content information, but CFRBM extracts the hidden relation which is the structural information of social network. Hence, as a feature selector, CFRBM promotes the performance of LR model.

The experimental results indicate that the structural information extracted by CFRBM can apparently promote the performance of basic classifier, and community information can improve the accuracy of distribution represented by CFRBM.

7   Conclusion

In this paper, we predict users' retweeting behaviors by collaborative filtering RBM in two ways. Firstly, when detailed information of dataset is unavailable, retweeting predicting problem is treated as a collaborative filtering task and CFRBM is employed as an independent classifier. Secondly, if the detail of users and microblogs is available, we apply CFRBM as a feature selector to extract the hidden features of similarity between users for the content-related LR model. Considering the community property of social network, we improve the performance of CFRBM by using information of communities detected by our proposed community detection algorithm CDRD. The LR model with extra feature of similarity obtains 3.79 % promotion in terms of F1-Score comparing with itself with basic features. Additionally, the hidden features produced by CFRBM with community can further enhance the performance of LR model. The experimental results indicate that CFRBM extracts hidden similarity of users well and notably promotes the effectiveness of basic classifier.

In the future, we would try to extract more information by CFRBM, such as hidden features of microblogs. We employ Jensen-Shannon divergence to compute the difference between extracted hidden features, so we could try more types of measurements to represent the difference. Additionally, we could try more dimensions reduction models like RBM to detect more types of features.

References

1. Tang, J., Sun, J., Wang, C., Yang, Z.: Social influence analysis in large-scale networks. In: Proceedings of the 15th ACM SIGKDD International Conference on Knowledge Discovery and Data Mining, pp. 807–816. ACM (2009)
2. Belák, V., Lam, S., Hayes, C.: Cross-community influence in discussion fora. In: ICWSM (2012)
3. Goyal, A., Bonchi, F., Lakshmanan, L.V.: Learning influence probabilities in social networks. In: Proceedings of the Third ACM International Conference on Web Search and Data Mining, pp. 241–250. ACM (2010)

4. Zhang, J., Liu, B., Tang, J., Chen, T., Li, J.: Social influence locality for modeling retweeting behaviors. In: Proceedings of the Twenty-Third International Joint Conference on Artificial Intelligence, pp. 2761–2767. AAAI Press (2013)

5. Hong, L., Dan, O., Davison, B.D.: Predicting popular messages in twitter. In: Proceedings of the 20th International Conference Companion on World Wide Web, pp. 57–58. ACM (2011)

6. Suh, B., Hong, L., Pirolli, P., Chi, E.H.: Want to be retweeted? large scale analytics on factors impacting retweet in twitter network. In: 2010 IEEE Second International Conference on Social Computing (Socialcom), pp. 177–184. IEEE (2010)

7. Boyd, D., Golder, S., Lotan, G.: Tweet, tweet, retweet: Conversational aspects of retweeting on twitter. In: 2010 43rd Hawaii International Conference on System Sciences (HICSS), pp. 1–10. IEEE (2010)

8. Yang, Y., Tang, J., Leung, C., Sun, Y., Chen, Q., Li, J., Yang, Q.: Rain: Social role-aware information diffusion. In: Proceedings of the 29th AAAI Conference on Artificial Intelligence (AAAI 2015) (2015)

9. Chen, C., Yin, H., Yao, J., Cui, B.: Terec: A temporal recommender system over tweet stream. Proc. VLDB Endowment **6**(12), 1254–1257 (2013)

10. Hinton, G.E., Salakhutdinov, R.R.: Reducing the dimensionality of data with neural networks. Science **313**(5786), 504–507 (2006)

11. Teh, Y.W., Hinton, G.E.: Rate-coded restricted boltzmann machines for face recognition. In: Advances in Neural Information Processing Systems, pp. 908–914 (2001)

12. Larochelle, H., Bengio, Y.: Classification using discriminative restricted boltzmann machines. In: Proceedings of the 25th International Conference on Machine Learning, pp. 536–543. ACM (2008)

13. Smolensky, P.: Information processing in dynamical systems: Foundations of harmony theory (1986)

14. Georgiev, K., Nakov, P.: A non-iid framework for collaborative filtering with restricted boltzmann machines. In: Proceedings of the 30th International Conference on Machine Learning (ICML-13), pp. 1148–1156 (2013)

15. Liu, Y., Tong, Q., Du, Z., Hu, L.: Content-boosted restricted boltzmann machine for recommendation. In: Wermter, S., Weber, C., Duch, W., Honkela, T., Koprinkova-Hristova, P., Magg, S., Palm, G., Villa, A.E.P. (eds.) ICANN 2014. LNCS, vol. 8681, pp. 773–780. Springer, Heidelberg (2014)

16. Hinton, G.E.: Training products of experts by minimizing contrastive divergence. Neural Comput. **14**(8), 1771–1800 (2002)

17. Salakhutdinov, R., Mnih, A., Hinton, G.: Restricted boltzmann machines for collaborative filtering. In: Proceedings of the 24th International Conference on Machine Learning, pp. 791–798. ACM (2007)

18. Forbes, P., Zhu, M.: Content-boosted matrix factorization for recommender systems: experiments with recipe recommendation. In: Proceedings of the fifth ACM Conference on Recommender Systems, pp. 261–264. ACM (2011)

19. Blei, D.M., Ng, A.Y., Jordan, M.I.: Latent dirichlet allocation. J. Mach. Learn. Res. **3**, 993–1022 (2003)

20. Heinrich, G.: Parameter estimation for text analysis. Technical report, Technical report (2005)

Cross-Collection Emotion Tagging
for Online News

Li Yu[1], Xue Zhao[1], Chao Wang[1], Haiwei Zhang[1,2(✉)], and Ying Zhang[1,2]

[1] College of Computer and Control Engineering, Nankai University, Tianjin, China
{yuli,zhaoxue,wangchao,zhanghaiwei,zhangying}@dbis.nankai.edu.cn
[2] College of Software, Nankai University, Tianjin, China

Abstract. With the rapid development of Internet and social media, online news has become an important type of information that attracts millions of readers to express their emotions. Therefore, it is of great significance to build an emotion classifier for online news. However, it largely relies on the collection with sufficient labeled news to build an emotion classifier and the manually labeling work can be quite labor intensive. Moreover, different collections may have different domains such as politics or entertainment. Even in the same domain, different collections require different classifiers, since they have different emotion labels and feature distributions. In this paper, we focus on the task of cross-collection emotion tagging for online news. This task can be formulated as a transfer learning problem which utilizes a source collection with abundant labeled data and a target collection with limited labeled data within the same domain. We proposed a novel method to transfer knowledge from source collection to help build an emotion classifier for target collection. Experimental results on four real datasets show that our method outperforms competitive baselines.

Keywords: Online news · Sentiment tagging · Cross-collection · Transfer learning

1 Introduction

With the explosive growth of the Internet, a large number of social media have been produced and its influence continues to increase. Online news is an important form that attracts millions of readers to express their feelings among kinds of social media. Readers often express their subjective emotions such as sadness, surprise and happiness by voting emotion tags after reading the news. Figure 1 shows an example from a popular Chinese news website (i.e., Society Channel of Sina News).

Emotion tagging for online news is an application of the research area of opinion mining and sentiment analysis which can be formulated as a sentiment classification problem. Traditional emotion tagging problems mainly focus on determining the emotions of the person who writes the news, while few focus on readers' perspective. Readers' emotion prediction aims to explore what

© Springer International Publishing Switzerland 2016
B. Cui et al. (Eds.): WAIM 2016, Part II, LNCS 9659, pp. 225–237, 2016.
DOI: 10.1007/978-3-319-39958-4_18

Voted by 6710 users

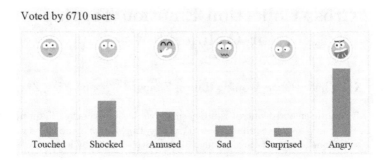

| Touched | Shocked | Amused | Sad | Surprised | Angry |

Fig. 1. An example of emotion tags voted by 6,710 users for a Sina news article.

most people think [1]. Also, most of the previous work just detect the polarity (e.g., positive or negative). In practice, readers' emotions can be divided into multiple categories (e.g., happy, sad, angry, surprised, etc.).

Traditional supervised learning methods have been applied to the problem of emotion tagging for online news. The performance of these methods highly depends on the availability of a collection with relative large amount of manually tagged news. However, the labeling work is often labor intensive for obtaining sufficient training data. Moreover, different collections may have different domains such as politics or entertainment, which leads to the distribution differences of domains [2]. Even in the same domain, different collections require different classifiers, since they have different emotion labels and word distributions.

This paper focuses on the task of cross-collection emotion tagging, which utilizes a source collection with abundant labeled data and a target collection with limited labeled data within the same domain. This paper proposes a novel transfer learning approach by applying latent discriminative model [3] on a source collection with abundant labeled news and a target collection with limited labeled data in the same domain and transferring knowledge across collections. Given a latest news, the model can use the knowledge from target collection and knowledge transferred from source collection and classify the news into an emotion tagging from the readers' perspectives. An extensive set of experiments is conducted with four real collections crawled from popular online news services in two groups. Our proposed model significantly outperforms alternatives in the task of emotion tagging for online news when the source collection has sufficient labeled data and target collection has limited labeled data. To the best of our knowledge, this is the first piece of research that focuses on modeling cross-collection emotion tagging for online news. Our proposed model can also be generalized and applied to the other cross-collection applications.

The rest of the paper is organized as follows. Section 2 reviews some related work. Section 3 proposes our approach for the scenario when the source collection has abundant labeled data and the target collection has limited labeled data. Experimental results are discussed in Sect. 4. The last Sect. 5 provides conclusions and points out potential work in the future.

2   Related Work

Sentiment analysis [4] has become an important subfield of information management. Pang *et al.* [5] exploited the positive or negative sentiment contained in movie reviews by applying support vector machine with unigram features. Mishne *et al.* [6] classified blog text according to the emotion reported by its author during the writing using SVM. Many machine learning techniques have been applied on sentiment classification, such as unsupervised learning techniques (e.g., [7]), supervised learning techniques (e.g., [5]) and semi-supervised learning techniques (e.g., [8]). For emotion tagging of online news, Zhang *et al.* [9,10] proposed a classification approach using information from heterogeneous sources to predict readers' emotions. The research work in [11] tries to predict upcoming news' polarity before news are issued. Quan *et al.* [3] proposed a latent discriminative model which incorporates latent variables into the LR model.

However, many previous works on emotion tagging assume that the labeled training data are sufficient. When a target collection has few labeled data, we have to transfer knowledge from a source collection with sufficient labeled data. Pan *et al.* [12] utilized transfer learning techniques to reuse data from an existing domain and thus required less labeled data from the new collection. Blitzer *et al.* [13,14] proposed the structural correspondence learning (SCL) algorithm to exploit domain adaptation techniques for sentiment classification. SCL is motivated by a multi-task learning method using alternating structural optimization (ASO) [15]. In addition, Li *et al.* [16] proposed to transfer common lexical knowledge across domains via matrix factorization techniques. Shu *et al.* [17] build a weakly-shared deep transfer networks to share knowledge between text and image. Zhang *et al.* [18] proposed research on knowledge transfer in sentiment analysis with auxiliary data from related sources. Zhang *et al.* [2] proposed a novel approach for cross-domain comments of online news using transfer learning and it is a baseline that will be tested in Sect. 4. In light of these considerations, this paper focuses on the transfer learning problem of cross-collection emotion tagging for online news.

3   Cross-Collection Emotion Tagging for Online News

This section focuses on the task of cross-collection emotion tagging for online news. We define our research problem and propose the approach for the scenario when the source collection has sufficient labeled data and the target collection in the same domain has limited labeled data.

3.1   Problem Definition

First of all, we define an online news collection containing a list of news as $\mathcal{D} = \{d_1, d_2, ..., d_{|\mathcal{D}|}\}$, while its emotion category set is defined as $\mathcal{E} = \{e_k\}(k = 1, \ldots, K)$. In each news, the news article can be described as feature vector $\mathbf{x}$ and its emotion label can be defined as $\mathbf{y} = \{y_k\}(k = 1, \ldots, K)$. y_k represents the user emotion voting probability of emotion e_k, and $\sum_{k=1}^{K} y_k = 1$. Thus an online news collection $\mathcal{D}$ can be denoted as $\{(\mathbf{x}_1, \mathbf{y}_1), (\mathbf{x}_2, \mathbf{y}_2), \ldots, (\mathbf{x}_{|\mathcal{D}|}, \mathbf{y}_{|\mathcal{D}|})\}$.

In our problem setting of cross-collection emotion tagging, we define a source collection $\mathcal{D}_S = \{d_{S_1}, d_{S_2}, \ldots, d_{S_m}\}$ and a target collection $\mathcal{D}_T = \{d_{T_1}, d_{T_2}, \ldots, d_{T_n}\}$ set within the same domain, where m is the size of source collection and n is the size of target collection. Since there are limited labeled data in the target collecion and abundant labeled data in the source collection, $0 < n \ll m$. Two emotion category sets can be defined as $\mathcal{E}_S = \{e_{S_k}\}(k = 1, \ldots, K_S)$ and $\mathcal{E}_T = \{e_{T_k}\}(k = 1, \ldots, K_T)$ respectively. We have a set of labeled data from the source collection $\mathcal{D}_S$ as $\{(\mathbf{x}_{S_1}, \mathbf{y}_{S_1}), (\mathbf{x}_{S_2}, \mathbf{y}_{S_2}), \ldots, (\mathbf{x}_{S_m}, \mathbf{y}_{S_m})\}$ and the target collection $\mathcal{D}_T$ as $\{(\mathbf{x}_{T_1}, \mathbf{y}_{T_1}), (\mathbf{x}_{T_2}, \mathbf{y}_{T_2}), \ldots, (\mathbf{x}_{T_n}, \mathbf{y}_{T_n})\}$.

In particular, we use not only emotion words in each collection, but also intersection words and union words of both collections to describe a news article:

- Emotion words in each collection are more likely to convey the emotions. Emotion words features are written as r_S and r_T(Note they have different feature space).
- Intersection words of $\mathcal{D}_S$ and $\mathcal{D}_T$ can be the relationship between collections and they can have the consistent distribution between source collection and target collection within the same domain[proved in Sect. 4.2]. Intersection words features are written as f_S and f_T(Note they have the same feature space).
- Union words of $\mathcal{D}_S$ and $\mathcal{D}_T$ show the distribution difference between collections which can be used to transfer knowledge across collections. Union words features are written as v_S and v_T(Note they have the same feature space).

Finally, the news article in both collections can be written as $\mathbf{x}_S = \{r_S, f_S, v_S\}$ and $\mathbf{x}_T = \{r_T, f_T, v_T\}$ respectively.

The task of cross-collection emotion tagging for online news is to learn an emotion classifier, predicting the emotion of unlabeled news in the target collection $\mathcal{D}_T$ based on the labeled data of both collections $\mathcal{D}_S$ and $\mathcal{D}_T$.

3.2   Latent Discriminative Model on Target Collection

In this section, latent discriminative model [3] is applied to target collection by adding an intermediate layer. Given a news d_{T_i} in target collection, the probability of evoked emotion being e_{T_k} is estimated using Bayes theorem as:

$$P(e_{T_k}|d_{T_i}) = \sum_{z=1}^{Z} P(e_{T_k}, l_z|d_{T_i})$$
$$= \sum_{z=1}^{Z} P(l_z|d_{T_i}) P(e_{T_k}|l_z, d_{T_i}) \tag{1}$$

where l_z denotes the latent variable and Z is the number of latent variables. This equation decomposes the original problem into two new subproblems: one is to build the relationship between latent variables l_z and the given news d_{T_i} and the other is to calculate the probability of emotion e_{T_k} given the combination of d_{T_i} and l_z.

$P(e_{T_k}|l_z, d_{T_i})$ could be simplified to $P(e_{T_k}|l_z)$ by assuming that the generation of emotions is entirely dependent on the latent variables. Thus it can be expressed in a soft-max function as the following formula:

$$P(e_{T_k}|l_z, d_{T_i}) = \frac{\exp(\omega_{zk} r_{T_i})}{\sum\limits_{j=1}^{K_T} \exp(\omega_{zj} r_{T_i})} \tag{2}$$

where r_{T_i} (proposed in Sect. 3.1) is a feature of d_{T_i}, conveying emotions. ω_{zk} is the vector of parameters corresponding to l_z and e_{T_k}. K_T is the number of emotion labels in target collection.

$P(l_z|d_{T_i})$ is also estimated by a soft-max function:

$$P(l_z|d_{T_i}) = \frac{\exp(\alpha_z f_{T_i})}{\sum\limits_{j=1}^{Z} \exp(\alpha_j f_{T_i})} \tag{3}$$

where f_{T_i} (proposed in Sect. 3.1) is another feature of d_{T_i}, using intersection words of both collections. α_z is the vector of parameters corresponding to l_z and d_{T_i}.

Finally, the probability of the evoked emotion being e_{T_k} in target collection can be written as:

$$P(e_{T_k}|d_{T_i}) = \sum_{z=1}^{Z} \frac{\exp(\alpha_z f_{T_i})}{\sum\limits_{j=1}^{Z} \exp(\alpha_j f_{T_i})} \frac{\exp(\omega_{zk} r_{T_i})}{\sum\limits_{j=1}^{K_T} \exp(\omega_{zj} r_{T_i})} \tag{4}$$

3.3 Latent Discriminative Model on Source Collection

Since the training data from the target collection are limited to make accurate prediction, it is necessary to utilize training data from source collection to help build a more accurate classifier. In particular, f_{T_i} and f_{S_i} may have the consistent distribution within the same domain, we set the same parameter α_z in the source collection, denoting the same weight. Therefore, the knowledge of source collection is shared on the same intermediate layer to help the target collection to build an emotion classifier.

Given a news article in source collection d_{S_i}, the probability of evoked emotion being e_{S_k} can be written as:

$$P(e_{S_k}|d_{S_i}) = \sum_{z=1}^{Z} \frac{\exp(\alpha_z f_{S_i})}{\sum\limits_{j=1}^{Z} \exp(\alpha_j f_{S_i})} \frac{\exp(\theta_{zk} r_{S_i})}{\sum\limits_{j=1}^{K_S} \exp(\theta_{zj} r_{S_i})} \tag{5}$$

where r_{S_i} is a feature of d_{S_i}, consisting of the emotion words in source collection. f_{S_i} is another feature of d_{S_i}, denoting the relationship between two collections. θ_{zk} is the vector of parameters corresponding to l_z and e_{S_k}.

3.4   Loss Function Definition

We use the negative log-likelihood as loss function. In the target collection, the loss of ith news d_{T_i}, referred as ξ_{T_i}, can be defined as:

$$\xi_{T_i} = -\log \prod_{k=1}^{K_T} P(e_{T_k}|d_{T_i})^{y_{T_{ik}}} = -\sum_{k=1}^{K_T} y_{T_{ik}} \log P(e_{T_k}|d_{T_i}) \tag{6}$$

where $y_{T_{ik}}$ is the kth user emotion voting probability of the ith news d_{T_i}. In a similar manner, the loss of ith news d_{S_i} in source collection, referred as ξ_{S_i}, can be defined as:

$$\xi_{S_i} = -\sum_{k=1}^{K_S} y_{S_{ik}} \log P(e_{S_k}|d_{S_i}) \tag{7}$$

By incorporating the auxiliary training data from the source collection, the final loss function can be defined as:

$$\underset{\omega,\theta,\alpha}{argmin} \frac{\lambda_1}{n} \sum_{i=1}^{n} \xi_{T_i} + \frac{\lambda_2}{m} \sum_{i=1}^{m} \beta_i \xi_{S_i} + \lambda_3 R(\omega) + \lambda_4 R(\theta) + R(\alpha) \tag{8}$$

where β_i re-weight the probability ξ_{S_i} from source collection to target collection which is introduced in the next Sect. 3.5. $R(\omega)$, $R(\theta)$ and $R(\alpha)$ are the three regularization penalties for parameters respectively. $\lambda_1, \lambda_2, \lambda_3$ and λ_4 are trade-offs that explore the relative importance of classification results.

3.5   Distribution Differences Between Collections

Although two collections come from the same domain, we can not simply identify the two collections as exactly the same. Here transfer learning method has been applied to transfer knowledge from source collection to target collection by re-weighting instances from the source collection. Through calculating the marginal probability of union words feature v_S and v_T, we know how they differ from collection to collection.

In particular, β_i refers to $\frac{Pr_T(v_{S_i})}{Pr_S(v_{S_i})}$, which represents the radio between the target and source collection. To estimate β_i, we adopt an approach based on kernel density estimation as:

$$\beta_i = \frac{Pr_T(v_{S_i})}{Pr_S(v_{S_i})} \propto \frac{\frac{1}{n}\sum_{j=1}^{n} \exp(-\frac{||v_{S_i}-v_{T_j}||}{\sigma^2})}{\frac{1}{m-1}(\sum_{j=1}^{m} \exp(-\frac{||v_{S_i}-v_{S_j}||}{\sigma^2})-1)} \tag{9}$$

where feature v_{S_i} and v_{T_i} (proposed in Sect. 3.1) have the same feature space but is under different distributions. m and n are the amount of labeled news counted on target and source collection. σ is the bandwidth parameter for Gaussian kernel. The -1 factor in the denominator removes the effect of the instance itself (i.e.,(v_{S_i}, v_{S_i})) in the source collection for modeling the probability ratio.

Obviously, if a source instance is more similar to a target instance, it contributes more to the β_i. Through re-weighting emotion probability, the source training data can be adjusted to fit the target collection distribution.

4   Experiments

A series of evaluations are conducted in this section to evaluate the new model. As there are many news portals with emotion votes for news in Chinese but few in English, our corpora are crawled from two popular Chinese news portals, Sina and QQ.

4.1   Datasets

Two groups of collections are gathered and each group contains two collections from one domain.

Table 1. The statistics of labeled news on four collections

(a) Collections in Entertainment

Sina Entertainment		QQ Entertainment	
Category	Count	Category	Count
Touched	334	Touched	186
Angry	521	Angry	944
Amused	252	Amused	936
Sad	207	Sad	249
Support	1022	Sympathetic	166
Seductive	35	Surprised	9
speculation	145	Happy	3542
Bored	1572	Anxious	1468
Total	4088	Total	7500

(b) Collections in Society

Sina Society 2013		Sina Society 2011	
Category	Count	Category	Count
Touched	1545	Touched	610
Angry	2920	Angry	2332
Amused	1512	Amused	779
Sad	1134	Sad	348
Surprised	235	Surprised	168
Shocked	319	Sympathetic	169
		Bored	226
		Warm	22
Total	7665	Total	4654

The first group of collections are crawled from Sina Entertainment and QQ Entertainment, which are among the largest news portals in China. Both of these collections are gathered in the same period of time within six months of 2011. The attributes include the URL, the news article, the news title, the publish date and the user emotion votes. Although two collections have eight kinds of emotion categories, they are not exactly the same. Here we use the news article as the input, the user emotion voting probability as the output. The second group of collections are crawled from Sina Society in 2011 and Sina Society in 2013. The attributes of Sina Society in 2011 have eight kinds of emotion categories. However, Sina Society changed its emotion category to six kinds since 2013.

Each collection defines its own emotion labels which are automatically anno-tated in original websites, therefore we can tag news articles with built-in emotion categories. In both groups, each of the two collections are different in emotion categories, but come from the same domain. Basically, the article with small emotion voting scores do not truly reflect actual situation. Therefore, we discuss those articles with more than 20 votes in total. The statistics are summarized in Tables 1(a) and (b).

4.2   Experimental Setup

In this experiment, we not only use textual emotion features to convey the emotions(r_S, r_T), but also use intersection words and union words features (f_S, f_T, v_S, v_T). For all collections, we use a Chinese word segmentation soft-ware ICTCLAS to segment the news article into terms and extract the emotion terms by NTU Sentiment Dictionary.

Distribution Cross-Collection. Figure 2(a) shows the feature distributions between QQ collection and Sina collection from entertainment domain. The X axis represents the intersection term features (f_S and f_T) which are ordered according to the term frequencies in Sina Entertainment. Then the ln(term fre-quency) values of both two collections are plotted. The comparison of features distributions (f_S and f_T) demonstrate the consistent distribution across collec-tion within one domain. Ying [2] explored the distributions between two collec-tions from different domains which were shown in Fig. 2(b). The comparison of features demonstrates the distribution differences across domains.

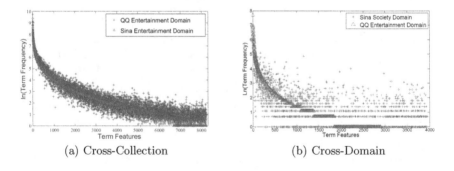

(a) Cross-Collection   (b) Cross-Domain

Fig. 2. Comparison of term feature distributions between collections

Evaluation Metrics. We adopt Accuracy(Accu), MRR and KL as the metrics. Assume that a collection contains N news. Given the ith news d_i, its user emo-tion voting probability y_i, its predicted emotion probability $\bar{y}_i$, we define three evaluation metrics:

1. Accuracy($Accu$) measures the validity of emotion prediction.

$$Accu = \frac{1}{N} \sum_{i=1}^{N} I(\bar{e}_i == e_i) \tag{10}$$

where e_i is its labeled emotion(truth) generated from the maximum probability of emotion label in y_i and $\bar{e}_i$ is its predicted emotion label generated from the maximum probability in $\bar{y}_i$. Also, $I(true) = 1$ and $I(false) = 0$.

2. Mean reciprocal rank(MRR) evaluates the average result of rank reciprocal.

$$MRR = \frac{1}{N} \sum_{i=1}^{N} \frac{1}{rank_i} \tag{11}$$

where $rank_i$ refers to the index of labeled emotion e_i in the sorted prediction emotion of $\bar{y}_i$.

3. Kullback Leibler divergence(KL) measures the distance between two probability distributions(y and $\bar{y}$).

$$KL = \frac{1}{N} \sum_{i=1}^{N} y_i \log \frac{y_i}{\bar{y}_i} \tag{12}$$

Comparison with Baselines. The following four methods are compared:

- **Probabilistic Model of Cross-Collection Emotion Tagging(P_CCET)** This approach is proposed in Sect. 3, which transfers knowledge from source collection with abundant labeled data to target collection with limited labeled data in a probabilistic way.
- **Probabilistic Model of Cross-Domain and Cross-Category Emotion Tagging (P_CDCCET).** This approach is proposed in [2], which models the relationship between different sets of emotion categories in different domains in a probabilistic way.
- **Latent Discriminative Model (LDM).** This approach is proposed in [3], which is based on observations and a joint distribution to capture the hidden structure in corpora of news. This method does not involve transfer learning techniques.
- **Emotion Tagging by Logistic Regression (ETLR).** This approach is based on multinomial logistic regression model. This method does not involve transfer learning techniques.

The performance of four methods is evaluated on the two groups of collections and each group has two settings by choosing either dataset as the target collection.

4.3 Results and Analysis

In this section, we do experiments to investigate the effectiveness of our proposed approach and baseline methods for the scenario when source collection

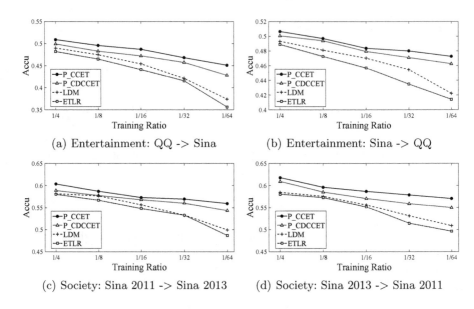

Fig. 3. The accu results on four collections.

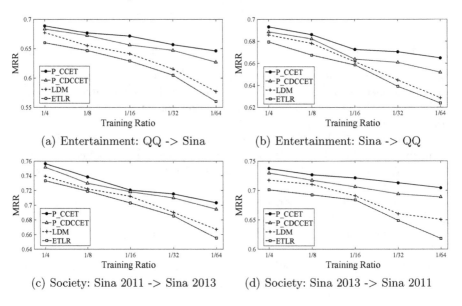

Fig. 4. The MRR results on four collections.

has sufficient data and target collection has limited data. We set four rounds of experiments: In each round of experiment, we specify one as target collection and the other as the source collection(e.g. Entertainment: QQ(source collection) -> Sina(target collection)).

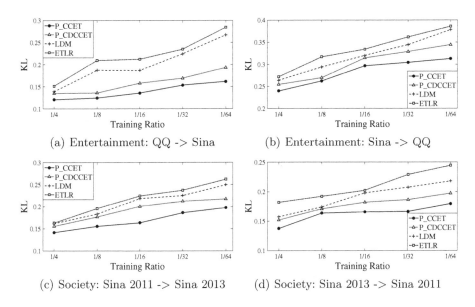

(a) Entertainment: QQ -> Sina   (b) Entertainment: Sina -> QQ

(c) Society: Sina 2011 -> Sina 2013   (d) Society: Sina 2013 -> Sina 2011

Fig. 5. The KL results on four collections.

In each round of experiment, we randomly selected 1/4, 1/8, 1/16, 1/32 and 1/64 of labeled data in target collection as the training data, while source collection is all selected as the training data. Remaining data in the target collection are used for testing. For all the methods, the trade-off parameters are set by ten fold cross-validations. The average experimental results of 100 independent runs are reported.

Figure 3(a) shows the performance of four methods on *Accu* when utilizing QQ entertainment with abundant labeled news to help the Sina entertainment with limited labeled news to build an emotion classifier under each size of training ratio. P_CCET outperforms competitive baselines, especially when Sina entertainment has only 1/64 of the original size.

In general, Figs. 3, 4 and 5 show the results of *Accu*, *MRR* and *KL* between four methods. Collections within the same domain may have the similar prediction probability. However, in most cases, prediction in society domain is more accurate than in entertainment domain under the same experimental setting.

Comparing four methods, it can be seen from these results that under most settings the P_CCET and P_CDCCET utilizing data from both the source and target collections beat LDM and ETLR using data only from the target collection, especially when training data from the target collection is not sufficient. The less training data the target collection has, the more benefit we gain from transfer learning. This fact clearly demonstrates the advantage of transferring knowledge from the source collection. The proposed approach P_CCET outperforms P_CDCCET in most cases. This is consistent with our expectation that sharing knowledge on intermediate layer between two collections helps to capture useful information across collections within the same domain.

Finally, significance tests using t-distribution are conducted for the results regarding from each run. Benefited from sufficient independent runs, such tests provide good discriminative power. P_CCET outperforms P_CDCCET, LDM and ETLR with 0.95 confidence level under all training ratios.

5   Conclusions and Future Work

This paper proposes a novel framework to address the task of predicting emotions for online news by modeling the relationship between related collections from the same domain. As far as we know, this is the first piece of research work to do cross-collection emotion tagging for online news. Our experimental results in four collections demonstrate the effectiveness of our proposed approach(P_CCET). For possible future research, we plan to design a better text representation scheme by combining full text representation with feature selection techniques. A more sophisticated modeling strategy for knowledge transfer may also improve the performance of cross-collection emotion tagging.

Acknowledgments. This work is partially supported by National 863 Project of China under Grant No. 2015AA015401, and National Natural Science Foundation of China under Grant No. 61402243. This work is also partially supported by Tianjin Municipal Science and Technology Commission under Grant No. 15JCTPJC62100, 13ZCZDGX01098, and 14JCYBJC15500.

References

1. Lin, K.H.Y., Yang, C., Chen, H.H.: Emotion classification of online news articles from the reader's perspective. In: Proceedings of the 2008 IEEE/WIC/ACM International Conference on Web Intelligence and Intelligent Agent Technology vol. 1, pp. 220–226. IEEE Computer Society (2008)
2. Zhang, Y., Zhang, N., Si, L., Lu, Y., Wang, Q., Yuan, X.: Cross-domain and cross-category emotion tagging for comments of online news. In: Proceedings of the 37th International ACM SIGIR Conference, pp. 627–636. ACM (2014)
3. Quan, X., Wang, Q., Zhang, Y., Si, L., Wenyin, L.: Latent discriminative models for social emotion detection with emotional dependency. ACM Trans. Inf. Syst. (TOIS) **34**(1), 2 (2015)
4. Pang, B., Lee, L.: Opinion mining and sentiment analysis. In: FTIR (2008)
5. Pang, B., Lee, L., Vaithyanathan, S.: Thumbs up?: sentiment classification using machine learning techniques. In: ACL (2002)
6. Hu, Y., Duan, J., Chen, X., Pei, B., Lu, R.: A new method for sentiment classification in text retrieval. In: Dale, R., Wong, K.-F., Su, J., Kwong, O.Y. (eds.) IJCNLP 2005. LNCS (LNAI), vol. 3651, pp. 1–9. Springer, Heidelberg (2005)
7. Turney, P.: Thumbs up or thumbs down?: semantic orientation applied to unsupervised classification of reviews. In: ACL (2002)
8. Sindhwani, V., Melville, P.: Document-word co-regularization for semi-supervised sentiment analysis. In: ICDM (2008)

9. Zhang, Y., Fang, Y., Quan, X., Dai, L., Si, L., Yuan, X.: Emotion tagging for comments of online news by meta classification with heterogeneous information sources. In: SIGIR (2012)

10. Zhang, Y., Su, L., Yang, Z., Zhao, X., Yuan, X.: Multi-label emotion tagging for online news by supervised topic model. In: Cheng, R., Cui, B., Zhang, Z., Cai, R., Xu, J. (eds.) Web Technologies and Applications. LNCS, vol. 9313, pp. 67–79. Springer, Heidelberg (2015)

11. Jakic, B., Weerkamp, W.: Predicting sentiment of comments to news on reddit (2012)

12. Pan, S., Yang, Q.: A survey on transfer learning. TKDE (2010)

13. Blitzer, J., Dredze, M., Pereira, F.: Biographies, bollywood, boom-boxes and blenders: Domain adaptation for sentiment classification. In: ACL (2007)

14. Blitzer, J., McDonald, R., Pereira, F.: Domain adaptation with structural correspondence learning. In: EMNLP, ACL (2006)

15. Ando, R., Zhang, T.: A framework for learning predictive structures from multiple tasks and unlabeled data. In: JMLR (2005)

16. Li, T., Sindhwani, V., Ding, C., Zhang, Y.: Knowledge transformation for cross-domain sentiment classification. In: SIGIR, ACM (2009)

17. Shu, X., Qi, G.J., Tang, J., Wang, J.: Weakly-shared deep transfer networks for heterogeneous-domain knowledge propagation. In: Proceedings of the 23rd Annual ACM Conference on Multimedia Conference, pp. 35–44. ACM (2015)

18. Zhang, D., Si, L., Rego, V.: Sentiment detection with auxiliary data. In: JIR (2012)

Online News Emotion Prediction
with Bidirectional LSTM

Xue Zhao[1], Chao Wang[1], Zhifan Yang[1], Ying Zhang[1,2]([⊠]), and Xiaojie Yuan[1,2]

[1] College of Computer and Control Engineering, Nankai University, Tianjin, China
{zhaoxue,wangchao,yangzhifan,zhangying,yuanxiaojie}@dbis.nankai.edu.cn
[2] College of Software, Nankai University, Tianjin, China

Abstract. Recent years have brought a significant growth in the volume of user generated data. Sentiment analysis is a crucial tool in the mining of such data, which is of great value for both improving particular services and assisting organizations' decision making process. Existing research focuses on identifying sentiment polarity on subjective text, such as tweets and product reviews. Sentiment analysis on news still remains a challenge: identifying effective emotion-differentiated features automatically from the more objective content, and modeling the longer document semantically. In this paper, we tackle this problem by improving document representations. From the word level, we implemented skip-gram model to learn the word representations with rich contextual information. Moreover, we propose two document representation approaches based on neural networks. We first introduce bidirectional long short-term memory (BLSTM) neural network to capture the complete contextual information in long news articles. In order to extract more salient information from document, we integrate a convolutional neural network with BLSTM to augment the document representations. Extensive experiments show the proposed model outperforms the other baseline methods.

Keywords: Sentiment analysis · Document representation · Bidirectional LSTM

1 Introduction

Emerging social media services have allowed the public to easily express their emotions, sentiments, opinions, and attitudes. Capturing this information is of great value for governments, companies and other organizations in regard to their policy, marketing and decision making process. Sentiment analysis on such user-generated documents has recently received increasing attention, for which it is used to identify sentiment polarity (*positive, neutral* or *negative*) or emotion categories (such as *happy, sad,* and *angry*) of a given text. In this paper, we focus on predicting the possible emotions triggered by news articles, which is a document-level finer-granularity sentiment analysis task. Sentiment analysis typically focuses on the subjective documents, for example, twitter, product comments, and movie reviews. Such documents are usually expressed with honest

B. Cui et al. (Eds.): WAIM 2016, Part II, LNCS 9659, pp. 238–250, 2016.
DOI: 10.1007/978-3-319-39958-4_19

opinions and emotions straightforwardly, whereas news articles attempt to stay objective. The emotions conveyed in news articles are less concrete and expressed much less explicitly. Although some journalists attempt to resort to other means to express their opinions and feelings, like embedding statements in a more complex discourse, and omitting or highlighting some facts, they still refrain from using emotional vocabulary. Another big difference of news and other social media is the length of documents. News has more sentences, some of which are uninformative and even add complexity. In this scenario, it is rather difficult to analyze sentiment on news articles. It is therefore necessary to represent word and the document semantically. However, early word representation simply maps the words to vectors by word occurrence, such as bag-of-words model. It suffers from data sparsity problem that the corresponding parameters of rare words cannot be properly estimated. It also ignores the context information around the word. In light of these problems, we applied skip-gram word embedding model to encode the words with real-value vectors which represent the rich context information. Moreover, most of research associates words to sentiment labels with hand-crafted features. Such feature engineering work is unable to define some latent discriminative factors and also labor-intensive.

Inspired by the success of neural networks in natural language processing (NLP) field, we take advantage of neural networks to learn more effective features automatically, with which we generate the document vector representations. Long short-term memory (LSTM) neural networks have become very popular in many sequence classification tasks. They are explicitly designed to overcome the gradient vanishing problem of traditional RNNs and then able to remember information for a longer period of time. Bidirectional LSTM (BLSTM) contains two LSTM layers running in the opposite directions. It takes into account both left and right context in making decisions at any point of word sequence, which makes it capable of modeling complete, sequential information of words in a long distance. In addition, convolutional neural networks (CNNs) have been widely used for document representation, because the convolution and pooling operations can extract local and global contextual features for every word in the document.

Analyzing the multiple emotions on news articles is quite challenging due to the characteristics of news. After the revival of deep learning, no recent literature has addressed this task with advanced neural networks. In this paper, we present a BLSTM neural network to learn the document representation for news emotion prediction. Furthermore, we explore the performance of the augmented feature set which consists of the document representations from BLSTM and CNNs neural networks. We conduct extensive experiments on a real world data set to prove the effectiveness of our proposed methods.

2   Related Work

2.1   Sentiment Analysis on Social Media

Social media are used daily to express personal thoughts and feelings, and allow researchers to gain valuable insight into the opinions of a large number of individuals. As a result, sentiment analysis is commonly applied to twitter [16], product comments [7], online news comments [19] and movie reviews [4]. However, after early attempts have been conducted on news [1], there are only a few research that put efforts to analyzing sentiments on objective news wire text.

The sentiment analysis methods can be broadly categorized into two main groups: machine-learning and lexical-based methods. Pang et al. [12] led the field with machine learning techniques by experimenting with various classifiers such as maximum entropy, Naive Bayes and SVM, using standard features such as unigram/bigrams, word counts/occurrence, word position and part-of-speech tagging. Bao et al. [1] proposed an emotion-topic model for social emotion classification by introducing an intermediate layer into Latent Dirichlet Allocation (LDA) [3]. On the other hand, lexical-based method is to make use of predefined lists of words, in which each word is associated with a specific sentiment. Very recently, Rao et al. [13] proposed lexicon-based emotion classification methods for online news. In particular, they built a word-emotion dictionary where each word is associated with the distribution on a series of emotions, as well as a topic-emotion dictionary where each topic is correlated with social emotions. Their experimental results show that their proposed methods outperform emotion-topic model in [1]. More details of sentiment analysis field are covered in the review of Liu [10].

2.2 Document Representation for Sentiment Analysis

Document level sentiment classification is a fundamental problem which aims at identifying the sentiment label of a document. Bag-of-words models are widely used for this task [12], as they can be composed into other effective features according to the specific document type. However, when they are used to encode documents into numerical vectors, they will inevitably cause huge dimensionality or sparse distribution problems. Feature engineering techniques can solve these problems, to a certain extent. Some representative features include word n-grams [17], syntactic relations [18], and sentiment lexicon features [8]. However, these feature engineering methods are unable to extract the discriminative information from data and quite labor intensive [15].

With the revival of deep learning, neural network becomes a more effective way to learn the vector representation of the word and the document. Earlier word representation methods rely on the word occurrence pattern, which makes them suffer from feature sparsity or feature explosion problem. Word embedding is an effective way to represent words semantically. The skip-gram word embedding model [11] has gained a lot of attention, and provides state-of-the-art word vectors. It follows the word distributional hypothesis that words in similar contexts have similar meanings. The words sharing similar contexts then can be mapped into close vectors, which contain rich context information. In this paper, we apply skip-gram word embedding to convert the words into fixed-length real-value vectors and then feed to our emotion prediction neural networks.

The document level composition needs more efforts on modeling the context semantically. LSTM is one of the state-of-the-art semantic composition models for sentiment classification [15]. BLSTM contains two LSTM in different direction that run in parallel: one on the input sequence and the other on the reverse of the input sequence. At each time step, the hidden state of the BLSTM is the concatenation of the forward and backward hidden states. Thus it can exploit both historical and future context. Convolution neural networks are widely used in document representation tasks, since the convolutional layer and pooling layer can capture local and global semantic information automatically. CNNs utilize layers with convolving filters to extract local features [9] and have achieved excellent results in sentence modeling [6].

In this work, we make full use of BLSTM and CNNs to capture the semantic context information as much as possible. We utilize BLSTM to learn complete information from word sequences in both directions. Because CNNs can extract important local information from the word context, we then take advantage of CNNs to improve the document representations by concatenating the outputs from both CNNs and BLSTM. A *softmax* layer is applied as the last output layer, which calculates the possibility distribution over emotions for the given document.

3   Emotion Classification Model

3.1   Input Layer

Word embedding is to convert words into fixed-length real-value vectors, at the same time, to map the words with similar context to the closer vectors. As for the first input layer of neural networks, there are two popular strategies to input data to neural networks: fine-tuned word embeddings and pre-trained word embeddings. The fine-tuned word embeddings are trained together with other layers in neural networks, and embeddings are therefore finely tuned for the specific task. Different from fine-tuned word embeddings, pre-trained word embeddings are independent from the neural networks and trained separately on corpus beforehand. In our paper, we explore the performance of both embedding methods in our emotion prediction task. In particular, we add a regular fully connected layer as the input embedding layer, which has the same number of output as the length of the word vector. We initialize the word vectors with random values drawn from uniform distribution which is scaled by the square root of the number of inputs [9], and update the embeddings together with the whole neural network. On the other hand, we prepare the pre-trained word embeddings with skip-gram model in advance, and feed the word embeddings to the neural network without the first embedding layer.

3.2   Bidirectional LSTM Neural Network

Recurrent neural networks (RNNs) are able to process input sequences of arbitrary length via the recursive application of a transition function on a hidden

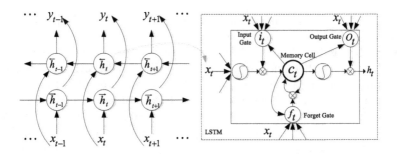

Fig. 1. Bidirectional long short-term memory (LSTM)

state vector h. Given an input sequence $x = (x_1, x_2, \ldots, x_T)$, RNNs compute the hidden vector sequence $h = (h_1, h_2, \ldots, h_T)$ and output vector sequence $F = (y_1, y_2, \ldots, y_T)$ by iterating the following equations from $t = 1$ to T [5].

$$h_t = \mathcal{H}(W_{ih}x_t + W_{hh}h_{t-1} + b_h) \tag{1}$$

$$y_t = W_{ho}h_t + b_o \tag{2}$$

where W denote weight matrices, b denote bias vector and $\mathcal{H}$ is the hidden layer activation function which is usually element-wise application of a sigmoid function. However, it has been observed that it is difficult to train RNNs to capture long-term dependencies since the backpropogated error will tend to either vanish or explode [2]. The most effective solution to this problem is the long short-term memory neural network (LSTM). Long short-term memory is a specific recurrent neural network with a special units called memory blocks, which contain one or more memory cells with self-connections. The memory blocks can store the temporal state of the network in addition to special multiplicative units called gates to control the flow of information. These gates allow the memory cell to store information over long periods of time and can avoid the vanishing gradient problem. Figure 1 illustrates a LSTM memory cell in a bidirectional recurrent structure. The hidden layer activation function $\mathcal{H}$ is implemented by the following composite function:

$$i_t = \sigma(W_{xi}x_t + W_{hi}h_{t-1} + W_{ci}c_{t-1} + b_i) \tag{3}$$

$$f_t = \sigma(W_{xf}x_t + W_{hf}h_{t-1} + W_{cf}c_{t-1} + b_f) \tag{4}$$

$$c_t = f_t c_{t-1} + i_t \tanh(W_{xc}x_t + W_{hc}h_{t-1} + b_c) \tag{5}$$

$$o_t = \sigma(W_{xo}x_t + W_{ho}h_{t-1} + W_{co}c_t + b_o) \tag{6}$$

$$h_t = o_t \tanh(c_t) \tag{7}$$

where σ is the logistic sigmoid function, and i, f, o and c are respectively the *input gate*, *forget gate*, *output gate* and *cell activation vectors*, all of which are the same size as the hidden vector h. the weight matrix subscripts are obvious, for example, W_{hi} is the hidden-input gate matrix. b is the bias term. As illustrated in Fig. 1, Bidirectional LSTM (BLSTM) is trained on input sequence in both forward and backward hidden layers, $\overrightarrow{h}$ and $\overleftarrow{h}$. The output sequence y is the combination of $\overrightarrow{h}$ and $\overleftarrow{h}$:

$$\overrightarrow{h_t} = \mathcal{H}(W_{x\overrightarrow{h}}x_t + W_{\overrightarrow{h}\overrightarrow{h}}\overrightarrow{h}_{t-1} + b_{\overrightarrow{h}}) \tag{8}$$

$$\overleftarrow{h_t} = \mathcal{H}(W_{x\overleftarrow{h}}x_t + W_{\overleftarrow{h}\overleftarrow{h}}\overleftarrow{h}_{t+1} + b_{\overleftarrow{h}}) \tag{9}$$

$$y_t = W_{\overrightarrow{h}y}\overrightarrow{h}_t + W_{\overleftarrow{h}y}\overleftarrow{h}_t + b_y \tag{10}$$

As a result, BLSTM takes into account both left and right context, which makes it capable of modeling complete sequential information in a longer distance. We build a BLSTM layer containing two separate LSTM layers running in different directions and feed the word vectors into it, the output of BLSTM layer is the representation for each document.

3.3 Convolutional Neural Network

CNNs have been successfully applied to image recognition field, due to the ability to exploit the spatially local correlation presented in natural images. In NLP task, the convolution operation works as feature extraction process within each sliding window over the word sequence, where each word is encoded into the fixed-length vectors. Figure 2 presents the complete CNNs structure. The input matrix s consists of the word vectors in the document, where each column corresponds to a d-dimensional word vector v_{w_i}.

$$s = [v_{w_1}, v_{w_2}, \ldots, v_{w_n}] \tag{11}$$

There are two dimensions of the input: word sequence dimension in s and word vector dimension in v. The words of the document are continuous over time and thus spatially correlated, whereas the dimensions in a word vector v are actually the separate features of words and context. Therefore, we apply a one-dimensional convolution layer together with a pooling layer in the word sequence

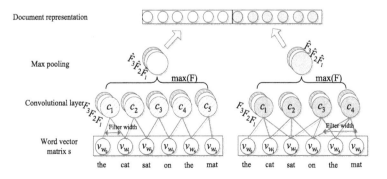

Fig. 2. A one-dimensional convolutional neural network

dimension, which contains more meaningful local contextual information. We consider the word and the context around the word w_i in a filter with size m. The convolution operation will be conducted over the window vector to obtain the context feature vector. Specifically, in convolution layer, the dot product is taken between a filter vector $w \in \mathbb{R}^{md}$ and a vector $s_{i:i+m-1} \in \mathbb{R}^{md}$ of m concatenated columns in d-dimensional word vector, in order to generate another feature sequence c. The i-th feature $c_i \in \mathbb{R}$ is generated as follow:

$$c_i = f(w \cdot s_{i:i+m-1} + b) \tag{12}$$

where b is a bias term and f is a rectified linear unit (ReLU) activation function in our model. $s_{i:i+m-1}$ refers to columns from i to $i+m-1$ word vectors of d. The outcomes of the convolution operation are basically the local features extracted for every single word. A feature map $F_j \in \mathbb{R}^{n+m-1}$ then can be defined as:

$$F_j = [c_1, c_2, \ldots, c_{n+m-1}] \tag{13}$$

However, we cannot simply utilize F to form a fixed-length document representation due to the varying length of input documents. Therefore, we use a *max pooling* layer on top of convolution layer, since it attempts to find the most salient semantic factors from the features produced by the convolutional layer. The value $\hat{F}_j = \max(F_j)$ is the feature corresponding to a particular filter. The output of pooling layer is the representations of documents. We use different convolutional filters whose widths are 2 and 3 to encode the semantics of bigrams and trigrams in the documents.

3.4   Concatenation Layer and Softmax Layer

Now we have two neural networks to obtain document representations. BLSTM can capture the important information around the word in both directions and in longer distance, and CNNs are designed for exploiting most useful contextual features from the document. Each dimension in the document representation is actually a document feature related to the contextual information. In order to obtain a more complete feature set, we augment the document representation by integrating both document representations of BLSTM and CNNs. Specifically, we concatenate the output vectors from BLSTM and CNNs together as the final document representation. We feed it to a linear layer whose output length is the class number, and add a *softmax* layer as the output layer to generate conditional probability of each emotion label. Formally, for K emotion categories, the probability of k-th category in the output is calculated as follows.

$$softmax_k = \frac{exp(x_k)}{\sum_{k'}^{K} exp(x_{k'})} \tag{14}$$

3.5   Neural Network Training

During the training process, since the output of *softmax* layer is a distribution over labels, we utilize the categorical cross entropy as our objective funtion.

$$H(p,q) = -\sum_x p(x) \log(q(x)) \tag{15}$$

where p is the true distribution over labels and q is the predicted distribution. Regularization is vital for good performance with neural networks, as their flexibility makes them prone to overfitting. Here we use Adagrad as our optimizer and introduce dropout technique to avoid overfitting [14]. The number o hidden layer is set as 50, based on our experiment results.

4   Experiments

In this section, we first introduce the experimental setups including data set, comparison baselines and evaluation metrics. We then report an extensive set of experiment results of the proposed and baseline methods. Analysis and discussions are presented based on the results.

4.1   Experimental Setups

Data Sets. we conducted on a large news article data set with eight emotion categories. Since there is no available data set of English news with multiple emotion labels, we can only use Chinese news to evaluate our model, but the proposed model is language independent. This data set is used in [13] for sentiment analysis. It is crawled from Sina news sites, between August 2009 to April 2012. The data contains segmented news documents and the user ratings over eight emotions of touching, empathy, boredom, anger, amusement, sadness, surprise and warmness. It contains 40,897 valid news articles with 445-word average length.

The statistics of the data set are summarized in Table 1. We assign the emotion label with the highest number of votes to each news article and count the total votes for each emotion category over all the news articles. As shown in Table 1, the largest category is 'anger', and there are 18,287 news articles having the highest number of votes over that emotion. The smallest one is 'warmness' with only 388 news articles. The category distribution is quite imbalanced and makes the emotion analysis on news more difficult.

Table 1. The statistics of top ranked emotion label sets.

Emotion label	# of articles	# of rating
touching	5,672	254,488
empathy	2,676	149,883
boredom	2,603	151,599
anger	18,287	830,927
amusement	6,812	320,824
sadness	2,888	212,316
surprise	1,571	105,585
warmness	388	58,196

Experiment Design. We conduct the following experiments to explore the performance of our emotion prediction neural networks. First, we explore two commonly used word embedding methods by feeding the vectors to neural networks.

1. Skip-gram word embedding (W2V). This popular word embedding model has been subsequently used in many natural language processing applications.
2. Fine-tuned word embedding (FW). It is trained by an extra regular fully connected embedding layer, in order to learn the task-specific word embedding.

The following methods are implemented to prove the effectiveness of proposed neural networks in emotion prediction for online news:

1. W2V+Conv-GRNN. It is a document representation method proposed in [15] which consists of convolution layer and gated recurrent neural network layer to learn sentence composition and document composition, respectively. The method outperforms various of traditional machine learning methods, for example, using SVM and text features, such as n-grams, character n-grams, and sentiment lexicon features.
2. Word-Emotion method(WE). It is proposed by [13] to classify emotions by building a word-level emotion dictionary. It is shown to outperform the other top-performing emotion classification algorithms, such as Emotion Topic Model, SWAT system, and ELDA. We pick the WE-max method which has achieved the best result.
3. BLSTM. It is the proposed method for emotion prediction task. The initialization function for the bias of the forget gate is set as one.
4. Multiple CNNs (mCNNs). They have been widely used to extract n-gram features from documents. We followed [15] and implemented three convolutional filters with width of 1, 2, 3 to capture the semantics of unigrams, bigrams and trigrams. We used average pooling to capture global semantic information.
5. BLSTM+CNNs. It is the proposed method where CNNs and BLSTM are designed to concatenated to build a more effective feature set and generate a more concrete document representation for emotion classification.

All of these neural networks are implemented with dropout technique to avoid the possible overfitting. We set the same number of hidden layers as 50, which achieves the best result in our most of the experiment. Besides these comparison experiments, we also explore different numbers of hidden layers of BLSTM to improve performance of our model.

Evaluation Metrics. Our interest is to predict the emotions of general public aroused by news articles, these emotions have the highest number of votes over all the other emotion categories. Thus, we regard the emotion with highest number of votes as the emotion label of the document. We use micro-averaged F1 measure ($MicroF1$) to evaluate the predicted emotion category. Micro-averaged F1 is the same as the accuracy at predicting top 1 emotion label. It weights equally all the documents in test data set. We avoid using averaged-F1 for each category or macro-averaged F1 for all categories due to the imbalanced distribution of news articles over emotion categories.

On the other hand, some other emotions can also have high score of votes, the ability of predicting other possible emotions triggered by news can also prove the effectiveness of the model. We applied averaged Pearson's correlation coefficient (AP) to measure the divergences between predicted emotion labels and ground truth emotion labels. Specifically, for each document d, the predicted and ground truth probability of emotions e on document d (i.e., X and Y), P_d is calculated as:

$$P_d = \frac{\sum_{i=1}^{E}(X_i - \overline{X})(Y_i - \overline{Y})}{(E-1)\sigma_X \sigma_Y} \tag{16}$$

Table 2. Experimental results on $MiccroF1$ and AP.

Methods	AF	AP
WE	56.70 %	54.65 %
FW + BLSTM	58.59 %	56.46 %
FW + BLSTM+CNNs	58.55 %	56.25 %
FW + Conv-GRNN	58.21 %	55.74 %
W2V+Conv-GRNN	58.96 %	57.75 %
W2V+GRNN	56.87 %	54.79 %
W2V+mCNNs	58.60 %	57.07 %
W2V+BLSTM	**59.15 %**	**55.53 %**
W2V+BLSTM+CNNs	**59.74 %**	**57.89 %**

Result and Analysis. We investigate the effectiveness of the proposed and baseline methods on the emotion prediction task for online news. Table 2 summarizes the evaluation results. Based on these results, we can make the following observations:

(1) Methods based on neural networks generally outperform WE in our task. This illustrates that news articles are objective documents which contain less straightforward emotion vocabularies, which makes lexicon-based methods weak compared to neural networks, which can exploit latent discriminative features from context information automatically.

(2) The word embeddings trained from raw corpus achieve better result compared to the fine-tuned word embeddings, which are trained for specific tasks, together with the neural networks. This presents that word vectors trained from skip-gram model contain more context information and thus can make big contribution to the success of the proposed neural networks. However, the fine-tuned word embeddings failed to associate the emotion labels with word context, due to the lack of sentiment vocabularies in news articles.

(3) W2V+Conv-GRNN achieves a very high prediction accuracy but is still weaker than BLSTM and the combination of BLSTM and CNNs. This proves that BLSTM has a strong ability to model long document and generate document representations.

(4) Multiple CNNs are good at exploiting local information from the document, this helps improve the emotion prediction AP result.

(5) CNNs doesn't provide much help to predict the top 1 emotion category, due to the fact that the single BLSTM has already achieved a very high result, this explains that the single BLSTM neural network can learn a very concrete feature set for the given documents. However, CNNs largely improve the result of AP, this means that CNNs can generally grasp more details of the documents compared to BLSTM, it is therefore able to predict the other possible emotions.

(6) GRNN performs worse than expectation, it may be because it is a successful model applied on shorter documents, such as tweets and comments, instead of long news articles.

In addition, a paired t-test shows that all improvements of proposed methods over the baselines are statistically significant with p-value < 0.001.

Tuning neural networks is a very tricky task due to the various of parameters and the possibility of overfitting. Here we use skip-gram word embeddings and BLSTM to explore the influence of hidden layers of BLSTM. As shown in Fig. 3, the varying number of hidden layers strongly affect the performance of emotion prediction. When the hidden layer is set as 50, both the micro-averaged F1 measure and averaged Pearson's correlation achieve the best results. Thus, in this paper, we set the number of hidden layer as 50.

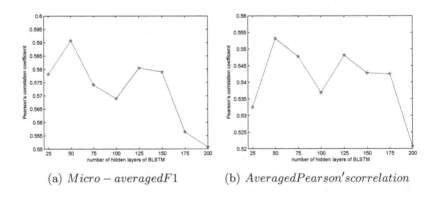

(a) $Micro - averagedF1$ (b) $AveragedPearson's correlation$

Fig. 3. $Micro - averagedF1$ and AP with different numbers of hidden layer.

5   Conclusions and Future Work

This paper proposes two neural network models based on BLSTM and CNNs to tackle the task of predicting general reader's emotion for online news. Specifically, we utilize skip-gram word embedding model to generate word-level representations, which contain the rich contextual information. We then feed the word vectors in BLSTM and a composition neural network of CNNs and BLSTM. BLSTM can model complete sequential information of words in a long distance, and CNNs can extract local and global contextual features for every word in the document. Experiment results prove that both of the proposed methods outperform the baseline methods in online news emotion prediction tasks. This work is an initial step towards a promising research direction. We will explore more about predicting emotions of online news with neural networks.

Acknowledgment. This work is supported by National 863 Project of China (Grant No. 2015AA015401), National Natural Science Foundation of China (Grant No. 61402243), Tianjin Municipal Science and Technology Commission (Grant No. 15JCT-PJC62100 and 13ZCZDGX01098)and Research Fund for the Doctoral Program of Higher Education of China (Grant No. 20130031120029).

References

1. Bao, S., Xu, S., Zhang, L., Yan, R., Su, Z., Han, D., Yu, Y.: Mining social emotions from affective text. IEEE TKDE **24**(9), 1658–1670 (2012)
2. Bengio, Y., Simard, P., Frasconi, P.: Learning long-term dependencies with gradient descent is difficult. Neural Netw. **5**(2), 157–166 (1994)
3. Blei, D.M., Ng, A.Y., Jordan, M.I.: Latent dirichlet allocation. J. Mach. Learn. Res. **3**, 993–1022 (2003)
4. Diao, Q., Qiu, M., Wu, C.Y., Smola, A.J., Jiang, J., Wang, C.: Jointly modeling aspects, ratings and sentiments for movie recommendation (jmars). In: Proceedings of the 20th ACM SIGKDD, pp. 193–202. ACM (2014)
5. Graves, A., Jaitly, N.: Towards end-to-end speech recognition with recurrent neural networks. In: Proceedings of the 31st International Conference on Machine Learning (ICML-14), pp. 1764–1772 (2014)
6. Kalchbrenner, N., Grefenstette, E., Blunsom, P., Kartsaklis, D., Kalchbrenner, N., Sadrzadeh, M., Kalchbrenner, N., Blunsom, P., Kalchbrenner, N., Blunsom, P.: A convolutional neural network for modelling sentences. In: Proceedings of the 52nd Annual Meeting of the Association for Computational Linguistics, pp. 212–217. Association for Computational Linguistics (2014)
7. Kim, S., Zhang, J., Chen, Z., Oh, A.H., Liu, S.: A hierarchical aspect-sentiment model for online reviews. In: AAAI (2013)
8. Kiritchenko, S., Zhu, X., Mohammad, S.M.: Sentiment analysis of short informal texts. J. Artif. Intell. Res. **50**, 723–762 (2014)
9. LeCun, Y., Bottou, L., Bengio, Y., Haffner, P.: Gradient-based learning applied to document recognition. Proc. IEEE **86**(11), 2278–2324 (1998)
10. Liu, B.: Sentiment analysis and opinion mining. Synth. Lect. Hum. Lang. Technol. **5**(1), 1–167 (2012)
11. Mikolov, T., Chen, K., Corrado, G., Dean, J.: Efficient estimation of word representations in vector space. In: Proceedings of Workshop at ICLR (2013)
12. Pang, B., Lee, L., Vaithyanathan, S.: Thumbs up? sentiment classification using machine learning techniques. In: Proceedings of the ACL Conference on EMNLP, vol. 10, pp. 79–86. ACL (2002)
13. Rao, Y., Lei, J., Wenyin, L., Li, Q., Chen, M.: Building emotional dictionary for sentiment analysis of online news. World Wide Web **17**(4), 723–742 (2014)

14. Srivastava, N., Hinton, G., Krizhevsky, A., Sutskever, I., Salakhutdinov, R.: Dropout: a simple way to prevent neural networks from overfitting. J. Mach. Learn. Res. **15**(1), 1929–1958 (2014)

15. Tang, D., Qin, B., Liu, T.: Document modeling with gated recurrent neural network for sentiment classification. In: Proceedings of EMNLP, pp. 1422–1432 (2015)

16. Tang, D., Wei, F., Yang, N., Zhou, M., Liu, T., Qin, B.: Learning sentiment-specific word embedding for twitter sentiment classification. In: Proceedings of the 52nd Annual Meeting of the Association for Computational Linguistics. vol. 1, pp. 1555–1565 (2014)

17. Wang, S., Manning, C.D.: Baselines and bigrams: Simple, good sentiment and topic classification. In: Proceedings of the 50th Annual Meeting of the Association for Computational Linguistics: Short Papers, vol. 2. pp. 90–94. Association for Computational Linguistics (2012)

18. Xia, R., Zong, C.: Exploring the use of word relation features for sentiment classification. In: Proceedings of the 23rd International Conference on Computational Linguistics: Posters, pp. 1336–1344. Association for Computational Linguistics (2010)

19. Zhang, Y., Zhang, N., Si, L., Lu, Y., Wang, Q., Yuan, X.: Cross-domain and cross-category emotion tagging for comments of online news. In: Proceedings of SIGIR Conference, pp. 627–636. ACM (2014)

Learning for Search Results Diversification in Twitter

Ying Wang[(✉)], Zhunchen Luo, and Yang Yu

China Defense Science and Technology Information Center, Beijing 100142, China
suneony@gmail.com, zhunchenluo@gmail.com, fengqy911@163.com

Abstract. Diversifying the results retieved is an effective approach to tackling users' information needs in Twitter, which typically described by query phrase are often ambiguous and have more than one interpretation. Due to tweets being often very short and lacking in reliable grammatical sytle, it reduces the effectiveness of traditional IR and NLP techniques. However, Twitter, as a social media, also presents interesting opportunies for this task (for example the author information such as the number of statuses). In this paper, we firstly address diversitication of the search results in Twitter with a learning method and explore a series of diversity features describing the relationship between tweets which include tweet content, sub-topic of tweet and the Twitter specific social information such as hashtags. The experimental results on the Tweets2013 datasets demonstrate the effectiveness of the learning approach. Additionally, the Twitter retrieval task achieves improvement by taking into account the diversity features. Finally, we find the sub-topic and Twitter specific social features can help solve the diversity task, especially the post time, hashtags of tweet and the location of author.

1 Introduction

Twitter provides a platform to allow users to post text messages known as tweets to update their followers with their findings, thinking and comments on some topics [1]. Twitter users often conduct search on the posted tweets to fulfill their information needs. As of 2014, the number of queries submitted to Twitter per day is reported to be more than two billion [2]. However, users' information needs, typically described by keyword based queries, are often ambiguous and have more than one interpretation. For example, as for the query "dreamliner battery", there are three relative tweets[1]:

- $tweet_1$: *Boeing 787 battery fire was difficult to control: An investigation of a battery fire aboard a Boeing 787*
- $tweet_2$: *Boeing 787 Dreamliner battery was miswired, Japan says - CTV News*
- $tweet_3$: *Rockford-Area News Boeing proposes battery fix to FAA for 787 Dreamliner planes*

---

[1] Tweets come from Tweets2013 corpus.

© Springer International Publishing Switzerland 2016
B. Cui et al. (Eds.): WAIM 2016, Part II, LNCS 9659, pp. 251–264, 2016.
DOI: 10.1007/978-3-319-39958-4_20

The $tweet_1$ is an introduction to the dreamlinear battery accident while the $tweet_2$ analyzes the reason of the accident and the $tweet_3$ refers to the solution of the accident. The three tweets refer to different aspects of the query and it is difficult to disamiguate the user intent without further information.

The typical web search also suffers this problem. Search results diversification has attracted considerable attention [3]. The key idea is to provide a diversified results list, in the hope that different users will find some results that can cover their information needs. Different methods on web search results diversification have been proposed in literature [4–6]. While, search results diversification in Twitter can be harder than typical web search largely due to tweet being often very short and lacking in reliable grammatical style and quality [7]. Morever, queries in Twitter are even shorter (1.64 words on the average). These factors reduce the effectiveness of traditional IR and NLP techniques.

Fortunately, Twitter also presents interesting opportunities. The rich environment presents us with a myriad of social information over-and-above just using terms in a post all of which potentially can improve on search results diversification performance in Twitter. For example, people usually use hashtags to indicate the topics of the tweet and the hashtags can be used for diversifying tweets.

In this paper, we firstly address the diversification of results in Twitter search with a learning method and propose a series of features including tweet content, the sub-topic of tweet and Twitter specific social information such as hashtags. Benefited from the learning method, we can easily incorporate aforementioned features into a ranking model easily. With each of these features describing the property of tweet in one perspective, we can diversify the tweets from multiple aspects. Obviously, the pure content is meaningful for diversity. Additionally, different tweets may associate with different sub-topics and two tweets may refer to the same sub-event if their sub-topics are similar. Hence, the sub-topics may help diversify the tweets. The rich social informations in Twitter can also reflect the differences between tweets. We investigate the effects of these features and produce a diverse ranking system [6].

With a series of experiments, we demonstrate the effectiveness of the learning method in tackling the search results diversification problem in Twitter. Our approach achieve comparable performance with the traditional diversification approaches [8] in the evaluations of most measures. Furthermore, we find the Twitter retrieval task achieves improvement by taking into account the diversity features which consider the relationship between tweets in the evaluations of α-nDCG, Precision-IA and Subtopic Recall. Finally, we find sub-topic and the Twitter specific social features can help solve the search results diversification, especially the post time, hashtags of tweet and the location of author.

The contributions of this paper can be summarized as follows:

(1) The proposal of a new learning method for search results diversification in Twitter.
(2) The exploration of a series of diversity features.
(3) The verification of the effectiveness of the proposed approach based on public datasets.

2   Related Work

We review related works on three main areas: Twitter search, diversifying web search results and diversifying Twitter search results.

2.1   Twitter Search

Relevance to the search query is the major ranking criteria in most of the work on tweet. Jabeur, Tamine, and Boughanem model the relevance of a tweet to a query by a Bayesian network that integrates a variety of features [9]. Zhang et al. train machine learning models for ranking tweets against a query [10]. Luo et al. introduce Twitter Building Blocks and their structural combinations, they use this structural information as features into a learning to rank scenario for Twitter retrieval [11]. Luo et al. integrate social and opinion-atedness information for tweets opinion retrieval [12].

TREC 2011 introduced the Microblog Track which addressed one single pilot task, entitled real-time search task, where the user wished to see the most recent but relevant information to the query. A total 59 groups participated in the track from across the world, with 184 submitted runs. The experimental results indicate the large gap between the best and medians evaluation score (e.g., MAP value) per-topic for 59 participated groups. It shows that Tweets retrieval is far from being a solved problem.

All these methods are simple ad-hoc retrieval approaches that only consider the relevance between tweet and query, while not taking into account the relationship between tweets.

2.2   Diversifying Web Search Result

There are many existing search results diversification methods in web search which can be mainly divided into two categories: implicit approaches and explicit approaches.

The implicit methods assume that similar documents cover similar aspects and model inter-document dependencies. For example, Maximal Marginal Relevance (MMR) method proposes to iteratively select a candidate document with the highest similarity to the user query and the lowest similarity to the already selected documents, in order to promote novelty [4].

The explicit methods explicitly model aspects of a query and then select documents that cover different aspects. The aspects of a user query can be achieved with a taxonomy [3], top retrieved documents [13], query reformulations [14], or multiple external resources [15].

All these methods are non-learning methods and utilize a heuristic predefined utility function. Zhu et al. address search results diversification as a learning problem where a ranking function is learned for diverse ranking [6].

None of the aforementioned methods are aimed at search results diversification in Twitter but the traditional way.

2.3   Diversifying Twitter Search Result

Tao, Hauff, and Houben create a microblog-based corpus (Tweets2013) for search result diversification experiments and a comprehensive analysis of the corpus showed its suitability for this purpose [16]. In this paper, we use this corpus to evaluate the performace of our approach.

Ozsoy, Onal, and Altingovde present an empirical analysis of a variety of search result diversification methods adopted from the text summarization and web search domains for the task of tweet ranking. Their experiments revealed that the implicit diversification methods outperform a popular explicit method, xQuAD, due to the vocabulary gap between the official query sub-topics and tweets [8].

Nevertheless, social information of Twitter was not taken into account by those methods.

3   Problem Definition

The diversification problem in Twitter can be naturally stated as a tradeoff between finding relevant tweets and diversifying the retrieved results:

Given an initial ranking R for a query q, find a re-ranking S that has the maximum coverage and the minimum redundancy with respect to the different aspects underlying q.

4   Overview of Our Approach

We adopt a learning approach to learn a diverse ranking function for tweets and use the tweet content, sub-topic of tweet and Twitter specific social information as features for diversifying tweets.

4.1   Learning to Rank Framework

Learning to rank is a data driven approach which effectively integrates a bag of features in the model. Figure 1 shows the paradigm of learning for tweet diverse ranking.

Firstly, a set of queries Q with related tweets and the ground-truth of the ranking in the form of a vector of ranking scores or a ranking list were used as training data. A bag of features related to the tweets is extracted to form a feature vector. Then a learning to rank algorithm is used to train a diverse ranking model. For a new query, their related tweets, which extract the same features to form feature vector, can be ranked by the diverse ranking function based on this model. The ranking performance of the model using a particular of feature sets in testing data can reflect the effect of these features for search results diversification in Twitter.

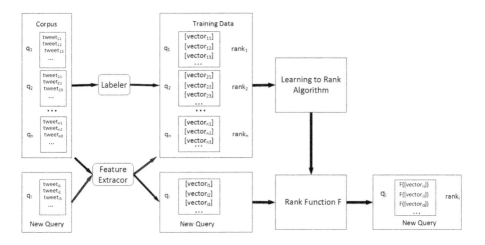

Fig. 1. Framework of learning to diverse rank in Twitter

4.2   Features for Diverse Ranking

We exploit three types of features for tweet diverse ranking:

(1) Tweet content feature refers to the one which describes the content diversity between tweets
(2) Sub-topic of tweet feature refers to the one that represents the sub-topic difference between tweets
(3) Twitter specific social features refer the ones that represent the particular characteristics of Twitter and the author's information.

In the next section, we will describe these three types of features in detail.

5   Feature Description

5.1   Tweet Content Feature

We use text dissimilarity as the content feature between tweets.

Text Dissimilarity. Obviously, text dissimilarity is meaningful for diversity. Two tweets may refer to the same sub-event if their texts are similar. We propose to represent it as the cosine dissimilarity based on weighted term vector representations, and define the feature as Eq. 1:

$$TT_1 = 1 - \frac{t_i \cdot t_j}{||t_i||\,||t_j||} \tag{1}$$

where t_i, t_j are the weighted tweet vector based tf*idf. The tf denotes the term frequency and idf denotes inverse document (tweet) frequency.

5.2   Sub-topic Feature

Sub-topics are the emerging topics related to an ongoing real-world topic. For instance, among all tweets[2] refer to the query "hillary clinton resign", we can discover "last, politician, popular, obama, clinton" and "look, woman, india, resign, speech" as different sub-topics which are characterized by a set of words. The sub-topic differs from sub-event in that sub-topics mostly happen across the same broad time interval, and with considerable everlap. However, intuitively, there are also some relationships between sub-topic and sub-event. Different tweets may associate with different sub-topics and two tweets may refer to the same sub-event if their sub-topics (characterized by a set of words) are similar. Therefore, we include the sub-topic of tweet as a feature for the diversity task and use Hierarchical Dirichlet Processes (HDP) to model implicit sub-topics distribution of candidate objects. For a pair of tweets, we define the implicit sub-topic feature as Eq. 2:

$$TT_4 = \sqrt{\sum_{k=1}^{m} (p(z_k|t_i) - p(z_k|t_j))^2} \tag{2}$$

where $p(z_k|t_i)$ is the probability distribution of $tweet_i$ on sub-topic z_k. We use hierarchical topic modelling to detect sub-topics in Twitter and conduct the sub-topics distribution using the HAC software[3]. Table 1 shows some sub-topics of tweets related to the query "hillary clinton resign" detected using HCA software and the probability distribution of tweet "Hillary Clinton tops Obama as most popular US politician" on the detected sub-topics.

Table 1. Sub-topics detected using HCA software and probability distribution of tweet on the detected sub-topics

subtopic	represented words	tweet
Topic1	last, politician, popular, obama, clinton	0.9807
Topic2	look, woman, india, resign, speech	0.0030
Topic3	run, go, want, hat, snap	0.0026
Topic4	justin, send, bieber, gonna, lift	0.0034
Topic5	clinon, dick, give, email, check	0.0017
Topic6	better, miss, bank, wemen, bill	0.0032
Topic7	first, john, visit, iraq, kabul	0.0002
Topic8	week, sworn, post, know, vote	0.0020
Topic9	support, former, samesex, syria, hurrah	0.0027

---

[2] Tweets come from Tweets2013 corpus.

[3] The software can be downloaded from http://mloss.org/software/view/527/.

5.3   Twitter Specific Social Features

Twitter has many special characteristics of social media. We exploit them and extract Twitter specific social features as follows:

Time Window Size. Two tweets may refer to the same sub-event if their post times are close. Time dissimilarity score between two tweets is based on the difference between their normalized timestamps (using Min-Max Normalization), and computed by Eq. 3:

$$TT_2 = |t_{norm}(t_i) - t_{norm}(t_j)| \tag{3}$$

where $t_{norm}(t_i)$, $t_{norm}(t_i)$ are the normalized timestamps of two tweets. For example, given the minimal timestamps *"Fri Feb 01 00:09:29 +0000 2013"* and the maximal timestamps *"Sun Mar 31 23:57:58 +0000 2013"*, the normalized timestamp of *"Tue Mar 25 14:45:00 +0000 2008"* is 0.387101.

Hashtags Dissimilarity. A hashtag refers to a word in the tweet that begins with the "#" character. It is used to indicate the topic of the tweet. Two tweets may refer to the same sub-event if they contain the same hashtags. For a given pair of tweets, we compute the Jaccard dissimilarity of the set of hashtags, which is defined as Eq. 4:

$$TT_3 = 1 - \frac{|Terms(t_i) \cap Terms(t_j)|}{|Terms(t_i) \cup Terms(t_j)|} \tag{4}$$

where $Terms(t_i)$, $Terms(t_j)$ are the sets of hashtags of two tweets.

Mentions Dissimilarity. In a tweet, people usually use "@" preceding a user name to reply other user (Mention). If two tweets related to one event mention the same person, two authors may talk about the same sub-event with the person they mention both. Therefore, we use a binary feature indicating whether two tweets contain the same "@username".

URL Dissimilarity. Sharing links in tweets is very popular in Twitter. Most tweets containing a link usually give an introduction to the links. Two tweets may refer to the same sub-event if they contain the same links. Therefore, we use a binary feature indicating whether two tweets contain the same links.

Author Dissimilarity. Twitter is a social network. The rich author information can also be used for the task. We use the following features related to the author of tweet: **location, language, verified, statuses, followers, friends, listed**. People living in the same area may care about the same event. Therefore, we consider the location information of authors and use a binary feature indicating whether two authors live in the same area. Intuitively, people use the same language are more likely to care about the same event than those use different languages. We use a binary feature indicating whether two authors use the same language. Additional information related the authority and activeness of users may also reveal the dissimilarity between events people care about. We use a binary feature indicating whether two authors are both verified users. As for the number of statuses, followers, friends and listed, we map them into an interval from 0 to 1 and calculate the dissimilarity between them.

6   Experiment

6.1   Dataset and Experimental Settings

For our evaluations, we use the Tweets2013 corpus[4] that is specifically built for tweet search results diversification problem. The dataset includes tweets collected between February 1, 2013 and March 31, 2013. There are forty seven query topics and each topic has 9 sub-topics on the average.

For learning method, we use the relational learning to rank algorithm proposed by [6]. Different from the traditional learning to rank algorithm where a ranking function is defined on the content of each individual document (tweet), the relational learning to rank algorithm considers both contents of individual document (tweet) and relationships between documents (tweets). Therefore, the diverse ranking function is defined as follows:

$$f_s(x_i, R_i) = w_r^T x_i + w^T h_s(R_i) \tag{5}$$

where x_i denotes the relevance features vector of the candidate tweet x_i, R_i stands for the matrix of relationships between tweet x_i and other selected tweets. For relevance, we use several features as summarized in Table 2.

Table 2. Relevance features for relational learning to rank algorithm

Feature	Description
$REL_{content}$	Content similarity between tweet and query
$REL_{hashtag}$	Whether a tweet contains a hashtag
$REL_{mention}$	Whether a tweet contains "@username"
REL_{url}	Whether a tweet contains links
$REL_{verified}$	Whether the author is a verified user
$REL_{statuses}$	The number of statuses
$REL_{friends}$	The number of friends
$REL_{followers}$	The number of followers
REL_{listed}	The times the author of a tweet has been listed

In particular, our experimentation focus on three main questions:

(1) Is the learning to rank approach effective on search results diversification problem in Twitter?
(2) Do the diversity features help in improving the Twitter retrieval?
(3) How does the individual feature effect on search results diversification in Twitter?

Five-fold cross validation is used in our experiments. We choose tweets of 27 queries as the training data. The remaining tweets are divided into evaluation data and validation data equally.

---

[4] The corpus is publicly available at http://wis.ewi.tudelft.nl/airs2013.

6.2   Evaluation Metrics

We evaluate diversification methods using the ndeval software[5] employed in TREC Diversity Tasks. We report results utilizing three popular metrics, namely, α-nDCG [17], Precision-IA [3], and Subtopic-Recall [18] at the cut-off values of 10 and 20, as typical in the literature.

6.3   Results

We first investigate whether the learning method is effective on search results diversification problem in Twitter. To evaluate the performance of learning method, we compare it with the state-of-the-art approaches [8], which are introduced as follows:

Max Marginal Relevance (MMR) is a classical implicit diversity method in the diversity research. It employs a linear combination of relevance and diversity [4].

eXplicit Query Aspect Diversification (xQuAD) is an explicit diversification method based on the assumption that aspects of a query can be known apriori [5].

Simple Yet (Sy) [16] present a framework for detecting duplicate and near-duplicate tweets and define a simple yet effective diversification method, so-called Sy [16].

While implementing these approaches in Twitter, [8] use three types of features, namely, the content, hashtags, time features for computing the similarity between tweets and the only content feature for computing the relevance between query and tweet. Therefore we choose to conclude those features in our learning method, marked as *RLTR*. The performaces of *RLTR* and baselines are shown in Table 3.

Table 3. Performance of Diverse Ranking Methods

Methods	α-**nDCG**		Prec-IA		ST-Recall	
	@10	@20	@10	@20	@10	@20
MMR	0.341	0.374	0.066	0.056	0.417	0.539
xQuAD	0.235	0.263	0.050	0.041	0.302	0.419
Sy	0.384	0.383	0.083	0.069	0.419	0.542
RLTR	0.382	0.417	0.058	0.051	0.351	0.483

From the results, we can see that our learning method achieves comparable performance with the traditional diversification approaches in the evaluations of most measures. Specially, in the evaluation of α-nDCG@20, the relative

---

[5] http://trec.nist.gov/data/web10.html.

improvement of our learning approach over the best of traditional diversification approaches is up to 8.9 %. The results demonstrate the effectiveness of the learning method in tackling the search results diversification problem in Twitter.

Next, we investigate whether the diversity features can help in improving the Twitter retrieval. To demonstrate the effectiveness of diversity features in Twitter retrieval, we made a contrast experiment. The first experiment only considers the relevance features between tweet and query, marked as *REL*, while the second takes into account all the diversity features on the basis of the first, marked as *REL+DIV*. The results of performace between *REL* and *REL+DIV* are shown in Table 4.

Table 4. Performance of Ranking Methods. A significant improvement over the baseline with bold respectively ($p < 0.05$)

Methods	α-**nDCG**		Prec-IA		ST-Recall	
	@10	@20	@10	@20	@10	@20
REL	0.282	0.322	0.045	0.043	0.273	0.445
REL+DIV	**0.491**	**0.510**	**0.074**	**0.069**	**0.441**	**0.578**

From the results, we can see that the *REL+DIV* outperforms the *REL* in the evaluations of all measures. In particular, the relative improvement of the Precision-IA@10 and Precision-IA@20 is up to 64.4 % and 60.4 %, the improvement of the α-nDCG@10 and α-nDCG@20 is up to 71.4 % and 58.4 %. The results demonstrate the effectiveness of the diversity features in Twitter retrieval.

Finally, we investigate the effects of sub-topic feature and each of individual Twitter specific social features. We use the learning method which takes into account the relevance features and leverages diversity feature, tweet content as a baseline, marked as *BASE*. We combine each individual feature with the *BASE*. Table 5 shows the performance of each ranking model.

From the results, we can see that sub-topic feature and the Twitter specific social features can help tackle the search results diversification task in Twitter, especially the post time, hashtags of tweet and the location of author.

Comparing with the baseline, the relative improvement of the model taking into account the sub-topic feature is up to 8.8 % and 10.9 % in the evaluations of α-nDCG@10 and α-nDCG@20 respectively. As for the sub-topic feature, different tweets may associate with different sub-topics and two tweets may refer to the same sub-event if their sub-topics are similar. For example, as for the query "North Korea nuclear", there are three relative tweets[6] and their distributions on sub-topics are shown as Table 6:

– *tweet$_4$: US and China agree on North Korea sanctions after nuclear test - Fox News: San Francisco ChronicleUS and China* http://t.co/9qs4CGruYb

---

[6] Tweets come from Tweets2013 corpus.

Table 5. Performance of Each Ranking Model. A significant improvement over the baseline with bold respectively (p < 0.05)

Methods	α-nDCG		Prec-IA		ST-Recall	
	@10	@20	@10	@20	@10	@20
BASE	0.417	0.428	0.056	0.049	0.349	0.465
+subtopic	**0.454**	**0.475**	**0.067**	**0.061**	**0.421**	**0.574**
+time	**0.445**	**0.463**	**0.064**	**0.057**	**0.393**	**0.536**
+hashtags	**0.431**	**0.459**	**0.063**	**0.059**	**0.401**	**0.519**
+location	**0.433**	**0.455**	**0.060**	0.051	**0.387**	**0.518**
+url	0.419	0.431	0.058	0.050	0.349	0.462
+verified	0.402	0.421	0.056	0.048	0.341	0.471
+language	0.413	0.420	0.054	0.049	0.339	0.466
+statuses	0.420	0.431	0.058	0.051	0.355	0.470
+friends	0.417	0.426	0.053	0.049	0.346	0.462
+followers	0.399	0.405	0.051	0.047	0.347	0.453
+listed	0.401	0.409	0.050	0.049	0.353	0.457
best	**0.499**	**0.514**	**0.075**	**0.068**	**0.447**	**0.580**

- $tweet_5$: *NEWSFLASH: Diplomats say US, China agree on new sanctions to punish North Korea for nuclear test. More as we get*
- $tweet_6$: *How Powerful Was #NorthKorea;s #Nuclear Test? #SouthKorea believes 7 kilotons; #Japan,10 & #Germany institute*

The $tweet_4$ and $tweet_5$ both refer to the "reaction of China to the nuclear test", while the $tweet_6$ refers to the "details of the third nuclear test". We can see from Table 6 that the distributions of the $tweet_4$ and $tweet_5$ are similar, while the $tweet_6$ is not. Therefore, the sub-topic of tweet can indeed help diversifying tweets.

We can also observe that the ranking model yield significant improvement over the baseline when taking into account time feature. The relative improvement is up to 6.7 % and 8.2 % in the evaluations of α-nDCG@10 and α-nDCG@20 respectively. As for the time feature, two tweets may refer to the same sub-event if their post times are close. For example, as for the query "hillary clinton resign", there are three relative tweets[7]:

- $tweet_7$: *TN China: Secretary of State Hillary Clinton formally resigns: Her resignation is effective upon the swearing[created_at:Fri Feb 01 20:39:04 +0000 2013]*
- $tweet_8$: *Secretary of State Hillary Clinton formally resigns: Her resignation is effective upon the swearing-in of John[created_at:Fri Feb 01 20:43:07 +0000 2013]*
- $tweet_9$: *Hillary Clinton: As Hillary Clinton leaves office after four years, John Kerry prepares to take over[created_at:Fri Feb 01 10:00:14 +0000 2013]*

---

[7] Tweets come from Tweets2013 corpus.

Table 6. Probability Distributions of Tweets on Sub-Topics

sub-topics	$tweet_4$	$tweet_5$	$tweet_6$
Topic1	0.0151	0.0212	0.0031
Topic2	0.0061	0.0868	0.0127
Topic3	0.8705	0.6907	0.0869
Topic4	0.0215	0.0248	0.0056
Topic5	0.0106	0.0140	0.0007
Topic6	0.0238	0.0141	0.0006
Topic7	0.0129	0.0133	0.0013
Topic8	0.0213	0.0203	0.8885
Topic9	0.0178	0.0648	0.0006

The $tweet_7$ and $tweet_8$ both refer to the "details of resignation" while the $tweet_9$ to "who follows Clinton as secretary of state". Obviously, the post times of the $tweet_7$ and $tweet_8$ are close, while the $tweet_9$ is far from them. Therefore, the time feature can improve the diversifying performance.

From Table 5, we can see that the model taking into account hashtags features outperforms the baseline in all settings. Especially, in the evaluations of α-nDCG@10 and α-nDCG@20, the relative improvement is up to 3.4 % and 9.6 %. As for the hashtags feature, two tweets may refer to the same sub-event if they contain the same hashtags. For example, as for the query "syria civil war", there are three relative tweets[8]:

- $tweet_{10}$: *RT @SyriaDayofRage: (02-21-13) #Damascus #Syria l Rebels Move Closer to Central Damascus as Clashes continue and Rebel Rockets hitting*
- $tweet_{11}$: *So far today 17 martyrs were reported in #Damascus and its suburbs, 3 in #Idlib, 3 in #Hama,3 in #Aleppo,2 in #Homs*
- $tweet_{12}$: *RT @lysdeschamps: #Syria 13 02 Killed: 231. 20 children, 98 women, 59 rebels. Aleppo 50 Idlib 33 Damascus 30 Daraa 21 Homs 18 Deir Azzo*

The $tweet_{10}$ and $tweet_{11}$ both refer to the "cities where the fighting is" while the $tweet_{12}$ refers to the "casualties of the war". we can see that the $tweet_{10}$ and $tweet_{11}$ both contain hashtag "#Damascus" which is a city, while the $tweet_{12}$ not. Therefore, the hashtag feature can also help tackling the task.

As for the location of author, the relative improvement of the model taking into it over the baseline is up to 3.8 % and 6.3 % respectively. It is due to that peoples who live in the same area may concern about the same sub-topic of event.

Finally we add all the features which can significantly improve the diversifying performance into a ranking model. They are **sub-topic**, **time**, **hashtag** and **location of author** features. Table 5 shows the **best** result of method which improves significantly over the baseline.

---

[8] Tweets come from Tweets2013 corpus.

7   Conclusion

To the best of out knowledge, we are the first to propose the learning method for search results diversification in Twitter. We explore a series of diversity features. The experimental results on the tweets2013 datasets demonstrate the effectiveness of our approach. Additionally, the Twitter retrieval task achieves improvement by taking into account the diversity features. Finally, we find the sub-topic and Twitter specific social features can help solve the diversity task, especially the post time, hashtags of tweet and the location of author.

References

1. Java, A., Song, X., Finin, T., Tseng, B.: Why we twitter: an analysis of a microblogging community. In: Zhang, H., Spiliopoulou, M., Mobasher, B., Giles, C.L., McCallum, A., Nasraoui, O., Srivastava, J., Yen, J. (eds.) WebKDD 2007. LNCS, vol. 5439, pp. 118–138. Springer, Heidelberg (2009)
2. Busch, M., Gade, K., Larson, B., Lok, P., Luckenbill, S., Lin, J.: Earlybird: Real-time search at twitter. In: IEEE 28th International Conference on Data Engineering (ICDE 2012), Washington, DC, USA (Arlington, Virginia), 1–5 April 2012, pp. 1360–1369 (2012)
3. Agrawal, R., Gollapudi, S., Halverson, A., Ieong, S.: Diversifying search results. In: Proceedings of the Second ACM International Conference on Web Search and Data Mining, pp. 5–14. ACM (2009)
4. Carbonell, J.G., Goldstein, J.: The use of MMR, diversity-based reranking for reordering documents and producing summaries. In: Proceedings of the 21st Annual International ACM SIGIR Conference on Research and Development in Information Retrieval, SIGIR 1998, August 24–28 1998, Melbourne, Australia, pp. 335–336 (1998)
5. Santos, R.L.T., Macdonald, C., Ounis, I.: Exploiting query reformulations for web search result diversification. In: Proceedings of the 19th International Conference on World Wide Web, WWW 2010, Raleigh, North Carolina, USA, 26–30 April 2010, pp. 881–890 (2010)
6. Zhu, Y., Lan, Y., Guo, J., Cheng, X., Niu, S.: Learning for search result diversification. In: The 37th International ACM SIGIR Conference on Research and Development in Information Retrieval, SIGIR 2014, Gold Coast, QLD, Australia, 06–11 July 2014, pp. 293–302 (2014)
7. Teevan, J., Ramage, D., Morris, M.R.: #Twittersearch: a comparison of microblog search and web search. In: Proceedings of the Forth International Conference on Web Search and Web Data Mining, WSDM 2011, Hong Kong, China, 9–12 February 2011, pp. 35–44 (2011)
8. Ozsoy, M.G., Onal, K.D., Altingovde, I.S.: Result diversification for tweet search. In: Benatallah, B., Bestavros, A., Manolopoulos, Y., Vakali, A., Zhang, Y. (eds.) WISE 2014, Part II. LNCS, vol. 8787, pp. 78–89. Springer, Heidelberg (2014)
9. Jabeur, L.B., Tamine, L., Boughanem, M.: Uprising microblogs: a bayesian network retrieval model for tweet search. In: Proceedings of the ACM Symposium on Applied Computing, SAC 2012, Riva, Trento, Italy, 26–30 March 2012, pp. 943–948 (2012)

10. Zhang, X., He, B., Luo, T., Li, B.: Query-biased learning to rank for real-time twitter search. In: 21st ACM International Conference on Information and Knowledge Management, CIKM 2012, Maui, HI, USA, 29 October–02 November 2012, pp. 1915–1919 (2012)

11. Luo, Z., Osborne, M., Petrovic, S., Wang, T.: Improving twitter retrieval by exploiting structural information. In: Proceedings of the Twenty-Sixth AAAI Conference on Artificial Intelligence, 22–26 July 2012, Toronto, Ontario, Canada (2012)

12. Luo, Z., Osborne, M., Wang, T.: Opinion retrieval in twitter. In: Proceedings of the Sixth International Conference on Weblogs and Social Media, Dublin, Ireland, 4–7 June 2012 (2012)

13. Carterette, B., Chandar, P.: Probabilistic models of ranking novel documents for faceted topic retrieval. In: Proceedings of the 18th ACM Conference on Information and Knowledge Management, CIKM 2009, Hong Kong, China, 2–6 November 2009, pp. 1287–1296 (2009)

14. Radlinski, F., Dumais, S.T.: Improving personalized web search using result diversification. In: SIGIR 2006: Proceedings of the 29th Annual International ACM SIGIR Conference on Research and Development in Information Retrieval, Seattle, Washington, USA, 6–11 August 2006, pp. 691–692 (2006)

15. He, J., Hollink, V., de Vries, A.P.: Combining implicit and explicit topic representations for result diversification. In: The 35th International ACM SIGIR Conference on Research and Development in Information Retrieval, SIGIR 2012, Portland, OR, USA, 12–16 August 2012, pp. 851–860 (2012)

16. Tao, K., Hauff, C., Houben, G.-J.: Building a microblog corpus for search result diversification. In: Banchs, R.E., Silvestri, F., Liu, T.-Y., Zhang, M., Gao, S., Lang, J. (eds.) AIRS 2013. LNCS, vol. 8281, pp. 251–262. Springer, Heidelberg (2013)

17. Clarke, C.L.A., Kolla, M., Cormack, G.V., Vechtomova, O., Ashkan, A., Büttcher, S., MacKinnon, I.: Novelty and diversity in information retrieval evaluation. In: Proceedings of the 31st Annual International ACM SIGIR Conference on Research and Development in Information Retrieval, SIGIR 2008, Singapore, 20–24 July 2008, pp. 659–666 (2008)

18. Zhai, C., Lafferty, J.D.: A risk minimization framework for information retrieval. Inf. Process. Manage. 42(1), 31–55 (2006)

Adjustable Time-Window-Based Event Detection on Twitter

Qinyi Wang[1], Jieying She[2], Tianshu Song[1], Yongxin Tong[1(✉)],
Lei Chen[2], and Ke Xu[1]

[1] SKLSDE Lab and IRC, Beihang University, Beijing, China
{wangqinyi,songts,yxtong,kexu}@buaa.edu.cn
[2] The Hong Kong University of Science and Technology, Hong Kong SAR, China
{jshe,leichen}@cse.ust.hk

Abstract. Twitter has become an important platform for reporting breaking news and instant events. However, it is almost impossible to detect events on Twitter manually due to the large volume of data and the noise in them. Though automatic event detection has been studied a lot, most works can only detect events in a fixed time window. In this paper, we propose an efficient system that can detect events in adjustable time windows. We detect terms with unusual frequency and group them into events. We further modify a segment tree data structure to support adjustable time window based event detection, which can efficiently aggregate statistics of terms of varied-sized time windows and is both space and time saving. We finally validate the effectiveness and efficiency of our proposed techniques through extensive experiments on real datasets.

1 Introduction

Twitter has become an important platform for fast reporting and broadcasting of news and events these days mainly due to it convenience and speed of spreading information. However, due to the large volume of data generated, monitoring and detecting events manually is infeasible and automatic detection is required. Although automatic event detection has been well studied in traditional media, i.e. the Topic Detection and Tracking (TDT) research program, event detection on Twitter is much more challenging due to its noisy data.

Some event detection techniques have been proposed recently. However, they fail to meet the following adjustable time window based event detection requirement. Suppose some users want to know what is happening around the world by monitoring Twitter in real-time. They would like to be reported breaking news or the hottest events during the past 120 min. When spotting a hot event in the past 120 min, they want to learn about how it developed and thus would like to check what happened during the past 60, 30 and 10 min.

Existing techniques cannot solve the above scenario due to the following reasons. First, they do not support detecting events in adjustable sizes of time windows and only divide the timeline into fixed time windows where events are

B. Cui et al. (Eds.): WAIM 2016, Part II, LNCS 9659, pp. 265–278, 2016.
DOI: 10.1007/978-3-319-39958-4_21

to be detected, which may not adapt to detecting different types of events or attract users desiring various degrees of real-time. Second, many works only focus on detecting category-specific or topic-specific events [8,12], which cannot attract users with broad interests. Third, some techniques, e.g. learning topic models [11], were too complex to support real-time application, which is important in case of breaking news and emergent events. Finally, some unsupervised techniques required several parameters or thresholds to detect events by identifying bursty terms [8], which could be sensitive to the parameter settings and hard to adapt to the dynamic environment of Twitter.

In this paper, we address the above problems with a solution with the following features. First, our solution supports real-time applications. We perform simple operations on the data with low latency, enabling timely event report. The time and memory consumed to detect events should not increase as time elapses and only recent statistics of data will be kept while obsolete ones will be discarded. Second, our solutions is completely unsupervised and no labeled data is required. Third, our solution is free of thresholds to identify bursty terms. Finally, we design a space-saving data structure that can efficiently detect events w.r.t. adjustable sizes of time windows.

Our approach achieves the goals by detecting unusually frequent words within an adjustable time window. We assume that an emerging event is indicated by a group of suddenly frequent terms, and we define a burst score for each term and group bursty terms based on their co-occurrence into candidate events. To enable adjustable time window based event detection, we modify the segment tree, Streaming Timeline Tree (ST-tree). More specifically, we first assume a minimum time window as a time unit, and an adjustable time window as several continuous time units. By properly storing statistics that we need to calculate scores for terms in the nodes and manage obsolete nodes and new ones, we are able to support adjustable time window based event detection.

The major contributions of our paper include:

- We propose a simple, effective and efficient method to detect open-domain emerging events.
- We design a data structure to support adjustable time window based event detection.
- We verify the effectiveness and efficiency of the proposed methods through extensive experiments on real datasets.

The rest of the paper is organized as follows. In Sect. 2, we formally define our problem. We then describe in detail our event detection algorithm in the Sect. 3. In Sect. 4, we present our ST-tree for flexible time frame event detection. Evaluation of our proposed methods is conducted in Sect. 5. We review previous works in Sect. 6 and conclude this work in Sect. 7.

2   Problem Formulation

We address our problem in a real-time environment, where we continuously receive and process streaming data from Twitter. A group of consecutive time windows (time frame) $f_{c-n+1}, \cdots, f_c$ are some non-overlapping time intervals of equal size on the timeline, where f_c is the most recent one. A time window is the basic unit which we identify events for. Note that "time frame" and "time window" are the same meaning and they are used interchangeably in this paper.

The major characteristic of an emerging event is that more and more people start to talk about it and some relevant terms, which will exhibit unusually higher frequency patterns than past records. We name them as abnormal terms. Three factors determine whether a term is abnormal or not: term frequency, the number of users using the term (i.e. user frequency), and its frequency statistics in previous time windows. More details will be described in the next section. In our work, we detect events with a sliding time window setting.

Definition 1 (Sliding time Window). *Given the current time window f_c with length L starting at time t_i and ending at $t_i + L$, the next sliding time window f_{c+1} begins at time $t_i + G$ and ends at $t_i + G + L$, where G is the sliding gap such that $G < L$ and L is divisible by G.*

Definition 2 (Event Detection). *The event detection problem is to identify groups of abnormal terms within the current time window and report the most abnormal groups as possible events in a sliding time window setting.*

We next define our adjustable time window based event detection problem.

Definition 3 (Time Unit). *A time unit is the minimal size of the time window available.*

Definition 4 (Adjustable Time Window). *An adjustable time window is one that consists of several consecutive time units, and the number of time units is specified by users.*

We limit the maximum number of time units an adjustable time window can consist. The reason is that a very long time window is uninteresting to Twitter users as mainstream media will catch up and report the events later. Though a larger time window may be applicable to event tracking or summarizing, we focus on detecting emerging events and thus define a maximum time window.

Definition 5 (Adjustable Time Window Based Event Detection). *The adjustable time window based event detection problem is to identify groups of abnormal terms within the current adjustable time window and report the most abnormal groups as possible events in a sliding time window setting with a sliding gap equal to the length of a time unit.*

3   Event Detection

3.1   System Overview

Our event detection system consists of two stages of processing. The first stage is to process each new tweet and store some statistics, and the second stage is to identify events at the end of the current time window.

3.2 Processing of Tweets

For each new tweet, we use unigrams as terms, since we find that unigrams outperforms n-grams in both effectiveness and efficiency. Details will be presented in Sect. 5. We then update the statistics of the current time window f_c: the number of tweets cd_{f_c} within f_c and for each term w a set of ids $tweet\_set_{f_c,w}$ referring to the tweets that contain the term and its frequency $cw_{f_c,w}$. For each term, we further maintain $uw_{f_c,w,u}$ for each user u, which is the number of tweets published by u during f_c that contain w, and $uwt_{f_c,w}$:

$$uwt_{f_c,w} = \sum_u (2 - \frac{1}{2^{uw_{f_c,w,u}-1}})(uw_{f_c,w,u} > 0) \tag{1}$$

We observe that around 30 % of tweets are retweets, many of which are generated by followers of celebrities, which makes the corresponding cw and uwt extremely large and thus results in false alarms. Note that breaking news could also trigger a large number of retweets, but it has longer retweeting chains and more origins (i.e. the original authors). Thus, when dealing with a tweet published by u_r and retweeted from u_o, we update uw_{f_c,w,u_o} instead of uw_{f_c,w,u_r} to reduce false alarms caused by retweets.

3.3 Identification of Abnormal Terms

Identification of abnormal terms is performed at the end of the current time window. We define significance score $ss_{f_c,w}$, abnormality score $as_{f_c,w}$ and abnormal terms as follows. Note that when calculating the average scores $ss_{avg,w}$ over past time windows, we use the most recent past non-overlapping time windows.

$$ss_{f_c,w} = \frac{cw_{f_c,w}}{cd_{f_c}} * \frac{uwt_{f_c,w}}{cd_{f_c}} \tag{2}$$

$$ss_{avg,w} = \frac{cw_{avg,w}}{cd_{avg}} * \frac{uwt_{avg,w}}{cd_{avg}} \tag{3}$$

$$as_{f_c,w} = \frac{ss_{f_c,w}}{ss_{avg,w}} \tag{4}$$

Definition 6 (Abnormal Term). *A term with abnormality score larger than 1 is abnormal.*

Notice that we simply regard all terms with scores higher than historical records as abnormal, as any threshold is sensitive to the fast changing data and it is impossible to define a universal threshold to effectively select terms under our unsupervised scheme.

3.4   Clustering Abnormal Terms

We then group terms into events. The main idea is to cluster two terms based on how often they co-occur as words that co-occur often are more likely to refer to the same event. We introduce a threshold *cluster_thres* to determine whether two terms should be grouped together, which controls how cohesive a cluster is.

We maintain a set $term\_set_{f_c,m}$ for each tweet m containing abnormal terms, which consists of all the abnormal terms mentioned in m. Such sets can be easily created from *tweet_set* of abnormal terms after we identify them. And we have the following lemma stating under what condition two terms co-occur.

Lemma 1. *Given a term w_i and $tweet\_set_{f_c,w_i}$, any term $w_j(\neq w_i) \in \cup_{m_i \in tweet\_set_{f_c,w_i}} term\_set_{f_c,m_i}$ must co-occur with w_i for at least once. Any other term $w_k(\neq w_i) \notin \cup_{m_i \in tweet\_set_{f_c,w_i}} term\_set_{f_c,m_i}$ does not co-occur with w_i.*

Then during clustering, we find highly co-occurring terms for each term w_i by scanning each term in $\cup_{m_i \in tweet\_set_{f_c,w_i}} term\_set_{f_c,m_i}$ and counting the frequency of the terms. The process is illustrated in Algorithm 1.

Algorithm 1. Clustering Abnormal Terms

input : Abnormal terms $\{w_1, \cdots, w_n\}$, $\{tweet\_set_{f_c,w_i}\}$, $\{term\_set_{f_c,m_i}\}$
output: Clusters $\{C_1, C_2, ..., C_m\}$

1 Mark each term as one cluster containing only itself;
2 **foreach** $i \leftarrow 1$ to $n - 1$ **do**
3   **foreach** $m_i \in tweet\_set_{f_c,w_i}$, $w_j \in term\_set_{f_c,m_i}$ **do**
4   **if** w_i and w_j are already in the same cluster **then**
5           $\lfloor$ continue;
6   **if** $w_j \notin candidate$ **then**
7           $\lfloor$ Add w_j to candidate, $cnt_{w_j} \leftarrow 1$;
8   **else**
9           $\lfloor$ $cnt_{w_j} \leftarrow cnt_{w_j} + 1$;
10   **foreach** $w_j \in candidate$ **do**
11   **if** $cnt_{w_j} / \min(|tweet\_set_{f_c,w_i}|, |tweet\_set_{f_c,w_j}|) > cluster\_thres$ **then**
12          $\lfloor$ Combine the clusters containing w_i and w_j;

13 **return** *Clusters* $\{C_1, C_2, ..., C_m\}$

3.5   Ranking Events

Finally, we return the top-k ranked clusters based on their abnormal terms:

Definition 7 (Rank of Clusters). *Given two clusters C_i and C_j, C_i ranks higher than C_j if and only if $as_{f_c,w_{i_1}} > as_{f_c,w_{j_1}}$, or $as_{f_c,w_{i_1}} = as_{f_c,w_{j_1}}$ and $as_{f_c,w_{i_2}} > as_{f_c,w_{j_2}}$, where w_{i_1}, w_{i_2} are the two terms of C_i with the highest and second-highest abnormality scores respectively, and w_{j_1}, w_{j_2} are the two of C_j with the highest and second-highest abnormality scores respectively.*

3.6   Complexity Analysis

The major time-consuming component is clustering, whose complexity is $O(n *$
$M * K)$, where n is the number of abnormal terms, M is the maximum size of
a *tweet_set*, and K is the maximum size of a *term_set* (relatively small). We
maintain six statistics for the current time window, and three for the past few
time windows, whose number is $V * L/G$, where V is the number of past time
windows. Thus, the total space complexity is $O(2W(VL/G+1)+UW+2MW)$,
where W is the number of terms, and U is the maximum number of users.

4   Adjustable Time Window Event Detection

In this section, we introduce the ST-tree structure. Let U_L denote the length of
a time unit, then $L_M * U_L$ is the maximum length of time window supported.

4.1   Description of ST-tree

Basically, ST-tree is almost identical to segment tree that allows efficient query
for statistics of an arbitrary interval. We update ST-tree in a different way. We
first present some basic concepts of ST-tree in this subsection.

ST-tree is a binary tree. Each leaf represents a time unit, and each parent
is the union of the time intervals its children refer to. We have $(V + 1) * L_M$
time units(leaves) in ST-tree, and where V is the number of time windows we
average on. Suppose the current time unit is tu_c, the root then covers the whole
time interval starting from the starting point of unit $tu_c - (V+1) * L_M + 1$, and
ending at the ending point of tu_c. Figure 1a shows an example of ST-tree.

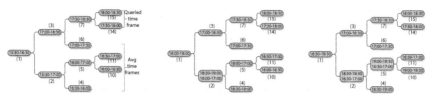

(a) The Original ST-tree   (b) After the 1st Update   (c) After the 2nd Update

Fig. 1. An example of ST-tree with time units of length 30 min ($L = V = 2$).

Algorithm 2. Build ST-tree

---

input: Current node x, st and ed
1 Mark x as representing $[st, ed)$;
2 **if** *Length of $[st, ed) > U_L$* **then**
3   Create children x_{child_left} and x_{child_right} of x;
4      $middle \leftarrow st + \lfloor \frac{(ed-st)}{2U_L} \rfloor U_L$;
5   Run 2 with input x_{child_left}, st and $middle$;
6   Run 2 with input x_{child_right}, $middle$ and ed;

Algorithm 2 presents how the structure of ST-tree is built during initialization, starting from the root given the initial interval $[st, ed]$.

Each node of ST-tree stores three statistics for the time interval it represents: cw, uwt and cd. When processing the current time unit, we could directly update the three statistics stored at the responding leaf by creating index for the leaves and visit them directly as we process the leaves from left to right in order. Updates for the other nodes are initiated when we arrive at the end of the current time unit. The detailed procedure will be described soon.

Besides the ST-tree, we also store $tweet\_set$ in the most recent L_M time units. Merging of the tweet sets of the most recent L time units for each term will be performed given a query for events in time window of length $L*U_L(1 \leq L \leq L_M)$.

4.2 Update

4.2.1 Calculation of Statistics

When at the end of the current time unit, we update ancestors of the current leaf. The statistics stored at a parent representing time interval f_{parent} are following (5) to (7), where f_{child_left} and f_{child_right} are the time intervals represented by its children.

$$cw_{f_{parent},w} = cw_{f_{child\_left},w} + cw_{f_{child\_right},w} \tag{5}$$

$$uwt_{f_{parent},w} = uwt_{f_{child\_left},w} + uwt_{f_{child\_right},w} \tag{6}$$

$$cd_{f_{parent}} = cd_{f_{child\_left}} + cd_{f_{child\_right}} \tag{7}$$

The true value of $uwt_{f_{parent},w}$ should not be the sum of $uwt_{f_{child\_left},w}$ and $uwt_{f_{child\_right},w}$ following the original definition of uwt. However, since storing a user set for each term in each node is too space-consuming, we instead store a single value and calculate the value of uwt by (6). In this way, we reduce our space and time consumption by relaxing the constraints on term's user frequency.

The whole process of updating is as follows: starting from the parent of the current leaf, we visit each node in the path from the current leaf to the root and update the corresponding statistics following (5) to (7).

4.2.2 Dealing with Obsolete Nodes

When processing the first $(V + 1)L_M$ time units, all we have to do is to fill in the values of the nodes. However, starting from the $(V + 1)L_M + 1$ time unit, we should keep the most recent statistics and discard the obsolete ones, as the statistics of unit $tu_c - (V + 1)L_M + 1$ will no longer be used when we proceed to the next unit $tu_c + 1$. We also need to find a node to store statistics of $tu_c + 1$.

Removing or adding nodes to the original tree will destroy its balanced structure. Thus, we propose to use the leaf representing $tu_c - (V + 1)L_M + 1$ to store statistics of $tu_c + 1$, so the oldest leaf becomes the newest leaf after update. Figures 1b and c show how the ST-tree changes after two updates of the original one in Fig. 1a.

Algorithm 3. Update of ST-tree

---

 input : ST and the current leaf x
 output: Updated ST
1   Mark each term as one cluster containing only itself;
2   **while** x *is not the root of* ST **do**
3     $x \leftarrow x$'s parent;
4   Update interval and statistics of x following (5) to (7);
5   **return** ST

Algorithm 4. Return Statistics of a Given Time Interval

---

 input : Current node x, ST and $[t_s, t_t)$
 output: $\{cw_{[t_s,t_t),w_i}\}$, $\{uwt_{[t_s,t_t),w_i}\}$ and $cd_{[t_s,t_t)}$
1   Initialize $\{now\_cw_{w_i}\}$, $\{now\_uwt_{w_i}\}$ and now_cd to 0;
2   **if** *The interval of $x \subset [t_s, t_t)$* **then**
3   **return** $\{cw_{w_i}\}$, $\{uwt_{w_i}\}$ *and cd stored at x*

4   **else**
5   **foreach** *Each child x_c of x* **do**
6   **if** *The interval of x_c overlaps with $[t_s, t_t)$* **then**
7   Run Algorithm 4 with input x_c, ST and $[t_s, t_t)$, aggregate the returned results
          to $\{now\_cw_{w_i}\}$, $\{now\_uwt_{w_i}\}$ and now_cd;

8   **return** $\{now\_cw_{w_i}\}$, $\{now\_uwt_{w_i}\}$ *and* now_cd

---

The process is illustrated in Algorithm 3. We use the oldest leaf to store statistics for the coming unit, and update its time interval. When arriving at the end of the newest time unit, we update its ancestors as usual. The time interval represented by a node may no longer be continuous.

4.3 Query

The query process is identical to that of our original problem, except that we first need to obtain the statistics for the queried time window and the past V time window from ST-tree, which is equivalent to the following problem: given a time interval $[t_s, t_t)$, return its statistics. The process is performed by running Algorithm 4 with input root of ST, ST, and $[t_s, t_t)$.

Note that Algorithm 4 applies to queries issued in the first $(V + 1)L_M$ time units as well as those issued afterwards. For queries issued after the first $(V + 1)L_M$ time units, the time intervals represented by the nodes may not be continuous. However, in such case, line 3 of Algorithm 4 will not be executed and we will keep exploring the children of such nodes until we reach at a node representing a continuous time interval that is subset of $[t_s, t_t)$.

Example 1. Suppose the current ST-tree is Fig. 1c, and we query for the statistics of 17:30-19:00. By running Algorithm 4, we will visit nodes 1, 2, 4, 3 and 7 in order, and aggregate the statistics stored in node 4 and 7.

4.4   Complexity Analysis

We compare ST-tree with two naive methods. One is Space-Severe (SS), which stores statistics for all possibly used time windows. Another one is Time-Severe (TS), which stores statistics only for time units and aggregates the results during querying. We only count the number of time windows/units/nodes.

Let $N = L_M * (V + 1)$ denote the number of leaves in ST-tree. The space complexity of ST-tree is $O(N)$. SS takes $O(NL_M)$ space, and TS takes $O(N)$ space. ST-tree takes $O(N)$ to build the tree and $O(logN)$ time to update. SS takes $O(L_M)$ time to update, and TS takes only $O(1)$ time. For query, the averaged time taken by ST-tree is $O(logN)$. SS takes $O(V)$ time, and TS takes $O((V + 1)L)$ time.

5   Experimental Study

5.1   Experimental Setup

We perform our experiments on machines with 2xIntel E5-2650 (8-core, 2 GHz, 20 M cache) CPU and 64G DDR3-1333 RAM. For all the experiments, we select the top-20 clusters detected in each time window and output the top-20 terms with the highest abnormality scores.

Twitter. We collected in total 11,625,484 tweets through Twitter's streaming API[1] continuously from Oct 31 to Nov 6, 2013. Only tweets written in English are used for evaluation. We use the data on October 31 to calculate past statistics, and detect events since Nov 1. The statistics are presented in Table 1.

Wikipedia. We refer to Wikipedia's current eventsportal, which keeps a list of daily events around the world with description of a few sentences, to see whether our algorithm can identify such events.

The Wikipedia data is processed as follows. For each event, we first extract non-stop words from its description as keywords. Second, we select 48 events that are ever mentioned in the Twitter dataset. Finally, we follow the links provided by Wikipedia to the background articles to check the exact time when mainstream media reported the events if available. When evaluating our algorithm, we do not count any cluster reported with delay of more than one day. To check whether a certain event is detected, we select clusters that contain at least two keywords of the event and then check manually whether the selected clusters contain highly relevant keywords of the events.

Table 1. Statistics of Collected Data

Number of retweets	3,488,124
Number of tweets per minute (on average)	1,153
Number of users	6,418,278
Number of different retweeted users	1,268,762

[1] https://dev.twitter.com/docs/api/1.1/get/statuses/sample.

5.2   Effectiveness Evaluation

We next evaluate our effectiveness in both aspects: how many events are detected and the impact of different parameter settings.

Accuracy. We compare with TwiCAL[11] and Twevent [7] in this part of evaluation. TwiCAL maintains a continuously updating demo[2], and thus we can directly obtain a list of events detected by TwiCAL. Since we follow the methods and settings of Twevent to detect events, we also set $V = 3$, $U_L = 10$ min and $clu\_thres = 0.8$ for our proposed algorithm.

We present the results in Table 2. The only event that is detected by TwiCAL but not by the others is that India launches a PSLV-XL rocket. The results indicate that our algorithm detects more events than both TwiCAL and Twevent do. One possible reason for the bad performance of TwiCAL is that it uses supervised technique and annotates only a very small number of tweets to extract event phrases, and thus may not adapt to the dynamic environment of Twitter when new concepts appear. Twevent filters out unnecessary events and thus has much fewer false alarms than our algorithm does. However, Twevent fails to detect many important events. One possible reason is that it introduces thresholds to identify bursty segments, which may not adapt well to various burst patterns of different types of events.

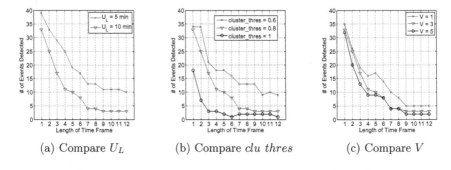

| (a) Compare U_L | (b) Compare $clu\ thres$ | (c) Compare V |

Fig. 2. Compare event detection results with different parameter settings.

Table 2. Number of events detected by different methods on different days

	Nov 1	Nov 2	Nov 3	Nov 4	Nov 5	Nov 6
TwiCAL	2	0	2	3	3	0
Twevent ($n = 1$)	0	0	0	0	1	0
Twevent ($n \leq 2$)	0	0	0	0	1	0
Twevent ($n \leq 3$)	1	0	0	0	1	0
Our Method	6	4	5	5	10	5

---

[2] http://statuscalendar.com.

Impact of Parameter Settings. We first compare the impact of U_L, clu_thres and V in Fig. 2. We set $U_L = 10$ min, $clu\_thres = 0.8$ and $V = 3$ if not specifying their varying values.

In Fig. 2a, we present the number of events detected with $U_L = 5$ min and 10 min. We can see that more events are detected with a shorter time unit. However, the drawback of using a shorter time unit is that we report events more frequently and thus may return more false alarm events.

We then compare how the clustering threshold affects the results. In Fig. 2b, we present the number of events detected with $clu\_thres = 0.6$, 0.8 and 1, respectively. The results show that a lower clu_thres yields better performance. However, we observe that the clusters generated by a lower clu_thres are more messed up with words referring to different events.

We finally evaluate how the number of past time windows we average on affects the results. The results in Fig. 2c indicate that our algorithm performs better with a smaller V. This may be that we could better adapt to the dynamic environment of Twitter when V is smaller.

5.3   Efficiency Evaluation

We finally evaluate whether our algorithms are suitable for real-time application. Particularly, the preprocessing component takes on average 0.8ms to process a tweet, which is far beyond the average arrival rate of tweets from Twitter's sampled streaming API (52 ms/tweet). We present the results as follows.

Identification of Abnormal Terms, Clustering and Ranking. The time consumed by identification of abnormal terms depends on the total number of terms in a queried time window. Our results indicate that our algorithm identifies abnormal terms in 2.7 s when we average over 5 past time windows and the time window is as large as 120 min, demonstrating that we could efficiently complete the task in this step of processing. The time consumed by ranking also depends on the length of the queried time window. Results show that we can return top-20 clusters of a 120-min time window in 0.25 s.

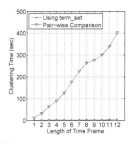

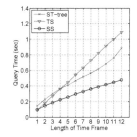

(a) Clustering time with $U_L = 10$ min, $clu\ thres = 0.8$, $V = 3$

(b) Query time of ST-tree, TS and SS with $U_L = 10$ min, $V = 5$

Fig. 3. Compare clustering time consumption and Query time of ST-tree.

We finally come to the time consumption of the clustering component. In Fig. 3a, we present how our strategy of using *term_set* improves the clustering efficiency. Notice that the naive pair-wise comparison method is extremely inefficient when the length of queried time window is large. However, our algorithm could finish clustering for a 120-min time window in 4 s.

ST-tree Specific. In this part, we evaluate the efficiency of ST-tree. Adjustable time window event detection differs from the original problem in three aspects: built-up of ST-tree, update and query of statistics. For the built-up of ST-tree, we can construct a complete ST-tree with 360 leaves ($L_M = 60$ and $V = 5$) in 0.032 s, showing that initialization of ST-tree is efficient.

Table 3. Update Time of ST-tree and SS (U_L=10 min)

Algorithm	L_M	V	# of Leaves	Update Time (sec)
ST-tree	6	5	36	0.0325
SS	6	5	-	0.4932
ST-tree	30	5	180	0.1764
SS	30	5	-	2.0573
ST-tree	30	10	330	0.3584
SS	30	10	-	2.2059

In Table 3, we present the update time of ST-tree and SS with different values of L_M and V. We do not compare with TS since TS only needs to update statistics of the current time unit, which is performed continuously in the preprocessing component. We can see that ST-tree takes much less time than SS to update statistics. Even when updating a ST-tree with 330 leaves, we finish the procedure in less than 0.4s, which is highly efficient.

In Fig. 3b, we present the query time of three algorithms w.r.t. various queried time window lengths L. We set the time unit as 10 min, and L_M as 12 time units, i.e. 2 h. The first observation is that SS is the most efficient in all lengths of queried time window due to its $O(V)$ query time complexity. Second, when the length of queried time window is small, i.e. less than 4, TS beats ST-tree. The reason is that TS simply aggregates statistics of L time units, while ST-tree may need to visit some deep nodes of ST-tree to obtain statistics of a short time window and thus result in a longer query time than TS.

6   Related Work

A comprehensive survey [2] on event detection on Twitter is proposed. We summarize major techniques of event detection on Twitter. Many previous works focused on detecting specific category of events [6,8,10,12], or required queries from users and assumed a topic-based stream [12]. Recently, some works focused on detecting topic-based events in crowdsourcing markets and social networks [16,17] and assisted task assignment in the markets and social networks [13,14,18,19]. In addition, [11] claimed to be the first to detect open-domain events on Twitter. Some others [5,20] could also detect unspecific events. However, [5,11,20] used time-demanding techniques and were not suitable for real-time applications on Twitter. Some works detected only geo-spatial events [1]. Furthermore, the issue of detecting experts on Twitter is also studied [3,4].

To detect emerging events, some works first identified bursty words [1,6,7,9]. However, they often introduced burst thresholds. For grouping terms/tweets into events, two types of techniques were developed: clustering [1,6,9] and graph partitioning [20]. Clustering of short-text tweets could be difficult, while graph partitioning based methods may be time-consuming.

Finally, adjustable time window event detection is not yet studied. [1,7] both used fixed time frames to detect bursty terms. [15] enabled timeline zooming.

7 Conclusion

In this paper, we study the problem of adjustable time window based event detection in Twitter. We design an abnormality score calculated by term frequency, user frequency and past averaged records to identify abnormal terms in a simple and free-of-threshold way. We then propose a highly efficient clustering algorithm to group co-occurring terms to form clusters, i.e. events. We further design ST-tree to adjustable time window based enable event detection. Extensive experiments on real dataset show that our algorithm can effectively and timely identify real-world events.

Acknowledgment. This work is supported in part by the National Science Foundation of China (NSFC) under Grant No. 61502021, 61328202, and 61532004, National Grand Fundamental Research 973 Program of China under Grant 2012CB316200, the Hong Kong RGC Project N_HKUST637/13, NSFC Guang Dong Grant No. U1301253, Microsoft Research Asia Gift Grant, Google Faculty Award 2013.

References

1. Abdelhaq, H., Sengstock, C., Gertz, M.: Eventweet: online localized event detection from twitter. PVLDB **6**(12), 1326–1329 (2013)
2. Atefeh, F., Khreich, W.: A survey of techniques for event detection in twitter. Comput. Intell. **31**(1), 132–164 (2013)
3. Cao, C.C., She, J., Tong, Y., Chen, L.: Whom to ask?: jury selection for decision making tasks on micro-blog services. PVLDB **5**(11), 1495–1506 (2012)
4. Cao, C.C., Tong, Y., Chen, L., Jagadish, H.V.: Wisemarket: a new paradigm for managing wisdom of online social users. In: SIGKDD 2013, pp. 455–463 (2013)
5. Cataldi, M., Di Caro, L., Schifanella, C.: Emerging topic detection on twitter based on temporal and social terms evaluation. In: MDMKDD 2010, p. 4 (2010)
6. Goorha, S., Ungar, L.: Discovery of significant emerging trends. In: SIGKDD 2010, pp. 57–64 (2010)

7. Li, C., Sun, A., Datta, A.: Twevent: Segment-based event detection from tweets. In: CIKM 2012, pp. 155–164 (2012)

8. Li, R., Lei, K.H., Khadiwala, R., Chang, K.C.: Tedas: a twitter-based event detection and analysis system. In: ICDE 2012, pp. 1273–1276 (2012)

9. Mathioudakis, M., Koudas, N.: Twittermonitor: trend detection over the twitter stream. In: SIGMOD 2010, pp. 1155–1158 (2010)

10. Popescu, A.M., Pennacchiotti, M.: Detecting controversial events from twitter. In: CIKM 2010, pp. 1873–1876 (2010)

11. Ritter, A., Etzioni, O., Clark, S., et al.: Open domain event extraction from twitter. In: KDD 2012, pp. 1104–1112 (2012)

12. Sakaki, T., Okazaki, M., Matsuo, Y.: Earthquake shakes twitter users: Real-time event detection by social sensors. In: WWW 2010, pp. 851–860 (2010)

13. She, J., Tong, Y., Chen, L.: Utility-aware social event-participant planning. In: SIGMOD 2015, pp. 1629–1643 (2015)

14. She, J., Tong, Y., Chen, L., Cao, C.C.: Conflict-aware event-participant arrangement. In: ICDE 2015, pp. 735–746 (2015)

15. Shou, L., Wang, Z., Chen, K., Chen, G.: Sumblr: continuous summarization of evolving tweet streams. In: SIGIR 2013, pp. 533–542 (2013)

16. Tong, Y., Cao, C.C., Chen, L.: Tcs: Efficient topic discovery over crowd-oriented service data. In: SIGKDD 2014, pp. 861–870 (2014)

17. Tong, Y., Cao, C.C., Zhang, C.J., Li, Y., Chen, L.: Crowdcleaner: Data cleaning for multi-version data on the web via crowdsourcing. In: ICDE 2014, pp. 1182–1185 (2014)

18. Tong, Y., She, J., Ding, B., Wang, L., Chen, L.: Online mobile micro-task allocation in spatial crowdsourcing. In: ICDE 2016 (2016)

19. Tong, Y., She, J., Meng, R.: Bottleneck-aware arrangement over event-based social networks: the max-min approach. World Wide Web Journal (to appear)

20. Weng, J., Lee, B.S.: Event detection in twitter. In: ICWSM 2011, pp. 401–408 (2011)

User-IBTM: An Online Framework for Hashtag Suggestion in Twitter

Jia Li and Hua Xu[✉]

State Key Laboratory of Intelligent Technology and Systems,
Tsinghua National Laboratory for Information Science and Technology,
Department of Computer Science and Technology, Tsinghua University,
Beijing 100084, China
callmecoding@sina.com, xuhua@tsinghua.edu.cn

Abstract. Twitter, the global social networking microblogging service, allows registered users to post 140-character messages known as *tweets*. People use the hashtag symbol '#' before a relevant keyword or phrase in their tweets to categorize the tweets and help them show more easily in a Twitter search. However, there are very few tweets contain hashtags, which impedes the quality of the search results and their applications. Therefore, how to automatically generate or recommend hashtags has become a particularly important academic research problem. Although many attempts have been made for solving this problem, previous methods mostly do not take the dynamic nature of hashtags into consideration. Furthermore, most previous work focuses on exploiting the similarity between tweets and ignores semantics in tweets.

In this paper, we regard hashtags as the underlying topics of the tweets. We first introduce an effective method for discovering the latent topics of streaming tweets, which uses the recently proposed incremental biterm topic model (IBTM). Then considering the personalized preferences, we propose a novel model, namely online Twitter-User LDA, to learn each Twitter user's dynamic interests. Finally, we design an online hashtag suggestion framework called User-IBTM by combining the online Twitter-User LDA and IBTM. As shown in the experimental results on real world data from Twitter, our designed framework outperforms several state-of-the-art methods and achieves satisfying performance (Code is available at https://github.com/worldcodingNow/UserIBTM/tree/master).

Keywords: Twitter · Hashtag suggestion · Online twitter-user LDA · Incremental biterm topic model

1 Introduction

As the number of available tweets explosively grows, the problem of managing tweets becomes extremely difficult, which can lead to information overload. To help categorize tweets, an efficient way called *hashtag*, the '#' symbol used before a relevant keyword or phrase, is introduced by Twitter.

© Springer International Publishing Switzerland 2016
B. Cui et al. (Eds.): WAIM 2016, Part II, LNCS 9659, pp. 279–290, 2016.
DOI: 10.1007/978-3-319-39958-4_22

Hashtags also attracted much attention in the academic area. Research [4, 8,9,17–20] shows that hashtags not only help users search tweets but also can be utilized in many research areas such as sentiment classification [9], breaking events detection [8], and hashtag-oriented spam detection [20].

Despite the importance of hashtags in actual use, large portions of tweets do not contain hashtags since they should be manually annotated. Zangerle et al. [24] find only 20 % of tweets in their datasets contain hashtags and Kywe et al. [13] report that less than 8 % of tweets contain tags. Therefore, a reliable hashtag recommendation system is needed to help users tag their new tweets.

To this end, many methods have been proposed to recommend hashtags for new tweets. An intuitive way is to directly classify new tweets with annotated hashtags such as Mazzia and Juett [16] utilize a Naive Bayes model as the classifier. More methods attempt to sovle this task based on traditional recommender systems [1,6,13,24]. However, these methods may not fare well in practice since they all use TF-IDF representations and ignore the underlying semantics in tweets. Recently a handful of methods based on topic models have been proposed such as Godin et al. [11] use standard latent Dirichlet allocation (LDA) to automatically recommend keywords as hashtags. As we can see, although many studies approach hashtag recommendation in Twitter, the dynamic nature here refers to the strong timeliness of hashtags, is ignored. It is clear that the number of the collected hashtags has become a bottleneck for those static methods, since there are very few hashtags in existing tweets and most hashtags only have a very short life span [13].

When a person tags a tweet, he/she typically finds a word or a phrase which has the similar meaning to the tweet. In other words, we think that good hashtags can be regarded as the low-dimensional theme representations of tweets. In this paper, we propose a novel online hashtag suggestion framework which considers the dynamic change of hashtags and the theme representations of tweets. As far as we know, our method is the first to consider the dynamic nature to suggest suitable hashtags. More precisely, we first introduce an online method for discovering the latent topics of streaming tweets, which uses the recently proposed incremental biterm topic model (IBTM) [7]. Then considering the personalized preferences, we propose a novel model, namely online Twitter-User LDA, for learning each user's dynamic interests. Finally, we design an online hashtag suggestion framework called User-IBTM by combining the online Twitter-User LDA and IBTM. As shown in the experimental results on real world data from Twitter, our designed method outperforms state-of-the-art methods and achieves satisfying performance.

The main contributions of our research are:

- Introduce an online topic model called IBTM to discover topic distribution of streaming tweet content.
- Design a novel model called online Twitter-User LDA, to capture Twitter users' dynamic interests.

– Propose an effective and efficient hashtag suggestion framework based on IBTM and online Twitter-User LDA. The work described here is, to the best of our knowledge, the first attempt at using online algorithms for suggestion hashtags in Twitter.

2   Related Work

In this section, we review the existing methods for hashtag recommendation in Twitter. Basically, these methods can be classified into two classes.

One class of these methods focuses on exploiting the similarity between tweets. Zangerle et al. [24] compare three ranking approaches to recommend hashtags based on the TF-IDF representations of tweets. Kywe et al. [13] suggest hashtags by combining hashtags of similar users and similar tweets. The TF-IDF method is used again in computing similarity of tweets. Otsuka et al. [19] propose the HF-IHU ranking scheme, which is a variation of TF-IDF, that also considers hashtag relevancy. Jeon et al. [12] use TF-IDF scheme to extract keywords from each tweet and then use a Naive Bayes classifier to map the keywords into pre-defined classes. They also consider the user interests but ignore the dynamic change of interests. The disadvantage of these methods is that they always recommend existing hashtags to new tweets. As noted by Kywe et al. [13], most hashtags have a very short life span and there are very few hashtags in existing tweets. Thus, the number of the collected hashtags has become a bottleneck and impedes the performance in practical use.

The other class of methods focuses on using conventional topic models to discover topic distribution of tweets. Godin et al. [11] first apply standard LDA to automatically recommend keywords as hashtags. Ma et al. [15] propose two PLSA-style topic models to model the hashtag annotation behavior. Feng and Wang [10] combine tweet content and explicit features to connect tweets with topics. However, because of the expensive computation, these methods may not apply to high volume streaming tweets. Considering trending effects, Lu and Lee [14] combine Topics-over-Time model and Mixed Membership Model to recommend hashtags. Like LDA, this model has to be trained repeatedly over the entire corpus if new tweets arrive. She and Chen [21] present a supervised topic model-based solution for hashtag recommendation. They treat hashtags as labels of topics and discover relationship among words, hashtags, and topics of tweets. Clearly, the scarcity of hashtags impedes the performance of this method.

As we can see, these methods either use static topic models or just consider the tweet similarity all ignore the dynamic nature, here refers to the strong timeliness of hashtags. Furthermore, most work does not take user interests into consideration. These weaknesses hinder the performance of previous methods in practical use.

3   Models and Hybrid Combination

In this section, we describe our architecture in detail for suggesting hashtags in Twitter. The whole framework is shown in Fig. 1. Facing streaming tweets,

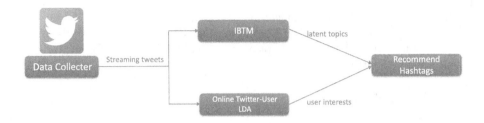

Fig. 1. The framework of hashtag recommendation.

IBTM and online Twitter-User LDA estimate new tweet's latent topic distribution and corresponding user's dynamic interests, respectively. Finally, combining outputs of the two models, recommended hashtags are produced for the new tweet.

In what follows, we first describe the IBTM proposed for online short texts. Then, we present the online Twitter-User LDA and show how it explores Twitter users' topics of interest dynamicly. In addition, we outline the hybrid method for hashtag suggestion in Twitter by combining IBTM and online Twitter-User LDA.

3.1   Incremental Biterm Topic Model (IBTM)

The key step in using topic models to suggest hashtags is to determine which topic the tweet addresses. Because of the shortness, noisiness, and sparsity of tweets, it is difficult to discover topics in a single tweet using conventional topic models (e.g. LDA). To solve this problem, a well-designed biterm topic model (BTM) [23] has been proposed recently for short texts. In this paper, we introduce an extension of BTM—incremental biterm topic model (IBTM) [7] by using online algorithm to learn topics of streaming tweets.

For completeness, we first briefly review the BTM [23]. Unlike LDA, BTM learns topics by modeling the generation of biterms (i.e., unordered co-occurring word pairs) in the collection, whose effectiveness is not affected by the length of documents, making it more appropriate for short texts. Gibbs sampling algorithm is used to estimate topics. More details about BTM are described in [23] and BTM's generative process is shown in Fig. 2.

IBTM is an online extension of BTM. It updates the model continuously whenever a biterm arrives, via a technique called incremental Gibbs sampler. In detail, when biterm b_i arrives, IBTM updates the model in two steps. Firstly, it draws the topic assignment of b_i from $P(z_i|z_{i-1}, B_i)$, where $z_{i-1} = \{z_j\}_{j=1}^{i-1}$ indicates all the previous topic assignments, and $B_i = \{b_1, b_2, ..., b_{i-1}\}$. Secondly, IBTM randomly chooses some previous biterms to construct a biterm sequence, called rejuvenation sequence $R(i)$, to resample their topic assignments. For each biterm $b_j \in R(i)$, it resamples the topic assignment z_j. The complete procedure of IBTM is outlined in Algorithm 1.

$$p(z_i = k|z_{-i}, B_i) \propto (n_{-i,k} + \alpha) \frac{(n_{-i,w_{i,1}|k} + \beta)(n_{-i,w_{i,2}|k} + \beta)}{(n_{-i,\cdot|k} + W\beta + 1)(n_{-i,\cdot|k} + W\beta)} \qquad (1)$$

Algorithm 1. Incremental BTM Algorithm

1: *Input:* K, α, β, biterm sequence $B = \{b_1, ..., b_N\}$
2: *Output:* Φ, Θ
3: **for** $i = 1$ to N **do**
4:　　Draw topic k from $P(z_i|z_{i-1}, B_i)$
5:　　Update n_k and $n_{w|k}$
6:　　Generate rejuvenation sequence $R(i)$
7:　　**for** $j \in R(i)$ **do**
8:　　　　Draw topic assignment k' from $P(z_i|z_{-j,i}, B_i)$
9:　　　　Update $n_{k'}, n_{w|k'}$
10:　　**end for**
11: **end for**
12: Compute Φ and Θ

A key concern of IBTM is how to generate the rejuvenation sequence $R(i)$. First, the length of $R(i)$ makes a trade-off between efficiency and effectiveness. The more sequence rejuvenated, the better approximation of the posterior distribution. Technically, if $R(i)$ is equal to B_i, the incremental Gibbs sampler closely approximates to the batch Gibbs sampler. Therefore, convergence is guaranteed as the number of biterms is resampled goes to infinity. Second, the choice of $R(i)$ affects the contribution of biterms received at different time in updating model. To make the model more sensitive to the dynamic change of topics, in our work, $R(i)$ is generated from a uniform distribution over a fixed-size sliding window covering the latest biterms.

3.2　Online Twitter-User LDA

People use Twitter in different ways, some as a news source, some as a platform to meet new people with same interests, and some as a microblog for updating people about their personal lives. Technically, people tweet personal lives, random thoughts, or anything that they are interested in.

Based on this observation, we think that the content of tweets can implicitly represent the interests of users. In addition, considering the temporal sequence of tweets, we have an intuitive hypothesis that user interests dynamicly change over time. However, previous research in modeling user interests did not address this issue explicitly, as most of them simply built a "bag-of-word" document containing his/her published tweets for each user [5,22]. We extend this model by using a computationally inexpensive online algorithm, namely incremental Gibbs sampler [3], which immediately updates estimations of the topics as each tweet is observed. That is to say, we propose the online Twitter-User LDA to learn users' dynamic interests.

Details about LDA can be found in [2].

Like LDA, our model assumes that underlying the streaming tweets, there are K topics. Each topic k is represented by a topic word distribution ϕ_k. Each document d is associated with one user and contains his/her L latest tweets. Here L is the same for each document.

$$p(z_{di}|rest) \propto \frac{(n_{kd}^{-di} + \alpha_k)(n_{kw}^{-di} + \beta_w)}{n_k^{-di} + \bar{\beta}} \qquad (2)$$

Facing streaming tweets, when a new tweet t arrives, online Twitter-User LDA first determines its sender. Without loss of generality, we assume that tweet t belongs to user u and u associates with document d. Then online Twitter-User LDA updates u's topics of interest in following steps:

1. For each word index i in tweet t:
   · sample a topic value by Eq. (2),
   · generate fixed-length rejuvenation sequence $R(i)$ from document d,
   · for each word in $R(i)$, resample a topic value by Eq. (2),
2. Discard the first tweet in document d,
3. Add tweet t to the end of document d.

After generating rejuvenation sequence, online Twitter-User LDA resamples the topic variables of some previous words. If these rejuvenation samplings are performed often enough and the length of $R(i)$ is long enough, the model closely approximates the posterior distribution $P(z_i|w_i)$ infinitely [3]. Like IBTM, we also generate $R(i)$ from a uniform distribution over a fixed-size sliding window covering the recent tweets.

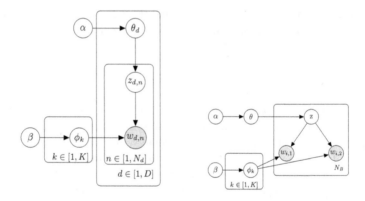

Fig. 2. Graphical representation of LDA (left) and BTM (right).

3.3 Online Framework: User-IBTM

Here we show the details about our proposed online framework: User-IBTM. In this paper, we utilize the ***problem decomposition*** technique to divide hashtag suggestion into two independent subtasks:

(a) determining the topic distribution of each new tweet;
(b) revealing the corresponding user's current topics of interest.

Subtasks (a) and (b) can be solved by IBTM and online Twitter-User LDA, respectively.

The key components of User-IBTM are two models and one hybrid method. Two models are IBTM and online Twitter-User LDA and their details are shown in Sects. 3.1 and 3.2. The hybrid method works as follows. When a new tweet t without hashtags is posted by user u, IBTM and online Twitter-User LDA first determine the topic distribution of t and u's current interests, respectively. Then according to the discovered topic distribution of t, it finds the topic k_{max} which has the maximum probability and selects N_t top words from k_{max}. Similarly, the hybrid method selects N_u top words from u's maximum probability topic of interests. Finally, the $N_t + N_u$ keywords are used as candidate hashtags to suggest for tweet t. As we can see the hybrid method looks simple but our experimental results show the method is very effective.

Although there is the *"cold start"* problem for capturing the new user's current interests in practical use, we can use the hybrid characteristic of User-IBTM to avoid it. When a new tweet without hashtags arrives, User-IBTM checks how many tweets the corresonding user has posted. If he/her has published more than L tweets, then User-IBTM works as mentioned in last paragraph, and if not, then only IBTM generates $N_t + N_u$ keywords from k_{max} as the suggested hashtags. In Sect. 4.4 we can see that IBTM also obtains effective performance.

4 Experiments

In this section, we show the effectiveness of User-IBTM that combines online Twitter-User LDA and IBTM to recommend hashtags on real world Twitter dataset.

4.1 Dataset and Preprocess

We used the public Twitter streaming API[1] to collect 6,600,669 tweets from January 2015 to July 2015. After observing the raw data, we carefully designed a series of steps to preprocess them. First, using regular expressions to remove non English tweets, which here were Chinese and Japanese tweets, and there were 5,002,842 pieces of tweet left. To get pure text contents, we removed URLs, punctuations, mentions, and HTML entities. Then we utilized an English stop word list to remove meaningless stopwords.

Considering the tweets sparsity, words that occurred less than 5 were removed. Furthermore, IBTM requires that each tweet contains at least two words. Therefore, those tweets that did not satisfy this condition were pruned away.

After preprocessing, there are 1,385,425 tweets posted by 2,094 users ranging between 2015/01/01 and 2015/07/31. In order to clarify the comprehensive suggestion performance of our method, we design two test sets. For the first one, we randomly select 50 tweets from each month that don't contain hashtags. For the second set, we randomly select 50 tweets from each month that contain hashtags. The rest of tweets constructed the training set and we removed hashtag symbol in training set.

---

[1] https://dev.twitter.com/streaming/overview.

4.2   Models and Parameter Settings

It is important to note that our method is unsupervised. Thus we do not compare with any supervised methods. As far as we know, there are very few unsupervised methods to suggest hashtags. In this paper, we compare our method against three models: LDA, BTM and IBTM. LDA as the most common probabilistic topic model, it is also the first topic model used in hashtag recommendation. BTM is designed specially for short texts to discover topics. IBTM is the extension of BTM by using online algorithm to detect topics for streaming short texts.

We set parameters $\alpha = 0.1$, $\beta = 0.01$ for LDA, BTM, IBTM, and online Twitter-User LDA. However, there is no single best or standard way to determine the topic numbers K. We tried values for K of 50, 75, 100, 150, and 200 topics to train LDA model. Although when K was fixed at 200, the trained model had the largest log-likelihood. By getting top 50 words with maximum probability for each topic, we found the model was more explanatory when $K = 150$. Finally, we set $K = 150$ for LDA and BTM. As for IBTM and online Twitter-User LDA, 150 was too large, we set $K = 50$. Moreover, Gibbs sampling was run for 500 iterations. Considering the trade-off between efficiency and effectiveness, we set $\{L, R(i)\} = \{50000,1000\}$ for IBTM and $\{L, R(i)\} = \{150,10\}$ for online Twitter-User LDA. All methods were implemented in Java and all experiments were run on a 2.83 GHz-CPU 16 GB-RAM Ubuntu 14.04 64 bit machine.

4.3   Evaluation Methods

As far as we know, there isn't a natural objective evaluation metric for unsupervised methods to suggest hashtags. In our experiments, we compared the performance of different approaches by asking 3 volunteers (represented by p_1, p_2, p_3) to independently hit one recommended hashtag if it could possibly be a hashtag for the testing tweet. Testing tweets in test set 2 all have hashtags. However, there may be different hashtags to describe one concept. Therefore, we asked the participants to hit a suggested hashtag as if its meaning is similar to the original hashtag.

Technically, We define the $hitrate_{p_i}@N$: given N suggested hashtags for each testing tweet, if p_i thinks one could be used as a hashtag, he would hit it. Hence, $hitrate_{p_i}@N$ = (number of hitting tweets)/(number of testing tweets). The average **hitrate@N** = (average number of hitting tweets)/(number of testing tweets). N$\in \{5, 10\}$.

4.4   Experimental Results

Table 1 lists the hitrates of all the approaches on two test sets. We have the following experimental results: (1) User-IBTM achieves the highest hitrate@5 and hitrate@10, which significantly outperforms other approaches. This demonstrates that our method based on dynamic user interests and tweet content can suggest more accurate hashtags; (2) IBTM falls behind User-IBTM but performs

Table 1. Hashtag recommendation performance on two test sets.

Approach	hitrate@5	hitrate@10
(a) Hitrate results on test set 1 (without hashtags)		
LDA	50.5%	57.0%
BTM	53.0%	59.5%
IBTM	67.5%	74.5%
User-IBTM	**70.5%**	**77.0%**
(b) Hitrate results on test set 2 (with hashtags)		
LDA	33.5%	36.0%
BTM	35.5%	39.5%
IBTM	57.0%	62.0%
User-IBTM	**64.5%**	**67.5%**

much better than standard LDA and BTM. LDA and BTM couldn't recommend satisfactory hashtags under two conditions. This is because static topic models cannot capture temporal change of topics and bursty topics. They just tend to discover common topics; (3) BTM outperfoms standard LDA in our Twitter corpus. LDA models each document, here refer to each tweet, as a mixture of topics and naturally suffers from the sparsity problem when tweets are extremely short; (4) Under the condition of test set 2, all of the methods have worse hitrates than themselves in test set 1. A reasonable explanation is that tagging is a subjective task. There isn't the only hashtag for a specific tweet.

To give a better sense of outcomes of different methods, we illustrate five test tweets and their corresponding recommended hashtags in Table 2. We can see that: (1) Tweets 1 and 2 are both about *charleston church shooting*. The differences are that tweet 1 has a hashtag #charleston and they are posted at different timestamps. LDA and BTM cannot capture the difference of post time. Therefore, they both respectively suggest the same hashtags to tweets 1 and 2. On the contrary, IBTM and User-IBTM recommend different hashtags; (2) Tweet 3 is about the *2015 Wimbledon Championships*. In our training set, there are only few tweets about this event. LDA and BTM do not detect the bursty topic. In contrast, IBTM and User-IBTM accurately capture this event; (4) Tweet 4 and tweet 5 are not about certain public event. They relate to the user's personal life and work. LDA, BTM, and IBTM all output an equalized topic distribution. Hence, in this condition, it is hard to suggest hashtags only based on tweet content. Note that tweet 4 and tweet 5 are posted at different timestamps. User-IBTM basically captures the dynamic change of interests.

The experimental results show that User-IBTM is also capable of suggesting new hashtags since it can detect the bursty events. Figure 3 illustrates some bursty topics captured by User-IBTM.

Table 2. Example tweets and corresponding recommended hashtags. Appropriate hashtags are highlighted. For User-IBTM, the first four tags are generated by IBTM and the last one is generated by online Twitter-User LDA.

#charleston, my heart hurts.	LDA	family, church, shooting, kill, victim
	BTM	police, shooting, kill, death, **charleston**
	IBTM	**charleston**, shooting, church, police, kill
timestamp: 2015-06-18-14-37-45	**User-IBTM**	**charleston**, shooting, church, police, love
who are the victims of the charleston shooting?	LDA	family, church, **shooting**, kill, **victim**
	BTM	police, **shooting**, church, police, kill
	IBTM	church, **charleston**, **shooting**, suspect, **charlestonshoot**
timestamp: 2015-06-19-21-14-09	**User-IBTM**	church, **charleston**, **shooting**, suspect, love
congratulations to,ambassador,on winning his 3rd #wimbledon title &9th grand slam victory!	LDA	win, world, final, cup, game
	BTM	win, world, cup, final, usa
	IBTM	**wimbledon**, win, **federer**, **title**, final
timestamp: 2015-07-13-05-56-51	**User-IBTM**	**wimbledon**, win, **federer**, **title**, college
our partners, do exactly this- negotiating lower vaccine prices & working with partners like, to reach every child.	LDA	happy, love, friend, gift, family
	BTM	love, girl, happy, friend, father
	IBTM	time, year, love, work, week
timestamp: 2015-05-21-00-23-53	**User-IBTM**	time, year, love,work, **child**
15m people are now receiving life-savinghiv treatment. incredible progress.	LDA	happy, love, friend, gift, family
	BTM	love, girl, happy, friend, father
	IBTM	time, year, love, work, week
timestamp: 2015-07-23-11-04-16	**User-IBTM**	time, year, love, work, **hiv**

Fig. 3. Examples of new topics discovered by User-IBTM.

5   Conclusions and Future Work

Hashtag recommendation in Twitter is an important problem. However, because of the diversity and short life spans of hashtags, it's difficult to recommend suitable hashtags. In this paper, we first propose the online Twitter-User LDA to reveal Twitter users' dynamic interests. Then we apply incremental biterm topic model (IBTM) to discover latent topics of streaming tweet content. Finally, we propose an online hashtag recommendation method based on online Twitter-User LDA and IBTM. Our proposed User-IBTM can handle streaming tweets with a long time span, as well as dynamic user interests. Experimental results demonstrate the proposed User-IBTM significantly outperforms standard LDA and BTM and performs better than IBTM. It is an effective and efficient unsupervised method to suggest accurate hashtags.

A potential extension to our proposed method could be to incorporate other attributes of tweets such as mentions, following relations, and URLs. Another future direction to suggest more suitable hashtags lies in using named entity recognition in pre-processing step to detect meaningful phrases as keywords.

Acknowledgments. This work is supported by National Basic Research Program of China (973 Program) (Grant No: 2012CB316301).

References

1. Abel, F., Gao, Q., Houben, G.-J., Tao, K.: Analyzing user modeling on twitter for personalized news recommendations. In: Konstan, J.A., Conejo, R., Marzo, J.L., Oliver, N. (eds.) UMAP 2011. LNCS, vol. 6787, pp. 1–12. Springer, Heidelberg (2011)
2. Blei, D.M., Ng, A.Y., Jordan, M.I.: Latent dirichlet allocation. J. Mach. Learn. Res. **3**, 993–1022 (2003)
3. Canini, K.R., Shi, L., Griffiths, T.L.: Online inference of topics with latent dirichlet allocation. In: International Conference on Artificial Intelligence and Statistics. pp. 65–72 (2009)
4. Chang, H.C.: A new perspective on twitter hashtag use: Diffusion of innovation theory. Proc. Am. Soc. for Inf. Sci. Technol. **47**(1), 1–4 (2010)
5. Chen, J., Nairn, R., Nelson, L., Bernstein, M., Chi, E.: Short and tweet: experiments on Recommending Content from Information Streams. In: Proceedings of the SIGCHI Conference on Human Factors in Computing Systems. pp. 1185–1194. ACM (2010)
6. Chen, K., Chen, T., Zheng, G., Jin, O., Yao, E., Yu, Y.: Collaborative personalized tweet recommendation. In: Proceedings of the 35th International ACM SIGIR Conference on Research and Development in Information Retrieval. pp. 661–670. ACM (2012)
7. Cheng, X., Yan, X., Lan, Y., Guo, J.: Btm: Topic modeling over short texts. IEEE Transactions on Knowledge and Data Engineering 26, 2928–2941 (2014)
8. Cui, A., Zhang, M., Liu, Y., Ma, S., Zhang, K.: Discover breaking events with popular hashtags in twitter. In: Proceedings of the 21st ACM International Conference on Information and Knowledge Management. pp. 1794–1798. ACM (2012)
9. Davidov, D., Tsur, O., Rappoport, A.: Enhanced sentiment learning using twitter hashtags and smileys. In: Proceedings of the 23rd International Conference on Computational Linguistics: Posters. pp. 241–249. Association for Computational Linguistics (2010)
10. Feng, W., Wang, J.: We can learn your# hashtags: Connecting tweets to explicit topics. In: 2014 IEEE 30th International Conference on Data Engineering (ICDE), pp. 856–867. IEEE (2014)

11. Godin, F., Slavkovikj, V., de Neve, W., Schrauwen, B., van de Walle, R.: Using topic models for twitter hashtag recommendation. In: Proceedings of the 22nd International Conference on World Wide Web Companion. pp. 593–596. International World Wide Web Conferences Steering Committee (2013)

12. Jeon, M., Jun, S., Hwang, E.: Hashtag recommendation based on user tweet and hashtag classification on twitter. In: Chen, Y., Balke, W.-T., Xu, J., Xu, W., Jin, P., Lin, X., Tang, T., Hwang, E. (eds.) WAIM 2014. LNCS, vol. 8597, pp. 325–336. Springer, Heidelberg (2014)

13. Kywe, S.M., Hoang, T.-A., Lim, E.-P., Zhu, F.: On recommending hashtags in twitter networks. In: Flache, A., Jager, W., Liu, L., Tang, J., Guéret, C., Aberer, K. (eds.) SocInfo 2012. LNCS, vol. 7710, pp. 337–350. Springer, Heidelberg (2012)

14. Lu, H., Lee, C.: The topic-over-time mixed membership model (tot-mmm): A twitter hashtag recommendation model that accommodates for temporal clustering effects (2015)

15. Ma, Z., Sun, A., Yuan, Q., Cong, G.: Tagging your tweets: A probabilistic modeling of hashtag annotation in twitter. In: Proceedings of the 23rd ACM International Conference on Conference on Information and Knowledge Management. pp. 999–1008. ACM (2014)

16. Mazzia, A., Juett, J.: Suggesting hashtags on twitter. EECS 545m, Machine Learning, Computer Science and Engineering. University of Michigan (2009)

17. Mehrotra, R., Sanner, S., Buntine, W., Xie, L.: Improving lda topic models for microblogs via tweet pooling and automatic labeling. In: Proceedings of the 36th International ACM SIGIR Conference on Research and Development in Information Retrieval. pp. 889–892. ACM (2013)

18. Meng, X., Wei, F., Liu, X., Zhou, M., Li, S., Wang, H.: Entity-centric topic-oriented opinion summarization in twitter. In: Proceedings of the 18th ACM SIGKDD International Conference on Knowledge Discovery and Data Mining, pp. 379–387. ACM (2012)

19. Otsuka, E., Wallace, S.A., Chiu, D.: Design and evaluation of a twitter hashtag recommendation system. In: Proceedings of the 18th International Database Engineering & Applications Symposium, pp. 330–333. ACM (2014)

20. Sedhai, S., Sun, A.: Hspam14: A collection of 14 million tweets for hashtag-oriented spam research. In: Proceedings of the 38th International ACM SIGIR Conference on Research and Development in Information Retrieval, pp. 223–232. ACM (2015)

21. She, J., Chen, L.: Tomoha: Topic model-based hashtag recommendation on twitter. In: Proceedings of the Companion Publication of the 23rd International Conference on World Wide Web Companion, pp. 371–372. International World Wide Web Conferences Steering Committee (2014)

22. Weng, J., Lim, E.P., Jiang, J., He, Q.: Twitterrank: Finding topic-sensitive influential twitterers. In: Proceedings of the Third ACM International Conference on Web Search and Data Mining, pp. 261–270. ACM (2010)

23. Yan, X., Guo, J., Lan, Y., Cheng, X.: A biterm topic model for short texts. In: Proceedings of the 22nd International Conference on World Wide Web. pp. 1445–1456 (2013)

24. Zangerle, E., Gassler, W., Specht, G.: Recommending#-tags in twitter. In: Proceedings of the Workshop on Semantic Adaptive Social Web (SASWeb 2011). CEUR Workshop Proceedings. vol. 730, pp. 67–78 (2011)

Unifying User and Message Clustering Information for Retweeting Behavior Prediction

Bo Jiang, Jiguang Liang, Ying Sha$^{(\boxtimes)}$, Lihong Wang, Zhixin Kuang,
Rui Li, and Peng Li

Institute of Information Engineering, Chinese Academy of Sciences,
Beijing 100093, China
{jiangbo,liangjiguang,shaying,kuangzhixin,lirui,lipeng}@iie.ac.cn

Abstract. Online social networks have been recently increasingly become the dominant platform of information diffusion by user's retweeting behavior. Thus, understanding and predicting who will be retweeted in a given network is a challenging but important task. Existing studies only investigate individual user and message for retweeting prediction. However, social influence and selection lead to formation of groups. The intrinsic and important factor has been neglected for this problem. In the paper, we propose a unified user and message clustering based approach for retweeting behavior prediction. We first cluster users and messages into different groups based on explicit and implicit factors together. Then we model social clustering information as regularization terms to introduce the retweeting prediction framework in order to reduce sparsity of data and improve accuracy of prediction. Finally, we employ matrix factorization method to predict user's retweeting behavior. The experimental results on a real-world dataset demonstrate that our proposed method effectively increases accuracy of retweeting behavior prediction compared to state-of-the-art methods.

Keywords: Retweeting behavior · Social networks · Matrix factorization · User clustering · Message clustering

1 Introduction

With the advent of social network platforms such as Twitter, Facebook and Weibo, thousands of millions of users have used these sites to share opinions and ideas with each other, and to engage in interesting activities about all kinds of topics and hot events. Social networks encourage connections, interactions and relationships between people. Thus, social network services allow a user to follow other users forming social link. On this basis, as message is forwarded from user to user, large cascades of reshares can be formed. As a result, information dissemination power has a unprecedented improvement via user's retweeting behavior. Retweeting has been considered as a key mechanism of information diffusion in Twitter [15]. Hence, understanding the retweeting effect factors from

© Springer International Publishing Switzerland 2016
B. Cui et al. (Eds.): WAIM 2016, Part II, LNCS 9659, pp. 291–303, 2016.
DOI: 10.1007/978-3-319-39958-4_23

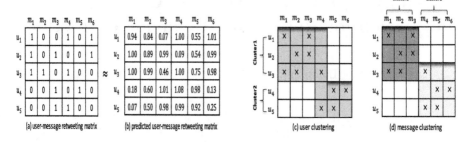

Fig. 1. Predicting unobserved retweetings based on observed interaction entities.

Fig. 2. Clustering for users and messages from different dimensions.

user's social footprints and predicting the hidden mechanism underlying diffusion are a critical but challenging task.

A number of research efforts have been performed towards investigating the factors that might affect a user to retweet messages of other users based on user survey [1,2,11,14], statistical analysis [15,17]. Meanwhile, various methods have also been proposed for predicting user's retweeting behavior from different perspectives, such as classifier-based method [7,10], influence-based method [9,19], graph-based method [17]. However, a common property of all the above mentioned models is that users and messages are assumed to be independent each other, respectively. In the real-world scenarios, as social activities grow, social influence and selection lead to formation of different groups, namely users belong to different groups due to the difference of individual preference, and messages belong to different groups due to the difference of referred topic. As a result, we can reach the conclusion that users are more likely to similar each other within the same group than those users who belong to other groups. Messages have the same property. Meanwhile, we argue that the users who belong to the same group are more likely to influence retweeting behavior each other due to their similar interests than these user who belong to other groups. For example in Twitter, a user can create groups of friends, relatives, coworkers and acquaintances that he post and forward on a regular basis. We also investigate that models which take homophily or similarity into account predicts social behavior much better than other more general models which do not take this into account.

Inspired by this, we propose a unified social clustering framework based on matrix factorization method through incorporating user and message clustering information to improve the accuracy of user's retweeting behavior prediction. Specifically, we factorize the user-message retweeting matrix into two intermediated latent matrices: latent user feature matrix and latent message feature matrix. The predicted user-message retweeting matrix is approximated as the product of user feature matrix and message feature matrix under some constraints, as illustrated in Fig. 1. Moreover, we employ clustering information in retweeting prediction to reduce sparsity of data and by doing so to improve accuracy of prediction, as illustrated in Fig. 2. We have conducted experiments on real social network dataset from Weibo. The results show incorporating cluster information from users and messages can reduce the data sparsity, and our method greatly outperforms the baseline methods by a large margin.

The main contributions of this work can be summarized as follows.

- We formulate the retweeting prediction problem as a predicting missing value task based matrix factorization, namely given the sets of users and messages, our goal is to find who will be retweeted based on partially observed entities.
- We exploit user and message clustering information as regularization terms to constrain objective function to reduce the sparsity of data and improve the performance of prediction.
- With extensive experiments on a real world dataset collected from Weibo, we empirically show the effectiveness and efficiency of our approach. Our approach outperforms state-of-the art methods with a significant margin.

The rest of the paper is organized as follows. Related work is introduced in Sect. 2. Our retweeting prediction model is proposed in Sect. 3 and experimental results are reported in Sect. 4. Conclusion comes in Sect. 5.

2   Related Work

There have been significant interests in algorithms for predicting retweeting behavior in social networks [4,5,7,9,12,13,18,20]. Here, we only summarize some representative investigations. For example, Yang et al. [17] proposed a factor graph model to predict user's retweeting behavior by analyzing influence that user, information, and time had on retweeting behavior. Luo et al. [10] employed a learning to rank based framework to discover the users who are most likely to retweet a specific post. Zhang et al. [19] demonstrated the existence of influence locality in social network and predicted user's retweeting behavior based on social influence locality via a logistic regression classifier. Jiang et al. [7] explored a wide range of features, such as user-based, content-based, relationship-based, and time-based, and then used classifier model as the solution to predict retweeting behavior. Most of the above methods are typically based on the effectiveness of leveraging the extracted of features for retweeting prediction. Choosing an appropriate feature set is the most critical part of these algorithms. However, some of these features may be computationally expensive for large social networks.

Recently, some works using matrix factorization for retweeting behavior prediction have been proposed. As far as we know, Wang et al. [16] utilized nonnegative matrix factorization to predict retweeting behavior from user and content dimensions by employing strength of social relationship to constrain objective function. However, this approach does not consider clustering information of user preferences and message referred topics. Hence social relationship undergo the data sparsity and limit the contribution of social regularization. Jiang et al. [6] proposed centroid-based and similarity-based message clustering retweeting prediction models which improve the prediction accuracy. It does not take into account influence from user clustering information.

Therefore, in our work, we give consideration to user and message clustering information from explicit and implicit dimensions, and integrate these clustering factors into matrix factorization model to reduce the data sparsity and improve the performance of retweeting prediction.

3   Social Clustering Prediction Model

In this section, we first present a formulation of the problem, and then introduce social clustering information from users and messages to reduce the data sparsity and improve the prediction performance. Finally, we give a unified framework for retweeting prediction, named SCRP (Social Clustering Retweeting Prediction).

3.1   Problem Formulation

We first formally define the problem of retweeting behavior prediction from the perspective of matrix factorization. Suppose that we are given M users and N messages, where the ith user denotes as u_i and the jth message denotes as m_j. The behaviors of users retweeting messages are represented in an $M \times N$ user-message retweeting matrix $\mathbf{R} = [\mathbf{r}_1, \cdots, \mathbf{r}_N]$, in which each row corresponds to a user and each column corresponds to a message. Meanwhile, whether user decide to retweet a message or not is a binary value task, hence the $(i, j)^{th}$ entry with $\mathbf{R} \in \mathbb{R}^{M \times N}$ can be represented as

$$R_{ij} = \begin{cases} 1 & \text{if } u_i \text{ retweeted } m_j \\ 0 & \text{otherwise} \end{cases} \tag{1}$$

Then we can model the problem of retweeting behavior prediction as a matrix completion task, where the unobserved entries in matrix $\mathbf{R}$ can be predicted based on the observed retweeting behaviors and other social factors.

Let $\mathbf{U} \in \mathbb{R}^{M \times K}$ be the latent user feature matrix, and $\mathbf{V} \in \mathbb{R}^{K \times N}$ be the latent message feature matrix, where K $(K \ll M, N)$ is the number of the latent features. We also assume that each row $U_i = [U_{i1}, U_{i2}, \cdots, U_{iK}]^T$ in $\mathbf{U}$ corresponds to a user and each column $V_j = [V_{1j}, V_{2j}, \cdots, V_{Kj}]^T$ in $\mathbf{V}$ corresponds to a message in latent feature space, respectively.

Now, the retweeting matrix $\mathbf{R}$ can be approximated by the product of two matrices: the latent user feature matrix $\mathbf{U}$ and the latent message feature matrix $\mathbf{V}$, i.e. $\mathbf{R}_{ij} \approx \mathbf{U}_i\mathbf{V}_j$. To learn the optimal latent feature matrices $\mathbf{U}$ and $\mathbf{V}$, we minimize the following objective function based on unobserved entries and observed entries.

$$\min_{U,V} \mathcal{J}(R, U, V) = \frac{1}{2} \sum_{i=1}^{M} \sum_{j=1}^{N} I_{ij}(R_{ij} - U_iV_j)^2 + \frac{\gamma}{2}\|U\|_F^2 + \frac{\lambda}{2}\|V\|_F^2 \tag{2}$$

where $\|\cdot\|_F$ denotes the Frobenius norm fitting constraint, γ and λ are regularization parameters. In order to focus more on the observed entries, we introduce an indicator function I_{ij} that is equal to 1 if u_i retweeted m_j and equal to 0 otherwise. The last two regularization terms are added to avoid overfitting.

Due to the severe sparsity of the retweeting matrix $\mathbf{R}$, it is impossible for directly learning the optimal latent spaces for users and messages by relying solely on observed retweeting entries. To alleviate the sparsity problem and improve the accuracy of prediction, we employ user and message clustering information to constraint the objective function.

3.2   User Clustering Factor

Users from social network are more likely to form a cohesive group due to social influence and selection. From a social and anthropological standpoint, people who have a common interest preference or a similar lifestyle are probably held together. We also argue that the users from the same group are more likely to similar each other than these users from the other groups. Hence user's interests and behavior pattern can be better represented by other users from the same group in the context of data sparsity.

Based on the above observation, we have the following assumptions that (1) the similar taste preference among users in observed spaces are consistent with the latent spaces; (2) users belonging to the same group should lie close to each other in the latent space; (3) each user can be represented by a linear combination of other users from the same group in the latent space.

In order to reduce the data sparsity and improve the accuracy of prediction, we perform K-means algorithm on the set of users $\mathcal{U}$ before predicting. More precisely, the set of users $\mathcal{U}$ can be divided into $\mathcal{U}_1, \mathcal{U}_2, \cdots, \mathcal{U}_p$, where $\mathcal{U}_i \cap \mathcal{U}_j = \varnothing$ and p is the number of users clustering. To formulate this, we introduce a user clustering sharing matrix $\mathbf{G} \in \mathbb{R}^{M \times M}$ with its $(i, j)^{th}$ entry defined as

$$g_{ij} = \begin{cases} 1 & \text{if } C_{u_i} = C_{u_j} \\ 0 & \text{otherwise} \end{cases} \tag{3}$$

where C_{u_i} and C_{u_j} are the clustering labels of users u_i and u_j, respectively. Then, to minimize the latent difference between users u_i and u_j who belong to the same group, we impose a social regularization term

$$\mathcal{J}_1 = \sum_{i=1}^{M} \sum_{j=1}^{M} g_{ij} S_u(i, j) \|U_i - U_j\|_F^2 \tag{4}$$

where $S_u(i, j)$ represents the similarity between u_i and u_j.

The similarity can refer to different dimensions. Here, we not only consider the similarity of user's taste preferences, but also take into account the similarity of user's interaction behaviors. The former can be profiled in the content of messages posted by user, and the latter can be reflected in user's social actions (e.g., posting, forwarding, commenting, etc.) which adopt the same message among users may have similar interests. Therefore, the similarity among users can be calculated based on the combine of taste preferences and social behaviors.

In order to calculate explicit taste preferences, we exploit LDA [8], which learns fixed-length feature representations from texts, to learn the vector representations of messages on the collection of user's messages. Then we calculate the taste preferences similarity between users u_i and u_j as following:

$$S_{taste}(i, j) = \frac{I(i)I(j)}{\|I(i)\| \, \|I(j)\|} \tag{5}$$

where $I(i) = \frac{1}{|D(i)|} \sum_{a \in D(i)} T_a$, $D(i)$ is the set of messages posted by user u_i, T_a is the learned vector representations for message a.

We also have the behavior footprint information of social users, and the behavior similarity between two users can be calculated by measuring the adopted interaction of these two users. To quantitatively measure the behavior similarity, we opt to choose Pearson Correlation Coefficient (PCC) [3], which is proposed to solve this problem that different users have different social action styles.

$$S_{behavior}(i, j) = \frac{\sum_{f \in I(i,j)} (R_{if} - \overline{R}_i) \cdot (R_{jf} - \overline{R}_j)}{\sqrt{\sum_{f \in I(i,j)} (R_{if} - \overline{R}_i)^2} \cdot \sqrt{\sum_{f \in I(i,j)} (R_{jf} - \overline{R}_j)^2}} \tag{6}$$

where $I(i, j)$ denotes the set of messages adopted by both u_i and u_j, $\overline{R}_i$ represents the average adopt of user u_i. Due to $S_{behavior}(i, j) \in [-1, 1]$, we also employ a sigmod function to map behavior similarities into $[0, 1]$.

Finally, the similarity between users u_i and u_j is calculated as following:

$$S_u(i, j) = \rho S_{taste}(i, j) + (1 - \rho) S_{behavior}(i, j) \tag{7}$$

where ρ is employed to control the contribution of each factor.

3.3   Message Clustering Factor

As is mentioned above, messages posted by various users have different structure styles and referred different topics. Therefore, messages can be divided into different groups based on structural and semantic information of texts. We also have the following assumptions that (1) the similar among messages in observed spaces are consistent with the latent spaces; (2) messages are more likely to similar within the same group compare to different latent groups; (3) each message can be represented by a linear combination of other messages from the same group in the latent space. Similarly, we also consider two dimensions of similarity for messages: structural information and semantic information.

Similar to user clustering, we also use K-means algorithm to perform messages clustering. More precisely, the set of messages $\mathcal{M}$ can be grouped into $\mathcal{M}_1, \mathcal{M}_2, \cdots, \mathcal{M}_q$, where $\mathcal{M}_i \cap \mathcal{M}_j = \varnothing$ and q is the number of messages clustering. To formulate this, we also construct a message clustering sharing matrix $\mathbf{H} \in \mathbb{R}^{N \times N}$ with its $(i, j)^{th}$ entry defined as

$$h_{ij} = \begin{cases} 1 & \text{if } C_{m_i} = C_{m_j} \\ 0 & \text{otherwise} \end{cases} \tag{8}$$

where C_{m_i} and C_{m_j} are the clustering labels of messages m_i and m_j, respectively. Then, to minimize the latent difference between messages m_i and m_j which belong to the same group, we impose a social regularization term

$$\mathcal{J}_2 = \sum_{j=1}^{N}\sum_{j=1}^{N} h_{ij} S_m(i,j)\|V_i - V_j\|_F^2 \tag{9}$$

where $S_m(i,j)$ represents the similarity between m_i and m_j which can be calculated by the combine of structure and semantic vector of two messages.

There is a wealth of evidence to suggest that structure features from messages are significantly associated with user's retweetability [15]. Here, we extract hashtag, URL, mention as a feature set in our proposed model. Specifically, we use a feature vector $\mathcal{V}_{structure}(j)$=(#hashtag, #URL, #mention) to represent the set of these features, where #hashtag/#URL/#mention denote the number of hashtag/URL/mention occurred for message m_j, respectively. Moreover, user's retweeting behavior is strongly correlated with the content of messages. Hence, we also use LDA method to measure the semantic information for messages. For a message m_j, we use $\mathcal{V}_{semantic}(j)$ to denote m_j's semantic feature vector. Now, we can combine structural vector $\mathcal{V}_{structure}(j)$ and semantic vector $\mathcal{V}_{semantic}(j)$ for message m_j into a compound vector $\mathcal{V}(j)$. In addition, we also use the two vectors mentioned above to calculate the similarities among messages. More specifically, the similarity between messages m_i and m_j is calculated as

$$S_m(i,j) = \lambda S_{structure}(i,j) + (1-\lambda) S_{semantic}(i,j) \tag{10}$$

where λ is the parameter controlling the contribution of each factor. $S_{structure}(i,j)$ and $S_{semantic}(i,j)$ are cosine similarities based on structural and semantic vectors, respectively.

3.4   Unified Prediction Model

Based on the above discussed, we demonstrate how to construct user clustering regularization and message clustering regularization, respectively. Now, we solve the optimization problem by combining $\mathcal{J}_1, \mathcal{J}_2$ with $\mathcal{J}$:

$$\min_{U,V} \mathcal{J}(R,U,V) = \frac{1}{2}\sum_{i=1}^{M}\sum_{j=1}^{N} I_{ij}(R_{ij} - U_i V_j)^2$$
$$+ \frac{\alpha}{2}\sum_{i=1}^{M}\sum_{k=1}^{M} g_{ik} S_u(i,k)\|U_i - U_k\|_F^2$$
$$+ \frac{\beta}{2}\sum_{j=1}^{N}\sum_{l=1}^{N} h_{jl} S_m(j,l)\|V_j - V_l\|_F^2$$
$$+ \frac{\gamma}{2}\|U\|_F^2 + \frac{\eta}{2}\|V\|_F^2 \tag{11}$$

where $\alpha \geq 0$ and $\beta \geq 0$ are the parameters controlling user clustering regularization and message clustering regularization on U_i and V_j, respectively.

A local minimum of the objective function given by Eq. (11) can be found by employing gradient descent method in feature vectors U_i and V_j, respectively.

$$\frac{\partial \mathcal{J}}{\partial U_i} = \sum_{j=1}^{N} I_{ij}(U_iV_j - R_{ij})V_j + \gamma U_i + \alpha \sum_{k=1}^{M} g_{ik}S_u(i,k)(U_i - U_k)$$

$$\frac{\partial \mathcal{J}}{\partial V_j} = \sum_{i=1}^{M} I_{ij}(U_iV_j - R_{ij})U_i + \eta V_j + \beta \sum_{l=1}^{N} h_{jl}S_m(j,l)(V_j - V_l)$$

(12)

4   Experimental Analysis

4.1   Dataset Description

We use a publicly available dataset released by [19] to evaluate the performance of our model. The dataset was collected from Weibo, which allows users to follow other users and receive messages from followed users. Like Twitter, it also provides retweeting function to encourage users to spread information. Specifically, in this paper, we randomly sample 10,000 messages retweeted by 690,787 users from the above dataset. Since the dataset doesn't contain the messages published by retweeters, we also collect messages posted by retweeters in order to calculate similarities among retweeters. Table 1 lists statistics of the dataset used in this paper.

Table 1. Retweeting data statistics

Dataset	#Users	#Retweeter's tweets	#Tweets	#Retweets	Sparseness
Weibo	690,787	131,129,186	10,000	1,435,720	0.02%

4.2   Experimental Settings

For the above dataset, we randomly sample 80% of the retweetings from user-message retweeting matrix as the training data to predict the remaining 20% of retweetings. The corresponding entries in $\mathbf{R}$ of positive instances for testing data are set to 0. We determine the number of clusters in the proposed model using rule of thumb: $k \approx \sqrt{n/2}$ with n as the number of users/messages. Meanwhile, we empirically set the number of topics to 100 and parameters $\rho = \lambda = 0.5$.

4.3   Comparative Algorithms

We implement the following baselines for comparison with our social clustering based retweeting prediction model (SCRP).

- **Naive Bayes**: The retweeting predication can be considered as a binary classification task, where each message is labelled either positive or negative instance to represent whether it will be retweeted or not.
- **LRC-BQ**: The method proposes a notion of social influence locality based on pairwise influence and structural diversity, and then uses a logistic regression classifier to predict user's retweeting behavior [19].
- **MNMFRP**: This method utilizes nonnegative matrix factorization to predict retweeting behavior from user and content dimensions, respectively, by using strength of social relationship to constrain objective function [16].
- **CRPM & IRPM**: The two methods use the clustering relationships of messages to predict retweeting behavior based on matrix factorization [6]. These models don't take into account clustering information from users.
- **SCRP-U**: This method only considers user clustering information in our proposed retweeting prediction model.
- **SCRP-M**: This method only utilizes message clustering information for user retweeting prediction model.

4.4   Evaluation Measures

To quantitatively evaluate the performance of the proposed model, we divide the constructed data set into training and test data, and perform 10-fold cross validation to alleviate the effects of random selection. We evaluate the performance of retweeting prediction in terms of Precision, Recall, F_1-score, and Accuracy.

4.5   Parameter Settings

In this section, we will investigate the effect of different parameter settings for our proposed model, including tradeoff parameters, dimension of latent features, and number of projected gradient iterations, on the performance.

Tradeoff Parameters: In our proposed method in this paper, the tradeoff parameters α, β, γ and η play the role of adjusting the strengths of different terms in the objective function. They control how much our method should incorporate the clustering information for retweeting prediction model. Taking the scales of U and V into account, we scan orders of magnitude and try different combinations of parameters as shown in Table 2. The results in Table 2 show that the parameter set $\alpha = \beta = \gamma = \eta = 10^{-4}$ produce the best performance. In our following experiments, we just use this parameter setting.

Number of Latent Features: To find a K-dimensional joint latent space for users and messages, we train U and V using gradient descent method. More specifically, we conduct extensive experiments with K from 2 to 80 on the constructed dataset. The results are shown in Fig. 3, from which we can see conclude that with the latent feature number K increasing, F_1-score increases gradually. We can also observe that F_1-score grow more slowly when $K > 50$. Considering the computation efficiency and storage cost, we choose $K = 50$ as the latent

Table 2. Tradeoff parameters on Weibo dataset (50 Hidden Features and 50 Iterations)

α	β	γ	η	F_1-score	Accuracy
10^{-6}	10^{-6}	10^{-6}	10^{-6}	0.808	0.768
10^{-5}	10^{-5}	10^{-5}	10^{-5}	0.835	0.808
10^{-4}	10^{-4}	10^{-5}	10^{-5}	0.839	0.825
10^{-4}	10^{-4}	10^{-4}	10^{-4}	**0.847**	**0.831**
10^{-4}	10^{-4}	10^{-3}	10^{-3}	0.823	0.788
10^{-3}	10^{-3}	10^{-3}	10^{-3}	0.816	0.780

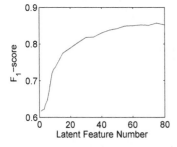

Fig. 3. Latent feature number on Weibo dataset (50 Iterations)

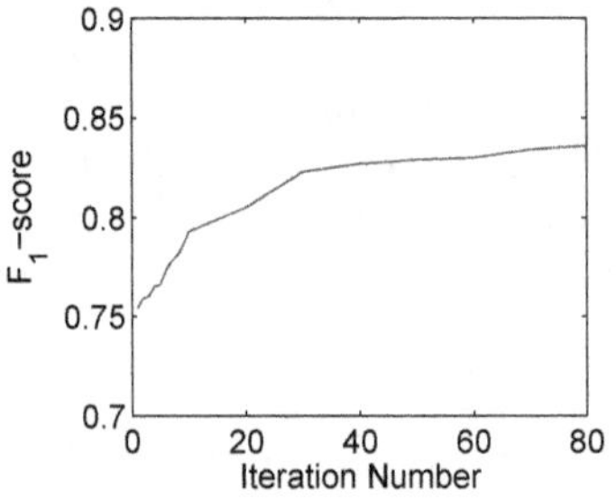

Fig. 4. Iteration number on Weibo dataset (50 Hidden Features)

space dimension in our experiments. Although it is not the perfect one, the following experiments demonstrate it is adequate.

Number of Iterations: When using gradient descent method to solve the objective function, we need to predefine a proper number of updating iterations to get a good performance while avoid overfitting. Figure 4 illustrates the impacts of the number of iterations on F_1-score. Considering the trade-off between the computational efficiency and the accuracy of prediction, we conduct 50 iterations for the solution in our experiments.

4.6 Effect of Sparseness

Based on the above parameter settings, we further exploit different training data sets to test the sensibility of the proposed model on constructed dataset. For example, training data 80 % means we randomly select 80 % of the retweeting behavior instances from user-message retweeting matrix as the training data to predict the remaining 20 % of retweeting entities. The overall performance of our proposed approach with different training set is illustrated in Fig. 5. From these figures, we can see conclude that the performance of our proposed SCRP method improves gradually as the number of training positive instances increase.

Moreover, we have also the following observation that the performances of our model change within a narrow range in the dataset with different sparseness which shows our model have good robustness. In general, social clustering based retweeting prediction performs better when observed retweeting instances are relatively more in the training data. This indicates that each user/message can be better represented by a linear combination of other users/messages from the same group in the latent space.

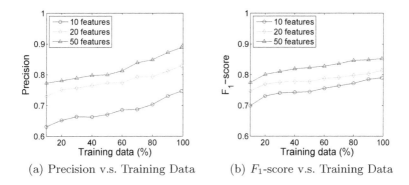

(a) Precision v.s. Training Data   (b) F_1-score v.s. Training Data

Fig. 5. Different training data settings to test our proposed SCRP model.

4.7   Prediction Performance

Our goal is to find who will be retweeted based on partially observed retweeting instances. Therefore, in this section, we will demonstrate the prediction performance of the proposed method, and compare it with other methods. Specifically, we set the optimal parameters when running the baselines. Then all experiments are performed 5 runs with the 50 dimensions to represent the latent features. We list the average results of each method in Table 3. Noted that both LRC-BQ model [16] and MNMFRP model [19] use the same original dataset with us. From these results, we can observe the following conclusions: (1) The proposed SCRP model, which incorporates user and message clustering factors together, significantly outperforms the baseline methods in our experimental results; (2) The prediction performance of SCRP is better than (CRPM &IRPM), which reveals that user clustering information is effectiveness of factor for retweeting prediction; (3) The comparison between SCRP-U v.s. MNMFRP reveals that the strategy of incorporating user clustering information to predict missing entities in the objective function is more effective compared with considering the strength of social relationship. In general, incorporating user and message clustering information can reduce the sparsity of data and improve the performance of prediction.

Table 3. Performance of retweeting behavior prediction.

Method	Precision	Recall	F_1-score	Accuary
Naive Bayes	0.562	0.555	0.558	0.555
LRC-BQ	0.698	0.770	0.733	0.719
MNMFRP	0.796	0.791	0.793	N/A
CRPM	0.814	0.833	0.823	0.821
IRPM	0.817	0.833	0.825	0.823
SCRP-U	0.846	0.809	0.827	0.809
SCRP-M	0.847	0.811	0.829	0.811
SCRP	**0.863**	**0.831**	**0.847**	**0.831**

5   Conclusion

In this paper, we propose a novel method, which incorporates the users and messages clustering information together, to predict user's retweeting behavior. The proposed model measures the similarities among users and messages using an ensemble from explicit and implicit dimensions, and then utilizes matrix factorization method to predict unobserved retweeting behaviors by employing cluster information of users and messages to constrain objective function. Experimental results demonstrate that the proposed method can achieve better performance than state-of-the-art methods.

Acknowledgments. This work was supported by National Key Technology R & D Program(No.2012BAH46B03), and the Strategic Leading Science and Technology Projects of Chinese Academy of Sciences(No.XDA06030200).

References

1. Abdullah, N.A., Nishioka, D., Tanaka, Y., Murayama, Y.: User's action and decision making of retweet messages towards reducing misinformation spread during disaster. J. Inf. Process. **23**(1), 31–40 (2015)
2. Boyd, D., Golder, S., Lotan, G.: Tweet, tweet, retweet: Conversational aspects of retweeting on twitter. In: HICSS, pp. 1–10 (2010)
3. Breese, J.S., Heckerman, D., Kadie, C.: Empirical analysis of predictive algorithms for collaborative filtering. In: UAI, pp. 43–52 (1998)
4. Can, E.F., Oktay, H., Manmatha, R.: Predicting retweet count using visual cues. In: CIKM, pp. 1481–1484 (2013)
5. Feng, W., Wang, J.: Retweet or not?: personalized tweet re-ranking. In: WSDM, pp. 577–586 (2013)
6. Jiang, B., Liang, J., Sha, Y., Wang, L.: Message clustering based matrix factorization model for retweeting behavior prediction. In: CIKM, pp. 1843–1846 (2015)
7. Jiang, B., Sha, Y., Wang, L.: A multi-view retweeting behaviors prediction in social networks. In: Cheng, R., Cui, B., Zhang, Z., Cai, R., Xu, J. (eds.) Web Technologies and Applications. LNCS, vol. 9313, pp. 756–767. Springer, Heidelberg (2015)

8. Blei, D.M., Ng, A.Y., Jordan, M.I.: Latent dirichlet allocation. J. Mach. Learn. Res. **3**, 993–1022 (2003)
9. Liu, L., Tang, J., Han, J., Jiang, M., Yang, S.: Mining topic-level influence in heterogeneous networks. In: CIKM, pp. 199–208 (2010)
10. Luo, Z., Osborne, M., Tang, J., Wang, T.: Who will retweet me? Finding retweeters in Twitter. In: SIGIR, pp. 869–872 (2013)
11. Metaxas, P., Mustafaraj, E., Wong, K., Zeng, L., O'Keefe, M., Finn, S.: What do retweets indicate? results from user survey and meta-review of research. In: ICWSM, pp. 658–661 (2015)
12. Naveed, N., Gottron, T., Kunegis, J., Alhadi, A.C.: Bad news travel fast: A content-based analysis of interestingness on twitter. In: WebSci, p. 8 (2011)
13. Petrovic, S., Osborne, M., Lavrenko, V.: RT to win! Predicting message propagation in Twitter. In: ICWSM (2011)
14. Recuero, R., Araujo, R., Zago, G.: How does social capital affect retweets? In: ICWSM (2011)
15. Suh, B., Hong, L., Pirolli, P., Chi, E.H.: Want to be retweeted? large scale analytics on factors impacting retweet in twitter network. In: SOCIALCOM, pp. 177–184 (2010)
16. Wang, M., Zuo, W., Wang, Y.: A multidimensional nonnegative matrix factorization model for retweeting behavior prediction. Math. Probl. Eng. **2015**, 1–10 (2015)
17. Yang, Z., Guo, J., Cai, K., Tang, J., Li, J., Zhang, L., Su, Z.: Understanding retweeting behaviors in social networks. In: CIKM, pp. 1633–1636 (2010)
18. Zaman, T.R., Herbrich, R., Van Gael, J., Stern, D.: Predicting information spreading in twitter. In: NIPS, pp. 17599–601 (2010)
19. Zhang, J., Tang, J., Li, J., Liu, Y., Xing, C.: Who influenced you? Predicting retweet via social influence locality. ACM TKDD **9**(3), 25 (2015)
20. Zhang, Q., Gong, Y., Guo, Y., Huang, X.: Retweet behavior prediction using hierarchical dirichlet process. In: AAAI, pp. 403–409 (2015)

KPCA-WT: An Efficient Framework for High Quality Microblog Extraction in Time-Frequency Domain

Min Peng$^{(\boxtimes)}$, Xinyuan Dai, Kai Zhang, Guanyin Zeng, Jiahui Zhu, Shuang Ouyang, Qianqian Xie, and Gang Tian

School of Computer, Wuhan University, Wuhan, China
{pengm,tracy_day,gunnyzeng,zhujiahui,ziyun,xieq,
tiang2008}@whu.edu.cn
zhangkaishiu@163.com

Abstract. Massive social event relevant messages are generated in online social media, which makes the filtering and screening a great challenge. In order to obtain massages with high quality, a high quality information extraction framework based on kernel principal component analysis and wavelet transformation (KPCA-WT) is proposed. First, based on multiple features fusion, we design an algorithm to extract the microblogs of high quality, which transforms the features into wavelet domain to capture the detailed differences between the feature signals. Then the weights of the features are evaluated by EM algorithm and fused further to get a comprehensive value of each message. In addition, to reduce the effect of noisy features and speed up the operation, these features are processed through kernel principal component analysis before transforming into wavelet domain. Experimental results show that the proposed framework can extract information with higher quality, less redundancy, and greatly reduce the time consumption.

Keywords: Information extraction · Feature fusion · Wavelet transformation · EM algorithm · KPCA

1 Introduction

In recent years, microblogging services have brought users to a new era of knowledge dissemination and information retrieval. There are billions of online users on Twitter who exchange information of interest every day in the form of short messages (no more than 140 characters), called tweets. According to statistics, over 400 million tweets are generated on Twitter per day[1]. This kind of social

---

This research is supported by the Natural Science Foundation of China under contract No. 61472291, and Natural Science Foundation of Hubei Province, China under contract No. ZRY2014000901.

[1] http://www.199it.com/archives/101807.html.

B. Cui et al. (Eds.): WAIM 2016, Part II, LNCS 9659, pp. 304–315, 2016.
DOI: 10.1007/978-3-319-39958-4_24

media have developed into important online platforms for users to share and discuss everything, including news, jokes, moods, and so on.

The messages of social media always relate to certain events happening in real-world and can be classified into different of topic sets [1]. For each topic set, the qualities of messages posted by different users are quite different. Some describe a detailed process of an event or express user's opinion clearly, while others only contain incomprehensible texts, Internet catchphrases, blurred expressions or even nonsense advertisements. In order to get the hang of the event overview, users usually need to scan hundreds of relevant messages. However, for a social event or its topics, there are always various redundant messages in discussion. This situation shows the prospect of a high quality extraction system for microblogs. Therefore, in this paper, we focus on the problem of extracting high quality messages in microblogs, i.e., a small collection of microblogs that can highly describe an event while containing little redundant information. Because of the limited length of microblog content, we also consider other features to evaluate a microblog's quality, including user authority, text structure, information propagation, and so on.

However, there are still some challenges for extracting the high quality messages from an event set. We list two as follows:

(1) Microblogs are usually comprised of multiple features, including noise or redundant features. How to measure the composite impact of all these features is a burning question.
(2) Due to the huge amount of spatial data, another important challenge for extraction is how to achieve better time efficiency.

To tackle the challenges above, in this paper, we aim to create a more efficient high quality microblog extraction framework based on our previous work [2]. For a given event message set, we consider multiple features such as commented number, forwarded number, URL, content and follower number. Then we address this task with three steps. First, we reduce the feature dimensions through feature transformation. Second, we construct a K-dimension feature matrix as an extraction basis, and transform it into time-frequency domain to reduce the time consumption and improve the extraction quality. Third, we estimate each feature's contribution degree to the information quality based on EM algorithm [3]. The main contributions of this paper are listed as follows:

(1) We design a feature-based transformation to reduce the feature dimension. Based on kernel principal component analysis (KPCA) [4], we can reduce the feature dimension, improve the time efficiency, and alleviate the influence of noisy features.
(2) We construct a feature matrix of different features from microblog messages. This matrix is then transformed into a uniform time-frequency domain based on wavelet transformation to reduce the computational complexity.
(3) We employ the EM algorithm to estimate the importance of each feature in the wavelet domain. Hence we combine them with reconstruction algorithm to derive a comprehensive score of each message.

2   Related Work

2.1   Feature Selection

Due to the heterogeneity and large scale of the data, selecting the most salient contents for an event is a challenging task. The past microblogging researches mainly focus on textual content analysis, such as language model [5], term frequency [6], hybrid TF-IDF [7], time delay [8], and topic hierarchy [9]. Most of these above works only refer to one or two social media features.

However, social media information contains not only content such as texts, pictures and videos, but also additional social attributions and authority attributions. These non-textual features usually convey more useful information in microblogs. Chen et al. [10] proposed a high quality threads finding method with time series analysis and features selection. Xi et al. [11] used features from the thread trees of forums, authors, and lexical distribution within a message thread and then applied linear regression and support vector machine to train the ranking function. Ghose and Ipeirotis [12] studied several factors on assessing review helpfulness including reviewer characteristics, reviewer history, readability and subjectivity.

2.2   Feature Fusion

In order to obtain high quality messages, extensive researches have been done in the data fusion. One of the pioneering works to investigate fusing result in multiple runs was proposed by Fox et al. [13]. They combined results from pairs of different weighting schemes from the SMART system, and found that significant improvements could be obtained by combining two different weighting schemes. Ogilvie and Callan [14] analyzed the conditions for combining different document representations based on TREC collection, and they found that the hypotheses for meta-search on classical text collections did not necessarily hold in web environments. Besides, Fan et al. [15] used generic programming (GP) to automatically optimize a search engine ranking function.

All these algorithms substantially analyze features in the time domain. There are also many feature analysis methods working in the frequency domain. [16] applied discrete Fourier transformation (DFT), which converted the signals from the time domain into the frequency domain. Compared with DFT, wavelet transformation [17] has more desirable features. Wavelet refers to a quickly varnishing oscillating function. Unlike the sine and cosine used in the DFT, which are localized in frequency but extended infinitely in time, wavelets are localized in both time and frequency domain. In this paper, we transform the features into wavelet domain, because the multi-resolution property of wavelet transformation can help in detecting the features at different levels of details. Since wavelet transformation is very fast, our approach thus becomes more efficient on the whole.

3   Problem Definition and Feature Analysis

3.1   High Quality Microblogs

Since microblog messages always relate to some certain social events, they can be reasonably classified into kinds of topic sets. For each microblog in a topic set Γ, it contains content feature, user authority feature and microblogging behavioral feature. Qualities of messages posted by different users are quite different and there is a lot of redundant information in the topic set Γ. In this paper, we attempt to extract a high quality microblog subset from Γ. Thus this high quality microblog subset should possess five characteristics: (1) It is the most relevant subset of the total, and the size of this microblog subset is far less than the original topic set ; (2) It can describe the event or express the opinion clearly; (3) It can generate wide attentions among users; (4) It is posted by influential people; (5) It has little redundancy and is readable.

3.2   Feature Analysis

In order to display the quality of the microblogs from different views, we firstly define the features. The microblogs have many features like commented number, forwarded number, URL, content, follower number, etc. The features used in our study include:

(1) *Length of a microblog*: Longer microblog may be more informative as each microblog has been limited to up to 140 characters.
(2) *Importance of content*: Each microblog has a score of TF-IDF to measure its importance.
(3) *Follower score*: Here we are concerned about those users who may publish more reliable microblogs, especially the authoritative users, because they always tend to publish high quality microblogs. Hence we utilize the number of followers to measure the authority of a user.
(4) *URL*: Users usually connect microblog and videos (or news websites) with URLs to present detailed description of event. For the ith microblog in topic set Γ, w_k is the frequency of the kth URL, then $URL = \sum w_k$ (if the kth URL is in the ith microblog).

3.3   Algorithm Overview

Based on the different features of the microblog, we formally define the task of high quality microblogs extraction as: Given an event set Γ, each microblog $d_i \in \Gamma (i = 1, 2, ..., N)$ has L features $F_l = (f_{i1}, f_{i2}, ..., f_{iL})$. Firstly, we reduce the dimension of the features and get the new K features through KPCA. Then, we construct a K-dimension feature matrix $\mathbf{F} = \{f_{ij} | i = 1, 2, ..., N; j = 1, 2, ..., K\}$ for these N microblogs. At last, we extract high quality information by fusing all these features with their corresponding weights and get the top-M items to constitute a high quality microblog subset.

4 Feature Dimension Reduction Based on KPCA

4.1 Pre-processing

The original data generally contains two aspects of the important information: (1) the difference of each index variation degree, embodied in the variation coefficient of each index (the ratio of the variance and mean); (2) the degree of interaction between various indexes. But for more groups of different dimensions or different order of magnitude, the data first should be dimensionless processed. Here we apply the method of equalization, where the mean value of each index is divided by their corresponding raw data. Assuming that the original sample data matrix is $\mathbf{X} = (x_1, x_2, ..., x_p)^T = (x_{ij})_{n \times p}$, let $y_{ij} = x_{ij}/\bar{x}_j (i = 1, 2, ..., n; j = 1, 2, ..., p)$, then we derive a matrix $\mathbf{Y} = (y_{ij})_{n \times p}$. Because the mean value of each column vector in $\mathbf{Y}$ is 1, we can deduce the diagonal elements of the covariance matrix as $s_{ij}/\bar{x}^2$, where s_{ij} is the covariance of the original data. That is to say, after equalization, the main diagonal elements of covariance matrix are the square of the index variation coefficients. We then set the raw data correlation coefficient between the various indexes as r_{ij}, and r'_{ij} after equalization, as shown in formula (1):

$$r'_{ij} = \frac{\frac{s_{ij}}{\bar{x}_i \cdot \bar{x}_j}}{\sqrt{\frac{s_{ii}}{\bar{x}_i^2}}\sqrt{\frac{s_{jj}}{\bar{x}_j^2}}} = \frac{s_{ij}}{\sqrt{s_{ii}}\sqrt{s_{jj}}} = r_{ij} \tag{1}$$

4.2 Detailed Steps of KPCA

The steps of KPCA are described as follows:

(1) When given a dataset, we deal with it by dimensionless processing. In detail, for input samples $\mathbf{X} = [x_1, x_2, ..., x_l]$, by using a nonlinear function Φ, we map the input samples into a high-dimensional feature space $F : R^N \to F, x \to \Phi(x)$. Therefore we can get mapped samples $\{\Phi(x_1), \Phi(x_2), ..., \Phi(x_l)\}$.

(2) By formulating PCA in a way that only involves the mapped samples via dot product, we do not need to know the mapping Φ explicitly, as the dot product can be evaluated by a positive semi-definite kernel function $k(x, y) = (\Phi(x), \Phi(y))$. Here we choose radial basis kernel $k(x, y) = exp(-(||x - y||^2)/c)$, and obtain the kernel matrix $\mathbf{K} = (k_{ij})_{l \times l}$.

(3) Assuming that $\{\Phi(x_1), \Phi(x_2), ..., \Phi(x_l)\}$ have been mean-centered in $\mathbf{F}$, therefore the kernel matrix $\mathbf{K}$ and the kernel function $k(x, y)$ are replaced by their centered versions $\tilde{\mathbf{K}}$:

$$\tilde{\mathbf{K}} = \mathbf{K} - I_l\mathbf{K} - \mathbf{K}I_l + I_l\mathbf{K}I_l, (I_l)_{ij} = 1/l \tag{2}$$

(4) By applying eigenvalue decomposition to $\tilde{\mathbf{K}}$, we can obtain the orthonormal eigenvectors $\alpha_1, \alpha_2, ..., \alpha_l$ and the associated corresponding eigenvalues $\lambda_1 \geq \lambda_2 \geq ... \geq \lambda_l$.

(5) We utilize these eigenvalues to calculate the cumulative contribution rate. The cumulative contribution rate is usually about 85 %–95 %, and then the principal component is chosen accordingly.

5   Feature Fusion Based on Wavelet Transformation

In this section, we introduce the two kinds of fusion methods. The first is the K-dimension feature fusion based on wavelet transformation (KD-FF_WF), where the feature space is the original one. The second is KPCA feature fusion based on wavelet transformation (KPCA-WT), where the features are derived by KPCA. These two methods are quite similar, as the only difference is the feature space, therefore we present these two methods in a common structure (namely, we mainly focus on the KPCA-WT in the following sections). The overview of these two methods can be summarized in Fig. 1.

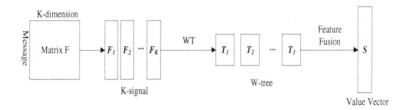

Fig. 1. Feature fusion framework based on wavelet transformation

5.1   Wavelet Transformation

After kernel principal component analysis for feature dimension reduction, a new K-dimension feature matrix $\mathbf{F} \in R^{N \times K}$ is obtained. Each dimension vector $F_k \in R^{N \times 1} (k = 1, 2, ..., K)$ is regarded as a one-dimension signal. We then transform the K-dimension feature into time-frequency domain via wavelet transformation to achieve feature fusion, select more significant features, and reduce computational complexity.

Wavelet transformation is good at time-frequency localization. It is a multi-resolution technique at different scale, i.e., inverse of frequency, to reflect different information [8]. Therefore, coefficients of wavelet transformation can reveal the multi-resolution analysis of features. What's more, the time complexity of the wavelet transformation is linear, namely $O(N)$, which is far below the computational complexity of the original feature space.

Wavelet has several types of basis functions that can be incorporated for feature analysis. Suppose that the wavelet basis function is h, fast wavelet transformation of each dimension results in a wavelet tree $T_k (k = 1, 2, ..., K)$ with L coefficient vector nodes. For each node $t_{kl} (l = 1, 2, ..., L)$, it has a value vector $W_{kl} (W_{kl} \in R^{n_l \times 1})$ to denote a set of corresponding wavelet/approximation coefficients, where n^l denotes the number of coefficients in l^{th} node of T_k, and all of the lth nodes in T_k have the same coefficient number n^l. In this paper, we choose the Daubechies wavelet, due to its nature of picking up details which might be lost by Haar wavelet. All these wavelet trees have a homogeneous structure, i.e., the coefficient set $T = \{W_{kl} | k = 1, 2, ..., K; l = 1, 2, ..., L\}$.

To facilitate the following discussion, all the coefficients W_{kl} in set T are also expressed as $C = \{C_l | l = 1, 2, ..., L\}$, where $C_l \in R^{nl \times 1}$ and $C_{lk} = W_{kl}$.

5.2   Feature Fusion in Wavelet Domain

According to some case studies, microblog feature vectors may share some commons in expressing the data. So when each feature is transformed into a tree with L nodes in wavelet domain, we need to fuse these K-dimension features by integrating the K-coefficient vectors of the lth nodes in each tree T_k respectively. Here, we incorporate EM algorithm to estimate the contribution degree of each coefficient vector C_{lk}. EM algorithm and its variants are appropriate to perform parameter inference, especially when there exist some latent variables in the probability distributions.

For each node set $N_l = \{t_{kl} | k = 1, 2, ..., K\}$, our goal is to determine the contribution degree α_{lk} of each dimension in the occurrence of all the coefficient matrix $C_l(C_l \in R^{n_l \times K})$.

Usually, the wavelet coefficients of each feature dimension contain quite a number of zero elements. So if these wavelet coefficients are sparse, they can be composed of a large number of small coefficients or a small number of large coefficients [18]. In this way, the Gaussian mixture distribution with mean zero is fit effectively for the actual distribution of the wavelet coefficients. So we assume that the observed coefficient distribution C_l can be described as a K-component Gaussian mixture distribution. Based on this mixture model, the occurrence probability of a fused coefficient c_{li} can be written as:

$$f(c_{li}) = \sum_{k=1}^{K} \alpha_{lk} f_{lk}(c_{lki}) \tag{3}$$

where $\alpha_{lk} \geq 0$, $\sum_{k=1}^{K} \alpha_{lk} = 1$, and $f_{lk}(c_{lki})$ denotes the probability density function of the kth component distribution $N(u_{lk}, \sigma_{lk}^2)$, α_{lk} denotes the contribution degree of the kth component in the lth node set N^l.

Then we set $\Phi_{lK} = \{\Theta_l; \Lambda_l\} = \{\alpha_{l1}, ..., \alpha_{lK}; \mu_{l1}, ..., \mu_{lK}, \sigma_{l1}^2, ..., \sigma_{lK}^2\}$ as the parameter vector of this mixture distribution. Λ_l is the distribution parameter vector (i.e., all the means and variances) of the lth node set. Then, the log-likelihood function over the entire l^{th} node set N^l is given as:

$$l(\Phi_l) = \ln(L(\Phi_l)) = \sum_{i=1}^{n^l} \ln\left(\sum_{k=1}^{K} \alpha_{lk} \frac{1}{\sqrt{2\pi\sigma_{lk}^2}} \exp\left\{\frac{(c_{lki} - u_{lk})^2}{2\sigma_{lk}^2}\right\}\right) \tag{4}$$

Therefore, the contribution degree can be updated with EM as:

$$\alpha_{lki}^{(s)} = \frac{\alpha_{lk}^{(s-1)} f_{lk}(c_{lki}, u_{lk}, \sigma_{lk}^2)}{\sum_{t=1}^{K} \alpha_{lt}^{(s-1)} f_{lt}(c_{lti}, u_{lt}, \sigma_{lt}^2)} \tag{5}$$

Hence we derive a group of optimum parameters $\Theta_l = \{\alpha_{l1}, \alpha_{l2}, ..., \alpha_{lK}\}$, and use them to fuse the K-dimension coefficient matrix C_l into a one-dimension coefficient vector $C_l{}^*$ by linearly weighting with Θ_l as follows:

$$c_{li}^* = \sum_{k=1}^{K} \alpha_{lk} c_{lki} \qquad (6)$$

When considering the case of wavelet transformation, for each node set $N_l(l = 1, 2, ..., L)$, we get its one-dimension coefficient vector $C_l^* = \{c_{li}{}^* | i = 1, 2, ..., n^l\}$. Then, we recompose a new signal S with coefficient set $C^* = \{C_l^* | l = 1, 2, ..., L\}$ via wavelet inverse transformation. Each element s_i in S denotes a finial comprehensive value of microblog i.

5.3   Computational Complexity

Here we mainly focus on the computational complexity of the proposed KPCA-WT. Computing wavelet transformation takes $O(N * j)$ for each dimension, where j is the wavelet decomposition level. Besides, inferring parameter vector Θ_l takes $O(s^l * K^3 * n^l)$ for each node set N_l, where s^l is the number of iterations, K is the number of features, and n^l is the number of coefficients in the lth node of T_k. As each Θ_l is inferred independently, the EM process can be parallelly executed for all node sets $N_l(l = 1, 2, ..., L)$. So the EM process takes at most $O(L * s^l * K^3 * n^l)$. Therefore, the overall complexity of KPCA-WT is $O(N * j) + O(L * s^l * K^3 * n^l)$. However, $\max(n^l) \approx N/2$, and $O(N * j)$ is much smaller than $O(L * s^l * K^3 * n^l)$, so the complexity of KPCA-WT algorithm approximates to $O(L * s^l * K^3 * n^l))$.

6   Experiments

6.1   Dataset and Experimental Setting

In order to validate our high quality microblog extraction framework, we utilize microblogs about social events from Sina Weibo, the most popular Chinese microblog platform. Via Sina Weibo API[2], we derive a microblog dataset from January to May in 2013. This dataset includes 8 hot social events happened during those months. For each event, there are about 10,000 to 100,000 relevant microblogs with a similar number of the involved users. The description of the dataset is shown in Table 1.

In this paper, several methods are incorporated to compare with the proposed extraction methods KPCA-WT and KD-FF_WT. The baselines includes (1) Most Recent (MR) tweet algorithm [19]: microblog extraction is based on timestamp reverse order; (2) Most Tweeted URL-based (MTU) tweets [8] algorithm: count the number of all URL occurrence for each event set, and then calculate the URL value of each microblog. Microblog extraction is based on URL value reverse order; (3) TF-IDF-based algorithm [20], commonly

---

[2] http://open.weibo.com/.

Table 1. Microblogs number and involved users number

Topic	Microblog	User
Two Sessions	73,337	59,257
Chinese First Lady Liyuan Peng	9,603	8,571
New Real Estate Policy	26,519	20,072
H7N9 Bird Flu	78,814	68,760
Ya'an Earthquake	101,978	87,577
Beijing Mist	32,613	30,201
Dead Pig	19,282	17,176
Funeral of Mrs. Thatcher	20,421	18,670

used in microblog summarization and extraction. We evaluate the effectiveness of our algorithm with time consumption, information quantity, and content redundancy. Specifically, we define the information quantity as $H(N) = \sum_{i=1}^{N} s_i \log_2(s_i)$, where s_i is the content score of the ith microblog, and utilize the variance (var $= \frac{1}{n-1} \sum_{i=1}^{N} (s_i - \overline{s_i})$) to measure the effect of redundancy reduction.

Table 2. Running time

Hot issues	KPCA-WT(s)	KD-FF_WT(s)
Two Sessions	75.77	104.14
Chinese First Lady Liyuan Peng	6.56	19.26
New Real Estate Policy	24.97	46.40
H7N9 Bird Flu	81.13	114.24
Ya'an Earthquake	54.84	80.70
Beijing Mist	32.66	58.55
Dead Pig	17.58	36.81
Funeral of Mrs. Thatcher	19.13	36.80

6.2 Experimental Results

(1) Comparison of Time Consumption. Using KPCA, the feature dimensions can be ultimately reduced by nearly a half. We compare the overall time consumption of KPCA-WT framework against with KD-FF_WT algorithm. The time consumption of KPCA-WT including: KPCA running time, wavelet decomposition and reconstruction time, EM training time. In Table 2, it is easy to see that the time consumption of KPCA-WT is less than the KD-FF_WT algorithm.

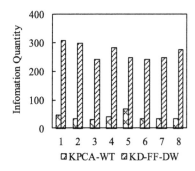

Fig. 2. Global information and its distribution

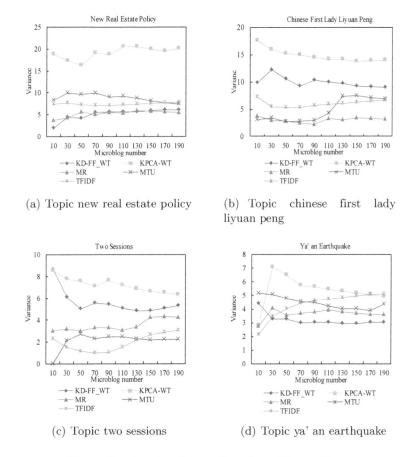

(a) Topic new real estate policy

(b) Topic chinese first lady liyuan peng

(c) Topic two sessions

(d) Topic ya' an earthquake

Fig. 3. Results of variance value of top-N microblogs

(2) Information Quantity. Through comprehensive signal S we can get top-N microblogs. Based on calculating the information quantity of these top-N microblogs, we can obtain the global information quantity $H(N)$, the lower the better. Figure 2 shows the global information quantity of the top-150 microblogs in eight event sets extracted by KPCA-WT framework and KD-FF_WT algorithm. We can see, through KPCA-WT framework, the global information quantity is considerable, often with abundant information.

(3) Effect of Reducing Content Redundancy. We perform the information redundancy statistics of top-N microblog results from different methods. As the most important feature of high quality information is the textual content, we calculate variance of the top-N microblogs, and higher variance denotes lower content redundancy. Taking four topics (New Real Estate Policy, Chinese First Lady Liyuan Peng, Two Sessions and Ya'an Earthquake) as examples, the results are displayed in Fig. 3.

From Fig. 3, we can see, in some event sets, without feature dimension reduction (KD-FF_WT algorithm), the extracted microblog content redundancy is larger. This is an evidence that noisy and redundant features do exist, and they would weaken the weights of the effective features and therefore influence the quality of extraction. During the experiments, the best performance is from the proposed KPCA-WT framework in our work, as its extracted microblogs obviously have higher variance than others. That is to say, our KPCA-WT framework is more effective in redundant content reduction.

7   Conclusions

In this paper, we propose a high quality information extraction framework to cater to this issue based on the idea of multiple features fusion. First, we formulate a high quality information extraction problem and analyze the internal relationships among different features. Then, we devise a framework of KPCA feature fusion based on wavelet transformation (KPCA-WT). Finally, along with a host of baselines, we evaluate the items generated by our framework on Sina Weibo dataset. With a set of eight hot events containing 300,000 microblog items, we evaluate the proposed framework through content redundancy, time consumption, information redundancy and subjective evaluation. Compared with some state-of-the-art methods, the proposed KPCA-WT achieves better performance in reducing redundant content and cutting down almost half of the running time. All in all, our work provides an effective strategy to study the microblog content selection and presentation by extracting high quality information from the original data resource. There are still a number of interesting directions for future work on high quality information extraction. For example, how could other implicit features impact on high quality extraction? And how could our method function in the heterogeneous microblog stream that is usually with multiple topics and multiple features?

References

1. Peng, M., Zhu, J., Li, X., et al.: Central topic model for event-oriented topics mining in microblog stream. In: CIKM 2015, pp. 1611–1620 (2015)
2. Peng, M., Huang, J., Fu, H., Zhu, J., Zhou, L., He, Y., Li, F.: High quality microblog extraction based on multiple features fusion and time-frequency transformation. In: Lin, X., Manolopoulos, Y., Srivastava, D., Huang, G. (eds.) WISE 2013, Part II. LNCS, vol. 8181, pp. 188–201. Springer, Heidelberg (2013)
3. Dempster, A.P., Laird, N.M., Rubin, D.B.: Maximum likelihood from incomplete data via the EM algorithm. J. Roy. Stat. Soc. **39**(1), 1–38 (1977)
4. Scholkopf, B., Smola, A., Mller, K.R.: Kernel principal component analysis. In: ICANN 1997, pp. 583–588 (1997)
5. O'Connor, B., Krieger, M., Ahn, D.: Tweetmotif: exploratory search and topic summarization for twitter. In: ICWSM 2010, pp. 384–385 (2010)
6. Yang, X., Ghoting, A., Ruan, Y., et al.: A framework for summarizing and analyzing twitter feeds. In: KDD 2012, pp. 370–378 (2012)
7. Sharifi, B., Hutton, M.A., Kalita, J.K.: Experiments in microblog summarization. In: SocialCom 2010, pp. 49–56 (2010)
8. Takamura, H., Yokono, H., Okumura, M.: Summarizing a document stream. In: Clough, P., Foley, C., Gurrin, C., Jones, G.J.F., Kraaij, W., Lee, H., Mudoch, V. (eds.) ECIR 2011. LNCS, vol. 6611, pp. 177–188. Springer, Heidelberg (2011)
9. Zhu, J., et al.: Coherent topic hierarchy: a strategy for topic evolutionary analysis on microblog feeds. In: Li, J., Sun, Y., Yu, X., Sun, Y., Dong, X.L., Dong, X.L. (eds.) WAIM 2015. LNCS, vol. 9098, pp. 70–82. Springer, Heidelberg (2015). doi:10.1007/978-3-319-21042-1_6
10. Chen, Y., Cheng, X., Yang, S.: Finding high quality threads in web forums. J. Softw. **22**(8), 1785–1804 (2011)
11. Xi, W., Lind, J., Brill, E.: Learning effective ranking functions for newsgroup search. In: SIGIR 2004, pp. 394–401 (2004)
12. Ghose, A., Ipeirotis, P.G.: Estimating the helpfulness and economic impact of product reviews: Mining text and reviewer characteristics. TKDE **23**(10), 1498–1512 (2011)
13. Fox, E.A., Shaw, J.A.: Combination of multiple searches. In: NIST SP, pp. 243–243 (1994)
14. Ogilvie, P., Callan, J.: Combining document representations for known-item search. In: SIGIR 2003, pp. 143–150 (2003)
15. Fan, W., Gordon, M.D., Pathak, P.: A generic ranking function discovery framework by genetic programming for information retrieval. Inf. Process. Manage. **40**(4), 587–602 (2004)
16. He, Q., Chang, K., Lim, E.P.: Analyzing feature trajectories for event detection. In: SIGIR 2007, pp. 207–214 (2007)
17. Daubechies, I.: Ten Lectures on Wavelets. Society for Industrial and Applied Mathematic, Philadelphia (1992)
18. Chipman, H.A., Kolaczyk, E.D., McCulloch, R.E.: Adaptive bayesian wavelet shrinkage. J. Am. Stat. Assoc. **92**(440), 1413–1421 (1977)
19. Burstei, J., Wolska, M.: Toward evaluation of writing style: finding overly repetitive word use in student essays. In: EACL 2003, pp. 35–42 (2003)
20. Becker, H., Naaman, M., Gravano, L.: Selecting quality twitter content for events. In: ICWSM 2011 (2011)

Big Data Analytics

Active Learning Method for Constraint-Based Clustering Algorithms

Lijun Cai, Tinghao Yu$^{(\boxtimes)}$, Tingqin He, Lei Chen, and Meiqi Lin

College of Information Science and Engineering, Hunan University, Changsha 410082, China
{ljcai,yth,hetingqin,chenleixyz123,lmq}@hnu.edu.cn

Abstract. Semi-supervision clustering aims to improve clustering performance with the help of user-provided side information. The pairwise constraints have become one of the most studied types of side information. According to the previous studies, such constraints increase clustering performance, but the choice of constraints is critical. If the constraints are selected improperly, they may even degrade the clustering performance. In order to solve this problem, researchers proposed some learning methods to actively select most informative pairwise constraints. In this paper, we presents a new active learning method for selecting informative data set, which significantly improves both the Explore phase and the Consolidate phase of the Min-Max algorithm. Experimental results on the data set of UCI Machine Learning Repository, using MPCK-means as the underlying constraint-based semi-supervised clustering algorithm, show that the proposed algorithm has better performance.

Keywords: Active learning · Clustering · Pairwise constraints

1 Introduction

Data clustering is a method of dividing a data set into clusters to group similar data points together, which is traditionally viewed as an unsupervised method for data analysis [1]. Clustering is generally done on the basis of some initial assumptions such as distance metric, data structure, number of clusters, data distribution and so on. Clustering result greatly relies on the correspondence between these assumptions and the actual model of clusters [2]. In fact, in many real application domains, some background knowledge or side information are available, which improve clustering performance. Consequently, constrained clustering has become popular recently for the reason of that it takes advantage of side information when available.

The instance-level constraints which were proposed by Wagstaff and Cardie [2] have become one of the most studied types of side information. There are two types of instance-level constraints: must-link and cannot-link constraints. The must-link constraints specify that two points must belong to the same cluster and the cannot-link constraints specify that two points must not belong to the same cluster [3]. According to the previous studies, such constraints increase clustering performance [1, 3–5]. Although the inclusion of instance-level constraints allows the user to manually inspect the data points in question, and precisely state whether two data points should belong

B. Cui et al. (Eds.): WAIM 2016, Part II, LNCS 9659, pp. 319–329, 2016.
DOI: 10.1007/978-3-319-39958-4_25

to the same class or not without the need to state the actual cluster they belong to, the acquisition of the constraints can be consuming and costly. Moreover, if the constraints are selected improperly, they may even degrade the clustering performance [10]. In order to solve this problem, researchers proposed some learning methods to actively select most informative pairwise constraints [3, 4, 6–9].

Comparing to the active learning studied in other domains, the research on active learning of instance-level constraints for semi-supervised clustering is relatively limited [3]. Basu et al. proposed an active selection method of constraints using farthest first query mechanism, which is the first study on this topic. We refer to this algorithm as the Farthest First Query Selection (FFQS) algorithm. It is a two phase approach including Explore and Consolidate. The Explore phase incrementally selects points using the farthest first query mechanism to find k points all belong to different clusters as skeleton of the clusters, where k is the number of clusters. The Consolidate phase selects non-skeleton points randomly and queries it against the skeleton until a must-link is obtained. An improved version of Consolidate phase using min-max criterion was proposed by Mallapragada et al., which selects the data point whose largest similarity to the skeleton points is smallest instead of random point selection. We refer to this algorithm as the Min-Max algorithm. While these two methods may not be robust as they select the first point at random in the Explore phase, and the selection of pairwise constraints didn't take the current clustering assignment into account. The clustering result can be improved on the current classification model, which is analogous to supervised active learning [3, 12, 13].

Some active constrained clustering algorithms which take advantages of the current clustering assignment also have been investigated. Xu et al. proposed an active constrained clustering algorithm named ACCESS, which examines the eigenvectors derived from similarity matrix of data [9]. However, the method is only suitable for two clusters problem. Huang and Lam derived a framework which takes an iterative approach to measure pairwise uncertainty and makes the selection choice [7]. Similar to this method, Xiong et al. proposed a neighborhood based framework to measure the point uncertainty, which also considers the number of queries to improve the clustering performance [3]. Nevertheless, due to the using of supervised learning methods like random forest, the amount of computations are expensive.

As we can see, all the constrained clustering algorithms mentioned above focus on minimizing the cost of constraint acquisition and improving the clustering performance as far as possible. While most of the existing work suffer from problems of inefficient performance or extensive computations.

This paper presents a new active learning method for selecting informative data set, which significantly improves both the Explore phase and the Consolidate phase of the Min-Max algorithm. Our proposed method is very simple but effective. First, the current clustering assignment of data points and their neighbors have been used to introduce a utility function for measuring point uncertainty and getting an informative data set. Then the Explore phase and Consolidate phase start in their order, which is same with Min-Max algorithm. The main contribution of our method is that the Explore phase and Consolidate phase are work on the informative data set (as opposed to the whole data set), which improves the clustering performance significantly. Furthermore, our method will select the most informative data point as the first point in the Explore phase rather than select data point at random, which eliminate the randomness of the Min-Max algorithm.

We apply our method to the well known MPCK-means (Metric Pairwise Constrained K-means) semi-supervised clustering algorithm and compare with random query selection algorithm, the FFQS algorithm and the Min-Max algorithm in terms of the normalized mutual information. Experimental results on the data set of UCI Machine Learning Repository [14] show that the proposed algorithm is better than the three algorithm mentioned above.

2   The Pairwise Constraints

The pairwise constraints were proposed by Wagstaff and Cardie. Besides, they have successfully developed a k-means variant which takes advantages of side information in the form of pairwise constraints [1]. These constraints include must-link and cannot-link constraints. The must-link constraints specify that two points must belong to the same cluster and the cannot-link constraints specify that two points must not belong to the same cluster.

The must-link constraints are transitive. If x_i must link to x_j which must link to x_k, we then know that x_i must link to x_k. Although the cannot-link constraints are not transitive, the closure is also performed. Because if we know that x_i cannot link to x_j which must link to x_k, then we know that x_i cannot link to x_k. According to the analysis above, the constraints can propagated in the data set. We can ultimately get all pairwise constraints based on these rules. These rules can be describe as follows.

– The transitive of Must-link constraints:

$$\left(x_i, x_k\right) \notin M \,\&\, \left(x_i, x_j\right) \in M \,\&\, \left(x_j, x_k\right) \in M \Rightarrow \left(x_i, x_k\right) \in M$$

– The transitive of Cannot-link constraints:

$$\left(x_i, x_k\right) \notin C \,\&\, \left(x_i, x_j\right) \in C \,\&\, \left(x_j, x_k\right) \in M \Rightarrow \left(x_i, x_k\right) \in C$$

The supervision information we select is critical to the performance of semi-supervised clustering algorithm. If the constraints are improper, they may have negative inspect on the clustering performance [3, 11]. Moreover, most of pairwise constraints is generated by manually inspecting the data points, and the acquisition of the constraints can be consuming and costly. So we propose an active learning method which can select the most informative pairwise constraints.

3   Active Learning Algorithms

As we mentioned in the introduction of this paper, the major improvement of our method compared to the Min-Max algorithm is that the Explore phase and Consolidate phase work on the informative data set selected rather than the whole data set. Another difference is that the first point in the Explore phase is the most informative data point rather than the point selected randomly. The core question we need to address is how to actively get informative data set and select the most informative data point before the Explore phase and Consolidate phase.

Suppose that there are data set $X = \{x_1, x_2, x_3, \dots, x_n\}$ which contains n instances and every instance has p variables. We should group them into k clusters. Assume that k has been given. x_{ij} is defined as jth variable of instance x_i. And the Euclidean distance between instance x_i and instance x_j is defined as follow.

$$d_{ij} = \sqrt{\sum\nolimits_{a=1}^{p} \left(X_{ia} - X_{ja}\right)} i = 1, \dots, n; \; j = 1, \dots, n \tag{1}$$

We get the cluster assignments $C = \{c_1, c_2, c_3, \dots, c_n\}$ by using any constraint-based semi-supervised clustering algorithm. In our method, the current clustering assignment of data points and their neighbors is used to introduce a utility function to measure point uncertainty and get an informative data set. Inspired by the k-nearest neighbors algorithm [15], we believe that for a data point, if the major of its neighbors are assigned to the different clusters, then the clustering result of this point is doubtful. We can measure the uncertainly of the given instance x_i through the number of its neighbors which is not assigned to the same cluster with it. Just like the k-nearest neighbors algorithm, the number of neighbors m will have great influence on the performance of our method.

The best choice of m depends upon the data set, such as the distribution of data points, the size of data set and so on. Generally, larger values of m make boundaries between informative data points and non-informative data points less distinct. If values of m is large enough, the informative data set S we selected may contain all the data points in the data set X and our proposed method will run as Min-Max algorithm. In contrast, if the values of m is too small, little informative data points will be selected and the effect of our proposed method will be reduced also. In this paper, we didn't pay much effort on the choice of m. We defined m as $m = n \times$ percent, $0 < percent < 1$. In which, n is the size of data set. Then we just simply adjust the value of percent to get enough informative data points in our experiments and the experiment results show that the performance of our proposed method is better than Min-Max algorithm in most case.

According to the previous studies on the k-nearest neighbors algorithm, much research effort has been put into the choice of k in the k-nearest neighbors algorithm and various techniques have been proposed [16, 17]. We believe that we can use the similar techniques to choose values of m in our proposed method and this is an important direction of our future work. As we have got m nearest neighbors N_i for each instance x_i, we can compute the number of neighbors which is not assigned to the same cluster with the instance x_i easily. As it is pointed in [18], selecting the instance with the largest uncertainly in deciding cluster membership is one of important principles of active learning. We defined P_i as the the number of neighbors which is not assigned to the same cluster with the instance x_i. Then the most uncertainly instance q can be chosen as

$$q = \arg\min_i P_i \tag{2}$$

And we can get informative data set as $S = \{s_i\}_{i=1}^{t}$ follow.

$$S = \left\{ s_i \mid s_i \in X, P_i > 0 \right\} \tag{3}$$

After we have got the most uncertainly instance q and the informative data set $S = \{s_i\}_{i=1}^t$, the Explore phase and Consolidate phase start in order. Different from Explore phase in the Min-Max algorithm, we select instance q as the first data point. What's more, the Explore phase and Consolidate phase work on the informative data set $S = \{s_i\}_{i=1}^t$ rather than the whole data set which is the most important difference between our proposed method and the Min-Max algorithm. The proposed method to select the informative data set is summarized in Algorithm 1.

Algorithm 1 Informative(X,C)

Input: A set of data instances $X = \{x_1, x_2, x_3, \ldots, x_n\}$;
Output: Informative data set $S = \{s_i\}_{i=1}^t$;

1: $C = $ Semi-Supervised-Clustering(D);
2: Compute the distance d_{ij} between instance i and instance j using Eq.1;
3: **for** each $x_i \in X$ **do**
4:     Select m nearest neighbors N_i for each instance x_i ;
5:     Compute P_i for each instance x_i ;
6: **end for**
7: **Return** $S = \{x_i \in X \mid P_i > 0\}$

---

4   Experiments

In this section, we apply our method to the well known MPCK-means (Metric Pairwise Constrained K-means) semi-supervised clustering algorithm and compare with random query selection algorithm, the FFQS algorithm and the Min-Max algorithm in terms of the normalized mutual information.

4.1   Datasets

In our experiments, we use eight benchmark UCI datasets that have been used in previous studies on constraint-based clustering [3, 4, 8, 19]. Our datasets include iris, wine, statlog-heart, ecoli, breast [20], parkinsons [21], statlog-image segmentation and digits-389. The iris dataset and the digits-389 dataset are used to evaluate Min-Max in [8]. For the ecoli dataset, we removed the smallest three classes, which is dealt in the same way with [3]. The characteristics of the data sets are shown in Table 1.

4.2   Evaluation Criterion

We use Normalized Mutual Information (NMI) to evaluate the clustering assignments. NMI considers both the clustering assignment and the class label as random variables, and measures the mutual information between these two random variables. We computed NMI following the methodology present in [22]. If C is the random variable representing the

Table 1. Charateristics of datasets.

Datasets	Number of data points	Dimension	Number of class
Iris	150	4	3
Wine	178	13	3
Heart	270	12	2
Ecoli	327	7	5
Breast	683	9	2
Parkinsons	195	22	2
Segment	2310	19	7
Digits-389	3165	16	3

cluster assignments of instances, and K is the random variable representing the class labels of the instances, then the NMI is computed by the following equation:

$$NMI = \frac{2I(C;K)}{H(C) + H(K)}$$

Where $I(X;Y) = H(X) - H(X|Y)$ is the mutual information between random variables X and Y. H(X) is the entropy of X, and $H(X|Y)$ is the conditional entropy X given Y.

4.3   Main Experimental Result

We apply our method to the well known MPCK-means (Metric Pairwise Constrained K-means) semi-supervised clustering algorithm and compare with random query selection algorithm, the FFQS algorithm and the Min-Max algorithm in terms of the NMI. We follow the experimental setup in [3]. For each given dataset, we apply our proposed method and three other baseline methods mentioned above to select up to 150 pairwise constraints, starting from no constraint at all. Then we apply MPCK-means to the dataset with these pairwise constraints and their transitive closures.

Because random query selection algorithm, the FFQS algorithm and the Min-Max algorithm are randomness, we repeat this process for 100 independent runs and report the average performance using NMI. Besides, the value of percent we chose for our proposed method in our experimental are shown in Table 2.

Table 2. The value of percent for each datasets.

Datasets	Iris	Wine	Heart	Ecoli	Breast	Parkinsons	Segment	Digits-389
n	150	178	270	327	683	195	2310	3165
percent	0.37	0.14	0.02	0.3	0.12	0.1`	0.01	0.008
m	55	25	5	98	82	19	23	25

The result of the experiments are shown in Fig. 1. We can see that our proposed method outperforms random query selection algorithm, the FFQS algorithm and the Min-Max algorithm on all the eight datasets considered. Besides, we find that the random query selection algorithm actually degrades the clustering performance for the datasets iris, wine, breast and digit-389 when the constraints increased. Previous studies have reported similar result that improper constraints may have negative inspect on the clustering performance, which we also mentioned in the previous section. This further proves that selecting the right constraints is very important.

4.4 Further Analysis

As we pointed out in the introduction section, the main improvement of our method compared to the Min-Max algorithm is that our Explore phase and Consolidate phase work on the data informative data set we selected rather than the whole data set and another difference is that we select the most informative data point as the first point in the Explore phase. Someone may doubt whether both of two factors are contributing to the performance advantage of our proposed method. Is it possible we just selected a good initial data point by adjusting the value of percent? To answer this question, we consider a variant of our proposed method, namely selecting the first point at random in the Explore phase. We follow the experimental setup above and let the variant of our proposed method repeat the same process for 100 independent runs on the six datasets which have been considered above. Note that the value of percent we chose is also not changed.

The experimental results are shown in the Fig. 2. We can see that the variant of our proposed method also shows better performance than the other three baseline method on the five datasets we considered. As for the remaining dataset ecoli, the performance of the variant of our proposed method is similar to the Min-Max. This is precisely what we expect, which proves that letting the Explore phase and the Consolidate phase run on the informative data set we selected can do much contribution to the performance of our proposed method. Besides, we observe that the performance of our proposed method remains superior to the variant on all the six datasets we consider. It suggests that the initial data point we selected also leads to the better choice of the pairwise constraints. In conclusion, both two factors we mentioned above contribute to the better performance of our proposed method.

4.5 Time Complexity Analysis

In this section, we analyze the runtime complexity of our proposed algorithm. We only consider the three major phase: selecting informative data set, the Explore phase and the Consolidate phase.

Selecting informative data set is the core part of our active learning algorithm, namely Algorithm 1. In line 1, any constraint-based clustering algorithm could be used to get the clustering assignment of data points. So the complexity depends on the chosen clustering algorithm. Note that the MPCK-means algorithm is used in our experiments, the complexity for this phase is $O(ndk)$ operations, where n is the number of data points, d is the number of dimensions and k is the number of clusters. The Euclidean distance is used in this study to calculate the distance between every pair of all instances, which will take $O(n^2)$ operations. In order to find m nearest neighbors N_i for each instance x_i and compute P_i for each instance x_i, a priority queue with the size of m is used for every instance. The computational complexity of this step is $O(n \log m)$, where m is generally

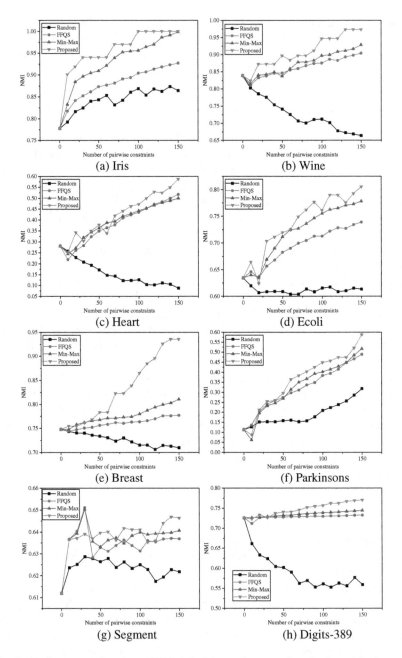

Fig. 1. Performance of random, FFQS, Min-Max and proposed method on eight datasets.

much smaller than n and can be viewed as a constant factor in this case. Therefore, the total computational complexity of this phase is $O(n^2)$.

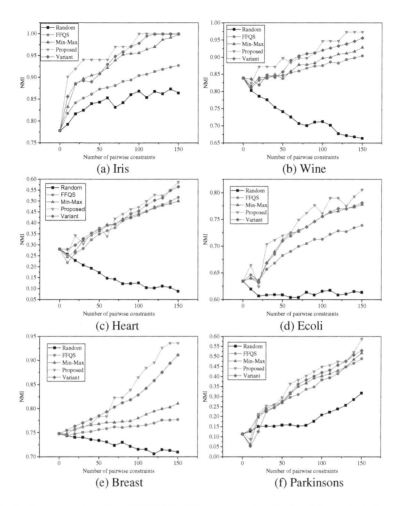

Fig. 2. Performance of random, FFQS, Min-Max, proposed method and the variant of proposed method on eight datasets.

As for the Explore phase and the Consolidate phase in our proposed method, their total runtime complexity is O(tf), where t is the size of informative data and f is the number of pairwise constraints. Obviously, $n^2 \gg tf$, the total computational complexity of our proposed method is $O(n^2)$. The complexity is generally less than most of the supervised learning method used in active learning methods.

5   Conclusions

This paper presents a new active learning method for constraint-based clustering algorithm which significantly improves the Min-Max algorithm. In our novel method, the current clustering assignment of data points and their neighbors is used to introduce a utility function to measure point uncertainty and get an informative data set. Then the Explore phase and Consolidate phase start in their order, which is same with Min-Max algorithm. Furthermore, our method will select the most informative data point as the first point in the Explore phase rather than select data point at random, which eliminate the randomness of the Min-Max algorithm. Experimental results on the data set of UCI Machine Learning Repository show that the proposed algorithm has better performance than its competitors.

There are a number of interesting directions to extend our work. In our proposed method, we need to choose a appropriate value of m, and research on how to obtain it. As we have mentioned in the previous section, we could try the similar techniques which have been used in the k-nearest neighbors algorithm. Besides, we consider a batch approach to select informative data points iteratively. Of course, our proposed method could be viewed as a general framework to improve the Min-Max algorithm also and we can try some other ways to select informative data set.

References

1. Wagstaff, K., Cardie, C., Rogers, S., et al.: Constrained K-means clustering with background knowledge. In: ICML, pp. 577–584 (2001)
2. Wagstaff, K., Cardie, C.: Clustering with instance-level constraints. In: Proceedings of the Seventeenth International Conference on Machine Learning, pp. 1103–1110 (2000)
3. Xiong, S., Azimi, J., Fern, X.Z.: Active learning of constraints for semi-supervised clustering. IEEE Trans. Knowl. Data Eng. **26**(1), 43–54 (2013)
4. Basu, S., Banerjee, A., Mooney, R.J.: Active semi-supervision for pairwise constrained clustering. In: Proceedings of 4th SIAM International Conference on Data Mining (SDM-2004), pp. 333–344 (2004)
5. Li, Z., Liu, J., Tang, X.: Pairwise constraint propagation by semidefinite programming for semi-supervised classification. In: International Conference on Machine Learning (2008)
6. Greene, D., Cunningham, P.: Constraint selection by committee: an ensemble approach to identifying informative constraints for semi-supervised clustering. In: Kok, J.N., Koronacki, J., Lopez de Mantaras, R., Matwin, S., Mladenič, D., Skowron, A. (eds.) ECML 2007. LNCS (LNAI), vol. 4701, pp. 140–151. Springer, Heidelberg (2007)
7. Huang, R., Lam, W.: Semi-supervised document clustering via active learning with pairwise constraints. In: ICDM IEEE Computer Society, pp. 517–522 (2007)
8. Mallapragada, P.K., Jin, R., Jain, A.K.: Active query selection for semi-supervised clustering. In: 19th International Conference on Pattern Recognition, ICPR 2008, pp. 1–4. IEEE (2008)
9. Xu, Q., desJardins, M., Wagstaff, K.L.: Active constrained clustering by examining spectral eigenvectors. In: Hoffmann, A., Motoda, H., Scheffer, T. (eds.) DS 2005. LNCS (LNAI), vol. 3735, pp. 294–307. Springer, Heidelberg (2005)
10. Kaufman, L., Rousseeuw, P.J.: Finding groups in data. An introduction to cluster analysis. J. Am. Stat. Assoc. **86**, 830–833 (1990)
11. Davidson, I., Wagstaff, K.L., Basu, S.: Measuring constraint-set utility for partitional clustering algorithms. In: Proceedings of the Tenth European Conference on Principles and Practice of Knowledge Discovery in Databases, pp. 115–126 (2006)
12. Guo, Y., Schuurmans, D.: Discriminative batch mode active learning. In: Advances in Neural Information Processing Systems, pp. 593–600 (2007)

13. Hoi, S.C.H., Jin, R., Zhu, J., et al.: Semi-supervised SVM batch mode active learning for image retrieval. In: 2013 IEEE Conference on Computer Vision and Pattern Recognition, pp. 1–7. IEEE (2008)
14. Blake, C., Keogh, E., Merz, C.J.: UCI Repository of Machine Learning Databases, Department of Information and Computer Science, University of California, Irvine (1998). http://www.ics.uci.edu/mlearn/MLRepository.html
15. Cover, T.M., Hart, P.E.: Nearest neighbor pattern classification. IEEE Trans. Inf. Theor. **13**(1), 21–27 (1967)
16. Nigsch, F., Bender, A., Van, B.B., et al.: Melting point prediction employing k-nearest neighbor algorithms and genetic parameter optimization. J. Chem. Inf. Model. **46**(6), 2412–2422 (2006)
17. Dhurandhar, A., Dobra, A.: Probabilistic characterization of nearest neighbor classifier. Int. J. Mach. Learn. Cybernet. **4**(4), 259–272 (2013)
18. Lewis, D.D., Catlett, J., Cohen, W., et al.: Heterogeneous uncertainty sampling for supervised learning. In: Machine Learning Proceedings, pp. 148–156 (1994)
19. Davidson, I., Wagstaff, K.L., Basu, S.: Measuring constraint-set utility for partitional clustering algorithms. In: Fürnkranz, J., Scheffer, T., Spiliopoulou, M. (eds.) PKDD 2006. LNCS (LNAI), vol. 4213, pp. 115–126. Springer, Heidelberg (2006)
20. Mangasarian, O.L., Wolberg, W.H.: Breast cancer diagnosis and prognosis via linear programming. Oper. Res. **43**(4), 570–577 (1970)
21. Little, M.A., Mcsharry, P.E., Roberts, S.J., et al.: Exploiting nonlinear recurrence and fractal scaling properties for voice disorder detection. Biomed. Eng. Online **6**, 23 (2007)
22. Strehl, A., Ghosh, J., Mooney, R.: Impact of similarity measures on web-page clustering. In: Workshop on Artificial Intelligence for Web Search (AAAI 2000), pp. 58–64 (2000)

An Effective Cluster Assignment Strategy for Large Time Series Data

Damir Mirzanurov, Waqas Nawaz[⊠], JooYoung Lee, and Qiang Qu

Institute of Information Systems, Innopolis University, Innopolis, Russia
{d.mirzanurov,w.nawaz,j.lee,qu}@innopolis.ru

Abstract. The problem of clustering time series data is of importance to find similar groups of time series, e.g., identifying people who share similar mobility by analyzing their spatio-temporal trajectory data as time series. YADING is one of the most recent and efficient methods to cluster large-scale time series data, which mainly consists of sampling, clustering, and assigning steps. Given a set of processed time series entities, in the sampling step, YADING clusters are found by a density-based clustering method. Next, the left input data is assigned by computing the distance (or similarity) to the entities in the sampled data. Sorted Neighbors Graph (SNG) data structure is used to prune the similarity computation of all possible pairs of entities. However, it does not guarantee to choose the sampled time series with lower density and therefore results in deterioration of accuracy. To resolve this issue, we propose a strategy to order the SNG keys with respect to the density of clusters. The strategy improves the fast selection of time series entities with lower density. The extensive experiments show that our method achieves higher accuracy in terms of NMI than the baseline YADING algorithm. The results suggest that the order of SNG keys should be the same as the clustering phase. Furthermore, the findings also show interesting patterns in identifying density radiuses for clustering.

1 Introduction

Time series are widely available from applications but often hard to be analyzed due to their high dimensionality [1–5]. Large volumes of high dimensional time series require new algorithms for complex tasks that need fast and scalable processing of time series data. Clustering, as one of the tasks, is extremely important for detecting and analyzing similar time series in a wide range of applications [1,6]. For instance, consider clustering time series generating by servers in terms of a set of metrics such as CPU usage [7], we can thus detect the servers with problems when they belong to the clusters containing problematic servers.

Recently, a fast time series clustering algorithm, YADING [1], which automatically groups large-scale time series data, is proposed. The algorithm has

Supported by RFBR grant 16-07-00202.

B. Cui et al. (Eds.): WAIM 2016, Part II, LNCS 9659, pp. 330–341, 2016.
DOI: 10.1007/978-3-319-39958-4_26

three main steps: (1) sampling the input time series, (2) clustering the samples, and (3) assigning the rest of the input data to the clusters generated on the samples. The last step is computationally intensive and ineffective, which is the most critical part of the algorithm especially when the input is large. Moreover, we observe that:

- In the assigning phase, YADING does not consider the order of SNG keys based on density radiuses, which could greatly influence the algorithm performance.
- In the assigning phase, YADING does not consider the distance values of the samples to the left data in identifying density radiuses in order to gain performance.

In this paper, we thoroughly studied the problem and propose a new strategy for assigning step, which significantly improve the performance in obtaining higher accuracy. In this paper, we find that, in general, the order of SNG keys based on density radiuses shall be the same as in the clustering phase where MULTI-DBSCAN is applied. MULTI-DBSCAN is a general DBSCAN algorithm [8] iterated through several density radiuses. We use the concept of inflection point that is the flattest part on k_{dis} curve of sample time series data. Note that k_{dis} of an object is the distance between this object and its k nearest neighbors. A k_{dis} curve is a list of k_{dis} values in descending order. Intuitively, the k_{dis} value of inflection point is the density radius. Our study suggests that finding inflection points depends on the boundaries of the k_{dis} curve. For example, the greatest difference between two points at the beginning leads to obtaining only one density radius. In summary, our contribution is threefold:

- We have thoroughly studied the problem of clustering large time series data, and propose a new strategy for assigning step of YADING in order to improve the performance.
- We have done extensive experiments on real-life datasets in addition to the original study of YADING. We have observed a series of findings including the identification of density radiuses for clustering, as well as the influence of descending and ascending orders of SNG keys.
- We have experimentally proved that the proposed scheme outperforms the baseline YADING by obtaining higher accuracy.

The remained sections are organized as follows. The related concepts are presented in Sect. 2. The literature is briefly reviewed in Sect. 3. The method is described in Sect. 4. The experimental study is shown in Sect. 5 followed by Sect. 6 as conclusion of this study.

2   Preliminaries

In this section, we briefly explain the basic concepts related to time series data clustering [1] for completeness.

SNG (Sorted Neighbors Graph): It is a data structure used for pruning unnecessary steps to reduce the overall computation between clustered and non-clustered data points [9–11]. SNG keys are sample entities that belong to arbitrary clusters. For each SNG key, we store distances from this entity to all other instances in an ascending order. K_{dis} *of an entity*: It refers to the distance between an entity and its kNN neighbors. The list of k_{dis} values in descending order creates k_{dis} curve. *Density radius*: It describes the density of a cluster, which can be defined as the most frequent k_{dis} value. It is used to estimate the density of clusters for input dataset. *Inflection point*: A point on a curve where its second derivative equals to zero. In another words, the points on the left and right sides of this point have the smallest differences. The neighboring points of the inflection point on k_{dis} curve have similar values.

3   Related Works

In this section, we briefly describe the existing methods in terms of measuring similarity, clustering, and data complexity reduction on large time series data. We also highlight one of the most recent time series clustering algorithm, i.e. YADING [1], which we extend to improve the results accuracy.

Similarity Measures of Time Series Clustering: Similarity of time series can be defined by multiple metrics, which may or may not depend on given time series data set. Similarity measures that depend on input data are used in clustering algorithms when time series instances are generated by probability distributions [12,13]. If time series entities are generated by similar distributions, then the time series entities should be similar. The aforementioned similarity measures that depend on input data help to improve clustering accuracy, however, the computation time complexity is usually high. Therefore, similarity measures [14,15] that do not depend on input data are preferred and based on Euclidean distance and two cross-correlation based distances, as described in the study [16].

Data Complexity Reduction Approaches: Data reduction is a commonly used technique to decrease execution time of clustering large-scale time series data. Data complexity reduction can be done in two ways; (1) Reducing time series dimensions, which can be described as vertical data reduction, where we reduce the number of dimensions for each data instance, (2) reducing time series data size through sampling, i.e. horizontal data reduction, that involves less number of data instances. Reducing the dimensionality of time series data can be conducted using following techniques: (a) Discrete Fourier Transform (DFT), (b) Discrete Wavelets Transform (DWT), (c) Singular Value Decomposition (SVD) and d) Piecewise Aggregate Approximation (PAA) [17]. PPA approach has lowest time complexity compared to other approaches. Sampling techniques are used to reduce size of input dataset. Different techniques for sampling are described in the literature, however, random sampling is a good decision due to its simplicity and low time complexity.

Clustering Methods: There are two prominent clustering methods for time series data: (1) partitioning methods and (2) density-based methods. In partitioning methods, we need to specify the number of clusters, k. Well-known examples of such methods are: *k-means* and *k-medoid*. CLARANS, an improved *k-medoid* method, is presented in the study [18]. In density-based algorithms, clusters are areas with high density that are divided by low density regions. DBSCAN is an example of density-based clustering algorithm, which is an efficient approach. However, DBSCAN has the following constraints: (a) it only handles datasets with one density; (2) it requires manual specification of the minimal number of neighbors and density radius.

Recently, an efficient technique [1] for time series data is presented, which relies on DBSCAN algorithm [8]. It has three key steps to cluster the time series data, i.e. data complexity reduction, sample data clustering, and labeling non-sampled data. However, the third step is not effective and incorrectly assigns labels to non-clustered data that results in reduction of accuracy. We suspect it is due to the fact that the comparison among non-sampled data was done with non-sorted order whereas the clustering of sample data was conducted in ascending order of density radiuses. Consequently, we extend this approach by introducing an effective cluster assignment strategy, which we discuss in the subsequent section.

4   Methodology

In this section, we briefly explain the naïve clustering approach for time series data [1]. Subsequently, we discuss the proposed optimization strategy to overcome the limitation of existing method.

4.1   Naïve Approach

We explain the naïve approach for time series clustering. The clustering algorithm consists of three sequentially dependent steps that include data dimensionality reduction with sampling, clustering of sampled data, and cluster assignment for non-sampled data. We provide a brief description for each step in the following paragraphs.

– **Data Dimensionality Reduction and Sampling:** We reduce the complexity of the original data through dimension reduction and sampling, i.e. reducing the number of dimensions for each time-series and total number of time-series respectively. A random sampling technique is used to extract the subset of the original data for efficiency, while we process the remaining data in cluster assignment phase. We use Piecewise Aggregate Approximation (PAA) algorithm [17] dimension reduction because of its simplicity and efficiency. This step yields a small subset of the original time-series data to be clustered in the next phase.

- **Clustering Sampled Data:** We perform multi-density based clustering on the sample data, where MULTI-DBSCAN is a general DBSCAN algorithm iterated through several density radiuses. We use the concept of inflection point that is the flattest part on k_{dis} curve of sample time series data. We denote k_{dis} of an object as the distance between this object and its kNN. A k_{dis} curve is a list of k_{dis} values in descending order. Intuitively, the k_{dis} value of inflection point is the density radius. After obtaining the density radius values and conducting MULTIDBSCAN, we get the sample clusters to assist the assignment phase.
- **Cluster Assignment of Non-Sampled Data:** In this phase, we assign labels to non-sampled data using clustered data from previous step. Inherently, it requires computing distances among every pair of time-series data, which has the highest time complexity compared with other steps. A pruning strategy is suggested based on unique data structure, called Sorted Neighbor Graph (SNG), to overcome this issue. For each time series in sample data, it computes distances for all the other non-sample time series data. The assignment criteria is as follows: if the distance between unlabeled and labeled time series smaller than density radius of labeled time series cluster, then unlabeled entity belongs to this cluster.

4.2   Optimization: Alternate Cluster Assignment Strategy

In naïve approach, we use SNG data structure to reduce the number of computations. However, it does not guarantee the accuracy of results. For instance, a single unlabeled time series is picked from non-sampled dataset and computed distance against each SNG key. If the resulting distance is less than the density radius of the cluster, to which clustered time series belongs to, then the unlabeled instance is considered to be in the same cluster. In such situations, there is high probability of high density clusters to swallow nearby small density clusters, which will result in lower accuracy. To avoid such scenario, we propose an ordering strategy for SNG keys based on density of the cluster they belong, as depicted in Algorithm 1. In general, we obtain the order of sorting same as it was during clustering phase. The detailed description of the algorithm is presented in Sect. 4.3.

In order to speedup the process, the naïve approach used the concept of *jump* to avoid the unnecessary computations for data points with minor differences for distances of density radiuses. However, this approach may neglect some legitimate points to be considered while jump-based scanning. To overcome this issue, we compare density radius of neighbour and the distance from a point to an object. This step introduces an additional check for density radious and distance to object, which is already computed, rather than computing the distance explicitly, as depicted in Fig. 1.

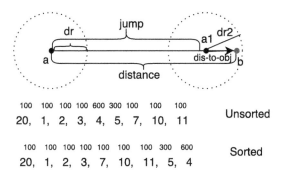

100	100	100	100	600	300	100	100	100	
20,	1,	2,	3,	4,	5,	7,	10,	11	Unsorted

100	100	100	100	100	100	100	300	600	
20,	1,	2,	3,	7,	10,	11,	5,	4	Sorted

Fig. 1. Pictorial description for density radius, distance and distance from neighbor to object along with the sorted and unsorted keys

4.3 Algorithm Details

We describe the proposed algorithm step-by-step in the following paragraphs. Initially, we have SNG keys and unlabeled objects as an input to this algorithm. First, for each object (*obj*) in list of unlabeled objects, we perform the following set of operations; (1) We have perform sorting of SNG keys in ascending order, based on density radius of each time series. (2) For each entity from sorted SNG keys, we pick an entity (*o*) given that it is not inspected and compute the L1 *distance* to object. If the *distance* is smaller than density radius of entity *o*, then we mark object with the same label as label of entity *o*, and break the loop for next iterations. Otherwise, continue and mark the entity *o* as inspected. At this step, we find the first suitable time series entities that the distance is less than density radius of the belonging cluster. The naïve approach had the unsorted list of SNG keys, therefore, it leads to choose an inappropriate cluster label. Consequently, it results in reduction of accuracy compared to our proposed approach. (3) The lines 16–31 describes the post-processing step for time series entity and does not comply with the aforementioned condition. It is the pruning stage to speeding up the performance.

We further highlight the changes introduced to the baseline algorithm in the subsequent discussion. The difference between the distance and the density radius of entity *o* is termed as the *jump* variable. We use the same searching mechanism, i.e. *BINARYSEARCH(SNG[o], jump)*, to find the index *i* of SNG key whose distance from entity *o* is equal or higher then *jump*. Additionally, we add marking of *neighbour* with index from *0* to *i-1*, and *i* should be greater then *0* as inspected, if its density radius is less than *dist-to-obj*. Distance to object (*dist-to-obj*) equals to the difference between total distance from *o* to *obj*, and distance from *o* to this *neighbour*. Then for all neighbors in *SNG[o]* with index greater then *i*, if the density radius of neighbour is less than *dist-to-obj* we consider this neighbour inspected. We repeat these steps until we either find the cluster label for each entity or mark it as noise.

Algorithm 1. Modified Algorithm for Cluster Assignment

1: /* uObj: the list of unlabeled objects */
2: ASSIGNMENT(SNG, $uObj$)
3: **for all** obj in $uObj$ **do**
4:      set the label of obj as noisy
5:      **/* Sorting the SNG keys based on densities */**
6:      $sorted\_sng\_keys \leftarrow$ SORT_density$\{keys\ of\ SNG\}$
7:      **for all** $o \in \{sorted\_sng\_keys\}$ **do**
8:          **if** o has been inspected **then**
9:              **continue**
10:         **end if**
11:         $dis \leftarrow$ L1 distance between o and obj
12:         **if** dis less than density radius of o **then**
13:             mark obj with same label of o **break**
14:         **end if**
15:         mark o as inspected
16:         $jump \leftarrow dis -$ density radius of o
17:         $i \leftarrow$ BINARYSEARCH($SNG[o]$, $jump$)
18:         **for all** $neighbor$ in $SNG[o]$ with index in $[0,i\text{-}1]$ /*If $i > 0$*/ **do**
19:             $dist\text{-}to\text{-}obj \leftarrow dis -$ distance from o to $neighbor$
20:             **if** density radius of $neighbor$ is less than $dist\text{-}to\text{-}obj$ **then**
21:                 mark $neighbor$ as inspected
22:             **end if**
23:         **end for**
24:         **for all** $neighbor$ in $SNG[o]$ with index greater than i **do**
25:             $dist\text{-}to\text{-}obj \leftarrow dis -$ distance from o to $neighbor$
26:             **if** density radius of $neighbor$ is less than $dist\text{-}to\text{-}obj$ **then**
27:                 mark $neighbor$ as inspected
28:             **else**
29:                 **break** /*this is a sorted list*/
30:             **end if**
31:         **end for**
32:     **end for**
33: **end for**

---

5   Experimental Study

We implement the YADING in three steps. First step is Data Sampling and Dimensionality Reduction phase. Second is the Sample Clustering. And finally, third step is Assignment clusters to remaining data set. We analyze the results of the proposed cluster assignment strategy with YADING as baseline approach.

5.1   Setup

To conduct experiments, we use two time series datasets from UCR library, i.e. Star Light Curves dataset with 8236 instances, 1024 dimensions, 3 cluster labels and Two Patterns dataset has 5000 instances, 128 dimensions, and 4 cluster labels. Each time series instance has one of the 3 cluster labels. As it will described below later, the cluster 1 and 3 is relatively close to each other. The hardware we use is standard workstation - with 8Gb RAM and Intel Core i5-3470 CPU @ 3.20 GHz 4 processor. We use Python language to both conduct YADING experiments and to plot the all graphs, e.g. clustered entities and k_{dis} curve).

5.2 Discussion

In this section, we analyze the base algorithm for each phase one by one through visual description of the clustering results over ground truth. Subsequently, we present the clustering results using proposed strategy for cluster assignment phase.

Data Sampling and Dimensionality Reduction Phase. For sampling strategy we use the random sampling, more precisely we specify the sampling size (s) as 2000, and take first s lines as sample. As dimensionality reduction strategy, we use Piecewise Aggregate Approximation (PAA) algorithm. But the key limitation of this algorithm is that the dimensionality of reduced data needs to be specified in advance. To overcome the limitation, it is proposed to automatically estimate the number of reduced dimensionality. Fast Fourier transform (FFT) and Inverse fast Fourier transform (IFFT) are used for this purpose. We found that 80 % percentile of

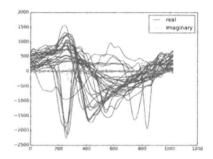

Fig. 2. Identifying Parameters through Fourier Transformation – Inverse Fast Fourier Transformation from 0 to 30 on sample data

all the sorted local minimums of auto-correlation curve is enough to estimate the highest dimensionality, because some of the time series can have extraordinary noise which can lead to potential instability. Figure 2 shows FFT and IFFT of a time series correspondingly. Based on IFFT we found first local min for each time series (ts) in sample dataset, and for sorted values we found 80 % as the estimated frame size (d). We obtained $d = 625.450585205 \geq 625$. Similarly, Fig. 2-(c) shows IFFT for time series from 0 to 30 on sample data.

Sample Data Clustering. In this step, we perform multi-density based clustering on sample data, where L1 distance measure is used for similarity estimation. We used multi-density based approach because of unavailability of data distribution information.

We applied the same algorithm as of base paper to find density radiuses. We have identified k_{dis} curve of sample dataset - list of all k_{dis} for sample dataset, in descending order. k_{dis} is a value between entity and his k-near neighbors.

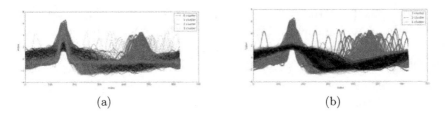

(a)   (b)

Fig. 3. Clusters of sample dataset (a) Obtained (b) Original

We specify $k = 4$. The density radiuses are the inflection points of this curve, i.e. flattest points where the neighboring points have the smallest differences. Figure 4 shows the k_{dis} curves of the sample data with the inflection points.

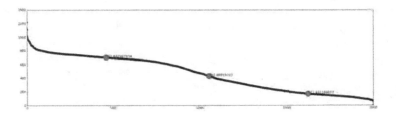

Fig. 4. k_{dis} curve of sample time series data with Inflection Points

Afterwards, we performed multi-density based clustering algorithm for each density, from lowest to highest, of the data. After finishing with one density, we removed labeled data instances from original data and repeat the process, that is how we used MULTI-DBSCAN. The result of this clustering are shown in Fig. 3, where we plot the clusters for better visual analysis.

Cluster Assignment for Non-Sample Data. In this step, we assigned labels to the remaining non-sample data based on obtained clusters in previous step. During assign phase, we have used SNG data structure, which is sorted neighbors graph, where it stores all his neighbours in an ascending order for each core point in the sample clusters. The results are shown in Fig. 5.

Proposed Cluster Assignment Strategy. In contrast to base paper strategy for cluster assignment, we ordered the SNG keys based on density values from lowest to highest. This sorting task has $O(N^2)$ complexity, were N refers to a number of instances in sample data. The sorting task is performed only once and has some impact on execution time. On the other hand, it improves the accuracy of results. The clustering results, using proposed cluster assignment strategy, are presented in Fig. 6. We can observe that the number of instances of black cluster reduces a lot. Therefore, it detects the instances of red cluster more precisely.

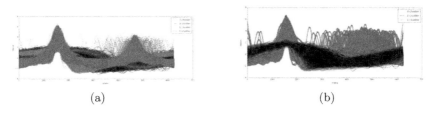

(a)   (b)

Fig. 5. Clusters of assigned data (a) Obtained (b) Original

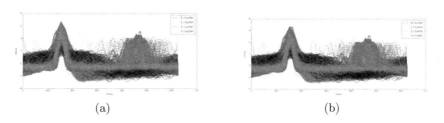

(a)   (b)

Fig. 6. Clusters of assigned and entire data using proposed strategy (a) Assigned Cluster Data (b) Entire Cluster Data

In previous case, we have validated the assumption that if we sort the keys of SNG based on density, from lower to higher, we will obtain the increase in accuracy. However, now we sorted the SNG keys in descending order in terms of density, and did the same experiments. We expect that the accuracy should decrease. The results in Fig. 7 proved this assumption. It happened because point always assigned to bigger clusters and consequently, we have less number of clusters for non-labeled data. We have analyzed the accuracy of clustering results with respect to each phase and varying the time series dataset. We observed that the proposed method archieved better accurancy compared with the base method, as shown in Fig. 8.

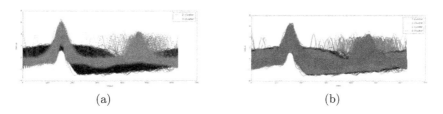

(a)   (b)

Fig. 7. Clusters of assigned and entire data using proposed strategy - with densities in descending order (a) Assigned Cluster Data (b) Entire Cluster Data

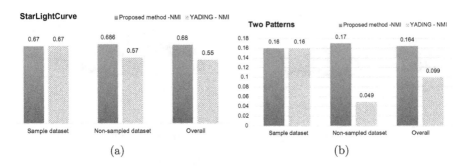

Fig. 8. Accuracy of the clustering results in terms of NMI measure (a) StarLightCurve Time Series Data (b) TwoPatterns Time Series Data

6   Conclusion

In this paper, we provide an experimental study and optimization for fast clustering algorithm, i.e. YADING, on time series data. We have analyzed the clustering results and observed the limitation of cluster assignment phase. Accordingly, we provided a unique mechanism, in extension to the original algorithm, by considering the order of SNG keys to overcome the aforementioned issue. The keys in SNG data structure are ordered that makes the algorithm more reliable in terms of accuracy. We sorted the candidates initially to look up distances based on their densities where values with smaller density are considered first. Furthermore, we investigated more general cases and found that the order of SNG keys should be the same as the clustering phase. We tested the base algorithm along with the proposed cluster assignment strategy on two time series datasets from UCR library, i.e. Star Light Curve and Two Patterns. We analyzed the accuracy of clustering results through NMI. The proposed strategy achieved better accuracy on Two Patterns dataset compared with Star Light Curve.

Nevertheless, the boundaries of k_{dis} curve plays a key role in finding density radiuses. It is an intereting problem to estimate such boundaries and regarded as our future work. As an extension to this idea, our work on improving the time efficiency of the overall algorithm by adapting constant time searching algorithm is in process.

References

1. Ding, R., Wang, Q., Dang, Y., Fu, Q., Zhang, H., Zhang, D.: Yading: fast clustering of large-scale time series data. Proc. VLDB Endowment **8**(5), 473–484 (2015)
2. Li, F., Li, H., Qu, Q.: Composite pattern query expression over medical data streams. In: BMEI, pp. 1–5 (2009)
3. Liu, S., Qu, Q., Chen, L., Ni, L.M.: SMC: A practical schema for privacy-preserved data sharing over distributed data streams. IEEE Trans. Big Data **1**(2), 68–81 (2015)

4. Qu, Q., Li, H., Wang, L., Miao, G., Wei, X.: Online constrained pattern detection over streams. In: FSKD, pp. 66–70 (2009)

5. Qu, Q., Liu, S., Jensen, C.S., Zhu, F., Faloutsos, C.: Interestingness-driven diffusion process summarization in dynamic networks. In: Calders, T., Esposito, F., Hüllermeier, E., Meo, R. (eds.) ECML PKDD 2014, Part II. LNCS, vol. 8725, pp. 597–613. Springer, Heidelberg (2014)

6. Liao, T.W.: Clustering of time series data–a survey. Pattern Recogn. **38**(11), 1857–1874 (2005)

7. Patterson, D.A., et al.: A simple way to estimate the cost of downtime. LISA **2**, 185–188 (2002)

8. Ester, M., Kriegel, H.P., Sander, J., Xu, X.: A density-based algorithm for discovering clusters in large spatial databases with noise. Kdd **96**(34), 226–231 (1996)

9. Eppstein, D., Paterson, M.S., Yao, F.F.: On nearest-neighbor graphs. Discrete Computat. Geom. 17(3) 263–282

10. Qu, Q., Qiu, J., Sun, C., Wang, Y.: Graph-based knowledge representation model and pattern retrieval. In: FSKD, pp. 541–545 (2008)

11. Zhu, F., Zhang, Z., Qu, Q.: A direct mining approach to efficient constrained graph pattern discovery. In: Proceedings of the ACM SIGMOD International Conference on Management of Data, SIGMOD 2013, New York, NY, USA, June 22–27, pp. 821–832 (2013)

12. Piccolo, D.: A distance measure for classifying arima models. J. Time Ser. Anal. **11**(2), 153–164 (1990)

13. Tran, D., Wagner, M.: Fuzzy C-means clustering-based speaker verification. In: Pal, N.R., Sugeno, M. (eds.) AFSS 2002. LNCS (LNAI), vol. 2275, pp. 318–324. Springer, Heidelberg (2002)

14. Yi, B.K., Faloutsos, C.: Fast time sequence indexing for arbitrary lp norms. In: VLDB (2000)

15. Yi, B.K., Jagadish, H., Faloutsos, C.: Efficient retrieval of similar time sequences under time warping. In: Proceedings of the 14th International Conference on Data Engineering, 1998, pp. 201–208, February 1998

16. Golay, X., Kollias, S., Stoll, G., Meier, D., Valavanis, A., Boesiger, P.: A new correlation-based fuzzy logic clustering algorithm for fmri. Magn. Reson. Med. **40**(2), 249–260 (1998)

17. Keogh, E., Chakrabarti, K., Pazzani, M., Mehrotra, S.: Dimensionality reduction for fast similarity search in large time series databases. Knowl. Inf. Syst. **3**(3), 263–286 (2001)

18. Raymond, T.N., Han, J.: Effecient and effictive clustering methods for spatial data mining. In: Proceedings of the 20th International Conference on Very Large Data Bases (1994)

AdaWIRL: A Novel Bayesian Ranking Approach for Personal Big-Hit Paper Prediction

Chuxu Zhang[1,2], Lu Yu[3], Jie Lu[4], Tao Zhou[5(✉)], and Zi-Ke Zhang[1(✉)]

[1] Alibaba Research Centre for Complexity Sciences, Hangzhou Normal University,
Hangzhou, China
cz201@cs.rutgers.edu, zhangzike@gmail.com

[2] Department of Computer Science, Rutgers University, New Brunswick, USA

[3] Alibaba Group, Hangzhou, China
coolluyu@gmail.com

[4] IBM Thomas J. Watson Research Center, Yorktown Height, USA
jielu@us.ibm.com

[5] Big Data Research Center,
University of Electronic Science and Technology of China, Chengdu, China
zhutouster@gmail.com

Abstract. Predicting the most impactful (big-hit) paper among a researcher's publications so it can be well disseminated in advance not only has a large impact on individual academic success, but also provides useful guidance to the research community. In this work, we tackle the problem of given the corpus of a researcher's publications in previous few years, how to effectively predict which paper will become the big-hit in the future. We explore a series of features that can drive a paper to become the big-hit, and design a novel Bayesian ranking algorithm AdaWIRL (**Ada**ptive **W**eighted **I**mpact **R**anking **L**earning) that leverages a weighted training schema and an adaptive timely false correction strategy to predict big-hit papers. Experimental results on the large ArnetMiner dataset with over 1.7 million authors and 2 million papers demonstrate the effectiveness of AdaWIRL. Specifically, it correctly predicts over 78.3 % of all researchers' big-hit papers and outperforms the compared regression and ranking algorithms, with an average of 5.8 % and 2.9 % improvement respectively. Further analysis shows that temporal features are the best indicator for personal big-hit papers, while authorship and social features are less relevant. We also demonstrate that there is a high correlation between the impact of a researcher's future works and their similarity to the predicted big-hit paper.

1 Introduction

The growing evolution of scientific research leads to an expanding body of literature year by year. Figure 1(a) reports the rapid increase of publication volume in the ArnetMiner[1] academic dataset of Computer Science. Figure 1(b) depicts

---

[1] https://aminer.org/AMinerNetwork.

© Springer International Publishing Switzerland 2016
B. Cui et al. (Eds.): WAIM 2016, Part II, LNCS 9659, pp. 342–355, 2016.
DOI: 10.1007/978-3-319-39958-4_27

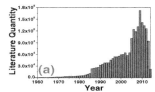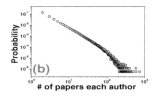

Fig. 1. (a) The growing volume of research literature. (b) Distribution of individual researchers' literature size.

the distribution of individual researchers' contribution in terms of the number of papers, showing more than 119,000 authors in the dataset having more than 10 publications each. Researchers are often recognized for their most influential (bit-hit) works. For example, in recognition of the landmark paper on Latent Dirichlet Allocation, the ACM Infosys Foundation Award in 2013 was presented to Dr. David Blei "for pioneering the area of topic modeling" [7]. Thus it is useful to continuously track the big-hit paper among the representative (e.g., first-author) publications of one's prior work as early dissemination of the big-hit work leads to significant attention or conducting future works related to the predicted big-hit paper can result in an ever larger reputation.

Traditionally, the citation count or the *h-index* [9] is used to measure the scientific impact. Accordingly, we consider the big-hit paper as the paper with the most citations.

Unlike previous works [21,24,25] that predict the exact citation value, we frame the problem as a ranking task to identify the paper that ranks first in the future for each researcher, which is inspired by the studies [16,17,27] in recommender systems that demonstrate predicting the users' preference ranking rather the rating score over items results the effective recommendation. As a pilot study for predicting personal big-hit work, our contributions are:

- We formalize a problem that predicts for each researcher the big-hit paper among one's previous few years' representative publications.
- We introduce various features that are correlated with papers' impact and design a novel Bayesian ranking algorithm AdaWIRL to solve the given problem.
- The experiment results on a large academic dataset show the effectiveness of our algorithm. AdaWIRL identifies the big-hit paper with over 0.783 accuracy and outperforms the traditional regression and ranking algorithms.

The rest of paper is organized as follows. Section 2 formally defines the personal big-hit paper prediction problem. Section 3 describes the algorithm proposed to solve the given problem. Section 4 presents experiments and analysis. Section 5 reviews related works. The paper concludes at Sect. 6.

2   Problem Definition

For ease of presentation, let A and L represent author and paper set respectively, and let a_r and l_i denote researcher r and paper i respectively. The corpus of a_r is represented by L_{a_r}. Letter c denotes papers' citation count and c_{l_i} is l_i's citation value. The true and predicted impact scores for l_i are symbolized by s_{l_i} and $\hat{s}_{l_i}$ respectively. Let $l^*{}_{a_r}$ and $\hat{l}^*{}_{a_r}$ represent the true and predicted most influential paper of a_r respectively. The problem is formalized as:

Problem 1 *Personal Big-Hit Paper Prediction: Given the publication corpus L_t before timestamp t and the selected first-author personal literature L_{a_r} of a_r in the previous Δt_b years before t, the task is to predict the big-hit paper $l^*{}_{a_r}$ among L_{a_r} for a_r after a given time period Δt_a years.*

Specifically, we use an example in the ArnetMiner dataset to illustrate the problem. During $2007 \sim 2009$, Dr. Jie Tang ($a_{1545692}$) published 9 first-author papers ($l_{1055861}, l_{1055867}, l_{1083734}, l_{1117023}, l_{1176930}, l_{1195999}, l_{1214702}, l_{1305623}$ and $l_{1318662}$) which have citation values of 4, 1, 96, 21, 21, 6, 89, 14 and 4 in 2014, respectively. By setting timestamp $t = 2010$, $\Delta t_b = 3$ and $\Delta t_a = 4$, we aim to find $a_{1545692}$'s big-hit work. The answer should be $l_{1083734}$ since it has the largest citation value.

3   Proposed Methods

3.1   Impact Ranking Model (IRM)

Inspired by previous works [16,17,27], we introduce the Bayesian pairwise ranking model into solution for the given problem. Let $(a_r, l_i, l_j) \in T$ represent a pair of papers of a_r and set T contains all cases. To capture the impact ranking of papers for each researcher, we maximize the posterior probability over the parameter space $\boldsymbol{\omega}$: $p(\boldsymbol{\omega}| >_{a_r}) \propto p(>_{a_r} |\boldsymbol{\omega}) \cdot p(\boldsymbol{\omega})$, where notation $>_{a_r} = \{l_i >_{a_r} l_j : ((a_r, l_i, l_j) \in T) \cap (s_{l_i} > s_{l_j})\}$ denotes the pairwise order structure for a_r and $p(\boldsymbol{\omega})$ is the prior probability. We use citation count to quantify the true impact score, i.e., $s_{l_i} = c_{l_i}$. Let set $T_>$ represent structure $>_{a_r}$ for all researchers and set $T_\le$ consists of the remaining cases not included in $T_>$. Without loss of generality, we assume that each $>_{a_r}$ is independent. Thus the likelihood function can be written as a product of single density for all researchers $a_r \in A$:

$$\prod_{a_r \in A} p(>_{a_r} |\boldsymbol{\omega}) = \prod_{(a_r, l_i, l_j) \in T} p(l_i >_{a_r} l_j |\boldsymbol{\omega})^{\delta((a_r, l_i, l_j) \in T_>)} \cdot (1 - p(l_i >_{a_r} l_j |\boldsymbol{\omega}))^{\delta((a_r, l_i, l_j) \in T_\le)}$$

Due to the antisymmetric nature of $>_a$, the log form of the objective is:

$$ln\, p(\boldsymbol{\omega}| >_a) = ln \prod_{(a_r, l_i, l_j) \in T_>} p(l_i >_{a_r} l_j |\boldsymbol{\omega}) \cdot p(\boldsymbol{\omega}) = \sum_{(a_r, l_i, l_j) \in T_>} ln\, p(l_i >_{a_r} l_j |\boldsymbol{\omega}) + ln\, p(\boldsymbol{\omega})$$

Let $p(l_i >_{a_r} l_j | \boldsymbol{\omega}) = \sigma(d_{(a_r,l_i,l_j) \in T_>}(\boldsymbol{\omega}))$, where σ is the sigmoid function: $\sigma(x) = \frac{1}{1+e^{-x}}$, and $d_{(a_r,l_i,l_j) \in T_>}(\boldsymbol{\omega})$ quantifies the impact difference between l_i and l_j. The predicted impact score $\hat{s}_{l_i}$ of l_i based on academic features $\boldsymbol{f}$ (described in Sect. 4.1) extracted from the corpus is formulated as $\hat{s}_{l_i} = \sum_{k=1}^{K} \boldsymbol{\omega}_k \cdot f_{ik}$, where f_{ik} denotes the k-th feature of l_i and K is the number of features. Intuitively, we define $d_{(a_r,l_i,l_j) \in T_>}(\boldsymbol{\omega}) = \hat{s}_{l_i} - \hat{s}_{l_j}$ and prior parameter distribution as $\boldsymbol{\omega} \sim \mathcal{N}(\boldsymbol{0}, \lambda_{\boldsymbol{\omega}} \boldsymbol{I})$. Therefore, the objective becomes:

$$IRM_{objective} \equiv \sum_{(a_r,l_i,l_j) \in T_>} ln \; \sigma(\hat{s}_{l_i} - \hat{s}_{l_j}) - \lambda_{\boldsymbol{\omega}} \cdot \|\boldsymbol{\omega}\|^2$$

where $\lambda_{\boldsymbol{\omega}}$ is the model specific regularization parameter.

3.2   Learning Algorithms

Impact Ranking Learning (IRL). The objective function of IRM derived above is differentiable. Thus we use stochastic gradient descent [26] for maximization. Specifically, the parameter $\boldsymbol{\omega}$ is randomly initialized according to $\boldsymbol{\omega} \sim \mathcal{N}(\boldsymbol{0}, \lambda_{\boldsymbol{\omega}} \boldsymbol{I})$. The IRL iteratively traverses each pairwise instance $(a_r, l_i, l_j) \in T_>$ and updates $\boldsymbol{\omega}$ by following the rule below until it meets the stopping criterion:

$$\boldsymbol{\omega} \leftarrow \boldsymbol{\omega} + \alpha \cdot \frac{\partial \, IRM_{objective}}{\partial \boldsymbol{\omega}} = \boldsymbol{\omega} + \alpha \cdot \{ \frac{e^{-(\hat{s}_{l_i} - \hat{s}_{l_j})}}{(1 + e^{-(\hat{s}_{l_i} - \hat{s}_{l_j})})} \cdot \frac{\partial(\hat{s}_{l_i} - \hat{s}_{l_j})}{\partial \boldsymbol{\omega}} - \lambda_{\boldsymbol{\omega}} \cdot \boldsymbol{\omega} \}$$

where α is the learning rate.

Weighted Training Schema (WIRL). The learning rate for each pairwise training case of IRL equals to α. However, we think the magnitude should be different for different training instances. Despite the multiplicative scalar $\Delta\phi_{a_r,l_i,l_j} = \frac{e^{-(\hat{s}_{l_i} - \hat{s}_{l_j})}}{(1+e^{-(\hat{s}_{l_i} - \hat{s}_{l_j})})}$ in each step, its value range $(0 < \phi_{a_r,l_i,l_j} < 1)$ can not fully capture diverse differences among all training cases. Thus we define an extra weight function $W(a_r, l_i, l_j)$ for updating model parameters:

$$W(a_r, l_i, l_j) = \begin{cases} e^{-(\hat{s}_{l_i} - \hat{s}_{l_j})} & for \; e^{-(\hat{s}_{l_i} - \hat{s}_{l_j})} \leq \tau \\ \tau & for \; e^{-(\hat{s}_{l_i} - \hat{s}_{l_j})} > \tau \end{cases}$$

The pairwise structure $l_i >_{a_r} l_j$ is well learned when $\hat{s}_{l_i} - \hat{s}_{l_j} > 0$ and we should update $\boldsymbol{\omega}$ slightly by setting $W(a_r, l_i, l_j) < 1$. Conversely, $\hat{s}_{l_i} - \hat{s}_{l_j} < 0$ means insufficient learning so $\boldsymbol{\omega}$ should be updated with a larger magnitude $W(a_r, l_i, l_j) > 1$. In addition, we cut off $W(a_r, l_i, l_j)$ by a threshold $\tau = 3$ to avoid the infinity value. With the weight function, we rewrite the iterative parameters updating step for weighted impact ranking learning (WIRL) algorithm as:

$$\boldsymbol{\omega} \leftarrow \boldsymbol{\omega} + \alpha \cdot W(a_r, l_i, l_j) \cdot \{ \frac{e^{-(\hat{s}_{l_i} - \hat{s}_{l_j})}}{(1 + e^{-(\hat{s}_{l_i} - \hat{s}_{l_j})})} \cdot \frac{\partial(\hat{s}_{l_i} - \hat{s}_{l_j})}{\partial \boldsymbol{\omega}} - \lambda_{\boldsymbol{\omega}} \cdot \boldsymbol{\omega} \}$$

Adaptive Timely False Correction (AdaWIRL). We presented the WIRL algorithm by integrating the weight function into the baseline learning. One more powerful trick that we introduce is adaptive timely false correction. Inspired by Nesterov's Accelerated Gradient Descent method [14], we add one compensation term to the parameter update framework but the addition step is only performed for false prediction instances after each iteration of WIRL. Thus the compensation addition step is: $\boldsymbol{\omega} \leftarrow \boldsymbol{\omega} + \beta \cdot \Delta\boldsymbol{\omega}_{\circlearrowleft}(a_r, l_i, l_j) \cdot \delta((\hat{s}_{l_i} - \hat{s}_{l_j}) < 0)$, where $\Delta\boldsymbol{\omega}_{\circlearrowleft}(a_r, l_i, l_j)$ is the compensation function and β is the magnitude adjusting parameter. The δ function guarantees $\hat{s}_{l_i} - \hat{s}_{l_j} < 0$. Then the key question is how to define $\Delta\boldsymbol{\omega}_{\circlearrowleft}(a_r, l_i, l_j)$ for pairwise structure $(a_r, l_i, l_j) \in T_>$. For the wrong prediction case, we have: $\hat{s}_{l_i} - \hat{s}_{l_j} = \sum_{k=1}^{K} \boldsymbol{\omega}_k \cdot (f_{l_i k} - f_{l_j k}) < 0$. In order to correct it, i.e., make $\hat{s}_{l_i} - \hat{s}_{l_j} > 0$, we set: $\Delta\boldsymbol{\omega}_{\circlearrowleft}(a_r, l_i, l_j) = \boldsymbol{f}_{l_i} - \boldsymbol{f}_{l_j}$. With suitable $\beta > 0$, $\Delta\boldsymbol{\omega}_{\circlearrowleft}(a_r, l_i, l_j)$ may be effective in correcting false prediction instances because:

$$\hat{s}_{l_i} - \hat{s}_{l_j} = \sum_{k=1}^{K} \{ \boldsymbol{\omega}_k + \beta \cdot \underbrace{(f_{l_i k} - f_{l_j k})}_{\equiv \Delta\boldsymbol{\omega}_{\circlearrowleft}(a_r, l_i, l_j)} \} \cdot (f_{l_i k} - f_{l_j k}) = \underbrace{\sum_{k=1}^{K} \boldsymbol{\omega}_k \cdot (f_{l_i k} - f_{l_j k})}_{<0} + \beta \cdot \underbrace{\sum_{k=1}^{K} (f_{l_i k} - f_{l_j k})^2}_{>0}$$

The second part of the above equation is the quadratic compensation term after adding $\Delta\boldsymbol{\omega}_{\circlearrowleft}(a_r, l_i, l_j)$, which is larger than 0. Thus the wrong prediction cases that satisfy $\beta \sum_{k=1}^{K}(f_{l_i k} - f_{l_j k})^2 > |\sum_{k=1}^{K} \boldsymbol{\omega}_k \cdot (f_{l_i k} - f_{l_j k})|$ will be corrected. The intuition behind this strategy is that we aim to adjust the model parameters to a proper extent and further correct false prediction cases. The corresponding algorithm AdaWIRL is illustrated by Algorithm 1.

Algorithm 1. WIRL with Adaptive Timely False Correction (AdaWIRL)

1: **procedure** : $\boldsymbol{\omega} \leftarrow AdaWIRL(T_>, \boldsymbol{f})$ // output $\boldsymbol{\omega}$: model parameters, input $T_>$: training set, input $\boldsymbol{f}$: features information
2:    Initialize $\boldsymbol{\omega} \sim \mathcal{N}(\boldsymbol{0}, \lambda_\omega \boldsymbol{I})$
3:    **repeat**
4:       randomly draw (a_r, l_i, l_j) from $T_>$
5:       compute $\hat{s}_{l_i} = \sum_{k=1}^{K} \boldsymbol{\omega}_k \cdot f_{l_i k}, \ \hat{s}_{l_j} = \sum_{k=1}^{K} \boldsymbol{\omega}_k \cdot f_{l_j k}$
6:       $\boldsymbol{\omega} \leftarrow \boldsymbol{\omega} + \alpha \cdot W(a_r, l_i, l_j) \cdot \{ \frac{e^{-(\hat{s}_{l_i} - \hat{s}_{l_j})}}{(1+e^{-(\hat{s}_{l_i} - \hat{s}_{l_j})})} \cdot \frac{\partial(\hat{s}_{l_i} - \hat{s}_{l_j})}{\partial\omega} - \lambda_\omega \boldsymbol{\omega} \}$
7:       compute $\hat{s}_{l_i} = \sum_{k=1}^{K} \boldsymbol{\omega}_k \cdot f_{l_i k}, \ \hat{s}_{l_j} = \sum_{k=1}^{K} \boldsymbol{\omega}_k \cdot f_{l_j k}$
8:       **if** $\hat{s}_{l_i} - \hat{s}_{l_j} < 0$ **then**
9:          $\boldsymbol{\omega} \leftarrow \boldsymbol{\omega} + \beta \cdot (f_{l_i} - f_{l_j})$
10:      **end if**
11:   **until** Convergence
12: **end procedure**

---

4   Experiments

4.1   Experimental Setup

Dataset. We use a real-world dataset from the ArnetMiner [19] - a well known platform for academic search. It contains 1,712,433 authors and 2,092,356 papers from major computer science venues for more than 50 years. Each paper contains information on title, abstract, authorship, publication year, publication venue and references. In total, we extract 4,258,615 edges from the collaboration network of authors and 8,024,869 edges in the citation network of papers. To conduct experiments, we sample all authors who published at least three first-author papers in past three years $2007 \sim 2009(\Delta t_b = 3)$ before $t = 2010$ and extract all information of their first-author papers. There are 29,158 authors and 148,893 papers in the sample set. Then we predict the big-hit work of these publications in 2013 or 2014 ($\Delta t_a = 3$ or 4) for each selected author.

Feature Engineering. In Sect. 3.1, the predicted impact of each paper is formalized as the combined influence of various factors. Authorship, social, venue, content and temporal features are used to represent those factors. Typically authorship and social features are "group-scale" which tend to have different values for different papers of each researcher. For example, three papers of a_r are to be ranked and we prefer maximum value among the authors' average citation counts to the first author's average citation count for each paper since the latter for the three papers to be ranked have the same value.

- *Authorship Features.* Papers' impact is naturally correlated with the authorship information. Previous works [5,24] studied the interplay between the citation count and authors' attributes. The spread of a paper's impact relies on the volume and citation information of authors' previous publications. We extract 6 correlated "group-scale" authorship features, as reported in Table 1.
- *Social Features.* Study [1] demonstrated that scientists tend to cite their co-authors' papers. We compute the author's co-authors number and the pagerank [15] (PR) values of authors in the weighted collaboration network (ACN) for each paper. In ACN, each edge denotes a collaboration relationship between two nodes (authors) and the weight of edges is defined as the frequency of collaboration between them. In addition, we think widely cited authors are authority researchers and define the corresponding PR value based on authors' "citing-cited" network (ACCN). Unlike ACN, the ACCN is weighted directed and each edge represents a citing-cited relationship between two authors. The 6 social features are illustrated in Table 1.
- *Venue Features.* Different venues have different reputations. In general, good papers in top venues tend to attract more attention than others. For example, in the data mining field, scientists pay much attention to publications in SIGKDD and prefer citing associated papers. In order to quantify the venue's influence, we compute the average citation value of all papers in each

venue. In addition, similar to the ACCN, we construct the weighted directed venues "citing-cited" networks (VCCN) and each edge represents a citing-cited link between two venues. The authority score of the venue is defined as the PR value. The 2 venue features are illustrated in Table 1.

- **Temporal Features.** Temporal features of literature have been found to be useful in modeling scientific impact [6,25]. We extract 3 temporal features of the paper including the publication time, citation increment in previous one year and average citation increment in previous years, as reported by Table 1.
- **Content Features.** Content is another factor that affects the paper's impact. As a widely used method for content analysis, topic modeling can be useful for predicting publications' impact. Like previous works [6,25], we run a 100-topics Latent Dirichlet Allocation [2]² (LDA) on the title and abstract of corpus with stop-words and high/low frequency words elimination, and it returns the probability distribution $p(z|l)$ over topics $z \in Z$ for each paper l. Intuitively, popular topics attract more attention and papers with popular topics may have a large impact. To quantify such an effect, we define topic popularity $popularity(z)$ as the citation count summation over corpus L_t before timestamp t: $popularity(z) = \sum_{l \in L_t} p(z|l) \times c_{lt}$ where c_{lt} is the citation value of l before t. Thus the popularity of target paper l^* is defined as: $popularity(l^*) = \sum_{z \in Z} p(z|l^*) \times popularity(z)$. Besides topic popularity, topic diversity [6,24,25] captures the topic range of the paper and a wider range means more attention from various research fields. The corresponding value of l^* is calculated by Shannon Entropy: $diversity(l^*) = \sum_{z \in Z} -p(z|l^*)logp(z|l^*)$. We further use the definition of author authority over topics in [6] here. The authority of researcher a on topic z is defined as: $authority(a, z) = \sum_{l \in L_{at}} p(z|l) \times c_{lt}$ where L_{at} is the researcher a's publications before t. Given target paper l^*, the consistence between author's authority and paper l^* is: $consistence(l^*, a) = \sum_{z \in Z} p(z|l^*) \times authority(a, z)$. The 4 content features are described in Table 1.

The values of above features are obtained from the corpus collected before $t = 2010$ and we normalize them by dividing the maximum value of each feature.

Evaluation Metric. In Sect. 3, set $T_>$ denotes all pairwise order instances. As a general machine learning task, we divide $T_>$ into two parts, one for training set $T_{>train}$ and the other for test set $T_{>test}$. The model is trained on $T_{>train}$ and then we evaluate its performances on $T_{>test}$. In this paper, we consider the recall accuracy of the big-hit paper for each researcher as the evaluation metric:

$$Pre@1 = \frac{1}{|A_{test}|} \sum_{a_r \in A_{test}} \delta(\hat{l}^*_{a_r} == l^*_{a_r})$$

where A_{test} is the set of researchers in $T_{>test}$. The true most-impactful (big-hit) paper $l^*_{a_r}$ is defined as the paper with the largest citation count of researcher a_r.

---

² http://radimrehurek.com/gensim/.

Table 1. Feature definition. Five groups of features extracted from each paper including authorship, social, venue, temporal and content features.

Category	Feature	Description
Authorship	$A\text{-}max\text{-}ave_{citation}$	The maximum value among all authors' average citation values.
	$A\text{-}ave\text{-}ave_{citation}$	The average value of all authors' average citation values.
	$A\text{-}sum\text{-}ave_{citation}$	The summation of each author' average citation value.
	$A\text{-}num\text{-}authors$	The authors number of the paper.
	$A\text{-}num\text{-}papers$	The summation of each author' previous papers number.
	$A\text{-}ave\text{-}num\text{-}papers$	The average value of all authors' previous papers numbers.
Social	$S\text{-}num\text{-}coauthors$	The summation of each author's previous co-authors number.
	$S\text{-}ave\text{-}ave_{citation}\text{-}co$	The average citation value of co-authors of the paper's authors.
	$S\text{-}max\text{-}PR_{ACN}$	The maximum pagerank value among the paper's authors in ACN.
	$S\text{-}ave\text{-}PR_{ACN}$	The average pagerank value of the paper's authors in ACN.
	$S\text{-}max\text{-}PR_{ACCN}$	The maximum pagerank value among the paper's authors in ACCN.
	$S\text{-}ave\text{-}PR_{ACCN}$	The average pagerank value of the paper's authors in ACCN.
Venue	$V\text{-}ave\text{-}citation$	The average citation value of the paper's venue.
	$V\text{-}PR_{VCCN}$	The pagerank value of the paper's venue in VCCN.
Temporal	$T\text{-}time$	The time/year lapse since the publication year of the paper.
	$T\text{-}one\text{-}\Delta citation$	The citation increment of the paper in previous one year.
	$T\text{-}ave\text{-}\Delta citation$	The average citation increment of the paper in previous years.
Content	$C\text{-}popularity_{LDA}$	The popularity of the paper.
	$C\text{-}diversity_{LDA}$	The topic diversity of the paper.
	$C\text{-}max\text{-}authority_{LDA}$	The maximum consistence between the authors' authorities and the paper.
	$C\text{-}ave\text{-}authority_{LDA}$	The average consistence between each author' authority and the paper

Comparison Methods. We compare the performances of our ranking algorithms with three categories of methods including:

- **Category I.** A baseline *Popular (Pop.)* method which predicts the big-hit papers according to the citation value at timestamp t.
- **Category II.** A series of machine learning regression methods[3], which identify the big-hit work according to the predicted impact score. It consists of *Logistic Regression, Naive Bayesian (NB), Random Forest (RF), Support Vector Machine (SVM)* and *AdaBoost Tree*. We only report the results of *NB*, *RF* and *SVM* because they have better performances.

---

[3] http://scikit-learn.org/.

Table 2. Performance comparison of different algorithms and methods. AdaWIRL outperforms the others, and achieves a $Pre@1$ score around 0.785.

Category			I	II			III		Our Solutions		
Metric	Δt_a	Training	Pop.	NB	RF	SVR	Co.Asc.	$Rank_{svm}$	IRL	WIRL	AdaWIRL
$Pre@1$	4	50 %	0.736	0.704	0.752	0.774	0.758	0.773	0.700	0.734	**0.786**
		40 %	0.736	0.702	0.748	0.773	0.754	0.772	0.697	0.733	**0.785**
	3	50 %	0.735	0.703	0.756	0.772	0.749	0.771	0.696	0.730	**0.783**
		40 %	0.735	0.703	0.752	0.772	0.749	0.771	0.694	0.739	**0.783**

- **Category III.** Various ranking algorithms[4,5] in Information Retrieval which includes *RankNet, RankBoost, AdaRank, LambdaMART, Coordinate Ascent (Co.Asc.)* and *RankSVM ($Rank_{svm}$)*. With proper parameter settings, *Co.Asc.* and *$Rank_{svm}$* achieve better performances and we report them in Table 2.

4.2 Experimental Results

Performance on Evaluation Metric. We construct $T_{>train}$ with 50 % or 40 % instances of $T_>$ and use it for model training. The performance is evaluated on the remaining 50 % or 60 % instances in $T_{>test}$. As described earlier, we set $t = 2010$, $\Delta t_b = 3$ and $\Delta t_a = 3$ or 4. The parameters in learning algorithms are fixed as: $\alpha = 0.0001$, $\lambda_\omega = 0.00001$ and $\beta = 0.001$. Table 2 reports the performances for various experiment settings. The results of ranking methods are averaged by 10 values during convergence range. Overall, all algorithms and methods perform well in $Pre@1$ (over 0.69) due to the well selected features. AdaWIRL has best score around 0.785 and outperforms the compared regression and ranking algorithms, with an average of 5.8 % and 2.9 % improvement respectively. In this work, AdaWIRL outperforms SVR and $Rank_{svm}$ because: (a) linear SVR may not achieve good prediction accuracy for non-linear paper citation distribution while AdaWIRL transforms the regression task to a pairwise classification problem; (b) unlike AdaWIRL, $Rank_{svm}$ can not capture the effect of parameter learning without using the current ranking information of papers in each iteration step.

Learning Convergence. We empirically study the learning curves of our methods and reports them in Fig. 2(a). Due to space limit, we only show the plot for one experiment setting (training = 50 % and $\Delta t_a = 4$) and the other cases have similar results. According to the figure, we find that our methods almost converge after a number of iterations. AdaWIRL has better performance and converges faster than IRL and WIRL. It indicates that the adaptive timely false correction strategy not only improves the prediction accuracy but also increases the convergence speed. As a result, a fixed number of training iterations is chosen as the stopping criteria of our ranking methods.

---

[4] http://www.cs.cornell.edu/People/tj/svm_light/svm_rank.html.
[5] http://people.cs.umass.edu/~vdang/ranklib.html.

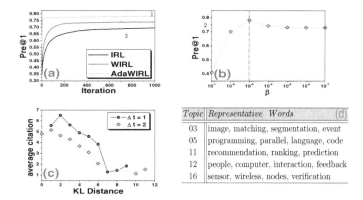

Fig. 2. (Best viewed in color.) (a) Learning curves of different ranking algorithms. (b) Impact of β for adaptive timely false correction strategy. (c) The relationship between average citation value of the selected author's future works and their KL distance from the predicted big-hit paper. (d) The selected hot-topics and their representative words. (Color figure online)

Impact of Hyperparameter. The parameter β plays an important role in AdaWIRL, as it determines how much compensation is added for each false prediction instance. On the one hand, if we use large β, $\boldsymbol{\omega}$ is changed significantly and may affect the accuracy of correct prediction cases. On the other hand, a tiny β adjusts $\boldsymbol{\omega}$ slightly and affects little in correcting false prediction samples. We investigate the relationship between AdaWIRL's performance and β, and report it in Fig. 2(b). The result meets what we expect. With the increment of β, $Pre@1$ increases at first. But when β goes beyond a certain threshold, $Pre@1$ decreases with the further increment of β. The existence of certain point of β (0.001) confirms that an appropriate integration of compensation term results in optimal performance. Given limited paper space, we report the impact of β in one experiment setting, i.e., training $= 50\%$ and $\Delta t_a = 4$.

4.3 Analysis and Discussion

Feature Contribution. Five groups of features are used in this work. We examine their contributions in two ways:

- **+Feature.** Maintain only one group of features to train the model.
- **-Feature.** Remove one group of features and keep the remaining groups of features for model training.

The prediction performances of AdaWIRL ($\Delta t_a = 4$) for different feature settings are reported in Table 3. Accordingly, the temporal features are shown as the best indication for big-hit papers and the score of AdaWIRL is over 0.77 with only temporal features while removing them leads to over 40 % loss. The temporal features capture the recent trends of different papers for each researcher,

Table 3. Contribution analysis of different groups of features. Temporal features are most influential while author and social related features have the least significance.

+/−	+					−					
Feature	Author	Social	Venue	Temp.	Content	Author	Social	Venue	Temp.	Content	All
50 % Train	0.259	0.034	0.413	**0.776**	0.176	0.784	0.783	0.775	**0.442**	0.781	0.786
40 % Train	0.269	0.051	0.414	**0.771**	0.135	0.783	0.782	0.774	**0.437**	0.782	0.785

and the "hot-papers" attract attention easily and have ever-growing impact. Meanwhile, the authorship and social features are least significant. Removing them results in tiny loss and using only those features has the poor performance with a below 0.27 accuracy. In feature engineering, although we prefer "group-scale" authorship and social features, it is common that those features of different papers of the same author are close. For example, Dr. Quanquan Gu ($a_{1037361}$) published 7 first-author papers during 2007 $\sim$ 2009 while all of them have the same value for A-max-$ave_{citation}$ because it is the average citation count of Dr. Jie Zhou, i.e., Dr. Quanquan Gu's advisor. The lack of variation for these groups of features may explain their low usefulness in predicting personal big-hit papers.

Future Works Impact Verification. Researchers may develop related future works based on content of the predicted bit-hit paper. We conduct experiment to reveal the relationship between citation values of the selected researcher's future works and their distance to the same researcher's predicted bit-hit paper. We run 100-topic LDA on corpus C_t published before $t = 2010$ and the first-author publications of selected researchers in Δt years after 2010, to get the topic distributions of these works. The distance between paper l_b and l_f is defined as the topic distributions' KL divergence [11]: $KL(p(Z|l_b), p(Z|l_f)) = \sum_{z \in Z} log \frac{p(z|l_b)}{p(z|l_f)} p(z|l_b)$. Figure 2(c) reports the relationship between average citation value of each selected author's future works in Δt years and their KL distance to the predicted big-hit papers. The big-hit papers of researchers are predicted by AdaWIRL with 50 % training cases and we only consider the results that at least 3 instances are in the given KL distance. According to Fig. 2(c), we find that with the increment of KL distance, the average citation value decreases, which meets our expectation: the higher the similarity, the larger the impact. Furthermore, we conduct 20-topic LDA on corpus $C_{big-hit}$ which contains the bit-hit papers of all selected authors and their most similar future works (with the smallest KL distance) in $\Delta t = 2$ years, to find the "hot-topics" of computer science corpus recently. Figure 2(d) shows five selected topics and their representative words. Obviously, topic 03, 05, 11, 12 and 16 are about compute vision, parallel programming, recommender systems, HCI and wireless communication, respectively.

5   Related Works

- ***Literature Scientific Impact Prediction.*** Scientific impact prediction has been extensively explored in recent years. Traditionally, the number of citations is used to quantify the scientific impact for individual paper or researcher thus many works focused on predicting future citation. Castillo et al. [5] estimated the author's citation value via using author reputation information. Yan et al. [24,25] used academic features for papers' citation prediction . In addition to citation count, the *h-index* [9] has been proved to measure both productivity and popularity of researchers. Yu et al. [7] investigated the correlation between researchers' *h-index* value and their collaboration signatures. Yu et al. [6] extracted various academic features for classifying the high impact papers that will contribute to improving the researcher's *h-index*.
- ***Pairwise Ranking Learning.*** The learning framework of AdaWIRL is based on Bayesian pairwise learning. Rendle et al. [16,17] proposed Bayesian personalized ranking (BPR) to explore users' preferences over items and make effective recommendation. Furthermore, BPR was improved by incorporating contextual information like social connections [27]. A boosting algorithm [12] for building ensemble BPR was proposed recently.

This paper is also related to other research tasks including collaboration link prediction [18,20], academic social network analysis [19], and relationship investigation for different influential nodes metrics [13]. Besides Bayesian pairwise ranking, there are many ranking methods in Information Retrieval from a pairwise [3,8,10] or list-wise perspective [4,22,23].

6   Conclusions

In this paper, we design AdaWIRL algorithm to predict the bit-hit paper for each researcher. The experiment on the ArnetMiner dataset shows the effectiveness of our method and AdaWIRL reaches over 0.783 accuracy in all experiment settings. In addition, we find that the temporal features are the best indication for big-hit work while the authorship and social information are least significant. Furthermore, we testify the correlation between the impact of researcher's future works and their similarity to the predicted big-hit paper. Overall, our work helps researchers to identify their big-hit papers in advance, which will be a useful guidance for disseminating impactful works or conducting related future work.

Acknowledgements. This work was partially supported by the National Natural Science Foundation of China (No. 11305043, No. 61433014), and the Zhejiang Provincial Natural Science Foundation of China (No. LY14A050001), the EU FP7 Grant 611272 (project GROWTHCOM) and Zhejiang Provincial Qianjiang Talents Project (Grant No. QJC1302001). Chuxu Zhang thanks to the assistantship of Computer Science Department of Rutgers University and Internship Experience of IBM Thomas J. Watson Research Center.

References

1. Bethard, S., Jurafsky, D.: Who should i cite: learning literature search models from citation behavior. In: CIKM 2010, pp. 609–618. ACM (2010)
2. Blei, D.M., Ng, A.Y., Jordan, M.I.: Latent dirichlet allocation. JMLR **3**, 993–1022 (2003)
3. Burges, C., Shaked, T., Renshaw, E., Lazier, A., Deeds, M., Hamilton, N., Hullender, G.: Learning to rank using gradient descent. In: ICML 2005, pp. 89–96. ACM (2005)
4. Cao, Z., Qin, T., Liu, T.-Y., Tsai, M.-F., Li, H.: Learning to rank: from pairwise approach to listwise approach. In: ICML 2007, pp. 129–136. ACM (2007)
5. Castillo, C., Donato, D., Gionis, A.: Estimating number of citations using author reputation. In: Ziviani, N., Baeza-Yates, R. (eds.) SPIRE 2007. LNCS, vol. 4726, pp. 107–117. Springer, Heidelberg (2007)
6. Dong, Y., Johnson, R.A., Chawla, N.V.: Will this paper increase your h-index? Scientific impact prediction. In: WSDM 2015, pp. 149–158. ACM (2015)
7. Dong, Y., Johnson, R.A., Yang, Y., Chawla, N.V.: Collaboration signatures reveal scientific impact. In: ASONAM 2015, pp. 480–487. ACM (2015)
8. Freund, Y., Iyer, R., Schapire, R.E., Singer, Y.: An efficient boosting algorithm for combining preferences. JMLR **4**, 933–969 (2003)
9. Hirsch, J.E.: An index to quantify an individual's scientific research output. PNAS **102**(46), 16569–16572 (2005)
10. Joachims, T.: Optimizing search engines using clickthrough data. In: KDD 2002, pp. 133–142. ACM (2002)
11. Kullback, S., Leibler, R.A.: On information and sufficiency. Ann. Math. Stat. **22**, 79–86 (1951)
12. Liu, Y., Zhao, P., Sun, A., Miao, C.: A boosting algorithm for item recommendation with implicit feedback. In: AAAI 2015, pp. 1792–1798. AAAI Press (2015)
13. Lü, L., Zhou, T., Zhang, Q.-M., Stanley, H.E.: The h-index of a network node and its relation to degree and coreness. Nat. Commun. **7**, 10168 (2016)
14. Nesterov, Y.: Smooth minimization of non-smooth functions. Math. Program. **103**(1), 127–152 (2005)
15. Page, L., Brin, S., Motwani, R., Winograd, T.: The pagerank citation ranking: bringing order to the web (1999)
16. Rendle, S., Freudenthaler, C.: Improving pairwise learning for item recommendation from implicit feedback. In: WSDM 2014, pp. 273–282. ACM (2014)
17. Rendle, S., Freudenthaler, C., Gantner, Z., Schmidt-Thieme, L.: Bpr: Bayesian personalized ranking from implicit feedback. In: UAI 2009, pp. 452–461. AUAI Press (2009)
18. Sun, Y., Han, J., Aggarwal, C.C., Chawla, N.V.: When will it happen? relationship prediction in heterogeneous information networks. In: WSDM 2012, pp. 663–672. ACM (2012)
19. Tang, J., Zhang, J., Yao, L., Li, J., Zhang, L., Su, Z.: Arnetminer: extraction and mining of academic social networks. In: KDD 2008, pp. 990–998. ACM (2008)
20. Wang, C., Han, J., Jia, Y., Tang, J., Zhang, D., Yu, Y., Guo, J.: Mining advisor-advisee relationships from research publication networks. In: KDD 2010, pp. 203–212. ACM (2010)
21. Wang, D., Song, C., Barabási, A.-L.: Quantifying long-term scientific impact. Science **342**(6154), 127–132 (2013)

22. Wu, Q., Burges, C.J., Svore, K.M., Gao, J.: Adapting boosting for information retrieval measures. Inf. Retrieval **13**(3), 254–270 (2010)
23. Xu, J., Li, H.: Adarank: a boosting algorithm for information retrieval. In: SIGIR 2007, pp. 391–398. ACM (2007)
24. Yan, R., Huang, C., Tang, J., Zhang, Y., Li, X.: To better stand on the shoulder of giants. In: JCDL 2012, pp. 51–60. ACM (2012)
25. Yan, R., Tang, J., Liu, X., Shan, D., Li, X.: Citation count prediction: learning to estimate future citations for literature. In: CIKM 2011, pp. 1247–1252. ACM (2011)
26. Zhang, T.: Solving large scale linear prediction problems using stochastic gradient descent algorithms. In: ICML 2004, p. 116. ACM (2004)
27. Zhao, T., McAuley, J., King, I.: Leveraging social connections to improve personalized ranking for collaborative filtering. In: CIKM 2014, pp. 261–270. ACM (2014)

Detecting Live Events by Mining Textual and Spatial-Temporal Features from Microblogs

Zhejun Zheng[1,2], Beihong Jin[1,2(✉)], Yanling Cui[1,2], and Qiang Ji[1,2]

[1] State Key Laboratory of Computer Sciences, Institute of Software,
Chinese Academy of Sciences, Beijing, China
jbh@otcaix.iscas.ac.cn
[2] University of Chinese Academy of Sciences, Beijing, China

Abstract. As microblogging services on the mobile devices are widely used, microblogs can be viewed as a kind of event sensor to perceive the dynamic behaviors in the city. In particular, detecting live events in microblogs, such as mass gathering, emergencies, etc., can help to understand what happened from the point of view of people who are present. For identifying the live events from a large number of short and noisy microblogs, the paper builds a generative probabilistic model named the ST-LDA model to cluster the microblogs whose semantics, time and space are similar into the same topic, and then determines the live events from the topics by an HMM-based method. The paper conducts the experiments on the real microblogs from *weibo.com*. Experimental results show that our method can detect live events more accurately and more completely than the LDA-based method and the TimeLDA-based method.

Keywords: Microblogs · Event detection · Topic model · Spatial-Temporal feature

1 Introduction

The top task of being smart cities is to comprehensively understand the events occurring in the cities, including crowd congregation, traffic accidents, and public safety incidents, etc. Only upon on this, can the urban operation mechanisms be analyzed and urban management strategies be decided. So far, some researches have been conducted on observing and analyzing the events in a city from various data. For example, considering that taxis are an important transportation means in cities, [1] estimates the social activeness of social activities in a city by analyzing the GPS-based trajectories of taxis and further detects social events, and [2] predicts whether a user will involve in urban activities by mining the relationship among different factors in event-based social networks. However, these are far from complete in the depth and breadth of sensing the events and cannot satisfy the real requirements.

Microblogging services provide a platform for sharing and exchanging information. Its users, i.e., microbloggers, can share with their followers activities they joined in, events they encountered as well as feelings they experienced anytime and anywhere. As microblogging services on the mobile devices are widely used, the number of

© Springer International Publishing Switzerland 2016
B. Cui et al. (Eds.): WAIM 2016, Part II, LNCS 9659, pp. 356–368, 2016.
DOI: 10.1007/978-3-319-39958-4_28

microbloggers continues to grow, further, more and more events happening in real life may be reflected in microblogs. Therefore, the microblogging services can be viewed as one kind of urban event sensor, which enable us to observe the city status and activities by discovering the events from microblogs.

From observations of the real world, we notice that a great number of people tend to post microblogs to share at that time what happened nearby (hereinafter referred to as live events), acting as a news reporter on the site. For example, they may report what congregation they are attending or report the accidents they encounter or see. Detecting these live events for knowing the crowd gathering and capturing emergencies from various levels and perspectives is imperative for city management. Therefore, this paper focuses on how to recognize live events effectively from a large number of microblogs, which can be helpful in identifying the city status and activities.

Compared to the traditional media (e.g., newspapers, magazines, journals, etc.), microblogs are short, noisy and large in quantity. Besides the text, a microblog usually contains some additional attributes such as the author, time and geographic location of the microblog. These characteristics of microblogs bring challenges as well as opportunities in the analysis and mining of microblogs. In order to recognize the live events, this paper improves the LDA (Latent Dirichlet Allocation) model [8] and proposes the ST-LDA (Spatial-Temporal LDA) model to recognize the topics of microblogs. The proposed model considers the text attributes, spatial-temporal attributes of microblogs, which ensures that the microblogs under the same topic not only have semantic similarity but also aggregate on both time and space. Furthermore, this paper builds a hidden Markov model (HMM) to discern the anomaly microblog topics and then determine the live events.

The remainder of this paper is organized as follows. Section 2 introduces the related work. Section 3 describes the ST-LDA model. Section 4 gives the method of identifying the live events from microblog topics. Section 5 shows the experimental evaluation. Finally, the paper is concluded in Sect. 6.

2   Related Work

In recent years, some researches pay attention to the social event detection from microblogs. As the early work, Lee et al. [3] present a geo-social event detection method through geo-tagged microblogs which are collected from Twitter. In [3], in order to detect unusual geo-social events with respect to an area, the ordinary status is first defined in terms of the number of microblogs posted from this location, the number of microbloggers and the number of moving microbloggers. Then, in a given area and a given time period, the sharp differences between current features and ordinary ones indicate that there are probably some events occurring. However, this method lacks thorough analysis in microblog contents, which leads to the detected events unreadable. Moreover, it is unable to handle noisy data properly. Sakaki et al. [4] select pre-defined event-related words from microblogs, and calculate the features, such as the frequencies of keywords and the positions of keywords in microblogs. Then, they employ the SVM classifier to pick up event-related microblogs for further analysis. While this method alleviates problems derived from data noise to some extent, the keyword selection will make a critical impact on the accuracy of event detection. In particular, sometimes keywords are hard to be selected in advance for a certain kind of event. Sankaranarayanan et al. [5] build a system called TwitterStand from Twitter, which can capture tweets that correspond to latest breaking news. In the TwitterStand, the news-related tweets are first recognized from the noisy dataset and an online clustering algorithm is performed on these news-related tweets. Then, each cluster is analyzed to describe the content of latest news and point out the location where the news happened. Becker et al. [6] explore a method to learn multi-feature similarity metrics for microblogs, and adopt the online clustering to aggregate the similar microblogs. Then, the clusters with comparatively large number of microblogs are regarded as the indication that there are some social events happening related to microblogs contents. The above detection methods are based on the sharp increase or decrease between current statistical indicators and previous ones, such as the number of microblogs or the number of people who post microblogs. In these methods, it is generally assumed that the great differences in a microblog dataset evidently indicate the critical clues to occurrences of social events. Becker et al. [7] give an event identification method by training a classifier to detect events from microblogs. They use an incremental, online clustering algorithm to cluster a stream of microblogs, then extract temporal features, social features, topical features and centric features from each cluster as the input of a SVM classifier which can decide whether each cluster of microblogs is related to a social event or not.

In essence, microblogs are a kind of short textual document, which are up to 140 characters long. Therefore, the event detection from microblogs can be started by identifying topics in microblogs which can be achieved by topic models such as the LDA model [8]. The LDA model is a generative probabilistic model for documents where documents are modeled via hidden Dirichlet random variables, namely topics, and these topics follow a probability distribution on a latent, low-dimensional topic space. While the LDA model is used to analyze microblogs, the differences between microblogs and traditional documents should be concerned. Compared to traditional documents, microblogs are relatively short. In many cases, a microblog message contains only one sentence, even one word. Therefore, instead of making the assumption that one word is generated by one topic in the original LDA model, for short documents like microblogs, it is practical and reasonable to assume that the whole document only contains one topic. In other words, all words from a microblog are generated by the same topic. Besides, the LDA generative process is modified to improve the accuracy of topic models in microblogs [9–11]. Specifically, Wang and McCallum [10] propose a modified model where each microblog's timestamp is generated from a certain topic in the same way as words in the microblog. As a result, topics deduced from this model have both semantic similarity among the words and the temporal concentration. Diao et al. [11] assume that microblogs which are posted during the same time period probably relate to the same topic. Moreover, they split a microblog dataset into several subsets according to their posted time, and put forward the TimeLDA model where topics of microblogs from the same time period conform to the same distribution. Meanwhile, analogous to the assumption for temporal feature, they also propose the UserLDA model where microblogs from the same author follow the same distribution. However, topic models proposed in [11] do not consider the spatial feature in microblogs, which makes these models hardly tell apart the events which happen in the same time period but at different locations. Zhou et al. [12] propose to use a filtering process plus a catego-rization process to explore events from tweets. The former process utilizes a well-trained classifier to separate event-related tweets from event-unrelated ones, and the latter process assigns each tweet to a certain event by a generative probabilistic model named LECM. In LECM, each event is characterized with semantic content, time and location. However, the location information here consists of the location-related words which are extracted from tweets. Such way of handling the locations may result in incorrect or imprecise description of detected events.

Compared to the existing work, our work is to identify live events, a special kind of social event, from microblogs. To do this, we build a generative probabilistic model which fully takes into account the characteristics of live events, and also design a HMM-based method to capture the live events.

3   ST-LDA Topic Model

3.1   Basic Idea

In the ST-LDA model, a microblog d is denoted as a triple $d = (c,t,g)$, where c is the textual content of microblog d, t and g are the posted time and geographical location of the microblog d, respectively.

The ST-LDA model is based on the following three assumptions:

- Each microblog is generated by a single topic, which is similar to the models proposed in [11]. At the same time, each word in a microblog is generated by either a topic allocated to the microblog or the background noise topic. The introduction of the noise topic is for handling noises in microblogs.
- Microblogs posted at the same time slice follow the same topic distribution. Based on this assumption, we split the microblog dataset into several parts according to their posted time.
- Microblogs that microbloggers post at the places close to each other could be more likely to be generated from the same topic than those posted from the places which are far away from each other. Therefore, we use a Gaussian distribution to describe the spatial distribution of a topic. In detail, for each topic, we use Gaussian parameter μ to denote the center position of this topic and use Gaussian parameter σ^2 to denote the extent to which the microblogs aggregate under this topic.

Table 1 lists the notations used in this paper.

The generative process of the ST-LDA model is described in Table 2, where K is the pre-defined number of topics, T is the number of time slices, D_t is the number of microblogs posted at the t-th time slice, N_d is the number of words in microblog d, N_g denotes the extent to which spatial feature of microblogs influences the topic distri-bution process. When N_g is set to a relatively big value in the model inference phrase, microblogs are more likely to be generated by the topics whose spatial centers are close to their posted locations. On the contrary, while a small value is set to N_g, microblogs are more likely to be generated by the topics that have high semantical similarity in their textual content.

Table 1. Notations

Symbol	Description
θ_t	multinomial parameter which microblogs posted at the tth time slice follow, $\theta_t \sim Dirichlet(\alpha)$
φ_k	multinomial parameter which words belonging to topic k follow, $\varphi_k \sim Dirichlet$ (β)
φ_b	multinomial parameter which words belonging to the noise topic follow, $\varphi_b \sim Dirichlet(\beta)$
z_d	topic of microblog d, $z_d \sim Multi(\theta_t)$
g_d	geolocation of microblog d, $g_d \sim Gaussian(\mu_{z_d}, \sigma_{z_d}^2)$
w_{dn}	the nth word in microblog d
x_{dn}	A variable which indicates whether w_{dn} is generated from noise topic or not, $x_{dn} \sim Bernoulli(\rho)$, $\rho \sim Beta(\lambda)$

Our generative process, which is responsible for allocating a topic for each microblog, is not only influenced by the similarity in textual contents, but also influenced by the temporal and spatial features embedded in the microblogs. Therefore, topics deduced from the ST-LDA model can be described with the textual feature plus the temporal and spatial features, which makes topics more understandable and more distinct with each other.

3.2 Model Derivation

There are two latent variables in the ST-LDA model. One is the topic of a microblog; the other is the variable which indicates whether the word in a microblog is generated by noise. Under the premise of using the ST-LDA model to generate microblogs, recognizing the topic of a microblog is to compute the conditional probabilities of the two latent variables, i.e., the probability that the topic of a microblog is k and the probability that a word is generated by noise, and further the topic of a microblog and whether a word is generated by noise can be obtained by Gibbs sampling.

In the ST-LDA model, the conditional probabilities of latent variables are expressed as follows. First, the probability that the topic of the microblog d is k is $p(z_d = k|z_{\neg d}, x, g, w, \alpha, \beta, \lambda, \mu, \sigma)$, which is calculated by (1), where z_d is the topic of d, x indicates whether the words are generated by noise, g is the geographical location of every microblog, w denotes all the words. Second, the probability that the n-th word in d is generated by noise is $p(x_{dn}|x_{\neg dn}, z, w, \alpha, \beta, \lambda, \mu, \sigma)$, which is calculated by (2). As we can see from (1) and (2), the conditional probability of latent variables can be expressed as the division of two joint probabilities.

$$p(z_d = k|z_{\neg d}, x, g, w, \alpha, \beta, \lambda, \mu, \sigma) \propto \frac{p(z, x, g, w|\alpha, \beta, \lambda, \mu, \sigma)}{p(z_{\neg d}, x_{\neg d}, g_{\neg d}, w_{\neg d}|\alpha, \beta, \lambda, \mu, \sigma)} \qquad (1)$$

$$p(x_{dn}|x_{\neg dn}, z, w, \alpha, \beta, \lambda, \mu, \sigma) \propto \frac{p(x, w|z, \alpha, \beta, \lambda, \mu, \sigma)}{p(x_{\neg dn}, w_{\neg dn}|z, \alpha, \beta, \lambda, \mu, \sigma)} \qquad (2)$$

Table 2. Generative process of the ST-LDA model

Generate multinomial parameter for noise topic from Dirichlet priori
$\varphi_b \sqcup Dirichlet(\beta)$

Generate Bernoulli parameter from Beta priori $\rho \sqsubset Beta(\lambda)$

for $t = 1$ to T do

   generate $\theta_t \sqcup Dirichlet(\alpha)$

end for

for $k = 1$ to K do

   generate $\varphi_k \sqcup Dirichlet(\beta)$

end for

for $t = 1$ to T do

   for $d = 1$ to D_t do

      generate $z_d \sqcup Multi(\theta_t)$

      for $n = 1$ to N_d do

         generate $x_{d,n} \sqcup Bernoulli(\rho)$

         if $x_{d,n} = 1$ then generate $w_{d,n} \sqcup Multi(\varphi_{z_d})$

         else generate $w_{d,n} \sqcup Multi(\varphi_b)$

      end for

      for $g = 1$ to N_g

         $g_d \sqcup Gaussian(\mu_{z_d}, \sigma_{z_d}^2)$

      end for

   end for

end for

From the generation method of the ST-LDA model, we have (3).

$$p(z, x, g, w | \alpha, \beta, \lambda, \mu, \sigma) = p(z|\alpha) \times p(x|\lambda) \times p(w|z, x, \beta) \times p(g|z, \mu, \sigma) \quad (3)$$

Therefore, we need to derive $p(z|\alpha)$, $p(x|\lambda)$, $p(w|z, x, \beta)$, and $p(g|z, \mu, \sigma)$.

First, we derive $p(z|\alpha)$ according to the conjugation between the Dirichlet distribution and the multinomial distribution. As shown in (4), N^{topic} is the number of topics, $N^{document}_{time=t,z=k}$ is the number of microblogs which are posted at time slice t and classified into topic k, $N^{document}_{time=t}$ is the number of microblogs in time slice t.

$$p(z|\alpha) = \prod_{t=1}^{T} \left(\frac{\Gamma(\alpha \cdot N^{topic})}{\Gamma(\alpha)^{N^{topic}}} \times \frac{\prod_{i=1}^{N^{topic}} \Gamma(\alpha + N^{document}_{time=t,z=k})}{\Gamma(\alpha \cdot N^{topic} + N^{document}_{time=t})} \right) \quad (4)$$

Next, we derive $p(x|\lambda)$. As shown in (5), $N^{word}_{x=1}$ is the number of words, which are generated by normal topics, in the microblog dataset, $N^{word}_{x=0}$ is the number of words, which are generated by noise, in the microblog dataset, and N^{word} is the number of words in the microblog dataset.

$$p(x|\lambda) = \frac{\Gamma(2\lambda)}{\Gamma(\lambda)^2} \times \frac{\Gamma(\lambda + N^{word}_{x=0}) \cdot \Gamma(\lambda + N^{word}_{x=1})}{\Gamma(2\lambda + N^{word})} \tag{5}$$

And, $p(w|z, x, \beta)$ is shown in (6), where $N^{vocabulary}$ is the number of different words in the dataset, $N^{word}_{x=1,z=k,w=j}$ is the number of j-th words generated by topic k, $N^{word}_{x=1,z=k}$ is the number of words generated by topic k, $N^{word}_{x=0,w=i}$ is the number of i-th words generated by noise.

$$p(w|z, x, \beta) = \frac{\Gamma(\beta \cdot N^{vocabulary})}{\Gamma(\beta)^{N^{vocabulary}}} \times \frac{\prod\limits_{i=1}^{N^{vocabulary}} \Gamma(\beta + N^{word}_{x=0,w=i})}{\Gamma(\beta \cdot N^{vocabulary} + N^{word}_{x=0})}$$

$$\times \prod\limits_{i=1}^{N^{topic}} \left(\frac{\Gamma(\beta \cdot N^{vocabulary})}{\Gamma(\beta)^{N^{vocabulary}}} \times \frac{\prod\limits_{j=1}^{N^{vocabulary}} \Gamma(\beta + N^{word}_{x=1,z=k,w=j})}{\Gamma(\beta \cdot N^{vocabulary} + N^{word}_{x=1,z=i})} \right) \tag{6}$$

Finally, we derive $p(g|z, \mu, \sigma)$ as shown in (7), where $N^{document}_{z=k}$ is the number of microblogs that are classified into topic k, $g_{document=j}$ is the geographical location of the j-th microblog, $\mu_{z=k}$ and $\sigma_{z=k}$ are the geographical location distribution parameters of the k-th topic.

$$p(g|z, \mu, \sigma) = \prod\limits_{i=1}^{N^{topic}} \prod\limits_{j=1}^{N^{document}_{topic=i}} Gaussian(g_{document=j}|\mu_{z=i}, \sigma_{z=i})^{N_g} \tag{7}$$

On the basis of Eqs. (4)–(7), we can calculate $p(z, x, g, w|\alpha, \beta, \lambda, \mu, \sigma)$. All the parameters in this joint probability are the statistic items of the dataset besides priori parameters. Similarly, the denominator $p(z_{-d}, x_{-d}, g_{-d}, w_{-d}|\alpha, \beta, \lambda, \mu, \sigma)$ in (1) can be obtained from the statistic items of all the microblogs except d. As thus, by simplification, we derive the probability that the topic of d is k, which is shown in (8).

$$p(z_d = k|z_{-d}, x, g, w, \alpha, \beta, \lambda, \mu, \sigma)$$

$$\propto (\alpha + N^{document}_{time=t,z=k} - 1) \times \frac{\Gamma(\beta \cdot N^{vocabulary} + N^{word}_{x=1,z=k} - N^{word}_{x=1,document=d})}{\Gamma(\beta \cdot N^{vocabulary} + N^{word}_{x=1,z=k})}$$

$$\times \prod\limits_{i=1}^{N^{vocabulary}} \frac{\Gamma(\beta + N^{word}_{x=1,z=k,w=i})}{\Gamma(\beta + N^{word}_{x=1,z=k,w=i} - N^{word}_{x=1,w=i,document=d})} \times Gaussian(g_{document=d}|\mu_{z=k}, \sigma_{z=k})^{N} \tag{8}$$

We can derive $p(x_{dn}|x_{-dn}, z, w, \alpha, \beta, \lambda, \mu, \sigma)$ in a similar way as shown in (9) and (10).

$$p(x_{dn} = 0 | x_{-dn}, z, w, \alpha, \beta, \lambda, \mu, \sigma) \propto \frac{(\lambda + N_{x=0}^{word} - 1)}{(2\lambda + N^{word} - 1)} \cdot \frac{(\beta + N_{x=0,w=dn}^{word} - 1)}{(\beta \cdot N^{vocabulary} + N_{x=0}^{word} - 1)} \tag{9}$$

$$
\begin{aligned}
p(x_{dn} = 1 | x_{-dn}, z, w, &\alpha, \beta, \lambda, \mu, \sigma) \\
&\propto \frac{(\lambda + N_{x=1}^{word} - 1)}{(2\lambda + N^{word} - 1)} \times \frac{(\beta + N_{x=1,z=z_d,w=dn}^{word} - 1)}{(\beta \cdot N^{vocabulary} + N_{x=1,z=z_d}^{word} - 1)} \\
&\times Gaussian(g_{document=d} | \psi_{z=z_d}^g)
\end{aligned}
\tag{10}$$

Then, according to the conditional probabilities of the two latent variables, we apply Gibbs sampling to estimate the parameters in the model. The sampling ends after enough iterations (500 iterations in the experiment). In the last iteration, the topic sampling result of a microblog is regarded as the topic of the microblog. We apply the maximum likelihood estimation to calculate θ_t and φ_k based on $N_{time=t,z=k}^{document}$, $N_{time=t}^{document}$, $N_{x=1,z=k,w=j}^{word}$, $N_{x=1,z=k}^{word}$ and $N_{z=k}^{document}$. At the same time, we apply the moment estimation to derive the geographical location distribution parameters (μ and σ^2) of every topic, which are shown in (11) and (12).

$$\mu_{z=k} = \frac{1}{N_{z=k}^{document}} \sum_{d=1}^{N_{z=k}^{document}} g_{document=d} \tag{11}$$

$$\sigma_{z=k}^2 = \frac{1}{N_{z=k}^{document}} \sum_{d=1}^{N_{z=k}^{document}} (g_{document=d} - \mu_{z=k})(g_{document=d} - \mu_{z=k})^T \tag{12}$$

Because φ_k represents the distribution of all the words in topic k, we can infer the semantic description of the topic according to φ_k. For example, if words "football, fighting, competition" account for a relatively large proportion in all the words which belong to a topic, then the semantics of the topic can be determined to be "football games".

4   Identifying Live Events from Anomaly Topics

By using the ST-LDA model, we can calculate the number of microblogs which are posted at the same time slice and have the same topic. Then, we build a hidden Markov model to describe the state changes of microblog topics.

Let X_n denote the state of microblog topic S at time slice n. Then, $\{X_n, n = 1, 2, \dots\}$ is modeled as a Markov chain that has two states, where state 0 indicates that the topic is normal at the time slice n, and state 1 indicates that the topic is abnormal at the time slice n. Let the probability distribution vector of the initial state of the Markov chain be $\pi = (p_0, p_1) = (P(X_1 = 0), P(X_1 = 1))$. We assume if topic S is normal at last time slice, then at the current time slice, the topic will be normal with a probability ρ, and abnormal with a probability $1-\rho$. Meanwhile, if topic S is abnormal at last time slice, then at the current time slice, the topic will be abnormal with a probability υ, and normal with a probability $1-\upsilon$. In general, ρ and υ can be obtained from historical data. As a result, we set the state transmission probability matrix A to $A = \begin{bmatrix} \rho & 1-\rho \\ 1-\upsilon & \upsilon \end{bmatrix}$. Because we can only observe the numbers of microblogs under topic S during T time slices, i.e., $\{c_i, i = 1,...T\}$ but cannot observe the states $\{X_i, i = 1,...T\}$ of the topic, i.e., the hidden state. Thus, we build a hidden Markov model, where A and π are mentioned above. B is an observation probability matrix, whose element $B_{ij} = P(c_i|X_j)$ denotes the probability that the observation is c_i given that the hidden state is X_j. Specifically, we adopt two Poisson distributions to describe the distribution of the numbers of the microblogs under the same topic in normal or abnormal states, respectively. The parameter of the Poisson distribution of the microblog number in normal state is set to the average number of microblogs which are classified into the topic S during T time slices, i.e., $\mu_0 = avg(c_1, c_2, \cdots c_T)$. Then, the parameter μ_1 of Poisson distribution of the microblog number in abnormal state is set to $\mu_1 = \mu_0 * 3$.

Now, given the number of microblogs which are posted during T time slices $(c_1, c_2, ..., c_T)$ and classified into one topic, we can predict the most possible topic state sequence $(X_1, X_2, ..., X_T)$ by using Viterbi algorithm. In the resultant state sequence, $X_i = 1$ indicates that the topic is abnormal at time slice i. Next, live events are determined as follows. The abnormal topic at time slice i is set to the topic of the live event. The median of time slice i is set to the occurring time of the live event. The geolocation center of the abnormal topic is set to the geographical location of the live event. Finally, the live events are sorted in descending order of the number of microblogs which are catalogued into the corresponding abnormal topic.

5   Evaluation

We evaluate the quality of live events detected from microblogs by our method (i.e., the ST-LDA-based method) via experiments. We adopt recall and precision as metrics. Recall is the ratio of the correctly-detected live events to all live events, and precision is the ratio of the correctly-detected live events to all detected live events.

We choose the other two methods as comparisons: one is based on the LDA model, the most widely-used topic model, and the other is based on the TimeLDA model. We extend them by adding the hidden Markov model in Sect. 4 so that they can detect anomalous events. In particular, in the LDA model which allocates one topic to one word, we cannot get the number of microblogs belonging to the corresponding event. So instead, we define hot degree for an event $e(topic = i, time = t)$ by (13).

$$H(e) = H(topic_{i,t}) = \sum_{d \in C_t} \frac{N^{word}_{z=i,document=d}}{N^{word}_{document=d}} \tag{13}$$

where $H(topic_{i,t})$ is the hot degree of topic i at time slice t. C_t is a collection of all microblogs at t. $N^{word}_{z=i,document=d}$ is the number of words in microblog d which are classified into topic i. $N^{word}_{document=d}$ is the number of words in microblog d. We call the two resultant methods the LDA-based method and the TimeLDA-based method, respectively.

The dataset used as the input of experiments are from *weibo.com*. It consists of 314,939 microblogs posted in March 2015, within the fifth ring road in Beijing city. After segmenting Chinese words, removing user names, stop words, punctuation, links, repost label, label "@", emoticon, high frequency words (e.g. we) and low frequency words, the dataset contains 117,043 different words.

Before experiments, parameters are set as follows. First, the time of day is divided into 4 slices, i.e., 0 am – 6 am, 6 am – 12 am, 12 am – 18 pm, 18 pm-24 pm. As thus, the time period of March 2015 is divided into 124 time slices. Then topic number K is set to 100. Next, similar to the parameter settings in [11], Dirichlet prior parameters are set as follows: $\alpha = 50/K, \beta = 0.01, \lambda = 1$. Finally, p_0, p_1, ρ, and v are set as follows: $p_0 = 0.5$, $p_1 = 0.5$, $\rho = 0.9$, $v = 0.6$. Different from the LDA-based method and the TimeLDA-based method, our method needs one additional parameter N_g which is set to 5.

In the experiments, we carry out our method and the other two methods to detect live events in the dataset and record the live events detected. For the ST-LDA-based method and the TimeLDA-based method, we calculate the number of microblogs which each topic contains while it is in the abnormal state and then sort the topics by the number of microblogs in descending order. The top-20 topics are regarded as the detected live events. For the LDA-based method, we sort the events detected in descending order according to (13), and get top 20 events as the resultant set.

We map the top 20 events detected by different methods to real events. We regard the union of three mapped event sets as the set of ground truth events. In particular, some events in the ground truth set are live events because they are characterized with locations pointing to places where these events happened. In detail, we find live events with Event IDs 1–8 belong to crowd gathering, and the events with Event IDs 9–11 are emergencies. We show all live events (with their Event ID) in Fig. 1, and show the remaining events in Table 3.

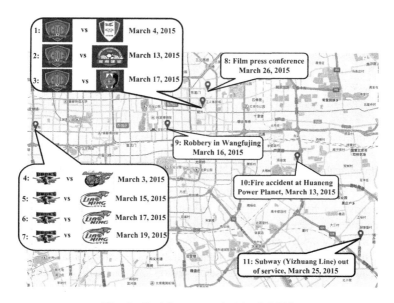

Fig. 1. Real live events in March 2015

Table 3. The other real events in March 2015

Event ID	Date (March 2015)	Event content
12	15th	CCTV (China Central Television) 3–15 party
13	2nd	Prince William's visit to China
14	5th	Lantern festival
15	8th	Women's day
16	14th	White valentine's day
17	22nd	CCTV CBA (Chinese Basketball Association) final

We give the top 20 live events detected by the ST-LDA-based method in Table 4. As shown in Table 4, our method can detect all the live events in March, 2015. Results in Table 4 also illustrate that our method can distinguish two events with same semantics which happen at the same time slice but at different places. For example, our method can detect two sports events on March 17, one with Event ID 6 is CBA game in MasterCard Center (Shou-gang basketball team vs. Liaoning basketball team) and the other with Event ID 3 is Asian Football Confederation (AFC) Champions League game in Beijing Workers' Stadium (Beijing Guoan FC vs. Urawa Red Diamonds). However, the TimeLDA-based method detects only one topic which contains the above two events, it cannot distinguish between the two events. In addition, top 5 live events detected by the ST-LDA-based method and their corresponding top 10 words are listed in Table 5. Live events detected by our method have more clear semantics because the spatial locations are regarded as one important dimension in our method.

Table 4. Top 20 events detected by the ST-LDA-based method

No.	Topic semantics	Event ID	No.	Topic semantics	Event ID
(1)	Basketball	(7)	(11)	Yizhuang line out of service	(11)
(2)	Soccer	(3)	(12)	Fire at Huaneng power plant	(10)
(3)	Soccer	(2)	(13)	Prince William's visit	(13)
(4)	Basketball	(6)	(14)	Emotional event	none
(5)	Basketball	(4)	(15)	Robbery on Wangfujing street	(9)
(6)	CCTV 3-15 party	(12)	(16)	Lantern Festival	(14)
(7)	Soccer	(1)	(17)	Film press conference	(8)
(8)	Emotional event	None	(18)	Emotional event	none
(9)	CCTV CBA final	(17)	(19)	Basketball	(5)
(10)	Emotional event	None	(20)	Emotion event	none

The recalls and precisions of three methods are shown in Fig. 2. Our method outperforms the other two methods in both precision and recall. The reason behind this is that the ST-LDA model can identify the topics of microblogs which aggregate on semantics, time and space, and these identified topics exactly contain the key constituent elements of live events. The same reason can be used for explaining the emotional events, which have no geographical clustering, are few in our result set (5 for our method, but 9 for the TimeLDA-based method, 10 for the LDA-based method).

Table 5. Top 10 words of top 5 live events detected by the ST-LDA-based method

Topic semantics	Top10 words
Basketball	Shou-gang, Beijing, Congratulation, amazing, championship, win, round, heer, final, game
Soccer	Guoan, soccer, fans, Workers' Stadium, Sanlitun, game, win, home court, AFC, VS
Subway derail	subway, line, train, Yizhuang, passengers, affect, debug, derail, test run, disaster
Huaneng power plant	Huaneng, power plant, happen, Chaoyang, fire, casualty, conflagration, helicopter, Wangsiying, disaster
Robbery	rob, Wangfujing Street, gun, masked man, taxi, criminal suspect

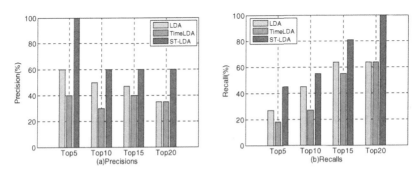

Fig. 2. Precisions and recalls

6   Conclusion

Live events recorded in microblogs provide the possibility of restoring the events (in particular, crowd gathering and emergencies) occurring in the city. Therefore, we present a microblog-oriented topic model, i.e., the ST-LDA model for analyzing the topics in the microblogs. By applying this model, the microblogs with the similar semantics and close spatial-temporal nature can be classified into a same topic. Then, we give an HMM-based method to discover the live events from topics. The experimental results show that our method of detecting live events has the higher recall and precision ratio than the LDA-based method and the TimeLDA-based method. In our next work, we will identify the live events from the stream of microblogs.

Acknowledgments. This work is supported by the National Natural Science Foundation of China under Grant No. 61472408 and No. 61379044.

References

1. Zhang, W., Qi, G., Pan, G.: City-scale social event detection and evaluation with taxi traces. Trans. Intell. Syst. Technol. Article No. 40 **6**(3). ACM (2015)

2. Du, R., Yu, Z., Mei, T.: Predicting activity attendance in event-based social networks: content, context and social influence. In: International Joint Conference on Pervasive and Ubiquitous Computing, pp. 425–434. ACM (2014)
3. Lee, R., Wakamiya, S., Sumiya, K.: Discovery of unusual regional social activities using geo-tagged microblogs. World Wide Web 14(4), 321–349 (2011)
4. Sakaki, T., Okazaki, M., Matsuo, Y.: Earthquake shakes Twitter users: real-time event detection by social sensors. In: International Conference on World Wide Web, pp. 851–860. ACM (2010)
5. Sankaranarayanan, J., Samet, H., Teitler, B.E.: Twitterstand: news in Tweets. In: ACM SIGSPATIAL International Conference on Advances in Geographic Information Systems, pp. 42–51. ACM (2009)
6. Becker, H., Naaman, M., Gravano, L.: Learning similarity metrics for event identification in social media. In: International Conference on Web Search and Data Mining, pp. 291–300. ACM (2010)
7. Becker, H., Naaman, M., Gravano, L.: Beyond trending topics: real-world event identification on Twitter. In: International Conference on Web and Social Media, pp. 438–441 (2011)
8. Blei, D.M., Ng, A.Y., Jordan, M.I.: Latent Dirichlet allocation. J. Mach. Learn. Res. 3, 993–1022 (2003)
9. Ramage, D., Dumais, S.T., Liebling, D.J.: Characterizing microblogs with topic models. In: International Conference on Web and Social Media, pp. 130–137 (2010)
10. Wang, X., McCallum, A.: Topics over time: a non-Markov continuous-time model of topical trends. In: ACM SIGKDD International Conference on Knowledge Discovery and Data Mining, pp. 424–433. ACM (2006)
11. Diao, Q., Jiang, J., Zhu, F.: Finding bursty topics from microblogs. In: Annual Meeting of the Association for Computational Linguistics: Long Papers-vol. 1, pp. 536–544. ACL (2012)
12. Zhou, D., Chen, L., He, Y.: An unsupervised framework of exploring events on Twitter: filtering, extraction and categorization. In: 29th AAAI Conference on Artificial Intelligence, pp. 2468–2474 (2015)

A Label Correlation Based Weighting Feature Selection Approach for Multi-label Data

Lu Liu, Jing Zhang$^{(\boxtimes)}$, Peipei Li, Yuhong Zhang, and Xuegang Hu

Hefei University of Technology, Hefei 230009, Anhui, China
liulu@mail.hfut.edu.cn, zhangjing@hfut.edu.cn

Abstract. Exploiting label correlation is important for multi-label learning, where each instance is associated with a set of labels. However, most of existing multi-label feature selection methods ignore the label correlation. Therefore, we propose a Label Correlation Based Weighting Feature Selection Approach for Multi-Label Data, called MLLCWFS. It is a framework developed from traditional filtering feature selection methods for single-label data. To exploit the label correlation, we compute the importance of each label in mutual information, and adopt three weighting strategies to evaluate the correlation between features and labels. Extensive experiments conducted on four benchmark data sets using two base classifiers demonstrate that our approach is superior to the state-of-the-art feature selection algorithms for multi-label data.

Keywords: Multi-label · Feature selection · Label correlation · Label weighting

1 Introduction

Traditional learning algorithms deal with problems where each instance is associated with a single label. In many real-world applications, however, one instance may associate with several labels simultaneously [1–3]. For example, an image can be tagged with a set of keywords [4], a document may belong to several predefined topics [5,6], and a gene may be related to multiple functions [7]. Hence, multi-label learning, such as image annotation, text categorization and protein function classification, had attracted much attention in the past years [8–11].

There are mainly two challenges in multi-label learning. First, like traditional single-label learning tasks, multi-label learning also suffers from the curse of dimensionality, which will cause serious problems when learning with high-dimensional data [8,12]. Feature selection is effective to solve the dimensionality excessively high problem. It aims to find a subset of relevant features that describes the data set as well as or even better than original feature set does [13,14]. Feature selection is able to remove redundant and irrelevant features, which can speed up learning algorithms and sometimes improve the performance, and make it possible to tackle the "curse of dimensionality" problem [11]. The second challenge is in multi-label learning, the instances are usually associated

© Springer International Publishing Switzerland 2016
B. Cui et al. (Eds.): WAIM 2016, Part II, LNCS 9659, pp. 369–379, 2016.
DOI: 10.1007/978-3-319-39958-4_29

with a set of labels which are typically interdependent and correlated [1], namely, the label correlation. For example, in image annotation, "desert" and "camel" tend to appear in the same image, while "desert" and "sea" are almost impossible to appear simultaneously, so it can be said that the correlation between "desert" and "camel" is stronger than that of "desert" and "sea". Although more and more feature selection algorithms have been proposed in recent years, most of them do not take the label correlation into account, which will result in the loss of useful information for learning algorithms.

Therefore, in this paper, we propose a label correlation based weighting multi-label feature selection approach, named as MLLCWFS. First, it is a framework built on traditional single-label filtering feature selection methods. This is because the filtering feature selection methods are faster while maintaining the effectiveness compared to other two feature selection approaches, namely, wrappers and embedded methods. We here choose a popular filtering feature selection method called mRMR (min-Redundancy and Max-Relevance) [15], because it can maximize the relevance between the selected feature subset and class labels, and meanwhile minimize the internal redundancy in feature subset. It enables effectively improving the performance of learning algorithms. Second, to exploit the label correlation, we compute the importance of each label in mutual information, and adopt three weighting strategies to evaluate the correlation between features and labels. Massive experiments conducted on four benchmark data sets using two base classifiers show that our approach outperforms to the state-of-the-art feature selection algorithms for multi-label data.

The rest of this paper is organized as follows. In Sect. 2, we review some related work. We give the details of our MLLCWFS approach and report the experiments in Sects. 3 and 4 respectively. Finally the conclusion in Sect. 5.

2   Related Work

In this section, we give a brief review of related work for multi-label learning. Existing methods can be simply classified into two categories, including the Problem Transformation (PT) approach and the Algorithm Adaption (AA) approach.

More precisely, the first category is to transform the multi-label data to single-label data in which a traditional feature selection method is then applied [11]. The representative work is the LP (Label Powerset) approach proposed by Tsoumakas et al. [16]. It treats each unique set of labels in the training set as a possible class of a traditional single-label classifier [17]. In other words, the original multi-label problem is transformed into a multi-class single-label problem. The main drawback of such approach is below: (1) No label correlation is taken into account. (2) It will cause the class-imbalance issue due to the excessive number of classes after the conversion of multi-label data into single-label data. Li and Zhang [18] proposed an Ensemble Multi-Label Feature Selection Algorithm Based on Information Entropy (called EMFSIE). It adopts the sum of information gain between the feature and each label in the label set as the information gain between the feature and the label set. Although the class-imbalance issue can be largely mitigated by this adoption, however, due to ignoring the label correlation, it still causes the loss of useful information.

The second category is the AA approach, it aims to tackle multi-label learning problem by adapting popular learning techniques to deal with multi-label data directly [3]. The AA approach can improve the information loss problem and the class-imbalance issue simultaneously, the representative work is the ML-kNN (Multi-Label k-Nearest Neighbor) algorithm [19]. It employs label prior probabilities gained from each instance's k nearest neighbors and uses maximum a posteriori (MAP) principle to determine labels [8]. The method overcomes the problem of imbalanced class distribution compared to the original kNN. Jungjit et al. [2] proposed two extensions to Multi-label Correlation-Based Feature Selection (called ML-CFS). It involves the label correlation and can get superior performance in Hamming Loss. In fact, the ML-CFS method is derived from the single-label CFS [20] method to deal with multi-label data. The basic idea of CFS method is to perform a search in the space of candidate feature subsets guided by a merit function, which evaluates the merit (quality) of each candidate feature subset.

However, both the PT approach and the AA approach are involved the single-label feature selection method. For example, the ML-CFS method requires improving the merit function which is more suitable for the single-label method. While the LP method has more options, such as mRMR, IG (Information Gain) [21], Relief and ReliefF [22] to name a few. Inspired by the thought in the EMFSIE algorithm and the extended ML-CFS algorithm, in this paper, we also consider the label correlation. However, contrary to these two algorithms, our approach is built on the traditional filtering feature selection algorithm, and it adopts three label weighting strategies to select the optimal feature subset.

3   Our Multi-Label Feature Selection Approach

In this section, we first give the formalization of our problem. Let D denote the m-dimensional instance space. And $D = \{d_i | d_i = (F_i, L_i), 1 \leq i \leq m\}$, where d_i is the i-th instance of the set D, $F = \{f_1, f_2, \cdots, f_d\}$ is a d-dimensional feature set, and $L = \{l_1, l_2, \cdots, l_d\}$ is a q-dimensional label set. Given an instance $F_i \in F$ the corresponding labels constitute a subset of L, which can be represented as a q-dimensional binary vector $L_i = \{l_{i1}, l_{i2}, \cdots, l_{iq}\}$, where $l_{ij} = 1$ if the instance has the label and otherwise $l_{ij} = 0$. The purpose of feature selection is to find a subset S of relevant features that describes the data set as well as or even better than original feature set F does, namely $\sum_{d_i \in D} |C_S(d_i) \in L_i| = \sum_{d_i \in D} |C_F(d_i) \in L_i|$, where $C(\cdot)$ is the classification model, L_i represents the true label of instance d_i, $C_F(d_i)$ and $C_{F_n}(d_i)$ represent the prediction labels of the classification model on the same instance d_i using the original feature set F and the selected feature subset F_n respectively.

Since most of existing multi-label feature selection methods ignore the label correlation, thus, we propose a multi-label feature selection approach named as

MLLCWFS. Contrary to existing multi-label feature selection methods, our app-roach is developed from traditional single-label filtering feature selection meth-ods. Meanwhile, we take into account the label correlation and use the label weighting mechanisms, because we believe the correlation between labels as well as the correlation between features and labels is different, thus, our approach combines the label correlation and label weighting to select the optimal fea-ture subset. Before giving the label weighting strategies, we first introduce our framework of multi-label feature selection as follows.

3.1   Framework of Our Approach

Measuring the correlation between features and labels is an important issue in feature selection method [12]. Regarding the traditional single-label problems, we can use mutual information or information gain to evaluate the correlation between features and labels, but there are some differences on the multi-label problem where each instance is associated with a set of labels. Thus, to deal with the multi-label feature selection, we need to find a method to measure the correlation between the feature and the label set. In this paper, we chose the mutual information as a base method rather than others such as the Pearson's correlation coefficient. This is because the former is effective in measuring the correlation between features and labels. Meanwhile, the label data are typically discrete, and it is suitable for measuring the correlation between two discrete variables in feature selection. Mutual Information is defined as follows.

Let f and l be two random variables with discrete values, to quantify how much information is shared by these two variables, a concept termed mutual information $I(f; l)$ is defined as shown in Eq. (1).

$$I(f; l) = \sum_{x \in f} \sum_{y \in l} \frac{p(x, y) \log (p(x, y))}{p(x)p(y)} \tag{1}$$

where $p(\cdot)$ is the probability density function. According to Eq. (1), we know that if f and l are closely related with each other, the value of $I(f; l)$ will be large; otherwise, f and l are completely independent and $I(f; l) = 0$.

As mentioned above, our approach is built on traditional single-label filtering feature selection methods and we chose mRMR as the base algorithm. Therefore, our framework of feature selection for multi-label data can be formalized below.

$$\Phi = \max(MR - mR), \tag{2}$$

$$MR(S, L) = \max \left\{ \frac{1}{|S|} \sum_{f_i \in S} R(f_i; L) \right\},$$

$$mR(S) = \max \left\{ \frac{1}{|S|^2} \sum_{f_i, f_j \in S} I(f_i; f_j) \right\}.$$

where $R(f_i, L)$ indicates the correlation between the feature f_i and the label set L, MR and mR represent the extended Max-Relevance criterion and min-Redundancy criterion respectively. And S is a feature subset, which jointly have the largest dependency on the label set [15]. In practice, incremental search methods can be used to find the near-optimal features defined by Φ. We can rewrite Eq. (2) into Eq. (3).

$$\Phi = \max_{f \in F-S} \left\{ R(f; L) - \frac{1}{|S|} \sum_{f_i \in S} I(f; f_i) \right\} \tag{3}$$

In this framework, our approach refines the mRMR algorithm in two dimensions, first, we chose mutual information to measure the correlation between the feature and the label set. Second, we give a definition to evaluate the importance of each label, that is, for a given label l_i, its importance can be represented as Eq. (4).

$$IMP(l_i) = \frac{1}{|L| - 1} \sum_{l_j \in L, j \neq i} I(l_i; l_j) \tag{4}$$

This is because the mRMR algorithm is based on the following two assumptions: First, the correlation between different features and labels is the same. Second, the label is independent of each other. However, in real-world tasks, these assumptions are unsustainable. Thus, we take into account the label correlation. And we propose the label weighting strategies, denoted as $W(l_i)$, which is a function relevant to the value of IMP. Therefore, we can define the correlation between the feature f and the label set L as shown in Eq. (5).

$$R(f; L) = \sum_{l_i \in L} I(f; l_i) W(l_i). \tag{5}$$

Details of our label weighting strategies are as follows.

3.2   Weighting Strategies

In this section, we introduce the weighting strategies proposed in this paper. There are three label weighting strategies, namely NCA (No Correlation Assignment) strategy, LCA (Large Correlation Assignment) strategy and SCA (Small Correlation Assignment) strategy.

Strategy 1. NCA: Assign the same weight to all labels, called the NCA strategy. The formalization of the NCA strategy can be seen as adding an equal weight to all labels, denotes as Eq. (6).

$$W(l_i) = \frac{1}{|L|} \tag{6}$$

Strategy 2. LCA: Assign a greater weight to the label with a larger IMP value as shown in Eq. (7), called the LCA strategy. The reason is because if a given label is highly correlated with other labels, namely the larger IMP value, it indicates the corresponding features correlated with the label are beneficial to the classification.

$$W(l_i) = \frac{IMP(l_i)}{\sum_{l_j \in L} IMP(l_j)} \qquad (7)$$

Strategy 3. SCA: Assign a greater weight to the label with a smaller IMP value as shown in Eq. (8), called the SCA strategy. This is because if a given label is weakly correlated with other labels, namely the smaller IMP value, it indicates it is hard to be predicted by other labels, and therefore the corresponding features correlated with this label are unbeneficial to the classification performance.

$$W(l_i) = \frac{1 - IMP(l_i)}{\sum_{l_j \in L}(1 - IMP(l_j))} \qquad (8)$$

The pseudo code of our MLLCWFS approach is given in Algorithm 1. More specifically, we use the incremental search methods to find the feature subset, suppose we already have S_{n-1}, the feature set with $n-1$ features. Our task aims to select the nth feature from the set $\{F - S_{n-1}\}$. This is done by selecting the feature that maximizes Eq. (3).

Algorithm 1. The MLLCWFS algorithm

Input: data set D, and selection ratio of features α
Output: selected feature subset S
  1: **procedure** MLLCWFS(D, n)
  2:      initialize $S \leftarrow \{\phi\}$ and $F = \{f_1, f_2, \cdots, f_d\}$
  3:      calculate the label importance $IMP(l_i)$ according to Eq. (4)
  4:      calculate the label weight $W(l_i)$ according to Eqs. (6), (7) and (8)
  5:      **while** $|S| \neq \alpha \cdot |F|$ **do**
  6:          find the feature $f(f \in F)$ maximizing Eq. (3)
  7:          set $S \leftarrow \{S \cup f\}$ and $F \leftarrow F \backslash f$
  8:      **end while**
  9:      **return** S
10: **end procedure**

---

4   Experiments

In our experiments, we compare our MLLCWFS approaches with three weighting strategies (including MLLCWFS$_{NCA}$, MLLCWFS$_{LCA}$ and MLLCWFS$_{SCA}$) with the EMFSIE feature selection algorithm for multi-label data. Our approaches adopt two base classifiers, such as ML-kNN and LP with the C4.5 decision tree [23], to make prediction using the 10-fold cross-validation. The parameters in ML-kNN is set to default values in the Mulan software [24]. Meanwhile, in our approaches, we select the top 15 % features of the data set as the final optimal feature subset, namely $\alpha = 0.15$.

4.1   Date Sets and Evaluation Measure

Experiments are conducted on four benchmark multi-label data sets, such as Enron, Medical, Genbase and Bibtex. Table 1 summarizes the statistics of these four data sets.

Table 1. Data sets used in our experiments

Name	Domain	#Instances	#Features	#Labels
Enron	text	1702	1001	53
Medical	text	978	1449	45
Genbase	biology	662	1186	27
Bibtex	text	7395	1836	159

To evaluate the feature subset selected in our approach, $MicroFMeasure$ [16] is adopted to evaluate the prediction performance of the classifiers. It is defined as:

$$MicroFMeasure = \frac{2 \times TP}{2 \times TP + FP + FN}. \tag{9}$$

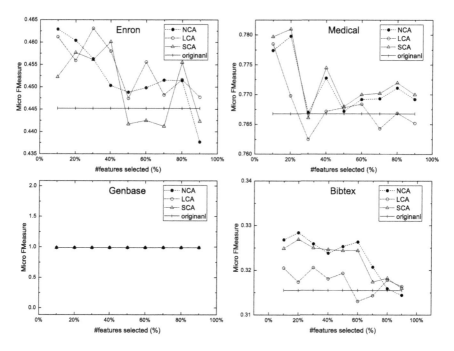

Fig. 1. Experimental comparison between our approaches in different strategies (using LP as the base classifier, called MLLCWFS + LP)

Here TP, FP and FN represent the number of True Positives, False Positives and False Negatives respectively. The larger the value of $MicroFMeasure$ is, the better the prediction result is.

4.2   Experimental

In this section, we first compare the three weighting strategies of our MLLCWFS approach. Second, we compare our MLLCWFS approach with the state-of-the-art feature selection algorithm EMFSIE for multi-label data.

Figures 1 and 2 show the results of the compared strategies on the benchmark four data sets, where LP and ML-kNN are used as the base classifier respectively. The "original" in figures indicates using the full data set rather than using the subset of the data set according to the selected subset of features. From the experimental results, we can see that with the same number of selected features, the performance of our approach with three strategies are generally better or comparable to the "original". Meanwhile, the performance of our approach using NCA and SCA strategies are very similar on all data sets in different base classifier, and most of the experimental results are better than the "original". For example, in both figures, the curves of NCA and SCA strategies on all data sets are almost coincident. These results validate the effectiveness of the three weighting strategies in our approach.

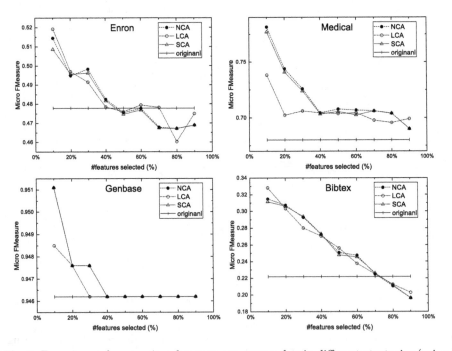

Fig. 2. Experimental comparison between our approaches in different strategies (using ML-KNN as the base classifier, called MLLCWFS + ML-KNN)

To examine the effectiveness of our proposed algorithm, Table 2 summarizes the experimental results of our approach with three weighting strategies compared to the state-of-the-art algorithm EMFSIE, where LP and ML-kNN are used as the base classifiers. In this table, we report the results of the average accuracy $\pm$ the standard deviation, the best or comparable results are highlighted in boldface. Meanwhile, we rank each compared approach, and give the average ranking score. From the experimental results, we can observe the following. There are at least two weighting strategies in MLLCWFS can beat the EMFSIE algorithm. For example, on the data set Bibtex, MLLCWFS$_{LCA}$ and MLLCWFS$_{SCA}$ approaches are ranked higher than EMFSIE. Meanwhile, our MLLCWFS$_{SCA}$ approach is the best approach. These data show that considering the label correlation and the weighting strategies is beneficial to improve the classification performance in our approach compared to the EMFSIE without consideration of the label correlation.

Table 2. Comparative results of different classifiers based on different feature selection methods (the larger the better)

Data Set	Methods	Base Classifiers		Avg. Rank
		LP	ML-kNN	
Enron	Original	0.4452±0.0281	0.4778±0.0226	5
	EMFSIE	**0.4596±0.0184**	0.4950±0.0236	2.5
	MLLCWFS$_{NCA}$	0.4501±0.0177	0.4969±0.0287	3.5
	MLLCWFS$_{LCA}$	0.4563±0.0224	**0.5111±0.0335**	**1.5**
	MLLCWFS$_{SCA}$	0.4512±0.0270	0.4990±0.0244	**2.5**
Medical	Original	0.7668±0.0301	0.6800±0.0401	5
	EMFSIE	0.7874±0.0280	0.7550±0.0344	2.5
	MLLCWFS$_{NCA}$	**0.7888±0.0327**	**0.7670±0.0315**	1
	MLLCWFS$_{LCA}$	0.7786±0.0262	0.7086±0.0366	4
	MLLCWFS$_{SCA}$	0.7813±0.0288	0.7587±0.0329	**2.5**
Genbase	Original	0.9868±0.0071	0.9462±0.0319	3
	EMFSIE	0.9815±0.0195	**0.9561±0.0202**	3
	MLLCWFS$_{NCA}$	**0.9868±0.0071**	0.9465±0.0281	2
	MLLCWFS$_{LCA}$	**0.9868±0.0071**	0.9476±0.0297	1.5
	MLLCWFS$_{SCA}$	**0.9868±0.0071**	0.9465±0.0281	2
Bibtex	Original	0.3155±0.0093	0.2218±0.0157	5
	EMFSIE	0.3218±0.0096	0.3254±0.0189	3
	MLLCWFS$_{NCA}$	0.3292±0.0116	**0.3286±0.0112**	1.5
	MLLCWFS$_{LCA}$	0.3222±0.0095	0.3212±0.0137	3.5
	MLLCWFS$_{SCA}$	**0.3335±0.0106**	0.3217±0.0121	2

5 Conclusion

In this paper, we proposed a multi-label feature selection approach called MLL-CWFS, whose framework is derived from traditional single-label filtering feature selection methods. Meanwhile, we used the label correlation as the label weight in three different ways to improve the classification performance. Experiments on four benchmark multi-label data sets show that our MLLCWFS approach outperforms a well-established multi-label feature selection algorithm.

However, how to select an optimal threshold in the selection of the number of features and how to apply our approach built on more traditional single-label feature selection methods are two significant and challenging issues in our future work.

Acknowledgments. This work is supported in part by the National 973 Program of China under grant 2013CB329604, the Program for Changjiang Scholars and Innovative Research Team in University (PCSIRT) of the Ministry of Education, China, under grant IRT13059, the Specialized Research Fund for the Doctoral Program of Higher Education under grant 20130111110011, the Natural Science Foundation of China under grants (61503112, 61273292, 61273297, 61229301, 61305063), and the Specified Research Fund for the Doctoral Program of HFUT under grant JZ2015HGBZ0461.

References

1. Gu, Q.-Q., Li, Z.-H., Han, J.-W.: Correlated multi-label feature selection. In: Proceedings of the 20th ACM International Conference on Information And Knowledge management, pp. 1087–1096 (2011)
2. Jungjit, S., Freitas, A.A., Michaelis, M., Cinatl, J.: Two extensions to multi-label correlation-based feature selection: a case study in bioinformatics. In: Proceedings of the IEEE International Conference on Systems, Man, and Cybernetics, pp. 1519–1524 (2013)
3. Zhang, M.-L., Zhou, Z.-H.: A review on multi-label learning algorithms. IEEE Trans. Knowl. Date Eng. **26**(8), 1819–1837 (2014)
4. Boutell, M.R., Luo, J.-B., Shen, X.-P., Brown, C.M.: Learning multi-label scene classification. Pattern Recogn. **37**(9), 1757–1771 (2004)
5. McCallum, A.K.: Multi-label text classification with a mixture model trained by EM. In: Proceedings of the AAAI 1999 Workshop Text Learning (1999)
6. Ueda, N., Saito, K.: Parametric mixture models for multi-labeled text. Adv. neural Inf. Process. Syst. **15**, 721–728 (2003)
7. Elisseeff, A., Weston, J.: A kernel method for multi-labelled classification. Adv. Neural Inf. Process. Syst. **14**, 681–687 (2002)
8. Zhang, Y., Zhou, Z.-H.: Multilabel dimensionality reduction via dependence maximization. ACM Trans. Knowl. Discov. Data **4**(3), 1–21 (2010). Article 14
9. Lee, J., Kim, D.-W.: Feature selection for multi-label classification using multivariate mutual information. Pattern Recogn. Lett. **34**(3), 349–357 (2013)
10. Huang, S.-J., Zhou, Z.-H.: "Multi-label Learning by Exploiting Label Correlations Locally. In: Proceedings of the 26th AAAI Conference on Artificial Intelligence, pp. 949–955 (2012)

11. Spolaôr, N., Cherman, E.A., Lee, H.D.: ReliefF for multi-label feature selection. In: Brazilian Conference on Intelligent Systems, pp. 6–11 (2013)
12. Zhang, Z.-H., Li, S.-N., Li, Z.-G., Chen, H.: Multi-label feature selection algorithm based on information entropy. J. Comput. Res. Dev. **50**(6), 1177–1184 (2013). (in Chinese)
13. Liu, H., Motoda, H.: Computational Methods of Feature Selection (Chapman & Hall/Crc Data Mining and Knowledge Discovery Series). Chapman & Hall/CRC, New York (2008)
14. Spolaôr, N., Monard, M.C., Lee, H.D., Tsoumakas, G.: Label construction for multi-label feature selection. In: Brazilian Conference on Intelligent Systems, pp. 247–252 (2014)
15. Peng, H.-C., Long, F.-H., Ding, C.: Feature selection based on mutual information criteria of max-dependency, max-relevance, and min-redundancy. IEEE Trans. Pattern Anal. Mach. Intell. **27**(8), 1226–1238 (2005)
16. Tsoumakas, G., Katakis, I., Vlahavas, I.: Mining multi-label data. In: Maimon, O., Rokach, L. (eds.) Data Mining and Knowledge Discovery Handbook, pp. 667–685. Springer, Heidelberg (2010)
17. Doquire, G., Verleysen, M.: Feature selection for multi-label classification problems. In: Advances in Computational Intelligence, pp. 9–16 (2011)
18. Li, S.-N., Zhang, Z.-H., Duan, J.-Q.: An ensemble multi-label feature selection algorithm based on information entropy. Int. Arab J. Inf. Technol. **11**(4), 379–386 (2014)
19. Zhang, M.-L., Zhou, Z.-H.: ML-kNN: A lazy learning approach to multi-label learning. Pattern Recogn. **40**(7), 2038–2048 (2007)
20. Hall, M.A.: Correlation-based feature selection for discrete and numeric class machine learning. In: Proceedings of the 17th International Conference on Machine Learning, ICML-2000, pp. 359–366 (2000)
21. Hoque, N., Bhattacharyya, D.K., Kalita, J.K.: MIFS-ND: A mutual information-based feature selection method. Expert Syst. Appl. **41**(14), 6371–6385 (2014)
22. Robnik-Šikonja, M., Kononenko, I.: Theoretical and empirical analysis of relief and RReliefF. Mach. Learn. **53**(1), 23–69 (2003)
23. Quinlan, J.R.: C4.5: Programs for Machine Learning. Morgan Kaufmann Publishers Inc., San Francisco (1993)
24. Tsoumakas, G., Spyromitros-Xioufis, E., Vilcek, J., Vlahavas, I.: Mulan: A java library for multi-label learning. J. Mach. Learn. Res. **12**, 2411–2414 (2011)

Valuable Group Trajectory Pattern Mining Directed by Adaptable Value Measuring Model

Xinyu Huang[1,2], Tengjiao Wang[1,2], Shun Li[3(✉)], and Wei Chen[1,2]

[1] School of Electronics Engineering and Computer Science,
Peking University, Beijing 100871, China
{xyhuang611,tjwang}@pku.edu.cn
[2] Key Laboratory of High Confidence Software Technologies,
Ministry of Education, Beijing 100871, China
[3] School of Information Science and Technology,
University of International Relations, Beijing, China
ls1977@gmail.com

Abstract. Group trajectory pattern mining for large amounts of mobile customers is a practical task in broad applications. Usually the pattern mining result is a set of mined patterns with their support. However, most of them are not valuable for users and it is difficult for users to go through all mined patterns to find valuable ones. In this paper, instead of just mining group trajectory patterns, we investigate how to mine the top valuable patterns for users, which has not been well solved yet given the following two challenges. The first is how to estimate the value of trajectory patterns according to users' requirements. Second, there are redundant information in the mined results because many mined patterns share common sub-patterns. To address these challenges, we define an adaptable value measuring model by leveraging multi-factors correlation in users' requirements, which is used to estimate the value of trajectory patterns. In order to reduce the redundant sub-patterns, we propose a new group trajectory pattern mining approach directed by the adaptable value measuring model. In addition, we extend and implement the algorithm as a parallel algorithm in cloud computing platform to deal with massive data. Experiments on real massive mobile data show the effectiveness and efficiency of the proposed approach.

Keywords: Trajectory pattern mining · Value measuring model · Multi-factors correlation

1 Introduction

With the development of Mobile Internet, large amounts of spatio-temporal trajectory data from moving objects are accumulated. Group trajectory pattern mining is to discover a group of moving objects moving together for a certain

This research is supported by the Natural Science Foundation of China (Grant No. 61572043, 61300003).

B. Cui et al. (Eds.): WAIM 2016, Part II, LNCS 9659, pp. 380–391, 2016.
DOI: 10.1007/978-3-319-39958-4_30

time period. These patterns can help the in-depth study of traffic surveillance, city planning and routes planning, etc.

Usually, to mine the group trajectory patterns, users will set the minimum length and support for the required patterns. Then the mining program will return a set of satisfied patterns, in which most of them are not valuable for users. It's necessary for users to go through all satisfied patterns to find the valuable ones, which is difficult because there are many satisfied patterns in the result. As a result, the valuable patterns may be buried in the massive result set. Therefore, instead of just mining group trajectory patterns, it's necessary to help users to find the top valuable patterns. Though this is a strategically important task in many applications, the problem has not been well solved yet given the following two challenges.

Firstly, it's difficult to estimate the value of the trajectory patterns because users have different mining requirements in different applications and there are lots of influential factors for a group trajectory pattern, such as trajectory length, time period, time granularity, etc. Take the green and blue group trajectory patterns in Fig. 1 as examples, it's easy to find that the green pattern with bigger support while the blue pattern with longer length. If users want to mine the longer group trajectory patterns which satisfy the minimum support, the users' requirement function can be described as Support-Unchange in Fig. 1. In this situation, the blue pattern is more valuable than the green pattern because the blue pattern is longer. However, in another situation, while users require longer length, they will also require smaller minimum support. Then the users' requirement function can be described as Inward-Curve in Fig. 1 and it's hard to determine which one is more valuable because it depends on the concrete requirement function. Therefore, if we only return the top frequent trajectory patterns as results, the results may not be valuable for all users. In this paper, to estimate the trajectory patterns' value, we define an adaptable value measuring model through leveraging the multi-factors correlation in users' requirements.

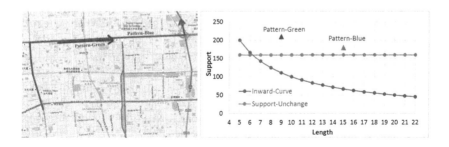

Fig. 1. Examples of group trajectory patterns with various factors

Secondly, if we just apply the traditional trajectory pattern mining approach and rank the mined patterns to find the valuable ones, there will be redundant information in the result set because many mined trajectory patterns share common sub-patterns. As Fig. 1 shows, if the blue pattern is more valuable than the green pattern, the green pattern will be redundant because it's sub-pattern of the blue pattern. Therefore, in order to reduce the redundant information, in this paper, we propose a new efficient group trajectory pattern mining approach directed by the adaptable value measuring(AVGPM) with linear time and space complexity. The proposed approach can mine the top valuable group trajectory patterns through maximizing the value of each satisfied pattern. The main contributions of this paper are summarized as follows:

1. We define an adaptable value measuring model to estimate the value of group trajectory patterns through leveraging the multi-factors correlation in users' requirements.
2. We present a configurable mobile trajectory building method to reduce the uncertainty of trajectories.
3. We propose a new group trajectory pattern mining approach directed by the adaptable value measuring model and validate the effectiveness and efficiency of the proposed approach.

The rest of the paper is organized as follows. Section 2 gives a structured overview of related works. We formalize the adaptable value measuring model in Sect. 3. An efficient algorithm to discover the group patterns directed by adaptable value measuring is shown in Sect. 4. Section 5 provides the experimental validation of the effectiveness and efficiency of the proposed approach, followed by conclusion in Sect. 6.

2   Related Work

In this section, we discuss the related work to our study including trajectory pattern mining, trajectory pattern recommending and trajectory uncertainty.

Trajectory Pattern Mining: Many research studies have investigated in trajectory pattern area, such as moving together patterns [1], trajectory clustering [2], periodic pattens [3] and frequent sequential patterns [4]. Moving together patterns mining is closely related to our task which is to discover a group of objects that move together for a certain time period, such as flocks, convoys, swarms and so on [5]. These patterns are distinguished by the following factors: the shape or density of a group, the number of the objects in a group, and the duration of a pattern. In this paper the proposed method focus on discovering convoy which requires a group of objects to be density connected during k consecutive time points. Jeung, H. et al. [6] formalized the concept of a convoy query using density-based notions, in order to capture groups of arbitrary extents and shapes. Different from the proposed method, our method aims to mine the top valuable trajectory patterns directed by the measurement defined by users.

Trajectory Pattern Recommendation: In recent years, there are many studies in recommending trajectory patterns [4,7–9]. Wei et al. [4] present a Route

Inference framework to construct the top-k routes traveling a given location sequence within a specified travel time according to user-specfied query. To help users to find interesting locations and classical travel sequences, Zheng et al. [7] proposed a HITS-based inference model by taking into account the factors of the locations. Yin et al. [8] investigate how to rank the trajectory patterns mined from the uploaded photos with geotags and timestamps. Instead of focusing on mining frequent trajectory patterns, they put more effort into ranking the mined trajectory patterns and diversifying the ranking results. To recommend a time-sensitive route, Hsieh et al. [9] devised a statistical model, to integrate four factors into a goodness function which aims to measure the quality of a route. Although these works are valuable in the specific scenario, they can't formulate users' requirements into a value measuring model and reduce the redundant sub-patterns based on the model.

Trajectory Uncertainty: Many trajectories have been recorded with a very low sampling rate, leading to object movement between sampling points uncertain. The research topics of trajectory uncertainty are studied in [10,11]. N. Pelekis et al. [10] study the effect of uncertainty and introduce a three-step approach to deal with it. However, this work focus on the trajectory uncertainty caused by measurement errors. Although K. Zheng et al. [11] investigated the problem of constructing popular routes sequentially passing the queried locations from uncertain trajectories, they use road network information to reduce the uncertainty of low sampled trajectories. Different from these works, we proposed a configurable mobile trajectory building method to reduce the trajectory uncertainty without using extra information like road network.

3 Adaptable Value Measuring Model

In this section, we first discuss the correlation of influential factors in trajectory patterns. We then give a formalized definition of the adaptable value measuring model to estimate the value of the trajectory patterns in Sect. 3.2.

3.1 Multi-factors Correlation

As the introduction mentions, there are lots of influential factors for a group trajectory pattern. Now we take the length and support as an example to discuss the multi-factor correlation in users' requirements as illustrated in Fig. 2(a):

1. For most applications such as traffic monitoring, while users require longer length for the trajectory pattern, they usually require smaller minimum support correspondingly. Therefore, the correlation between length and support can be formalized as the function whose length and support is inversely proportional as Linear Type in Fig. 2(a) or the function whose support decreases progressively along some curve as Inward Curve.

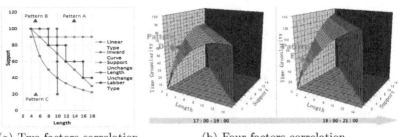

(a) Two factors correlation   (b) Four factors correlation

Fig. 2. Multi influential factors model case

2. For some applications such as urban planning, when the required length of trajectory pattern fluctuates to some extent, its required minimum support stay unchanged. For example, if the required length increase from 4 to 8, the required minimum support stay unchanged. However, if the required length increase from 8 to 9, the required minimum support will decrease. This kind of correlation function can be present as Ladder Type in Fig. 2(a).
3. For some particular applications, the correlation function may present two extreme cases: in one case(e.g. in group migrations application), when the length of the patterns increase, the required minimum support stay unchanged presented as Support-Unchange Type in Fig. 2(a); in the other case(e.g. in emergency events monitoring application), when the support of the patterns increase, the required length stay unchanged as Length-Unchange Type in Fig. 2(a).

3.2   Formalized Definition of the Adaptable Value Measuring Model

To estimate the value of the trajectory patterns, we propose a new adaptable value measuring model based on the muti-factors correlation in users' requirements discussed in Sect. 3.1.

Firstly, we define the adaptable correlation function based on multi-factors correlation.

Definition 1. *(Adaptable Correlation Function):* $w_n = f(w_1, w_2, \ldots, w_{n-1})$, $\forall 1 \leq i \leq n$, w_i *means the ith of factors in the pattern*

Take the application just considering length and support as an example, its adaptable correlation function is defined as $y = f(x)$, in which x means the length of trajectory pattern and y means the minimum support when the required length is x.

Observing all kinds of functions in Fig. 2(a), we assume that the value is negative for arbitrary point(x, y) like Pattern C, which is at the lower left of the function $y = f(x)$. The farther away from the function, the smaller the value of the pattern. Inversely, the applying value is positive for arbitrary point(x, y) like

Pattern B or Pattern A, which is at the upper right of the function $y = f(x)$. And the farther away from the function, the larger the applying value.

According to assumptions, *the value of pattern A > the value of pattern B > 0 > the value of pattern C* in Fig. 2(a). Obviously, the value of pattern C is negative which means this pattern is not valuable.

Therefore, if the trajectory pattern with length as X and support as Y, it's easy to find out that the size of area, which is formed by line $x = X$, line $y = Y$ and function $y = f(x)$, is proportional to the applying value for users. What's more, we can get a higher dimensional function $w_4 = f(w_1, w_2, w_3)$ by adding time factor and time granularity factor shown in Fig. 2(b). According to Fig. 2(b), given certain time as W_1 and time granularity as W_2, for the trajectory pattern whose length is W_3 and support is W_4, we can easily find that the area formed by $w_3 = W_3, w_4 = W_4$ and function $w_4 = f(W_1, W_2, W_3)$ is proportional to the the applying value of the trajectory pattern.

Take pattern D and pattern E as an example, supposing the respective corresponding section area integral are V_D and V_E in Fig. 2(b), we can find that V_D and V_E are negative and $V_D < V_E$. What's more, traffic flow is smooth at night but intensive during rush hours, the required support at night is usually smaller than the required support during rush hours. Therefore, for the pattern D during rush hour and pattern E at night, the applying value of pattern E is larger than pattern D, which correspond to the relationship between V_E and V_D.

Therefore, the adaptable correlation function can be used to describe the multi-factors correlation in the users' requirements. To estimate the value of the trajectory patterns, we define the adaptable value measuring model based on the adaptable correlation function.

Definition 2. (*Adaptable Value Measuring*): *For any trajectory pattern, suppose there are n influential factors , and for $\forall 1 \leq i \leq n, w_i$ means the ith influential factor of trajectory pattern and W_i means the value of the ith influential factor, then the adaptable value measuring based on multi influential factors of this trajectory pattern can be defined as following:*

$$F(W_1, W_2, \ldots, W_n) =$$

$$\int_{f_n^{-1}(W_1, W_2, \ldots, W_{(n-2)}, W_n)}^{W_{n-1}} |W_n - f_n(W_1, W_2, \ldots, W_{n-1}, w_{n-1})| dw_{n-1} \qquad (1)$$

$w_n = f(w_1, w_2, \ldots, w_{n-1})$ means adaptable correlation function of trajectory pattern based on multiple factors, $f_n^{-1}(W_1, W_2, \ldots, W_{n-2}, W_n)$ is the inverse function of $f_n(W_1, W_2, \ldots, W_{n-2}, w_{n-1}) = W_n$ Therefore, with the adaptable value measuring model, we can easily estimate the trajectory patterns' applying value by leveraging muti-factors correlation in users' requirements.

4 Group Trajectory Pattern Mining Directed by Adaptable Value Measuring(AVGPM)

In this section, we present the new group trajectory pattern mining approach directed by the adaptable value measuring model. The approach consists of three phases: trajectory building, closed group discovering, group trajectory pattern mining. In the first phase, we propose a configurable mobile trajectory building method to build users' trajectories. In the second phase, we introduce a closed group trajectory clustering algorithm to accelerate the group trajectory pattern mining phase. In the third phase, we present our group trajectory pattern mining algorithm directed by the adaptable value measuring model to reduce the redundant sub-patterns. What's more, we accelerate group trajectory pattern mining by applying filter-and-refinement paradigm in Sect. 4.4.

4.1   Configurable Mobile Trajectory Building (CMTB)

In order to study the similarity of moving groups, we need to build the mobile trajectory for each user. However, there is uncertainty for the trajectory data because objects move continuously while their locations can only be updated at discrete times, leaving the location of a moving object between two updates uncertain [5]. For the given timestamp sequence $T = < t_1, t_2, \ldots, t_m >$, it is necessary to calculate the user's position at each timestamp t_i according to collected trajectory data. Assuming the collected trajectory data sequence is $TP = < tp_1, tp_2, \ldots, tp_n >$, two strategies are adopted in order to build the mobile trajectories.

Firstly, we adopt the "nearest" assignment strategy. For any time point tp_i in TP, we find the closest timestamp t_k in T, and assign the monitoring value at time tp_i to the value at timestamp t_k. If there is another time point tp_u in TP which is closer to t_k, we reassign the monitoring value at tp_u to the value of t_k instead. This strategy guarantee the position in the built trajectory at each timestamp is definitely the latest monitoring value.

Secondly, for the missing value in the trajectory built by the above strategy, if there is no position data collected by the base station at near time, we use interpolation strategy based on adjacent speed to assign its value. The above assignment strategy could be used to construct a trajectory during some given timestamps from discrete and irregular data which are collected by the base stations. Moreover it guarantees the precision of the analog position.

4.2   Grid Cell Clustering(GCC)

An important preprocess phase for group trajectory pattern mining is clustering the trajectories at each timestamp. A straightforward approach is performing density-based clustering [12] on the trajectories of users at each timestamp in TP to detect all density-connected clusters.

However there may be ten millions of users per day in the mobile communication, the number of subspaces would be extremely large. A more efficient strategy is to check the desity-connected clusters through a 2-D space which divided into some grid cells with size of $\varepsilon/2 \times \varepsilon/2$. The distance between any two points in the same cell is at most ε. Therefore, any cell containing more than a minimum number of points is part of a cluster; such cells are called dense groups. First we merge the points which belong to sparse cells into its neighboring dense groups, then the remaining sparse cells will be ignored. Finally the DBSCAN proceeds to merge neighboring dense groups. Our Grid Cell Clustering algorithm, referred as GCC, shares many characteristics with the MC1 algorithm [13]. Unlike the MC1, we first filter the sparse cells and merge the points in these cells into their dense neighbors, which can highly accelerates the process of clustering by minimizing the useless candidates.

4.3   The Algorithm of AVGPM

In this section, we present the details of the group trajectory pattern mining algorithm directed by the adaptable value measuring model. Given a set of trajectory sequences S, a timestamp sequence T, a value measuring model based on users' requirements and some parameters $(K, \varepsilon, MinS)$, the algorithm aim to mine the top K valuable group trajectory patterns for users.

Algorithm 1 presents the detailed pseudocode of AVGPM. Line 2 performs trajectory building for the set of sequence S, and the cost is $O(|S|)$. Closed group clusters are detected at each timestamp (Lines 3–7), and the cost for this step is $O(|MT| \times |T| + |C|)$. The algorithm builds a cluster matrix, which uses the clusters represent moving objects' position at each timestamp, from the original trajectory matrix MT, and each cluster's support at each timestamp is counted (Lines 9–15), which would be executed for $|MT| \times |T|$ times. Furthermore, the non-frequent clusters are filtered, and the rows and columns which have no frequent clustering are also deleted, in order to reduce the candidates for the following steps (Lines 16–19), and the cost is also $O(|MT| \times |T|)$.

After minimizing the useless candidates, AVGPM will mine the top K valuable group trajectory patterns directed by the value measuring model VM (Lines 20–31). We apply an approach searching from a candidate cluster and then extend it to more valuable one to maximize the potential value according to VM(Line 28). For example, if a trajectory pattern A_1 with length as 5 and support as 100, AVGPM will extend the A_1 to trajectory pattern A_2 with length increase to 6 and support decrease to 90 if A_2 is more valuable than A_1 according to VM. However, if A_2 is less valuable than A_1, AVGPM will not extend the A_1 to A_2. To improve efficiency, AVGPM adopt greedy search paradigm to search for a local optimum instead of the exhaustive search for a global optimum for the trajectory patterns. Compared the greedy search and exhaustive search for global optimum on our real mobile dataset, we find about 79.2 % of the local optimum is also the global optimum, which prove the effectiveness of applying greedy search for a local optimum. The cost of AVGPM depends on the cost of the pattern extending step in Line 28, which is $O(K \times |MT|^2 \times |T|^2)$ if brute force algorithm is used.

4.4   Faster Group Trajectory Mining - AVGPM+

As the AVGPM performs expensive pattern extending at every time, in this section, we apply a filter-and-refinement paradigm to reduce the overall computational

Algorithm 1. AVGPM(SequenceSet S, real ε , int MinS, Timestamp T, Value-Model VM, int K)

1: $GP \leftarrow \emptyset;//$ Set of group patterns
2: $MT \leftarrow CMTB(S,T);//$Building the trajectories MT from S
3: **for** each $t \in T$ **do**
4:    $C \leftarrow$ *Build CellSet by assigning each point of MT to a cell;*
5:    $G \leftarrow GCC(C,\varepsilon,MinS);//$discover the closed groups
6:    $ClustersSet[t] \leftarrow G$
7: **end for**
8: //Build the cluster matrix from trajectory matrix MT
9: **for** each $t \in T$ **do**
10:    **for** each $mt \in MT$ **do**
11:       $cluster \leftarrow FindCluster(ClustersSet[t], mt[t]);$
12:       $Matrix[mt.user][t] \leftarrow cluster;//$Build cluster matrix
13:       $ClusterSup[t][cluster] \leftarrow ClusterSup[t][cluster] + 1;$
14:    **end for**
15: **end for**
16: **for** each $t \in T$ **do**
17:    delete the non-frequent clusters in Matrix;
18:    filter rows and columns contain no frequent cluster;
19: **end for**
20: **for** each $t \in T$ **do**
21:    find max cluster CE and support S in ClusterSup[t]
22:    $Maxs[t].cluster \leftarrow CE; Maxs[t].support \leftarrow S;$
23: **end for**
24: $foundCount \leftarrow 0;$
25: **while** $foundCount < K$ **do**
26:    $MaxCluster \leftarrow the\ first\ max\ cluster\ in\ Maxs;$
27:    //Extend maxCluster to be its most valuable pattern
28:    $gp \leftarrow Extend(MaxCluster, VM, T, Matrix)$
29:    Update gp's statistics in ClusterSup and Maxs;
30:    $GP \leftarrow GP \bigcup gPattern; foundCount \leftarrow foundCount + 1;$
31: **end while**
32: return GP;

---

cost. It accelerates the process of trajectory mining and brings higher efficiency by avoiding the redundant checks in pattern extending step.

Suppose that C is the group pattern candidate that is currently being extended. We first apply a refinement step to filter out all the useless rows from the trajectory matrix M which do not contain C in its timestamp. After this filter step, the remaining useful rows could be much smaller than the original rows, the cost is $O(|M|/|T|)$. Suppose the remaining matrix is M', the pointer matrix is built using M' which stores the pointer of the items that are in the same cluster(like a liked list), and the cost is $O(|M'|)$. Finally, a given value measuring model is used to discover the most valuable group patterns by using dynamic programming techniques , but unlike the dynamic programming strategy we store the minimum interim results which only requires $O(|M'|)$ of memory cost, and its time cost is $O(|M'| \times |T|)$. And the total cost of AVGPM+ is $O(K \times (|T| + 1) \times |M|)$.

However, at large scales, mining for valuable patterns faces great challenges in terms of data management as well as computation(the cost could be extremely high, even infinite). In order to meet the challenges, we store the data to a distributed computing cloud. For the most costly steps of AVGPM(closed groups clustering, cluster matrix building and non-frequent clusters filtering step) could be computed in parallel way for different clusters/areas(grid cells) at different timestamp, so AVGPM can obtain good speedups and expansibility. And for the pattern extending step, as its cost is small, it could be in-memory computed efficiently by a single machine.

5   Experimental Evaluation

In this section, we conduct extensive experiments to evaluate the effectiveness and efficiency of the proposed model and approach based on a real trajectory dataset, which is collected in a large-scale exposition held in Shanghai containing about 10–15 million of signal data and 40–60 million users per day on average.

We first compare the discovery with CuTS [6], an algorithm for traditional group trajectory pattern mining problem. We then analyze the speedups and expansibility of the AVGPM+. The experiments are conducted on a distributed computing cloud. The configuration for each compute node is: Dell PowerEdge 2970 server, AMD Opteron 2212 * 2 CPU, 2 G memory, 500 G disk driver.

To compare the discovery between CuTS and AVGPM+, the experiment takes traffic monitoring application as an example, which mainly considers two general influential factors, length and support. Generally speaking, in this application, while users require longer length, they will require smaller support correspondingly. Therefore, the adaptable correlation function according to users' requirements can be described as Inward Curve in Fig. 2(a). In the experiment, we set the required length as 5 and the minimum support as 50.

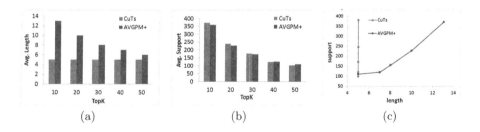

(a)   (b)   (c)

Fig. 3. The experiment result of effectiveness

Figure 3(a) shows the experiment result for CuTS and AVGPM+. We can find that the length for the trajectory patterns mined by CuTs without adaptable value measuring are constantly 5 while AVGPM+ with adaptable value

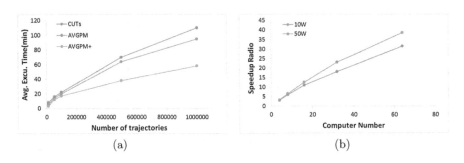

Fig. 4. The experiment result of efficiency

measuring have longer length. Figure 3(b) shows that TopK trajectory patterns mined by CuTS and AVGPM+ have similar support. From Fig. 3(c), we can find that with the similar support for the trajectory patterns mined by AVGPM+ have longer length, which means AVGPM+ can find the more valuable trajectory patterns for users.

From the experiment result, we can conclude the algorithm with the adaptable value measuring model can mined out the trajectory patterns with larger applying value. What's more, it can adapt to different value measuring defined by uses.

Figure 4(a) reports the execution times obtained by using three algorithms over datasets with increasing number of input trajectories. The curves show an almost linear scalability for the algorithms. And all the algorithms have very similar execution time when the number of trajectories is small. However, with the number of trajectories increasing, the AVGPM+ generate higher efficiency, even it considers more influential factors than CuTS. Since with the statistics created parallelly in the filter step, we only need a smaller cost to find the group patterns from these statistics and filtered matrix.

Figure 4(b) shows the speedups of the algorithm AVGPM using different number of trajectory and computing sites respectively. As shown, with the increase of data, the growth rate slows down. This is mainly because with the site increase, the data synchronization between computer nodes also increase. In addition, we can see from Fig. 4(b), our algorithm demonstrate greater advantages and higher performance in dealing with large-scale data sets.

6   Conclusion

Different from the traditional trajectory pattern mining just discovery the satisfied patterns, in this paper, we investigate to mine the top valuable trajectory patterns for users. In this study, we formally proposes the adaptable value measuring model based on the correlation of influential factors in users' requirements to estimate the value of group trajectory patterns. To reduce redundant information, we propose a new group trajectory pattern mining approach based on the measuring model to efficiently discover valuable patterns. The AVGPM+ approach accelerates the process of trajectory mining by efficiently reducing the amounts of data that need further processing and avoiding the redundant checks. In addition, we proposed a configurable mobile trajectory building technique to reduce the uncertainty of trajectories. At last we validate the effectiveness and efficiency of our proposals by conducting extensive experiments based on a real and large scale mobile data.

In the future, we plan to automatically generate the multi-factors correlation function through learning users' requirement samples and provide some correlation function as templates for users.

References

1. Zheng, K., Zheng, Y., Yuan, N.J., Shang, S., Zhou, X.: Online discovery of gathering patterns over trajectories. IEEE Trans. Knowl. Data Eng. **26**(8), 1974–1988 (2014)
2. Lee, J.G., Han, J., Whang, K.Y.: Trajectory clustering: a partition-and-group framework. In: Proceedings of the 2007 ACM SIGMOD International Conference on Management of Data, pp. 593–604. ACM (2007)
3. Li, Z., Wang, J., Han, J.: Mining event periodicity from incomplete observations. In: Proceedings of the 18th ACM SIGKDD International Conference on Knowledge Discovery and Data Mining, pp. 444–452. ACM (2012)
4. Wei, L.Y., Zheng, Y., Peng, W.C.: Constructing popular routes from uncertain trajectories. In: Proceedings of the 18th ACM SIGKDD International Conference on Knowledge Discovery and Data Mining, pp. 195–203. ACM (2012)
5. Zheng, Y.: Trajectory data mining: an overview. ACM Trans. Intell. Syst. Technol. (TIST) **6**(3), 29 (2015)
6. Jeung, H., Yiu, M.L., Zhou, X., Jensen, C.S., Shen, H.T.: Discovery of convoys in trajectory databases. Proc. VLDB Endowment **1**(1), 1068–1080 (2008)
7. Zheng, Y., Zhang, L., Xie, X., Ma, W.Y.: Mining interesting locations and travel sequences from gps trajectories. In: Proceedings of the 18th International Conference on World Wide Web, pp. 791–800. ACM (2009)
8. Yin, Z., Cao, L., Han, J., Luo, J., Huang, T.S.: Diversified trajectory pattern ranking in geo-tagged social media. In: SDM, SIAM pp. 980–991 (2011)
9. Hsieh, H.P., Li, C.T., Lin, S.D.: Exploiting large-scale check-in data to recommend time-sensitive routes. In: Proceedings of the ACM SIGKDD International Workshop on Urban Computing, pp. 55–62. ACM (2012)
10. Pelekis, N., Kopanakis, I., Kotsifakos, E.E., Frentzos, E., Theodoridis, Y.: Clustering trajectories of moving objects in an uncertain world. In: Ninth IEEE International Conference on Data Mining, ICDM 2009, pp. 417–427. IEEE (2009)
11. Zheng, K., Zheng, Y., Xie, X., Zhou, X.: Reducing uncertainty of low-sampling-rate trajectories. In: 2012 IEEE 28th International Conference on Data Engineering (ICDE). pp. 1144–1155. IEEE (2012)
12. Ester, M., Kriegel, H.P., Sander, J., Xu, X.: A density-based algorithm for discovering clusters in large spatial databases with noise. Kdd **96**, 226–231 (1996)
13. Kalnis, P., Mamoulis, N., Bakiras, S.: On discovering moving clusters in spatio-temporal data. In: Medeiros, C.B., Egenhofer, M., Bertino, E. (eds.) SSTD 2005. LNCS, vol. 3633, pp. 364–381. Springer, Heidelberg (2005)

DualPOS: A Semi-supervised Attribute Selection Approach for Symbolic Data Based on Rough Set Theory

Jianhua Dai[1,2(✉)], Huifeng Han[2], Hu Hu[2], Qinghua Hu[1], Jinghong Zhang[2], and Wentao Wang[2]

[1] School of Computer Science and Technology, Tianjin University,
Tianjin 300350, China
david.joshua@qq.com
[2] College of Computer Science and Technology, Zhejiang University,
Hangzhou 310027, China

Abstract. Rough set theory, supplying an effective model for representation of uncertain knowledge, has been widely used in knowledge engineering and data mining. Especially, rough set theory has been used as an attribute selection method with much success. However, current rough set approaches for attribute reduction are unsuitable for semi-supervised learning as no enough labeled data can guarantee to calculate the dependency degree. We propose a new attribute selection strategy based on rough sets, called DualPOS. It provides mutual function mechanism of multi-attributes, and generates the most consistent one as a candidate. Experiments are carried out to test the performances of classification and clustering of the proposed algorithm. The results show that DualPOS is valid for attribute selection in semi-supervised learning.

Keywords: Semi-supervised · Attribute selection · Rough set theory · Dual-dependence degree · DualPOS

1 Introduction

As data mining aims to address larger, more complex tasks, the problem of focusing on the most relevant information in a potentially overwhelming quantity of data has become increasingly important [1]. In such cases, attribute selection or feature selection is necessary.

Attribute selection has become the focus of much research in areas of applications, including text processing of internet documents, gene expression array analysis, and big data analysis. There are many potential benefits of attribute selection: facilitating data visualization and data understanding, reducing the measurement and storage requirements, reducing training and utilization times, defying the curse of dimensionality to improve prediction performance [2]. It is a process of choosing a subset of attributes from the original set of attributes forming patterns in a given dataset. The subset should be necessary and sufficient

© Springer International Publishing Switzerland 2016
B. Cui et al. (Eds.): WAIM 2016, Part II, LNCS 9659, pp. 392–402, 2016.
DOI: 10.1007/978-3-319-39958-4_31

to describe target concepts, retaining a suitably high accuracy in representing the original attributes. The importance of attribute selection is to reduce the problem size and resulting search space for learning algorithms [3].

Due to the abundance of noisy, irrelevant or misleading features, the ability to handle imprecise and inconsistent information in real world problems has become one of the most important requirements for attribute selection. Rough set theory [4,5] can handle uncertainty and vagueness, discover patterns in inconsistent data. Rough set theory has been an useful attribute selection method (also called attribute reduction). The rough set approach to attribute selection is to select a subset of attributes (or features), which can predict the decision concepts as well as the original attribute set. The optimal criterion for rough set attribute selection is to find shortest or minimal reducts while obtaining high quality classifiers based on the selected attributes [6]. Usually, there are a few reducts. But it has been shown that the problem of minimal reduct generation is NP-hard and the problem of generation of all reducts is exponential [7]. Therefore, heuristic approaches have to be considered.

In general, there are two kinds of rough set methods for attribute selection, hill-climbing (or greedy) methods and stochastic methods [8]. The hill-climbing approaches usually employ rough set attribute significance as heuristic knowledge. For example Hu et al. [9] gave a reduction algorithm using the positive region-based attribute significance as the guiding heuristic information. Hu [10] and Susmaga [11] et al. considered both indiscernibility and discernibility relations in attribute reduction. Dai et al. [12–15] studied attribute reduction methods for incomplete information and numerical information. Some researchers also use stochastic methods for rough set attribute selection [16–19]. However, most of these studies focus on labeled information. In many applications, only partly labeled information is available. Semi-supervised learning can readily use available unlabeled data to improve supervised learning tasks when the labeled data is scarce or expensive [20]. In this circumstance, the above reduction methods are no longer suitable.

In this article we address a new attribute selection algorithm based on rough set theory for semi-supervised learning, named Dual POSitive region method (denoted as DualPOS). It is based on the degree of dual-dependency, which can be described as a measure of interrelation among attributes. Meanwhile, we propose the consistency to a class label as a criterion of selecting a candidate attribute. Our goal is to utilize as much as information from datasets, especially under the condition of a lack of class labels.

The rest of the paper is organized as follows. In Sect. 2, we review some preliminary notions for this task. In Sect. 3, we propose the concept of dual-dependency. Based on the proposed dual-dependency, relevance between conditional attributes and decision, as well as redundancy between two conditional attributes, are defined. Consequently, a new attribute selection algorithm based on rough set theory for semi-supervised learning, named Dual POSitive region method is constructed. Experiments are conducted in Sect. 4. Section 5 concludes this paper.

2   Preliminaries

In this section, we will review some basic notions in rough set theory which can be found in [21–24].

In rough set theory, these objects are described in accordance with the format of the data table, in which rows are considered as objects and columns as attributes.

An information system (information table) is a quadruple $IS \doteq < U, A, V, f >$, where U is a non-empty finite set of objects called the universe; A is a non-empty finite set of attributes; V is the union of attribute domains, $V = \bigcup_{a \in A} V_a$, where V_a is the value set of attribute a, called the domain of a; $f : U \times A \rightarrow V$ is an information function which assigns particular values from domains of attributes to objects such as $\forall a \in A, x \in U, f(a, x) \in V_a$, where $f(a, x)$ denotes the value of attribute a for object x.

A decision system (decision table) can be defined as a quadruple $DS =< U, C \cup D, V, f >$, where C is the set of conditional attributes, while D is the set of decision attributes, and $C \cap D = \emptyset$. Each attribute $a \in C \cup D$ defines an information function $f : U \times C \cup D \rightarrow V$, where V is the union of both the conditional attributes domain and the decision attributes domain, $V = V_C \cup V_D = \{V_c | \forall c \in C\} \cup \{V_d | \forall d \in D\}$.

Unlike an information system, a decision system contains the decision attribute set D. Hence, a labeled data set is represented as a decision system. It should be also noticed that the attributes are supposed to be symbolic.

The concept of indiscernibility is central to rough set theory. Given a knowledge representation system $S =< U, A, V, f >$, for every set of attributes $B \subset A$, there is an associated equivalence relation $IND(B)$:

$$IND(B) = \{(x, y) | \forall a \in B, f(a, x) = f(a, y)\}. \tag{1}$$

$IND(B)$ is called the B-indiscernibility relation. If $(x, y) \in IND(B)$, then objects x and y are indiscernible from each other by attributes from B. An equivalence class of $IND(B)$ containing x is defined as $[x]_{IND(B)} = [x]_B = \{y | \forall (x, y) \in IND(B)\}$. The family of all equivalence classes of $IND(B)$ will be denoted by $U/IND(B)$, or simply U/B.

Let X be a subset of U, $\underline{B}(X)$ and $\overline{B}(X)$ are called B-lower approximation and B-upper approximation with respect to B.

$$\underline{B}(X) = \{x | \forall x \in U, [x]_B \subseteq X\}. \tag{2}$$

$$\overline{B}(X) = \{x | \forall x \in U, [x]_B \cap X \neq \emptyset\}. \tag{3}$$

Every subset defined through upper and lower approximations is known as Rough Set, such as the order pair $< \underline{B}, \overline{B} >$ is called a rough set of X with respect to $IND(B)$.

Based on lower approximation and upper approximation, the boundary region can be denoted as $BN_B(X) = \overline{B}(X) - \underline{B}(X)$. Additionally, $POS_B(X) = \underline{B}(X)$ is called the B-positive region of X.

The $\underline{B}(X)$ and $POS_B(X)$ are the set of these objects, which certainly belongs to the set X. The $\overline{B}(X)$ is the set of the objects, which possibly belongs to the set X. $BN_B(X)$ is an undecidable area of the universe, that is, none of the objects belonging to the boundary region can with certainty be classified into X or $-X$, where $-X$ denotes the complement of X in U, as far as the attributes B are considered.

In a decision system, the attribute reduction algorithm is traditionally based on the degree of dependency, called SinglePOS for convenience in this paper. Thus, we recall the degree of the dependency and the significance at first.

Given a decision system $DS =< U, C \cup D, V, f >$, the decision set D determines a partition $U/D = \{X_1, X_2, \ldots, X_r\}$ in U. For any $B \subseteq C$, the set $\underline{B}(X_1) \cup \underline{B}(X_2) \cup \ldots \cup \underline{B}(X_r)$ is called the B-positive region of DS:

$$POS_B(D) = \bigcup_{X \in U/D} \underline{B}(X). \tag{4}$$

It is a positive region of the partition U/D with respect to B, as well as the set of all elements of U that can be uniquely classified to blocks of the partition U/D, by means of B.

We say that the set of decision attributes D depends on B in a degree $k(0 \leq k \leq 1)$, denoted $B \Rightarrow_k D$,

$$k = \gamma(B, D) = \frac{|POS_B(D)|}{|U|} = \frac{|\bigcup_{X \in U/D} \underline{B}(X)|}{|U|}. \tag{5}$$

Obviously, if $k = 1$, D depends totally on B, and if $k < 1$, D depends partially (in a degree k) on B. The coefficient k expresses the radio of all elements of the universe, which can be properly classified to blocks of the partition U/D, employing attributes B and will be called the degree of the dependency.

By calculating the change in dependency when an attribute is added to the set of considered conditional attributes, a measure of the significance of the attribute can be obtained. The higher the change in dependency, the more significant the attribute is. If the significance is 0, then the attribute is dispensable. More formally, given $DS =< U, C \cup D, V, f >$, $B \subseteq C$, $a \in C$, and $a \notin B$, the degree of the significance of a to B with respect to D is

$$SIG(a, B, D) = \gamma(B \cup \{a\}, D) - \gamma(B, D) = \frac{|POS_{B \cup \{a\}}(D) - POS_B(D)|}{|U|}. \tag{6}$$

3   A Dual-dependency-based Reduction Algorithm (DualPOS)

In this section, we will propose a new reduction algorithm of rough-set theory in order to solve semi-supervised attribute selection problem. It is based on the degree of dual-dependency, called the DualPOS algorithm.

At first we set an expansible conception of the dependency degree, that is, dual-dependency degree. Given an information system $=< U, A \cup D, V, f >$, for any subset of attributes $A_i, A_j \subseteq A$, we define the dual-dependency degree between A_i, A_j as follows:

$$dualDep(A_i, A_j) = \frac{|POS_{A_i}(A_j)| + |POS_{A_j}(A_i)|}{2|U|} \qquad (7)$$

It is obvious that the dual-dependency performs better than the original one when the interrelation considered. Moreover in a decision system, it can generate a measure of inter-dependency among different conditional attributes, or between conditional attributes and the decision attribute. Then we take into account the relative redundance of an attribute to others and make a decline of the influence of the missing class labels. Those will be applied as a selection criteria in the DualPOS.

A decision system $semiDS =< U, A \cup D, V, f >$ is called a partially labeled decision system, if $U = U^L \cup U^U$, where U^L means the labeled object set and U^U means the unlabeled object set. In formal, let $semiDS =< U, C \cup D, V, f >$, $A_i, A_j \subseteq A$, and usually the set of decision attributes D will be transformed into the single attribute d. Then we can define the relavance between conditional attributes A_i and the decision attribute d as

$$Rel(A_i, d) = dualDep^L(A_i, d) \qquad (8)$$

where, $dualDep^L(A_i, d)$ means dual-dependency based on labeled objects.

The redundancy between conditional attributes A_i and A_j is denoted by

$$Red(A_i, A_j) = dualDep(A_i, A_j). \qquad (9)$$

So whether a candidate attribute a should be put into a reduct S or not depends on the comparison between the $Rel(S \cup \{a\}, d)$ and $Red(S, a)$. If the redundancy (between a and the selected attribute set S) is smaller than the relevance (between $S \cup \{a\}$ and the decision attribute d), a could be a member of the new reduct, otherwise it is not.

A conditional attribute and the decision attribute are consistent to a certain extent , if they mostly map to a similar value. It means the decision attribute is possibly derived from this attribute, which should be a candidate attribute of a reduct.

We can calculate the consistence in following way. Given a dataset $X = X^L \cup X^U$, for the labeled data X^L, the consistent rate of an attribute a_i to the decision attribute is

$$Con(a_i) = \frac{\sum_{x \in X^L} \sum_{y \in X^L} Cons_i(x, y)}{|X^L| \cdot |X^L|}, \qquad (10)$$

where

$$Cons_i(x, y) = \begin{cases} 1 & a_i(x) = a_i(y) \oplus d(x) = d(y) \\ 0 & \text{otherwise} \end{cases}$$

is a measure of consistence between the attribute a_i and the class label, in regard of two objects. $a_i(x) = a_i(y)$ represents x and y get the same value in a_i, $d(x) = d(y)$ represents they both have an identical label, and $\oplus$ is an exclusive-OR mark.

Intuitively, the more consistent a_i and d are, the higher $Con(a_i)$ is. Hence, the attribute with the highest consistence will be a candidate of a reduct.

The dual-positive region based attribute reduction algorithm (denoted as DualPOS), as shown in Algorithm 1, sorts the consistence of each attribute and selects the highest one as a candidate. The attribute should be added, if the relevance of the former is not smaller than the redundancy of the latter. The final reduct can be generated when every attribute is considered.

Algorithm 1. DualPOS algorithm

Input: A dataset $X = X^U \cup X^L$, the set of attribute A
Output: A selected feature S
 1: $S = \emptyset, Rest = A$
 2: $S = S \cup argmax_{a_i \in A} Cons(a_i)$
 3: $Rest = Rest - S$
 4: **while** $Rest \neq \emptyset$ **do**
 5:    $a_i = argmax_{a_i \in Rest} Cons(ai)$
 6:    **if** $Rel(S \cup \{a_i\}, d) \geq Red(S, a_i)$ **then**
 7:       $S = S \cup a_i$
 8:    **end if**
 9:    $Rest = Rest - a_i$
10: **end while**
11: **return** S

Through the DualPOS, we can achieve the attribute selection result of those datasets with part labels.

4   Experiment and Analysis

4.1   Dataset

The datasets for our experiments is from UCI (University of California Irvine). And all the objects are labeled with the class. More detail can be found in Table 1. *#Instances*, *#Attribute* and *#Classes* represent the amount of the objects, attributes and classes, respectively. The continuous attributes in these datasets are discretized by supervised MDL (Minimum Description Length) [25].

4.2   Evaluation Standards

The selected attributes will be evaluated by make classifications and clusterings. As for classification, we compare the accuracy (ACC) of both all features and the

Table 1. Dataset information

DataSet	#Instances	#Attribute	#Classes	Abstract
Colic	368	22	2	Horse Colic Data Set; Well documented attributes
Credit	690	15	2	This data concerns credit card applications; Good mix of attributes
Diabetes	768	8	2	From National Institute of Diabetes and Digestive and Kidney Diseases
Heart-cleveland	303	13	5	Heart Disease Data Set
Heart-hungarian	294	13	5	Heart Disease Data Set
Heart-statlog	270	13	2	Heart Disease Data Set
Hepatitis	155	19	2	From G.Gong: CMU; Includes cost data (donated by Peter Turney)
Ionosphere	351	34	2	Classification of radar returns from the ionosphere
Musk2	707	166	2	Musk (Version 2) Data Set
Promoters	106	57	2	E. Coli promoter gene sequences (DNA)
SPECT	374	21	2	Single Proton Emission Computed Tomography (SPECT) images
Voting	435	16	2	1984 United Stated Congressional Voting Records
WDBC	569	30	2	Breast Cancer Wisconsin (Diagnostic) Data Set

selected attributes, while the accuracy (ACC) as well as the normalized mutual information (NMI) for clustering [26].

Let x_i stand for an object, r_i be the result label of a classification or clustering algorithm, and s_i be the real label, then the ACC can be calculated by

$$ACC = \frac{\sum_i^n \delta(s_i, map(r_i))}{n}, \tag{11}$$

where n is the amount of the objects of a dataset, $\delta(x,y) = 1$ if $x = y$, $\delta(x,y) = 0$ otherwise, $map(r_i)$ is a mapping function of which the best is the Kuhn-Munkres Algorithm [27].

Let C represent the real partition of a dataset, C' be the result by means of a clustering algorithm, their mutual information (MI) can be defined as

$$MI(C, C') = \sum_{c_i \in C, c'_j \in C'} p(c_i, c'_j) \log_2 \frac{p(c_i, c'_j)}{p(c_i) \cdot p(c'_j)}, \tag{12}$$

where $p(c_i)$ is the probability of a object selected by c_i, while $p(c'_i)$ is by c'_j and $p(c_i, c'_j)$ is by c_i and c'_j simultaneously. Then the normalized mutual information (NMI) of C and C' can be denoted by

$$NMI(C, C') = \frac{MI(C, C')}{\max(H(C), H(C'))}, \tag{13}$$

in which $H(C)$ and $H(C')$ separately express the entropy of C and C'. It is clear that $NMI \in [0, 1]$. The two partitions are the same if $NMI = 1$, otherwise they are independent.

Table 2. The performance of classification

Datasets	#Features		NBC		C4.5		JRip		PART		CART	
	DualPOS	FullFeature	DualPOS	FullFeature	DualPOS	FullFeature	DualPOS	FullFeature	DualPOS	FullFeature	DualPOS	FullFeature
			ACC±Std	ACC±Std	ACC±Std	ACC±Std	ACC±Std	ACC±Std	ACC±Std	ACC±Std	ACC±Std	ACC±Std
Colic	6	22	81.47±0.23	81.47±0.35	80.49±0.62	84.84±0.35	81.25±0.72	84.35±0.53	78.97±0.89	80.92±1.09	80.49±0.52	84.46±0.59
Credit	10	15	87.01±0.17	86.17±0.26	86.99±0.47	86.61±0.51	85.71±0.58	86.58±0.49	85.51±0.61	85.88±0.67	84.87±0.24	84.61±0.44
Diabetes	5	8	77.89±0.25	77.84±0.11	75.70±0.77	77.50±0.92	77.01±0.47	77.32±0.56	75.52±0.62	77.06±0.76	74.77±0.79	76.74±0.80
Heart-cleveland	8	13	81.78±0.36	83.96±0.18	78.42±1.86	78.02±1.29	81.78±1.54	81.39±1.13	80.99±1.33	80.86±1.26	81.06±1.11	80.99±1.79
Heart-hungarian	7	13	83.88±0.19	84.69±0.24	79.86±1.39	80.48±0.85	80.34±1.06	79.73±0.89	80.20±1.21	81.02±0.91	79.18±1.30	78.98±1.55
Heart-statlog	8	13	81.33±0.33	83.78±0.41	82.44±0.89	82.07±1.33	83.78±1.40	83.78±1.40	84.59±2.00	83.63±1.86	81.70±1.75	82.59±0.79
Hepatitis	10	19	85.16±0.00	84.00±0.54	80.52±1.40	80.26±1.91	79.23±2.16	80.00±2.28	81.68±1.34	82.32±0.87	79.74±0.98	79.61±0.35
Ionosphere	9	34	91.51±0.13	90.77±0.38	89.34±0.59	89.74±0.67	89.91±0.72	91.51±1.16	89.86±0.38	89.91±0.32	87.75±0.94	90.31±0.83
Musk2	11	166	87.81±0.25	85.26±0.27	89.84±0.49	90.83±1.00	89.28±0.46	90.01±1.05	90.69±0.42	90.69±0.46	89.70±0.65	90.27±0.50
Promoters	4	57	90.57±0.94	90.38±0.79	81.89±1.40	79.81±1.58	81.78±1.54	81.70±2.27	81.13±2.58	86.04±1.81	83.02±1.49	74.34±2.86
SPECT	7	21	77.59±0.55	75.40±0.46	76.74±1.25	82.89±0.60	75.51±0.82	79.30±1.52	77.97±1.90	83.80±2.89	77.59±0.51	80.64±1.75
Voting	13	16	90.16±0.19	90.30±0.19	96.32±0.28	96.37±0.34	96.00±0.31	95.45±0.44	95.77±0.38	95.84±0.47	96.00±0.53	95.54±0.13
WDBC	13	30	95.89±0.44	95.85±0.10	94.31±0.24	95.85±0.10	94.73±0.51	96.34±0.19	95.82±0.47	95.99±0.23	94.45±0.55	94.13±0.92
Average	8.54	32.85	85.54	85.37	84.07	85.02	84.32	85.15	84.89	85.59	83.87	84.09

Table 3. The performance of clustering

Datasets	KModes		EM		Cobweb		FarthestFirst		CLOPE	
	DualPOS	FullFeature	DualPOS	FullFeature	DualPOS	FullFeature	DualPOS	FullFeature	DualPOS	FullFeature
	ACC±Std	ACC±Std	ACC±Std	ACC±Std	ACC±Std	ACC±Std	ACC±Std	ACC±Std	ACC	ACC
Colic	62.23±10.85	58.10±8.19	77.45±0.00	66.85±0.00	72.61±3.87	66.25±2.75	63.26±8.97	57.12±8.89	79.08	74.18
Credit	72.55±10.48	75.45±11.76	86.67±0.00	73.33±0.00	83.86±0.91	71.28±2.49	65.22±5.66	60.03±8.59	80.58	79.71
Diabetes	60.05±7.24	60.05±7.24	66.41±0.00	66.41±0.00	65.73±1.19	65.73±1.19	57.14±8.83	57.14±8.83	76.30	67.06
Heart-cleveland	79.27±2.83	79.41±2.74	81.52±1.38	80.26±1.61	78.35±1.06	80.46±2.19	77.10±2.32	75.38±5.60	83.83	58.75
Heart-hungarian	82.31±1.42	82.31±1.42	84.08±0.15	84.08±0.15	81.70±1.69	81.70±1.69	77.35±5.61	77.35±5.61	80.27	77.21
Heart-statlog	82.67±0.41	82.67±0.41	80.37±0.00	80.74±0.00	80.89±1.25	79.41±3.24	68.89±12.17	65.63±14.34	72.22	74.07
Hepatitis	73.03±5.19	69.81±11.39	83.23±0.00	75.23±0.58	80.13±1.06	79.48±0.29	79.61±5.19	71.87±11.81	85.16	83.87
Ionosphere	87.35±1.02	88.77±0.96	90.03±0.00	89.17±0.00	89.12±1.22	79.43±2.87	85.36±3.11	78.35±2.38	95.73	95.73
Musk2	73.86±6.07	61.67±11.30	82.60±0.00	53.61±0.00	82.18±0.00	82.18±0.00	77.54±13.20	57.00±9.66	86.28	87.98
Promoters	62.08±7.53	61.32±10.05	61.89±5.88	57.17±5.99	73.96±16.70	69.43±3.10	62.64±3.10	58.11±6.21	100.00	100.00
SPECT	60.64±9.50	62.57±5.09	73.80±0.00	54.55±0.00	67.81±4.20	61.28±4.10	60.32±8.91	62.14±6.42	63.90	58.29
Voting	87.22±0.50	87.03±0.72	87.82±0.00	87.82±0.00	81.98±6.78	82.90±3.44	86.90±2.17	86.11±2.63	83.91	82.76
WDBC	94.48±0.10	92.79±0.48	95.25±0.00	94.55±0.00	92.76±1.15	83.23±2.46	84.08±13.55	88.51±8.37	97.19	89.63
Average	75.21	74.00	80.85	74.14	79.31	75.60	72.72	68.83	83.42	79.17

Table 4. The normalized mutual information of clustering

Datasets	KModes		EM		Cobweb		FarthestFirst		CLOPE	
	DualPOS	FullFeature	DualPOS	FullFeature	DualPOS	FullFeature	DualPOS	FullFeature	DualPOS	FullFeature
	NMI±Std	NMI±Std	NMI±Std	NMI±Std	NMI±Std	NMI±Std	NMI±Std	NMI±Std	NMI	NMI
Colic	0.110±0.073	0.075±0.049	0.205±0.000	0.109±0.000	0.149±0.042	0.084±0.043	0.043±0.085	0.063±0.063	0.096	0.082
Credit	0.197±0.113	0.234±0.131	0.429±0.000	0.158±0.000	0.351±0.017	0.135±0.029	0.082±0.058	0.051±0.068	0.087	0.098
Diabetes	0.069±0.022	0.069±0.022	0.089±0.000	0.089±0.000	0.049±0.011	0.049±0.011	0.022±0.018	0.022±0.018	0.062	0.053
Heart-cleveland	0.173±0.023	0.156±0.017	0.199±0.008	0.191±0.001	0.114±0.011	0.132±0.019	0.127±0.026	0.130±0.052	0.126	0.046
Heart-hungarian	0.157±0.013	0.157±0.013	0.181±0.010	0.181±0.010	0.130±0.012	0.130±0.012	0.119±0.038	0.119±0.038	0.140	0.139
Heart-statlog	0.343±0.022	0.343±0.022	0.281±0.000	0.287±0.000	0.269±0.036	0.252±0.055	0.147±0.154	0.136±0.164	0.129	0.167
Hepatitis	0.150±0.016	0.159±0.052	0.300±0.000	0.187±0.005	0.175±0.059	0.143±0.042	0.103±0.079	0.077±0.109	0.109	0.175
Ionosphere	0.468±0.021	0.478±0.044	0.507±0.000	0.475±0.000	0.474±0.042	0.272±0.054	0.393±0.063	0.236±0.041	0.173	0.168
Musk2	0.139±0.063	0.051±0.033	0.206±0.000	0.026±0.000	0.093±0.023	0.039±0.027	0.168±0.084	0.040±0.023	0.078	0.092
Promoters	0.066±0.082	0.072±0.092	0.051±0.053	0.024±0.026	0.251±0.256	0.110±0.032	0.053±0.023	0.032±0.030	0.151	0.150
SPECT	0.055±0.084	0.044±0.032	0.223±0.000	0.006±0.000	0.087±0.040	0.036±0.035	0.048±0.058	0.040±0.030	0.073	0.043
Voting	0.474±0.016	0.465±0.034	0.486±0.000	0.486±0.000	0.375±0.098	0.376±0.068	0.460±0.047	0.439±0.074	0.320	0.276
WDBC	0.678±0.004	0.611±0.017	0.711±0.000	0.680±0.000	0.614±0.047	0.385±0.063	0.429±0.199	0.510±0.192	0.216	0.258
Average	0.237	0.224	0.298	0.223	0.241	0.165	0.169	0.146	0.135	0.134

4.3 Results and Analysis

Since our dataset are all labeled, partially labeled data can be simulated by removing some labels randomly. We set a percent of data labeled (Lpercent), and assign it from 0 to 0.5, with step size 0.05. The change of amount and the

Table 5. The time of attribute selection and classification

Datasets	FeatureSelection	NBC		C4.5		JRip		PART		CART	
	DualPOS	DualPOS	FullFeature	DualPOS	FullFeature	DualPOS	FullFeature	DualPOS	FullFeature	DualPOS	FullFeature
Colic	0.093	0.003	0.011	0.009	0.012	0.016	0.040	0.009	0.025	0.240	0.691
Credit	0.140	0.003	0.008	0.012	0.017	0.041	0.056	0.025	0.033	0.643	0.775
Diabetes	0.047	0.003	0.003	0.009	0.009	0.041	0.055	0.012	0.022	0.296	0.390
Heart-cleveland	0.031	0.003	0.002	0.003	0.008	0.019	0.023	0.012	0.014	0.259	0.353
Heart-hungarian	0.016	0.000	0.003	0.006	0.005	0.012	0.017	0.009	0.012	0.187	0.284
Heart-statlog	0.015	0.003	0.002	0.003	0.006	0.016	0.019	0.009	0.009	0.140	0.183
Hepatitis	0.016	0.003	0.000	0.000	0.005	0.009	0.009	0.003	0.009	0.078	0.128
Ionosphere	0.125	0.003	0.003	0.006	0.008	0.028	0.055	0.009	0.019	0.384	1.250
Musk2	2.824	0.003	0.033	0.013	0.117	0.059	0.464	0.025	0.382	0.618	9.185
Promoters	0.031	0.000	0.003	0.003	0.006	0.006	0.019	0.003	0.014	0.034	0.373
SPECT	0.046	0.000	0.003	0.006	0.014	0.025	0.059	0.013	0.033	0.165	0.479
Voting	0.063	0.003	0.002	0.003	0.006	0.016	0.014	0.006	0.009	0.206	0.254
WDBC	0.285	0.003	0.006	0.006	0.011	0.044	0.062	0.011	0.020	0.543	1.030
Average	0.287	0.002	0.006	0.006	0.017	0.026	0.069	0.011	0.046	0.292	1.183

Table 6. The time of attribute selection and clustering

Datasets	FeatureSelection	KModes		EM		Cobweb		FarthestFirst		CLOPE	
	DualPOS	DualPOS	FullFeature	DualPOS	FullFeature	DualPOS	FullFeature	DualPOS	FullFeature	DualPOS	FullFeature
Colic	0.093	0.003	0.014	0.040	0.081	0.072	0.115	0.001	0.000	0.219	0.213
Credit	0.140	0.004	0.007	0.085	0.085	0.156	0.215	0.001	0.000	0.297	0.335
Diabetes	0.047	0.007	0.002	0.069	0.084	0.072	0.079	0.000	0.000	0.172	0.101
Heart-cleveland	0.031	0.001	0.002	0.117	0.206	0.024	0.048	0.000	0.000	0.062	0.048
Heart-hungarian	0.016	0.000	0.003	0.106	0.149	0.031	0.034	0.000	0.000	0.016	0.038
Heart-statlog	0.015	0.003	0.002	0.022	0.029	0.023	0.026	0.000	0.000	0.048	0.016
Hepatitis	0.016	0.000	0.002	0.012	0.021	0.013	0.019	0.000	0.000	0.000	0.022
Ionosphere	0.125	0.004	0.008	0.042	0.110	0.059	0.220	0.000	0.000	0.499	1.948
Musk2	2.824	0.013	0.072	0.203	0.818	0.277	1.753	0.004	0.007	0.421	1.771
Promoters	0.031	0.003	0.007	0.066	0.103	0.067	0.110	0.000	0.000	0.172	0.344
SPECT	0.046	0.004	0.004	0.051	0.182	0.041	0.080	0.000	0.000	0.031	0.032
Voting	0.063	0.003	0.004	0.042	0.043	0.069	0.072	0.000	0.000	0.031	0.047
WDBC	0.285	0.006	0.011	0.070	0.157	0.166	0.297	0.001	0.001	0.793	0.873
Average	0.287	0.004	0.010	0.071	0.159	0.082	0.236	0.000	0.001	0.212	0.445

accuracy of the selected attributes are obtained by 10-folds cross validation with 10 times in CART (Classification And Regression Tree) algorithm.

When Lpercent = 50 %, we will compare the evaluation effect of classification and clustering with the attribute subset by DualPOS and universal set. Classification algorithms used contain NaiveBayes Classifier (NBC), C4.5, JRip, PART and CART. For clustering, K-Modes, EM, Cobweb, FarthestFirst and CLOPE are employed.

The experimental results are illustrated in Tables 2, 3 and 4. The accuracies are very close, which implies the selected features are able to remain most original information. On clustering, the performance of selected attributes by DualPOS is obvious better.

As Tables 5 and 6 show, the time costs of both classification and clustering in DualPOS are cut down, which indicates that our attribute selection of DualPOS is not only effective but also efficient.

5 Conclusion

In this paper, we have studied the attribute selection problem for partially labeled data based on rough set theory since existing studies are unsuitable. A concept of dual-dependency degree is introduced to measure the relation and redundancy, which is an extend from the well used degree of dependency. Besides, the definition of consistency is presented to evaluate a candidate attribute. Consequently, a new attribute selection algorithm based on rough set theory for semi-supervised learning, named dual positive region method is constructed. Experimental results demonstrate that the proposed attribute selection method is effective. The algorithm is able to handle attribute selection in semi-supervised learning.

Acknowledgements. This work was partially supported by the National Natural Science Foundation of China (No. 61473259, No. 61070074, No. 60703038), the Zhejiang Provincial Natural Science Foundation (No. Y14F020118), the National Science & Technology Support Program of China (2015BAK26B00, 2015BAK26B02) and the PEIYANG Young Scholars Program of Tianjin University (2016XRX-0001).

References

1. Blum, A.L., Langley, P.: Selection of relevant features and examples in machine learning. Artif. Intell. **97**(1C2), 245–271 (1997)
2. Guyon, I., Elisseeff, A.: An introduction to variable and feature selection. J. Mach. Learn. Res. **3**, 1157–1182 (2003)
3. Bae, C., Yeh, W.C., Chung, Y.Y., Liu, S.L.: Feature selection with intelligent dynamic swarm and rough set. Expert Syst. Appl. **37**(10), 7026–7032 (2010)
4. Pawlak, Z.: Rough sets. Int. J. Comput. Inform. Sci. **11**(5), 341–356 (1982)
5. Pawlak, Z.: Rough sets and fuzzy sets. Fuzzy Sets Syst. **17**(1), 99–102 (1985)
6. Revett, K., Iantovics, B.: A survey of electronic fetal monitoring: a computational perspective. Stud. Comput. Intell. **486**, 135–141 (2014)
7. Skowron, A., Rauszer, C.: The discernibility matrices and functions in information systems. In: Slowiński, R. (ed.) Intelligent Decision Support. Theory and Decision Library, vol. 11, pp. 331–362. Springer, Netherlands (1992)
8. Vafaie, H., Imam, I.F.: Feature selection methods: genetic algorithms vs. greedy-like search. In: Proceedings of the International Conference on Fuzzy and Intelligent Control Systems, pp. 39–43 (1994)
9. Hu, X., Cercone, N.: Learning in relational databases: a rough set approach. Comput. Intell. **11**(2), 323–338 (1995)
10. Hu, X.: Knowledge discovery in databases: an attribute-oriented rough set approach. Ph.D. thesis, Citeseer (1995)
11. Susmaga, R.: Reducts and constructs in attribute reduction. Fundamenta Informaticae **61**(2), 159–181 (2004)
12. Dai, J., Wang, W., Xu, Q.: An uncertainty measure for incomplete decision tables and its applications. IEEE Trans. Cybern. **43**(4), 1277–1289 (2013)
13. Dai, J., Wang, W., Tian, H., Liu, L.: Attribute selection based on a new conditional entropy for incomplete decision systems. Knowl.-Based Syst. **39**, 207–213 (2013)
14. Dai, J., Xu, Q., Wang, W., Tian, H.: Conditional entropy for incomplete decision systems and its application in data mining. Int. J. Gen. Syst. **41**(7), 713–728 (2012)
15. Dai, J., Xu, Q.: Attribute selection based on information gain ratio in fuzzy rough set theory with application to tumor classification. Appl. Soft Comput. **13**(1), 211–221 (2013)

16. Dai, J., Li, Y.X., Liu, Q.: Hybrid genetic algorithm for reduct of attributes in decision system based on rough set theory. Wuhan Univ. J. Nat. Sci. **7**(3), 285–289 (2002)
17. Dai, J., Chen, W., Gu, H., Pan, Y.: Particle swarm algorithm for minimal attribute reduction of decision data tables. In: Proceedings First International Multi-Symposiums on Computer and Computational Sciences (IMSCCS 2006), Hangzhou, China, I, pp. 572–575, April 2006
18. Bazan, J.G., Nguyen, H.S., Nguyen, S.H., Synak, P., Wróblewski, J.: Rough set algorithms in classification problem. In: Polkowski, L., Tsumoto, S., Lin, T.Y. (eds.) Rough Set Methods and Applications. Studies in Fuzziness and Soft Computing, vol. 56, pp. 49–88. Springer, Heidelberg (2000)
19. Wroblewski, J.: Finding minimal reducts using genetic algorithms. In: Proccedings of the 2nd Annual Join Conference on Infromation Science, pp. 186–189 (1995)
20. Zhu, X., Goldberg, A.B.: Introduction to semi-supervised learning. Synth. Lect. Artif. Intell. Mach. Learn. **3**(1), 1–130 (2009)
21. Pawlak, Z., Sowinski, R.: Rough set approach to multi-attribute decision analysis. Eur. J. Oper. Res. **72**(3), 443–459 (1994)
22. Pawlak, Z., Grzymala-Busse, J., Slowinski, R., Ziarko, W.: Rough sets. Commun. ACM **38**(11), 88–95 (1995)
23. Dai, J., Xu, Q.: Approximations and uncertainty measures in incomplete information systems. Inf. Sci. **198**, 62–80 (2012)
24. Dai, J., Wang, W., Xu, Q., Tian, H.: Uncertainty measurement for interval-valued decision systems based on extended conditional entropy. Knowl.-Based Syst. **27**, 443–450 (2012)
25. Fayyad, U., Irani, K.: Multi-interval discretization of continuous-valued attributes for classification learning. In: Proceedings of the 13th International Join Conference on Artificial Intelligence, pp. 1022–1027 (1993)
26. Jain, A., Zongker, D.: Feature selection: evaluation, application, and small sample performance. IEEE Trans. Pattern Anal. Mach. Intell. **19**(2), 153–158 (1997)
27. Zhu, H., Zhou, M.: Efficient role transfer based on kuhn-munkres algorithm. IEEE Trans. Syst. Man Cybern. Part A Syst. Hum. **42**(2), 491–496 (2012)

Semi-supervised Clustering Based on Artificial Bee Colony Algorithm with Kernel Strategy

Jianhua Dai[1,2(✉)], Huifeng Han[2], Hu Hu[2], Qinghua Hu[1], Bingjie Wei[1], and Yuejun Yan[1]

[1] School of Computer Science and Technology, Tianjin University,
Tianjin 300350, China
david.joshua@qq.com
[2] College of Computer Science and Technology, Zhejiang University,
Hangzhou 310027, China

Abstract. Artificial Bee Colony (ABC) algorithm, which simulates the intelligent foraging behavior of a honey bee swarm, is one of optimization algorithms introduced recently. The performance of the ABC algorithm has been proved to be very effective in many researches. In this paper, ABC algorithm combined with kernel strategy is proposed for clustering semi-supervised information. The proposed clustering strategy can make use of more background knowledge than traditional clustering methods and deal with non-square clusters with arbitrary shape. Several datasets including 2D display data and UCI datasets are used to test the performance of the proposed algorithm and the experiment results indicate that the constructed algorithm is effective.

Keywords: ABC · Semi-supervised clustering · Kernel method · Non-square cluster

1 Introduction

Clustering is an important unsupervised data mining technique which has many potential applications in different areas [1–4]. Major concern of clustering algorithm is to form groups such that the data in some groups have similar property and data in different groups are different to each other. The most popular class of clustering algorithms is the k-means algorithm which is a center based, simple and fast algorithm. However, k-means algorithm highly depends on the initial states and always converges to the nearest local optimum. To solve this problem, evolutionary computation methods such as Simulated Annealing, Genetic Algorithm, Particle Swarm Optimization can be used to avoid the local optimum to obtain a better result [5,6].

However, traditional clustering can not deal with semi-supervised learning where some labels of objects are given to wish a better result. Algorithms with the ability to obtain this goal is called semi-supervised clustering. Many semi-supervised clustering methods have been presented in recent years [7–9].

© Springer International Publishing Switzerland 2016
B. Cui et al. (Eds.): WAIM 2016, Part II, LNCS 9659, pp. 403–414, 2016.
DOI: 10.1007/978-3-319-39958-4_32

These algorithms can deal with constraints between objects such as must-link and cannot-link, where those links tell us some objects should belong to the same clustering and some objects should belong to different clusters. In many approaches, traditional clustering algorithms are modified so that pairwise constraints are used to guide the algorithms towards a more appropriate data partitioning. For example, Cop-Kmeans [7] is a semi-supervised variant of K-means, where the initial background knowledge provided in the form of constraints between instances in the dataset is used in the clustering process.

Swarm intelligence draw a lot of research interests to many research scientists of different fields in recent years. Swarm intelligence algorithms, such as Particle Swarm Optimization and Artificial Bee Colony algorithm, attempt to solve optimization problems by imitating the collective behavior of social insect colonies and other animal societies [10]. In this paper, Artificial Bee Colony Algorithm is introduced to deal with semi-supervised clustering problem.

The rest of this paper is organized as follows. Section 2 reviews some basic notions in ABC algorithm. In Sect. 3, a semi-supervised clustering method based on ABC and kernel strategy is constructed. Experiments on 2D display datasets and real world datasets are conducted in Sect. 4. Section 5 concludes the whole paper.

2   Artificial Bee Colony Algorithm

Artificial Bee Colony algorithm, proposed by Karaboga [11], is a simple and robust stochastic optimization algorithm based on swarm intelligence [12–15]. In ABC algorithm, each solution is viewed as a food source and bees try to find a best food source with best resource (fitness). The population of artificial bees consists of three components: employed bees, onlooker bees and scouts.

(1) Employed bees: these bees go to the food sources, and then they share the nectar and the position information of the food sources with onlooker bees in the dance area.
(2) Onlooker bees: which are waiting in the dance area and determine to choose a food source shared by employed bees. Food sources with better nectar will be chosen with a bigger probability.
(3) Scouts: the third kind of bees carry out random search to discover new sources.

In an ABC algorithm, there is a bijective mapping between employed bees and food sources. Hence, the number of employed bees must equal to the number of food sources which are also the number of onlooker bees. If a food source can not be improved by search around it's neighborhood, the corresponding employed bee will make a random search to generate a new food source. The process of ABC algorithm is shown in Algorithm 1.

In the process of the algorithm, an onlooker bee chooses a food source by a probability associated with the fitness of each food source. The probability is computed by:

An initial set of food sources $food_i, i = 1, 2, ...SN$ will be generated in the search space.

Evaluate fitness value fit_i for all pop_i in the population.

while $T < maxiter$ **do**

    foreach *Employed bees* **do**

        Produce a new food source based on the old one corresponding to the employed bees by Eq. (2).

        Evaluate the fitness value of the new food source. If the new food source has a better fitness, it will replace the old one.

    end

    foreach *Onlooker bees* **do**

        Select a food source $food_i$ according to probability computed by Eq. (1) of each food source with roulette wheel selection strategy.

        Produce a new food source based on the old one by Eq. (2).

        Evaluate the fitness value of the new food source. If the new food source has a better fitness, it will replace the old one.

    end

    If a food source can not be improved after a certain number of local searching made by employed bees or onlooker bees, then replace it with a new food source at a random value.

    Set $T = T + 1$

end

Algorithm 1: ABC

$$p_i = \frac{fit_i}{\sum_{i=1}^{SN} fit_i} \tag{1}$$

The fit_i is the fitness value of the ith food source and SN is the number of food sources.

For an employed bee or an onlooker bee, a new food source can be generated by the following method:

$$newFood_i = food_i + c(food_i - food_k) \tag{2}$$

where, $food_i$ is the old food source to be updated, c is a random number in $[-1,1]$, $food_k$ is a food source randomly selected from the set of other food sources, and j is a random number selected from 1 to D, D is the dimension of each food source.

3   Semi-supervised Clustering Based on ABC Combing with Kernel Strategy

Semi-supervised clustering is a kind of clustering with semi-supervised information. Normally, the information consists of must-link constraint or cannot-link constraint. The must-link constraint is described by the set of object pairs which must belong to the same cluster, and the cannot-link constrains is described by the set of object pairs which should belong to different clusters. In this section, we try to handle semi-supervised clustering problem with ABC algorithm combining with kernel strategy.

3.1 Objective Function for Semi-supervised Information

K-means is a popular clustering algorithm based on iterative relocation that partitions a dataset to k clusters, locally minimizing the average squared distance between the data points and the cluster centers. For a set of data points $X = \{x_1, x_2, \ldots, x_n\}$, $x_i \in R^d$, the k-means method generates a k-partitioning $\{C_j\}_j^k$ of X. If $\{\mu_1, \mu_2, \ldots, \mu_k\}$ denote the k partition centers, then the following objective function

$$\mathcal{J}_{kmeans} = \sum_{j=1}^{k} \sum_{x_i \in C_j} ||x_i - \mu_j||^2 \qquad (3)$$

is locally minimized.

However, traditional k-means algorithm can not deal with non-square cluster with arbitrary shape. To overcome this weakness, kernel method can be introduced to the objective function as follows:

$$\mathcal{J}_{kernel-kmeans} = \sum_{j=1}^{k} \sum_{x_i \in C_j} ||\Phi(x_i) - \Phi(\mu_j)||^2$$

$$= \sum_{j=1}^{k} \sum_{x_i \in C_j} ||\Phi(x_i) - \frac{1}{|C_j|} \sum_{l \in C_j} \Phi(x_l)||^2 \qquad (4)$$

Non-square clusters can be handled by this objective function. However, the objective function can not handle semi-supervised information. Hence, a new kind of objective function is constructed with the ability to make use of additional information such as must-link constraint and cannot-link constraint to get a better result.

$$\mathcal{J}_{obj} = \sum_{j=1}^{k} \sum_{x_i \in C_j} ||\Phi(x_i) - \Phi(\mu_j)||^2 + \sum_{(x_i, x_j) \in Must-link} \omega_{i,j} I(l_i \neq l_j)$$

$$+ \sum_{(x_i, x_j) \in Cannot-link} \overline{\omega}_{i,j} I(l_i = l_j) \qquad (5)$$

where l_i is the label assigned to x_i, I is the indicator function with $I(true) = 1$ and $I(false) = 0$ and ω_{ij} or $\overline{\omega}_{i,j}$ is a weight which can measure the important degree of the must-link or cannot-link between x_i and x_j.

In our method, ω_{ij} and $\overline{\omega}_{i,j}$ are both measured by the distance between the ith and jth objects in the kernel space, i.e. $||\Phi(x_i) - \Phi(x_j)||^2$. For the must-link constraint, the cost of violating is higher for the a pair of close points than a pair of points far apart. Hence ω_{ij} is set to $\alpha_{max} - ||\Phi(x_i) - \Phi(x_j)||^2$, where α_{max} is set to $\max_{(x_i, x_j) \in Must-link} ||\Phi(x_i) - \Phi(x_j)||^2$.

On the contrary, we think the cost of violating is smaller for a pair of close points than a pair of points far apart for cannot-link constraint. Hence, $\overline{\omega}_{i,j}$ is set to $||\Phi(x_i) - \Phi(x_j)||^2$.

3.2 Configuration of ABC Algorithm for Semi-supervised Clustering

In our ABC optimization algorithm, each solution is expressed by $food_i = (\mu_1, \mu_2..\mu_k)$ where μ_j is a vector denoting the center of the jth cluster in the original data space. Hence the distance between a point and a cluster can be defined as:

$$
\begin{aligned}
dis_\Phi(x_i, \mu_j) &= ||\Phi(x_i) - \Phi(\mu_j)||^2 \\
&= \Phi(x_i)\Phi(x_i)^T + \Phi(\mu_j)\Phi(\mu_j)^T - \Phi(x_i)\Phi(\mu_j)^T - \Phi(\mu_j)\Phi(x_i)^T \quad (6) \\
&= K(x_i, x_i) + K(\mu_j, \mu_j) - 2 \times K(x_i, \mu_j)
\end{aligned}
$$

Consequently, a label can be assigned to each object as:

$$
l_i = argmin_j dis_\Phi(x_i, \mu_j) \tag{7}
$$

After the assigning process, we can compute the fitness of each solution as follows:

$$
fit_i = \frac{1}{1 + \mathcal{J}_{obj}(food_i)} \tag{8}
$$

With these definition, the ABC optimization algorithm can be applied to the semi-supervised clustering problem directly.

4 Experiments

In this section, our algorithm is compared to traditional k-means, cop k-means, kernel k-means, cop kernel k-means algorithms on both 2D data and real-world data. In all the experiments, Gaussian kernel is used:

$$
K(x_i, x_j) = \exp \frac{-||x_i - x_j||^2}{2\sigma^2} \tag{9}
$$

4.1 Experiment Results on 2D Display Dataset

First of all, our algorithm is tested on 2D display dataset. The set up parameters and experiment results are shown in Figs. 1, 2, 3 and 4. From these figures, we can see that our method has the ability to deal with non-square shape clusters with better results.

The proportion of constraint for the experiments is set to $\alpha = 0.025$. The constraints are generated by following steps:

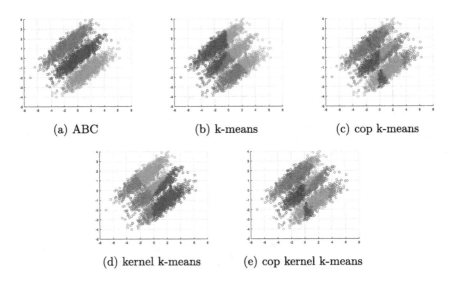

(a) ABC   (b) k-means   (c) cop k-means

(d) kernel k-means   (e) cop kernel k-means

Fig. 1. Mixed Gauss, $\sigma = 5$

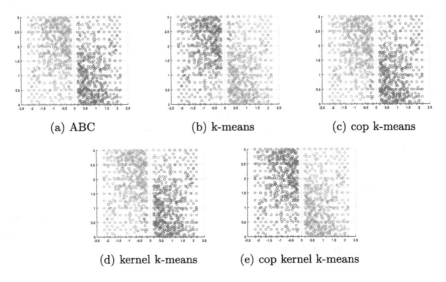

(a) ABC   (b) k-means   (c) cop k-means

(d) kernel k-means   (e) cop kernel k-means

Fig. 2. WingNut, $\sigma = 5$

(1) First of all, produce ($\frac{N^2-N}{2}$) pairs of potential constraints in the original dataset, where N is the object number of the original dataset. The label index of a potential constraint determine it belongs to must-link set or cannot-link set.

(2) Then α of constraints are randomly constructed from the potential constraints produced in Step 1.

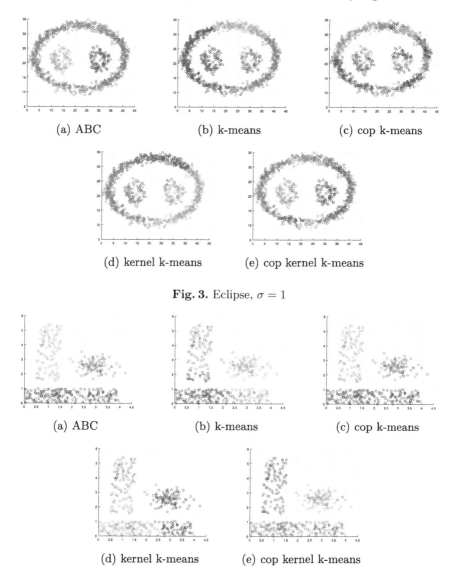

(a) ABC   (b) k-means   (c) cop k-means

(d) kernel k-means   (e) cop kernel k-means

Fig. 3. Eclipse, $\sigma = 1$

(a) ABC   (b) k-means   (c) cop k-means

(d) kernel k-means   (e) cop kernel k-means

Fig. 4. Lsun, $\sigma = 5$

Experiments on several well known 2D data are conducted. The results are shown in Figs. 1, 2, 3 and 4. From the figures, we find the proposed method outperform the comparative methods obviously.

4.2   Experiments on Real World Datasets

In this subsection, the experiments are conducted on several UCI datasets. The datasets are summarized in Table 1. In all the experiments, K is set to the number of real classes. The results are measured by the Accuracy (AC) index. The AC index is defined as:

$$r = \frac{\sum_{l=1}^{k} a_l}{n} \tag{10}$$

where a_l is the number of instances occurring in both lth cluster and its corresponding class and n is the number of instances in the data set.

Table 1. Dataset

Name	Number of attributes	Number of objects	Classes	$-\sigma^2-$
Iris	4	150	3	50
Fertility diagnosis	9	100	2	50
Planning relax	12	182	2	50
SPECTF	44	80	2	500
Glass	10	214	6	50
Haberman	3	306	2	250
ModelN	5	145	4	50
Wine	13	178	3	500
Foresttest	27	325	4	500

Table 2. Iris

Index	ABC	K-means	Cop K-means	Kernel K-means	Cop Kernel K-means
mean	0.9653	0.8697	0.9250	0.8510	0.8777
min	0.9400	0.5067	0.5267	0.5067	0.4133
max	0.9867	0.8933	0.9867	0.8933	0.9867
sd	0.0131	0.0855	0.0958	0.1155	0.1621

At first, experiments are conducted with certain proportion of constrains 0.025. The constraints are generated by the same method in the pervious subsection. 20 times experiments are conducted for each dataset and the results are shown in Fig. 5 and Tables 2, 3, 4, 5, 6, 7, 8, 9 and 10. In Fig. 5, x label is the Number of each experiment and y label is the AC index. Tables 2, 3, 4, 5, 6, 7, 8, 9 and 10 are statistics including mean, max, min and sd representing the average value, the maximal value, the minimum valued and the standard deviation of the corresponding accuracy results. From this experiment, we can see that our algorithm can lead to better results than others.

Moreover, we try to find the influence of semi-supervised information for these algorithms. Different proportions of constraints are tested on each dataset. For a certain proportion, 20 times of experiments are did and their average

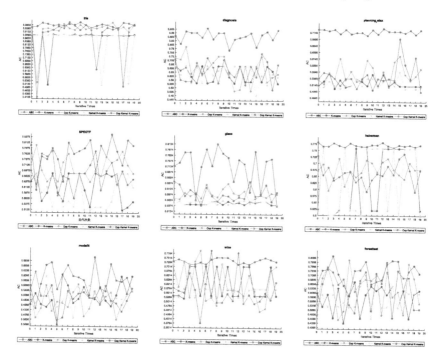

Fig. 5. Experiment on UCI dataset with AC index

Table 3. Fertility diagnosis

Index	ABC	K-means	Cop K-means	Kernel K-means	Cop Kernel K-means
mean	0.8585	0.6170	0.6470	0.6230	0.6380
min	0.7700	0.5600	0.5100	0.4800	0.5200
max	0.9100	0.6700	0.7900	0.6700	0.7500
sd	0.0362	0.0481	0.0792	0.0530	0.0665

Table 4. Planning relax

Index	ABC	K-means	Cop K-means	Kernel K-means	Cop Kernel K-means
mean	0.7184	0.5220	0.5569	0.5181	0.5473
min	0.7033	0.5000	0.5055	0.5000	0.4890
max	0.7308	0.5604	0.6264	0.5659	0.6923
sd	0.0077	0.0179	0.0363	0.0178	0.0457

accuracies are shown in Fig. 6. Here, the x label is the proportion of constraint and y label is the average AC index. On the whole, from the figure, we can see that our algorithm outperforms the other comparative algorithms.

Table 5. SPECTF

Index	ABC	K-means	Cop K-means	Kernel K-means	Cop Kernel K-means
mean	0.7606	0.6175	0.7006	0.5869	0.6744
min	0.6000	0.5125	0.6375	0.5125	0.5625
max	0.8500	0.6750	0.8000	0.6750	0.7750
sd	0.0630	0.0627	0.0472	0.0677	0.0628

Table 6. Glass

Index	ABC	K-means	Cop K-means	Kernel K-means	Cop Kernel K-means
mean	0.7105	0.5481	0.5879	0.5512	0.5902
min	0.5374	0.5421	0.5561	0.5421	0.5374
max	0.8131	0.5561	0.6355	0.5607	0.6449
sd	0.0806	0.0066	0.0205	0.0070	0.0304

Table 7. Haberman

Index	ABC	K-means	Cop K-means	Kernel K-means	Cop Kernel K-means
mean	0.7696	0.6480	0.6725	0.6337	0.6618
min	0.7614	0.5000	0.5948	0.5000	0.5915
max	0.7810	0.7582	0.7190	0.7582	0.7386
sd	0.0060	0.1251	0.0341	0.1279	0.0385

Table 8. ModelN

Index	ABC	K-means	Cop K-means	Kernel K-means	Cop Kernel K-means
mean	0.5883	0.5066	0.4797	0.5024	0.4990
min	0.4414	0.3585	0.3862	0.3931	0.3793
max	0.7034	0.6138	0.5655	0.6069	0.6276
sd	0.0648	0.0623	0.0546	0.0530	0.0537

Table 9. Wine

Index	ABC	K-means	Cop K-means	Kernel K-means	Cop Kernel K-means
mean	0.7356	0.6003	0.6008	0.5795	0.5604
min	0.7135	0.5169	0.4438	0.5225	0.3764
max	0.7535	0.7022	0.7921	0.7191	0.7921
sd	0.0135	0.0855	0.1037	0.0643	0.0973

Table 10. Foresttest

Index	ABC	K-means	Cop K-means	Kernel K-means	Cop Kernel K-means
mean	0.7192	0.5785	0.6328	0.5783	0.6343
min	0.5938	0.4338	0.4523	0.4615	0.4646
max	0.8185	0.6708	0.7969	0.7108	0.7846
sd	0.0618	0.0731	0.1039	0.0669	0.1062

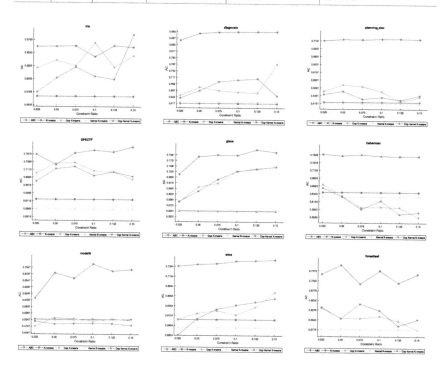

Fig. 6. Experiments of AC index at different proportions of constraints

5   Conclusion

In this paper, the Artificial Bee Colony algorithm combined with kernel strategy is introduced to deal with semi-supervised clustering problem with must-link and cannot-link constraints. The kernel function and constraints information are both used in our objective function. Our algorithm is compared to k-means and Cop k-means algorithms both in the original space and the kernel space on 2D display data and real world data. The results are very encouraging and verify the effectiveness of our algorithm.

Acknowledgements. This work was partially supported by the National Natural Science Foundation of China (No. 61473259, No. 61070074, No. 60703038), the Zhejiang Provincial Natural Science Foundation (No. Y14F020118), the National Science & Technology Support Program of China (2015BAK26B00, 2015BAK26B02) and the PEIYANG Young Scholars Program of Tianjin University (2016XRX-0001).

References

1. Kanungo, T., Mount, D., Netanyahu, N., Piatko, C., Silverman, R., Wu, A.: An efficient k-means clustering algorithm: analysis and implementation. IEEE Trans. Pattern Anal. Mach. Intell. **24**(7), 881–892 (2002)
2. Sander, J., Ester, M., Kriegel, H., Xu, X.: Density-based clustering in spatial databases: the algorithm GDBscan and its applications. Data Min. Knowl. Discov. **2**(2), 169–194 (1998)
3. Dhillon, I.S., Guan, Y., Kulis, B.: Kernel k-means: spectral clustering and normalized cuts. In: Proceedings of the Tenth ACM SIGKDD International Conference on Knowledge Discovery and Data Mining. ACM, pp. 551–556 (2004)
4. Zhang, R., Rudnicky, A.: A large scale clustering scheme for kernel k-means. In: Proceedings of the 16th International Conference on Pattern Recognition, vol. 4. IEEE, pp. 289–292 (2002)
5. Maulik, U., Bandyopadhyay, S.: Genetic algorithm-based clustering technique. Pattern Recogn. **33**(9), 1455–1465 (2000)
6. Van der Merwe, D., Engelbrecht, A.: Data clustering using particle swarm optimization. In: Proceedings of the 2003 Congress on Evolutionary Computation, vol. 1. IEEE, pp. 215–220 (2003)
7. Wagstaff, K., Cardie, C., Rogers, S., Schrödl, S.: Constrained k-means clustering with background knowledge. In: Proceedings of the Eighteenth International Conference on Machine Learning, pp. 577–584 (2001)
8. Basu, S., Banerjee, A., Mooney, R.J.: Semi-supervised clustering by seeding. In: Proceedings of the Nineteenth International Conference on Machine Learning, ICML 2002, pp. 27–34 (2002)
9. Kumar, N., Kummamuru, K.: Semisupervised clustering with metric learning using relative comparisons. IEEE Trans. Knowl. Data Eng. **20**(4), 496–503 (2008)
10. Bonabeau, E., Dorigo, M., Theraulaz, G.: Swarm Intelligence: From Natural to Artificial Systems. Oxford University Press, New York (2003)
11. Karaboga, D.: An idea based on honey bee swarm for numerical optimization. Technnical report, TR06, Erciyes University, Erciyes (2005)
12. Karaboga, D., Basturk, B.: Artificial bee colony (ABC) optimization algorithm for solving constrained optimization problems. In: Melin, P., Castillo, O., Aguilar, L.T., Kacprzyk, J., Pedrycz, W. (eds.) IFSA 2007. LNCS (LNAI), vol. 4529, pp. 789–798. Springer, Heidelberg (2007)
13. Karaboga, D., Ozturk, C.: A novel clustering approach: Artificial bee colony (ABC) algorithm. Appl. Soft Comput. **11**(1), 652–657 (2011)
14. Szeto, W., Wu, Y., Ho, S.C.: An artificial bee colony algorithm for the capacitated vehicle routing problem. Eur. J. Oper. Res. **215**(1), 126–135 (2011)
15. Karaboga, D., Akay, B.: A survey: algorithms simulating bee swarm intelligence. Artif. Intell. Rev. **31**(1–4), 61–85 (2009)

Distributed and Cloud Computing

HMNRS: A Hierarchical Multi-source Name Resolution Service for the Industrial Internet

Yang Liu[1(✉)], Guoqiang Fu[1], and Xinchi Li[2]

[1] China Academy of Information and Communication Technology, Beijing, China
{liuyang7,fuguoqiang}@caict.ac.cn
[2] China Telecom Corporation Limited Beijing Research Institute, Beijing, China
lixc@ctbri.com.cn

Abstract. The concept of Industrial Internet brings great potential for realizing intelligent products. In order to identify and locate information objects associated with these smart things, an identifier-locator mapping system is one of the most recognized technologies that enable the product information network. However, existing approaches lack enough attention to the locality issue. No matter aggregating information to the initial manufacturer or a random node calculated by some consistent hashing algorithms, all of these proposals will violate the administration of autonomous product information system as well as cause high latency and communication costs. To solve this issue, we propose a hierarchical multi-source name resolution service based on a two-level mapping system. Only the latest redirection is published on a global scale, while detailed mapping records are still maintained in each local subdomain. Furthermore, we propose two types of resolution mode based on this mapping system. A basic mode always originates the resolution request from an inter-domain name resolution, while the improved mode is a locality preference query mechanism. Through the theoretical analysis and simulation experiments, we find that if only the percentage of local queries exceed 20 %, the improved mode will behave much better for a trace query.

Keywords: Identifier and locator · Name resolution service · Locality · Industrial internet · Internet of things

1 Introduction

According to Meyer et al. [1], an ***intelligent product*** not only refers to the physical entity, but also includes its information-based representation. Through gathering and analyzing this information data, designers can make decision to accelerate the rate of product innovation, distributors can reasonably arrange scheduling to avoid long tail effect, and consumers can acquire necessary priori knowledge to better enjoy the service. This concept has become popular in the emergent ***Industrial Internet*** [2] paradigm, which will augment physical entities and devices with sensing, computing, and communication capabilities by means of the Internet of Things (IoT) technology. It connects otherwise isolated intelligent products and autonomous product information management systems [3] to form a collaborative product information network. Interoperability among these systems will greatly promote information sharing for collaborative business

© Springer International Publishing Switzerland 2016
B. Cui et al. (Eds.): WAIM 2016, Part II, LNCS 9659, pp. 417–429, 2016.
DOI: 10.1007/978-3-319-39958-4_33

partners. Once vast amounts of physical entities in the real world are seamlessly integrated into the virtual information network, how to identify and locate associate information objects becomes one of the most critical issues.

The key of this issue is designing an effective and efficient **name resolution service** **(NRS)**, which can provide mapping conversion from an identifier to its corresponding locators. Currently, various researches have been carried out to solve this problem from both academic and industry. These existing approaches either design the mapping service based on Domain Name Service (DNS), or based on Distributed Hash Table (DHT). While each of them has its pros and cons, a common weakness is that they do not pay enough attention to the locality issue. No matter aggregating mapping records from multiple information sources into the same place, or randomly distributing them throughout the overall network according to some consistent hashing algorithm, lack of attention to locality will violate the administration of autonomous product information system as well as cause high latency and communication costs.

In this paper, we propose a hierarchical multi-source name resolution service: HMNRS. Our basic idea is dividing a larger mapping system into multiple autonomous administrative subdomains, and connecting them together through product information links (PIL). Specifically, we design a two-level mapping system. The upper-level service platform always maintains a redirection to the tail of each product's PIL, and the lower-level autonomous subdomains manage detailed mapping records for all of the associated information objects which locate inside the subdomain. So only an abstract of mapping record is published on a global scale, while a major part of them is still maintained in each local subdomain. Based on such a mapping system, we propose two types of resolution mode. Considering the user pattern in product information network, it is unnecessary for the name resolution to always originate from an inter-domain NRS. So an improved resolution mode may first use an intra-domain NRS, which is much better if only the percentage of local queries exceed some extents.

2 Related Work

As mentioned before, from autonomous product information systems to collaborative product information network, the design of NRS is a critical step. Recently, many researchers have dedicated their efforts to developing such an identifier-locator mapping system.

(1) Centralized NRS. In a centralized approach, mapping records are aggregated into the same place, usually under the manufacture's control. For example, two kinds of mapping system are included in the EPCglobal network architecture [4]. The Object Name Service (ONS) [5] maps an EPC code to the address of the corresponding EPCIS repository. It is implemented based on DNS and uses type 35 Naming Authority Pointer (NAPTR) records [6]. ONS does not provide serial-level look up for individual objects, but only class-level product information, which is usually a pointer to the manufacture's EPCIS. Discover Service (DS) is designed to address this problem, which is also most similar to our work. It is expected to enable users to query dynamic real-time information from multiple sources. But until writing this paper, there is still no explicit resolution for DS.

(2) Semi-distributed NRS. In a semi-distributed approach, NRS is designed as a structured peer-to-peer overlay network. The most common type of structured P2P networks relies on the DHT technology, in which consistent hashing algorithm and its optimized variations (e.g., Chord [7], CAN [8], Pastry [9], Tapestry [10]) are used to assign resolution task to a particular peer. DHT technology has several very interesting properties, such as self-organization, robustness, and load balancing that are clearly desirable for a NRS. Several NRS designs based on the DHT mechanism have been proposed [11, 12].

(3) Fully-distributed NRS. In a fully-distributed approach, there is no global data structure restriction. Each node randomly forms connections to each other, and maintains a local NRS. In [13], the authors propose such an unstructured P2P network architecture for traceability application. It is a completely distributed manner without any centralized management node. Each node stores product information in its local repositories and maintains the hints of where the product is from and going to. Information query will be first processed in a local search engine according to local knowledge and security policy. If failed, query will be forwarded in the light of directional routing hint, in order to avoid aimless flooding query.

3 System Design

In this section, we describe in detail the designed system to provide a hierarchical multi-source name resolution service (HMNRS). We first give an overview of our approach. Subsequently, we will introduce each of its four components: network topology, naming schema, mapping record, and routing mechanism.

3.1 Overview

At the beginning of this paper, we have put forward the requirements for designing a hierarchical, scalable, effective and efficient NRS to query products information, which call for a fundamental shift from traditional NRS solutions such as DNS or DHT. To combines the advantages of both, we design a Hierarchical DHT structure to implement the HMNRS. In our approach, a two-level DHT overlay network is employed, consisting of an upper-level inter-domain NRS and numerous lower-level intra-domain NRS. The fundamental principle of our proposal is by means of the hierarchy to reduce latency and provides administrative control for each subdomain. The name resolution server located in the upper-level acts like a virtual gateway for product, which only keeps the latest index for redirecting the queries to the last subdomain. Meanwhile, the name resolution servers located in the lower-level maintain the local bindings between identifier and corresponding locators. To query all of needed information objects, we introduce the concept of product information link (PIL) to satisfy the special multi-source nature of NRS in product information network. We set two directional pointers for each information object, which can be used to connect the multiple information sources for each product.

3.2 Network Topology

A logical network topology to deploy the HMNRS system is depicted in Fig. 1, which is utilized so that the previously mentioned tasks can be fulfilled. The essential elements of HMNRS system are defined as following:

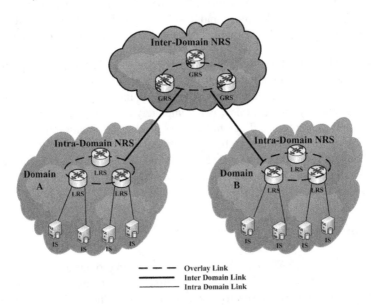

Fig. 1. Network topology of HMNRS.

Name Resolution Service Node. It refers a node that maintains mapping record and response name resolution request, which include two types: the Global Resolution Servers (GRS) and the Local Resolution Servers (LRS).

Information Service Node. It refers a node that maintains concrete information objects associated.

3.3 Naming Scheme

A name refers to a logical symbol for uniquely identifying an object. Accordingly, the name can be divided into identifier and locator. The former is a unique identity while the latter is a routable address to find the relative product information. In order to interconnect all relative information objects, at least, a global unique product identifier (UID) is needed to identify the physical object (e.g. Electronic Product Code – EPC). But for distinguishing different information objects associated with the same physical object and ensure the flexibility of local management, we name them with different local identifier (LID) generated from the UID. A LID is two-stage naming schema consisted of a string NameID and a binary NumericID. The special character "!" is used as a delimiter between these two parts. The NameID part is arranged by the owner of product information in the form of a hierarchical domain name, while the NumericID part is the hash value of its UID. Figure 2 uses an example to illustrate this naming schema. We assumed that the product's UID is "6900123456789" and the hash value of "6900123456789" equals to "101", then the latest information object's LID is "c.ucas.cn!101".

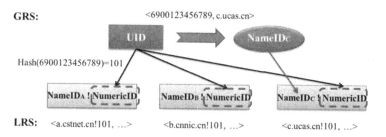

Fig. 2. Naming scheme of HMNRS.

3.4 Mapping Record

A mapping record keeps a binding relationship among different object names, which may also contain some other auxiliary information (e.g., record time, information publisher, scope, and et al.). It could map an identifier to several locators, or indirectly to another identifier.

In our proposal, in order to ensure administrative control for each subdomain, the upper-level GRS node only stores an abstract mapping record, which is a redirection from UID to the NameID part of LID. Its key is the identifier of the product, and value is the NameID of the last owner along the PIL. The lower-level LRS nodes will maintain the binding between identifier of information object and its locators. The primary key for this schema is the identifier of a product information object id, which includes a string NameID and a binary NumericID. The former is set as a unique hierarchical identifier (e.g., domain name) of the publisher who registers this product information, while the latter is the hash value of the unique product identifier. $from$ and to indicate the owner of previous and next subdomain for this product, which can be used further to construct the PIL. $scope$ restrict the propagation range of this mapping record. Multiple 4-touple map: $<l, t, a, p>$ are possibly included. l defines the locator where the concrete information object can be found, t records when the binding data is submitted, a defines an action that took place when the product information data is captured, and p identifies a possible container product. The so-called action determines the context in which a capture event took place, e.g. if an item was just observed to occur in a specific place, or it is packed/unpacked from a transport container.

3.5   Routing Mechanism

The upper-level NRS does not contain any trade secrets, but only plays a redirection function. So we do not restrict a particular DHT algorithm for the implementation of inter-domain routing mechanism. A public third party can be chosen to maintain this service. The lower-level NRS directly relates to the enterprises' confidential information. To ensure the locality of content and routing, our intra-domain routing mechanism for sending and receiving resolution messages is designed based on SkipNet. To illustrate our basic idea, we take an example which is shown in Fig. 3. To simplify the problem, we suppose that the nodes belongs to one company have the same NameID. While each node's random choice of ring memberships can be encoded as a unique binary NumericID. For example, node H and G have the same NameID "c.cnnic.cn", while node H has "110" as its NumericID but node G has "011" as its NumericID. We continue to use the sample data in Fig. 3, i.e., when the product "6900123456789" is transferred into this subdomain, it is maintained by owner who is identified as "b.cnnic.cn", so that its LID is "b.cnnic.cn!101". Then if the resolution request originates from node A, it will be first routed to node E according to NameID, and then routed to node F according to NumericID.

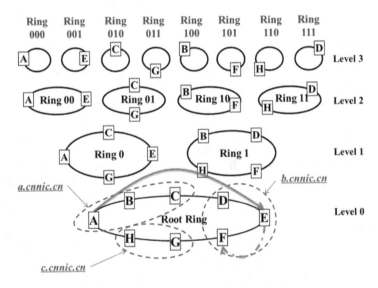

Fig. 3.   Intra-domain routing based on SkipNet.

4   Service Procedure

In this section, we describe the service procedure from three aspects: how to publish mapping records in the resolution system, how to update the dynamic mapping relationship, and how to resolve the identifier of a physical product into the locators of the associated information object.

(1) Registration. Unlike the Internet search engines can make use of hyperlinks between web pages to crawl the index information. Product information data itself does not have such a nature. So actively publishing the mapping records is necessary premise for information sharing. Whenever real-time product information is captured and submitted to the corresponding information server, registration will be triggered. Firstly, the binding between identifier and locator will be stored at a corresponding LRS node. Subsequently, it will be forwarded to an upper-level GRS node.

(2) Update. Until the product is transferred into a new resolution subdomain, its emergent mapping record will be registered in that subdomain. Correspondingly, the upper-level GRS mapping record should be updated either.

(3) Resolution.

Basic mode. A basic name resolution is a typical "top-down" operation, which originates from an inter-domain NRS. The requester first takes the UID of a product as the key to perform inter-domain resolution query. If successes, a GRS mapping record will be returned to indicate the tail of the product's PIL. According to this routing hint, resolution message will be redirect to the latest resolution subdomain. If it is a track query, response can be returned immediately. Otherwise, if it is a trace query, it will travel reverse along with the PIL until fetch all of needed information.

Improved mode. From the perspective of the operating sequence, basic resolution mode is reverse to registration. It seems that local and global is a paradox. If we want to make sure which domain has the answer of our question, we have to make an inter-domain name resolution firstly. But as a matter of fact, it is not necessary that resolution request must originate from the upper-level GRS query. A special nature of the product information network makes us rethink this problem, i.e., its user pattern. Who is concerned about particular product information? It is most likely to be one partner right on the path of product's PIL or in the vicinity. Just considering this situation, we set bidirectional pointer for each LRS mapping records. So in our improved resolution mode, resolution message will first be routed in the local subdomain. By means of PIL information, all subdomains which have information associated with the same product can be linked together. Only if requester is in the vicinity of one product information object, queries can be directly conducted forward and backward along with the PIL, needless to execute inter-domain NRS. We can take advantage of this nature to change the query starting point to effectively reduce the resolution latency as well as ensure the locality of NRS.

5 Analysis

So far we have described how HMNRS works. In this section, we present some analysis about its scalability, multi-source and locality. Next section, we will focus on evaluation the performance of our proposal.

5.1 Scalability

DHT based network topology makes HMNRS more scalable to support even with thousands or millions of nodes. With the incensement of products connected to the Internet of Things, each enterprise can create its own local subdomain to provide intra-domain NRS. Correspondingly, we can increase the number of upper-level GRS nodes to connect much more such resolution subdomain. The cost of a resolution operation grows as the logarithm of the number of nodes, so even very large systems are feasible. No parameter tuning is required to achieve this scaling. When some nodes join or leave this system at any time, the DHT based overlay network could make adaptive adjustments to maintain a normal query.

5.2 Multi-source

In host-centric network, the legacy NRS is designed to provide a "one-to-one" mapping service. As the most mature NRS for the Internet, DNS just translates human-readable domain names into machine-readable IP addresses. In information-centric network [14], named content probably has no less than one copy, so that a potential "one-to-many" mapping is adapted to shorten query time. Nevertheless, only the nearest copy is returned. But in our research scenario, i.e., a product-centric network, the situation is completely different. Throughout the entire products life cycle, several owners will publish and maintain relevant information associated with the same product. This information is not a copy of each other, but different object. If the user wants to trace query the product's logistics trajectory, any partial response is not enough, unless all of the relative information together could reconstruct the complete history of individual product.

Our proposal could effectively solve this problem without aggregating all mapping records but just the latest redirection. Mapping records, which may be a trade secret either, are still maintained where they are produced. Information owner could choose to modify or delete the local mapping records at any time. The communication overhead for redirection update is far less than that for uploading every piece of mapping data. Until some requesters initiate query for a particular product, a resolution message will be routed along the PIL of this product. Then, multiple nodes will collaborate temporarily to answer a query.

5.3 Locality

From performance's perspective, as many as possible local queries can dramatically reduce the query delay. Because the average query hops in a DHT-based overlay network directly depends on the number of nodes. Dividing a larger mapping system into multiple autonomous administrative subdomains makes the routing in each subdomain faster. Through evaluation in next section, we can find that even plus switching time between different levels, the total query delay is still reduced.

From security's perspective, information sharing must be guaranteed in a certain scope. Proper information sharing among cooperators contributes to discover much more meaningful knowledge. Contrary, uncontrolled information distribution may leak valuable trade secrets [15]. Recent research [16] shows that even a simple mapping record may disclose user privacy too. Just such a concern that attackers may use mapping records to infer sensitive business information leads partners reluctant to share product information. Based on our HMNRS system, information owners can flexibly set their local policies for inter-domain query. For example, Attribute Based Access Control (ABAC) technology [17] can be adapted to answer a legitimate query or just forward reluctant query.

6 Evaluation

In this section, we mainly evaluate the performance of HMNRS. We examine the effect of different parameters on the performance of query processing. Since the resolution latency is mainly determined by the hardware which the simulations performed. In our experiment, we measured the average query hops as the evaluation criteria. Through previous theoretical analysis, we can find that the average query hops is mainly influenced by these three factors: the resolution network scale (N), the percentage of local queries (p) and the average number of information sources who has associate information objects with the same product (s).

(1) Simulation Scenarios

Our simulation experiment has been performed using the simulator developed in Java based on PeerSim. All experiments were conducted on an Intel Core i7-2600 CPU 3.4 GHz system with 4 GB of RAM. For each experiment, we repeated the simulation for ten times and computed the average value as the final evaluation criteria.

- Considering most DHT algorithms take logarithmic number of hops, we adapt Chord to implement inter-domain name resolution service without loss of generality.
- The number of GRS nodes is constant set as $N_{GRS} = 1000$ during the simulation, because the inter-domain NRS act as a basic service infrastructure in the collative product information network, where churn is not a critical issue to deal with.
- In order to evaluate the effect of number of nodes, we will alter the value of N_{LRS}. We construct d subdomains ($d = 1000$), and set the number of nodes in each subdomain to be the same as N_{LRS}.
- In addition, prior to our measurement, we introduced $c = 1000$ products in to the network, and then registered some associate information objects in s random subdomain.

(2) Effect of p

We will first conduct a simple experiment to prove the effect of the percentage of local queries p on the average query hops. In order to observe the unique effect of parameter p, we restrict the number of LRS nodes in each subdomain to be the same as $N_{LRS} = 100$, and set the average number of information object associate with the same product as $s = 10$. Figure 4(a) shows that the improved mode is not suited for a track query. Because based on this improved mode, a local query will forward the resolution message according to its PIL's directional routing hints. On average, traversing along with the PIL to its end always takes *s/2* times intra-domain NRS query processing. But the basic mode could immediately forward the resolution message to the end. Figure 4(b) shows the situation for a trace query is opposite. Because traversing along with the PIL is inevitable for such a trace query. If the PIL's routing hints can be used directly, then an inter-domain NRS seems redundant. We can find that when the percentage of local queries is greater than 20 %, the improved mode already behaves better.

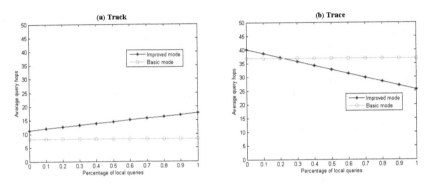

Fig. 4. Average query hops vs. different percentage of local queries, when s = 10, NLRS = 100.

(3) Effect of NLRS

Subsequently, we dynamically change the number of LRS nodes in each subdomain by adjusting the value from 100 to 500. To simplify the problem, we continue to set the average number of information object associate with the same product as *s* = 10.

Figure 5(a) and (b) present the relationship between the average query hops and network scale under different percentage of local queries. The x-axis indicates the percentage of local queries; the y-axis indicates the network scale, measured by the number of LRS nodes; the z-axis indicates average query hops. Generally speaking, as

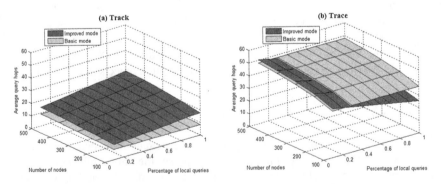

Fig. 5. Average query hops vs. network scale under different percentage of local queries, when s = 10 (Color figure online).

the number of nodes increases, the number of hops also increases. Figure 6(a) shows that the greater the scale of a subdomain is, the better a basic mode behaves for a track query. But the effect of parameter N_{LRS} is weaker than that of parameter p. This situation seems more apparent in Fig. 6(b). We can also find that the turning point where the improved model dominants is almost always 20 % regardless of the scale of the subdomain.

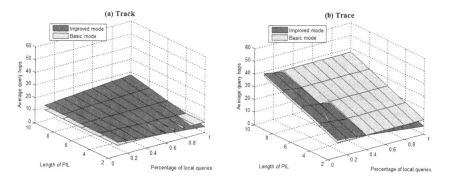

Fig. 6. The average query hops vs. the average length of PIL under different percentage of local queries, when NLRS = 100 (Color figure online).

(4) Effect of s

In many similar simulation works, the number of information object associate with the same product (this concept equals to what we mention in this paper as the length of a PIL) is usually arranged as an immutable value or even does not consider this problem. As a matter of fact, it is varied for different application scenarios. In order to simulate the real application environment for product information network as much as possible, we simulated different value of s to judge its effect on performance. We chose a fixed network topology by set $N_{LRS} = 100$ and alter the average number of registered information objects associate with the same product. Figure 6(a) shows that when the average length of PIL is smaller than 4, the dominant of the basic mode for a track query is no longer significant. Because a few times intra-domain NRS query processing is still less than a switching between the upper-level and lower-level NRS. As it is shown in Fig. 6(b), for a track query, the length of PIL also impact on the turning point where the improved model dominants. If s equals 10, such a turning point occurs when the percentage of local queries is greater than 20 %. But if s reduces to 6, such a turning point occurs until the percentage of local queries is greater than 40 %.

7   Conclusion and Future Work

In this paper, we present the design and evaluation of a hierarchical multi-source name resolution service (HMNRS). The proposal aims to identify and locate information object in a collaborative product information network. Based on a full consideration of locality issue, we construct a two-level DHT resolution network to balance between global and local view of mapping records. Two types of resolution mode have been studied in this mapping system. Through the theoretical analysis and simulation experiments, we can find that the basic mode is good for tracking the product's latest information object. However, to trace a comprehensive locator list, the improved mode, which is a locality preference query mechanism, behaves better than that of the basic mode.

Currently, many efforts have been made on product information network from both academic researchers and industrial practitioners, but there are still many research problem and implementation issues unsolved. For example, the privacy problem is always an obstacle to promote the implementation of a collaborative product information network. Not only information owners' privacy, but also the requesters' privacy is needed to be properly protected. Our approach provides a feasible solution to control information dissemination. Future work will focus on a flexible and cross-domain access control mechanism. In addition, products can be associated with each other because of containment (e.g. item is packed into a case) or assembly (e.g. an engine block is used for a car). So a recursive query mechanism is needed to study for acquiring a complete locator list.

References

1. Meyer, G.G., Främling, K., Holmström, J.: Intelligent products: a survey. Comput. Ind. **60**(3), 137–148 (2009)
2. Evans, P.C., Annunziata, M.: Industrial internet: pushing the boundaries of minds and machines. General Electric, p. 21 (2012)
3. Ranasinghe, D.C., Harrison, M., Främling, K., McFarlane, D.: Enabling through life product-instance management: solutions and challenges. J. Netw. Comput. Appl. **34**(3), 1015–1031 (2011)
4. EPCglobal: Object Name Service (ONS). Version 2.0.1. EPCglobal Standards (2013)
5. EPCglobal: The EPCglobal architecture framework. Version 1.5. EPCglobal Standards (2013)
6. Mealling, M., Daniel, R.: The naming authority pointer (NAPTR) DNS resource record. IETF Standard, RFC 2915 (2000)
7. Stoica, I., Morris, R., Karger, D., Kaashoek, M.F., Balakrishnan, H.: Chord: a scalable peer-to-peer lookup service for internet applications. In: Proceedings of SIGCOMM 2001, San Diego, CA, pp. 149–160, August 2001
8. Ratnasamy, S., Francis, P., Handley, M., Karp, R., Schenker, S.: A scalable content-addressable network. In: Proceedings of SIGCOMM 2001, San Diego, CA, pp. 161–172, August 2001
9. Rowstron, A., Druschel, P.: Pastry: scalable, decentralized object location, and routing for large-scale peer-to-peer systems. In: Guerraoui, R. (ed.) Middleware 2001. LNCS, vol. 2218, pp. 329–350. Springer, Heidelberg (2001)
10. Zhao, B.Y., Kubiatowicz, J.D., Joseph, A.D.: Tapestry: an infrastructure for fault-tolerant wide-area location and routing. Technical report CSD-01-1141, U.C. Berkeley, April 2001
11. Fabian, B., Gunther, O.: Distributed ONS and its impact on privacy. In: Proceedings of IEEE International Conference on Communications (ICC 2007), Glasgow, Scotland, pp. 1223–1228 (2007)

12. Xu, D.G., Qin, L.H., Park, J.H., Zhou, J.L.: ODSA: chord-based object discovery service architecture for the internet of things. Wirel. Pers. Commun. **73**, 1–22 (2013)
13. Cheung, A., Kailing, K., Schonauer, S.: Theseos: a query engine for traceability across sovereign, distributed RFID databases. In: Proceedings of the 23rd International Conference on Data Engineering (ICDE 2007), Istanbul, Turkey, pp. 1495–1496, April 2007
14. Ahlgren, B., Dannewitz, C., Imbrenda, C., Kutscher, D., Ohlman, B.: A survey of information-centric networking. IEEE Commun. Mag. **50**(7), 26–36 (2012)
15. Eurich, M., Oertel, N., Boutellier, R.: The impact of perceived privacy risks on organizations' willingness to share item-level event data across the supply chain. Electron. Commer. Res. **10**(3), 423–440 (2010)
16. Yan, Q., Li, Y., Deng, H.R.: Anti-tracking in RFID discovery service for dynamic supply chain systems. Int. J. RFID Secur. Crypt. **1**(1), 25–35 (2012)
17. Shi, J., Li, Y.J., He, W., Sim, D.: SecTTS: a secure track & trace system for RFID-enabled supply chains. Comput. Ind. **63**(6), 574–585 (2012)

Optimizing Replica Exchange Strategy for Load Balancing in Multienant Databases

Teng Liu[1], Qingzhong Li[1,2(✉)], Lanju Kong[1], Lei Liu[1], and Lizhen Cui[1]

[1] School of Computer Science and Technology, Shandong University, Jinan, China
liutengsd@gmail.com,
{lqz,klj,l.liu,clz}@sdu.edu.cn
[2] Dareway Software Co., Ltd, Jinan, China

Abstract. Resource sharing in multitenant databases is a challenge issue. The phenomenon can adversely affect a tenant's performance due to contending for shared resources among other tenants, and may cause a performance crisis. In this paper, a performance crisis is mitigated by a dynamic load balancing mechanism, which based on exchanging the roles between tenants' primary replicas and secondary replicas. The mechanism is composed of two parts: firstly, to balance resource utilization across servers, queries are dynamically allocated according to resource consumption; secondly, Improved Simulated Annealing Algorithm is developed to identify an optimal subset of tenants on overloaded servers, and for each tenant in the set, a suitable secondary replica can be selected to exchange roles with its primary replica to mitigate a crisis. Experimental results show significant reduction of service level objective violations compared to migration-based load balancing method and no load balancing method, respectively.

Keywords: Resource consumption · Load balancing · Replica exchange · Multitenant databases

1 Introduction

Consolidating fragmental workloads on a shared database is an efficient way of providing cost efficiency service. Database services in the cloud have successfully using multitenant databases which employs the consolidation strategy to reduce costs [1,2]. In the real world, a server with many tenants or tenants with heavy workload may result in a hotspot, i.e.,-an overloaded server [3,4]. In addition, an imbalanced usage of resources among different servers could cause resource waste and significantly degrade the overall performance [5]. As a result, load balancing in a multitenant database has attracted many research efforts. Mishima et al. proposed Madeus, a middleware approach conducts database live migration to efficiently mitigate hotspots [3]. Delphi is presented to solve a performance crisis by using a hill-climbing search algorithm to identify a new tenant placement strategy [4]. Later, Xiangyu et al. proposed an organized non-uniform replica placement strategy [6]. The aforementioned approaches are data

© Springer International Publishing Switzerland 2016
B. Cui et al. (Eds.): WAIM 2016, Part II, LNCS 9659, pp. 430–443, 2016.
DOI: 10.1007/978-3-319-39958-4_34

migration-based load balancing methods, which do not target the problem of load balancing without data movement. To this end, this paper presents a light-weight load balancing mechanism, which can be used to solve the imbalanced usage of resources, and address the hotspot problems in light of exchanging the roles between tenants' primary replicas and secondary replicas. The developed mechanism does not incur data movement, which may improve resource and time efficient. Existing works of using replica exchange for load balancing cannot be readily implemented in a practical system, for the reason that their works assume the tenant has only one primary replica and one secondary replica for fault tolerance [7]. In this paper, we extend the problem to a more general case, where each tenant has one primary replica with one or more secondary replicas. Additionally, secondary replicas in our developed mechanism are not only simply used as asynchronous replicas for high availability purposes, but also can serve 'read-only' queries.

In detail, the mechanism is composed of two parts: (1) it analyzes characteristics of queries, and dynamically allocates queries based on the resource consumption to balance the resource utilization across servers. Queries on the same server should have complementary resource consumption. For instance, co-locating a mix of I/O-heavy and CPU-heavy queries is probably better than co-locating multiple CPU-heavy queries; (2) once the system detects a performance crisis, Improved Simulated Annealing Algorithm (ISAA) is employed to identify an optimal subset of tenants that should be subject to the replica exchange. Both the primary and secondary replicas in the mechanism can serve 'read-only' queries, and meanwhile, the primary replica can serve 'write' queries as well. Hence, the primary replica receives a larger amount of workload compared with the secondary replica. After tenants perform the replica exchange, workload can be effectively moved from the primary replica to the secondary replica, and the hotspot is mitigated. Finally, the roles of the primary and the secondary replicas are exchanged and the 'write' queries are sent to the new primary replica, which was a secondary replica before the exchange. The contributions of this paper can be summarized as follows:

(1) A dynamic load balancing mechanism is developed to mitigate hotspots through exchanging the roles between tenants' primary replicas and secondary replicas. Unlike the traditional data migration-based load balancing method, the mechanism quickly finds and balances the load in a lightweight manner.
(2) The mechanism dynamically allocates queries based on the resource consumption, which makes full use of server resources and achieves the goal of minimizing the overall operating costs.
(3) ISAA algorithm is presented to identify an optimal subset of tenants. Different from the Hill climbing search algorithm which is good for finding a local optimum, ISAA is likely to find a good approximation to the global optimum of a given function in a large search space.

The remainder of this paper is organized as follows. Section 2 investigates the related work. In Sect. 3, we present the detailed implementation of allocating queries in terms of the resource consumption. Section 4 demonstrates how to mitigate hotspots with the developed mechanism. We present the experimental comparisons in Sect. 5. Section 6 concludes this paper.

2   Related Work

Several related approaches have been proposed for problems related to load balancing. Curino et al. [8] discussed migration-based load balancing in the context of multitenant DBs, suggesting an approach of on-demand migration, where the migration lazily moves the data as it is needed at the destination. Elmore et al. [9] proposed Zephyr, a technique to efficiently migrate a live database in a shared nothing transactional database architecture to load balancing. And, Hwang et al. [10] proposed Ursa, which scales to a large number of storage nodes and objects and aims to minimize latency and bandwidth costs during system reconfiguration. Toward this goal, Ursa formulates an optimization problem that selects a subset of objects from hotspots and performs topology-aware migration to achieve cost-effective load balancing. These approaches can mitigate a performance crisis by data migration-based load balancing. Since the mechanism presented in this paper does not incur data movement, it is significantly more efficient than these data migration methods. More recently, Shankar et al. [11] presented a framework to balance multi-tenancy with performance-based service level objectives (SLOs). It takes as input the tenant workloads, their performance SLOs, and the server hardware that is available to the provider, and outputs a cost-effective recipe that specifies how much hardware to provision and how to schedule the tenants on each hardware resource. However, their approach does not target the problem of tenants with variable and unknown workloads.

Previous works [12,13] have suggested an idea of using replica exchange for load balancing, but the details have not been explored. Later, Moon et al. [7] presented a load balancing method SWAT, using replica exchange to achieve a desired load balancing effect. They assume that a 'read-only' workload incurs zero loads on the secondary replicas, and secondary replicas just execute the update logs relayed from primary replicas. But in practical use, secondary replicas can also bear the load of 'read-only' queries and the load should not be neglected. In this paper, we design, implement, and evaluate replica exchange for the purpose of load balancing. And we provide the rule of selecting a suitable replica among tenants multiple secondary replicas to exchange roles with the primary replica, which is not mentioned in the existing methods.

3   Dynamically Query Allocation

Figure 1 provides an overview of the system architecture which we consider in this paper. When a query arrives, the Web-server forwards the query to the Master. The Master analyzes characteristics of the query and learns a load model which represents resource consumption of the query, then it dynamically directs the query to the right DB-server based on the workload level of servers. The Crisis detection engine continuously monitors the trade-off between efficient resource sharing among multiple tenants and their performance. The workload information of each server is collected by the Crisis detection engine and sent to the Scheduling algorithm engine. Once a performance crisis is detected, the Scheduling algorithm engine employs the ISAA algorithm to identify an optimal subset of tenants on the hotspot, and sends the set to the Scheduling executor. Finally, the executor performs the replica exchange, achieving the balanced load. Table 1 shows the important system parameters used in this paper.

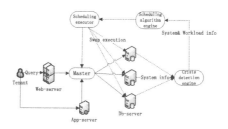

Fig. 1. Overview of the system architecture

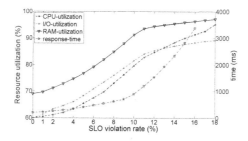

Fig. 2. Resource utilizations and average query response time with various SLO violation rates

3.1   Load Model

We use supervised learning techniques to learn characteristics of queries and train three load models: query load model, replica load model and server load model. The load models are representative of their resource consumption and have three important components: CPU consumed, disk I/O bandwidth consumed and main memory (RAM) consumed. Queries are classified into different types in terms of their resource consumption. In order to train the query load model, all types of queries are running in isolation on a database server.

Definition 1. *The load of a query is defined as the amount of CPU, I/O and RAM resources required to serve the query, as a percentage of the server capacity.*

Table 1. Notation of system parameters

Q_i	i-th query
T_i	i-th tenant
S_i	i-th server
T_i^p/T_i^s	T_i's primary replica / secondary replica
Q_j^p	A query set consists of a series of query which is executed on T_j^p
R_j	A replica set consists of a series of replica which is on S_j
$Load\,(T_i^p/T_i^s/S_i)$	The load of $T_i^p/T_i^s/S_i$
$Load_j\,(T_i^p/T_i^s)$	The load of T_i^p/T_i^s which is on S_j
$Load_j\,(Q_i)$	The load of Q_i which is executed on T_j^p/T_j^s

For example, $Q_1\left[10,\,20,\,10\right]$ means that the query Q_1 consumes 10 percent of CPU resources, 20 percent of I/O resources and 10 percent of RAM resources.

Definition 2. *We use a linear additive model for individual resource loads as in [14]. Therefore, the replica's load is defined as the sum of the load of queries which are executed on the replica T_j^p or T_j^s.*

$$Load(T_j^p) = \sum_{Q_i \in Q_j^p} Load_j(Q_i). \tag{1}$$

For instance, Q_1 and Q_2 are executed on the T_3^s. The query load models of Q_1 and Q_2 are $Q_1\left[10,\,20,\,10\right]$ and $Q_2\left[20,\,30,\,20\right]$, respectively. Hence, we can calculate the replica load model of T_3^s as $\left[30,\,50,\,30\right]$.

Definition 3. *Same as above, the load of a server is defined as the sum of the load of replicas which are on the server S_j.*

$$Load(S_j) = \sum_{T_i^p \in R_j} Load_j(T_i^p) + \sum_{T_i^s \in R_j} Load_j(T_i^s). \tag{2}$$

3.2 Training Server Performance Labels and Query Allocation

In this section, we train the server performance labels corresponding to the load level of servers, and dynamically allocate queries based on the resource consumption.

The server tends to be overloaded when its resource utilization reaches a critical point. To determine resource thresholds of the server, we want to find a certain point that indicates the overloaded situation. And we also intend to find a point where the server resources are fully utilized and the average query response time can be accepted. We run a controlled experiment involving a single server and use TPC suite to comprise different workload levels. Figure 2 shows the resource utilizations and the average query response time with various SLO violation rates. We observe that when the SLO violation rate is in the range of 1% to 10%, CPU utilization is greater than 60% as well as I/O, meanwhile, RAM utilization is greater than 70%, and the average query response time is in an acceptable range. The response time of queries will be exploded when the SLO violation rate begins to greater than 10%. We use the two points where the SLO violation rates are 1% and 10% to determine the resource thresholds, recorded as X,Y. Consequently, the thresholds of CPU and I/O are defined as $(60, 80)$ and the threshold of RAM is $(70, 90)$.

We now explain how to use the thresholds to train class labels (L) corresponding to the load level of servers and set $L = \{UNDER, GOOD, OVER\}$. If the resource utilization is below the lower bound (X), then the resource is labeled under-utilized $(UNDER)$. And if the resource utilization is above the upper bound (Y), the resource is marked as over-utilized $(OVER)$. Resource utilization in the range of X to Y is considered good utilization $(GOOD)$. As long as there is one resource marked as **OVER**, the server is in an overloaded condition and becomes a hotspot.

Definition 4. *Let the vector P represents the load level of servers and $P(S_i) = \{(M_i, N_i, U_i) \mid M_i, N_i, U_i \in L\}$, where M_i, N_i and U_i represent class labels of CPU, I/O and RAM, respectively.*

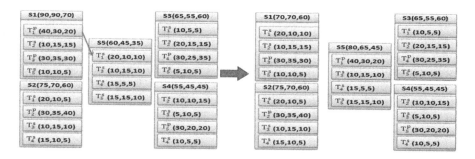

Fig. 3. Load balancing by exchanging the roles between T_1^p at S_1 and T_1^s at S_5

For example, $P(S_1) = (OVER, GOOD, GOOD)$ indicates that server S_1's CPU resource is labeled as **OVER**, I/O and RAM resources are marked as **GOOD**.

When a query Q_i submitted by tenant T_i arrives, the Master dynamically directs Q_i to the right DB server based on the load level of servers. As mentioned in the introduction, the 'write' query can be executed on the server which contains T_i^p. The execution of the 'read-only' query will be on any server that contains a replica of the tenant T_i, whereas how to choose the best one among T_i's replicas is a challenge. Each server at most contains one replica of tenant T_i. The problem of choosing a server S_j to execute Q_i can be formulated as:

$$P(S_j) \neq OVER \wedge P(S_j') \neq OVER \wedge Min(R \ resource \ utilization). \quad (3)$$

where $P(S_j) \neq OVER$ means that none of the three resources of server S_j are labeled as $OVER$, and $S_j{}' = S_j \cup Q_i$.

R represents the maximum resource needed to serve Q_i. If the result is not unique, we choose the server with the minimum R resource utilization. And finally, the Q_i is executed on S_j.

4   Crisis Mitigation

4.1   Identifying an Optimal Subset of Tenants

Recent works in solving a performance crisis often use hill-climbing search algorithm to identify a set of tenants [4]. Hill-climbing is good for finding a local optimum, but it is not necessarily guaranteed to find the best possible solution out of all possible solutions. We design the ISAA algorithm as outlined in **Algorithm 1**. Our system employs the ISAA algorithm to identify an optimal subset of tenants that should be subject to the replica exchange. The ISAA algorithm developed from simulated annealing is a generic probabilistic meta-heuristic for the global optimization problem. The ISAA algorithm finds the tenant set A through performing what-if analysis. For each primary replica T_i^p on the hotspot S_j, if the hotspot can be removed after exchanging the roles between tenant T_i's primary replica and secondary replica, then add T_i to the set A. Otherwise, we accept the tenant T_i by an acceptance probability function $f(T) = e^{\frac{dE}{T}}$. T is the initial temperature. And r is used to control the parameter T. The algorithm cannot random accept tenant T_i all the time because of time limit.

We illustrate the key idea by using the example in Fig. 3. There are five servers S_1 through S_5 and five tenants T_1 through T_5. Each tenant T_i has a primary replica T_i^p and three secondary replicas T_i^s. The number next to servers and replicas represent their respective load. The thresholds of CPU, I/O and RAM are $(60, 80)$, $(60, 80)$ and $(70, 90)$, respectively. As the server S_1's CPU and I/O resources are labeled as $OVER$, it becomes a hotspot. The ISAA algorithm is used to identify the tenant set A on the server S_1. One of the possible solutions is exchanging the roles between T_1^p at S_1 and T_1^s at S_5, which effectively moves the load from S_1 to S_5. The load of server S_1 changes from $[90, 90, 70]$ to $[70, 70, 60]$ and the performance crisis is mitigated. It is obvious that the solution is not the only one.

4.2   Performing the Replica Exchange

The ISAA algorithm is employed to identify an optimal subset of tenants on hotspots in Sect. 4.1. Then for each tenant in the set, a suitable secondary replica should be selected among the tenant's multiple secondary replicas to exchange roles with its primary replica.

We define each tenant T_i maintains two matrices A_i and B_i. Matrix A_i is composed of the load of servers which contain T_i's secondary replicas.

Algorithm 1. Improved Simulated Annealing Algorithm

Input:

$\{Query, replica, server\}$ load models; resource thresholds; server performance labels

Output:

Tenant set A

1: initialize $A = \emptyset$

2: **for** each hotspot S_j **do**

3:    **while** T_i^p on server S_j **do**

4:      exchange the roles between T_i's primary replica and secondary replica

5:      $S'_j = S_j - T_i^p + T_i^s$

6:      $dE = |Load\left(S'_j\right) - Load\left(S_j\right)|$

7:      **if** (hotspot S_j removed) **then**

8:        add T_i to the set A

9:        break

10:      **else if** ($e^{\frac{dE}{T}} > random\left(0,1\right)$) **then**

11:        add T_i to the set A

12:      **end if**

13:      $T = r \times T$

14:    **end while**

15: **end for**

16: **return** A

Each column in the A_i represents a server's load. A_i is a 3-by-N matrix, and N is the number of tenant T_i's secondary replicas. Matrix B_i is composed of the difference between $Load\left(T_i^p\right)$ and $Load_j\left(T_i^s\right)$, and B_i is also a 3-by-N matrix. We formulate the problem of selecting a suitable secondary replica among tenants' multiple secondary replicas as compute a binary variable assignment for a row vector X. The row vector X is composed of x_j, $(1 \leq j \leq N)$, where $x_j = 1$ means the jth column of A_i is selected, and $x_j = 0$ means otherwise, subject to

$$\forall x_j \in \{x_1, x_2, ..x_N\};\ x_j \in \{0,1\};\ x_1 + x_2 + .. + x_N = 1. \tag{4}$$

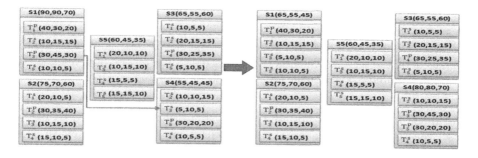

Fig. 4. Load balancing by exchanging the roles between T_3^p at S_1 and T_3^s at S_4

The performance crisis is mitigated means that after the replica exchange, all servers' loads will be below the threshold of $[80, 80, 90]$. The row vector X is computed by the following function:

$$A_i * X^T + B_i * X^T \leq L_H{}^T. \tag{5}$$

where $L_H = [80, 80, 90]$.

$$Max\,(\cos\theta)\,; \quad \cos\theta = \frac{Load_j(T_i^s) * Load\,(T_i^p)^T}{|Load_j\,(T_i^s)| \times |Load\,(T_i^p)|}. \tag{6}$$

For explanation, we use the example in Fig. 4. The ISAA algorithm identifies the tenants set $\{T_3\}$ on the hotspot S_1. Then we should select a suitable replica among the tenant T_3's multiple secondary replicas to perform the exchange. The server S_2, S_4 and S_5 contain a secondary replica of T_3. Therefore, A_3 and B_3 are calculated as follows:

$$A_3 = \begin{bmatrix} 75 & 55 & 60 \\ 70 & 45 & 45 \\ 60 & 45 & 35 \end{bmatrix} \quad B_3 = \begin{bmatrix} 20 & 25 & 20 \\ 30 & 35 & 30 \\ 20 & 25 & 20 \end{bmatrix}. \tag{7}$$

$$X = [x_1, x_2, x_3]\,; \; x_i \in \{0,1\}\,; \; x_1 + x_2 + x_3 = 1. \tag{8}$$

From Eq. (5), we can obtain $X = [0, 0, 1]$ or $X = [0, 1, 0]$. The first solution $X = [0, 0, 1]$ means that the roles of T_3^p at S_1 and T_3^s at S_5 are exchanged, the second solution $X = [0, 1, 0]$ means that the roles of T_3^p at S_1 exchange with T_3^s at S_4. The solution derived from Eq. (6) is $X = [0, 1, 0]$. The load of server S_1 changes from $[90, 90, 70]$ to $[65, 55, 45]$.

5   Experimental Evaluations

In this section, we apply the system architecture described in Sect. 3 to hypothetical SaaS scenarios, and illustrate the merits of the mechanism.

Systems. We deploy and evaluate our system on a cluster of ten machines with an Intel Xeon E312xx processor, 4G memory and 20 GB Disk. Nine of them are used as database servers. The other one is used as a master. MySQL 5.5 is used with InnoDB and a 1 GB buffer pool on CentOS generated by OpenStack.

Database and Tenants. We use the TPC $Benchmark^{TM}$ C (TPC-C) database, generated up to 100 tenants. Each tenant has one primary replica and one or more secondary replicas. The nine servers are in a random fashion while ensuring that a single server contains at most one replica of a tenant. The data size of tenants varied from 100 MB to 2 GB in the experiments.

Evaluation Metric. We focus on SLOs in terms of query response time as a mechanism for quantifying the quality-of-service of a multitenant DBMS. We use the SLO violation rates as our main performance metric to show the effectiveness of our load balancing mechanism in reducing the SLO violations.

5.1   Load Model and Server Performance Labels Evaluations

Load Model Evaluations. Before evaluating effectiveness and scalability of our mechanism, we briefly revisit the load model generation and provide a basic validation that the proposed models do capture relative resource consumption. There are three load models we have trained, the query load model, the replica load model and the server load model. Since all types of queries are executed on the server, respectively, and the required server resources are computed for each type of queries. We can conclude that the query load model do capture resource consumption of the query. The server load model is a linear additive model composed of several replica loads. The replica load model is a linear additive model as well. Therefore, we only need to evaluate the server load model.

We use TPC-C suite to comprise different workloads and pick one server from the nine servers as an example. We compare the CPU utilization obtained from the server load model referred to as trained-CPU with the actual CPU utilization of the server in Fig. 5(a). The line of CPU utilization is the average utilization of each 10 s interval. The results demonstrate that the two curves follow the same trend, and the trained-CPU is able to represent the actual CPU utilization. Figure 5(b) shows the similar results as above, where we compare the trained-I/O utilization against the actual I/O utilization. We also show the difference between the trained-RAM utilization and the actual RAM utilization in Fig. 5(c), and the good simulating results are obtained.

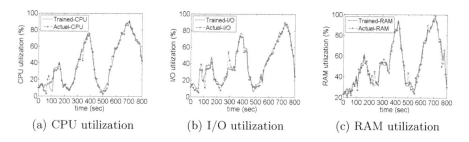

(a) CPU utilization   (b) I/O utilization   (c) RAM utilization

Fig. 5. Trace resource utilizations and performance of the server load model

Server Performance Labels Evaluations. To validate that the class labels are real representative of the actual load level of servers, we continuously monitor the performance of the servers. We use the rules described in Sect. 3.2 to label the server resources, and the resource utilizations are used for validation. Figure 6 shows the CPU, I/O and RAM utilizations with various labels:

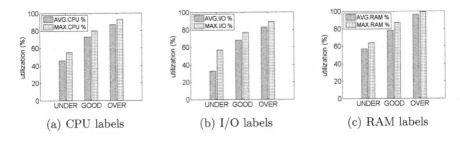

Fig. 6. Server resource utilization under different labels

UNDER, *GOOD*, and *OVER*. Figure 6(a) displays the average and the max CPU utilizations of the nine servers under the three labels. We observe that the average CPU utilization is 42 % and the max CPU utilization is 58 % with the label is *UNDER*. And when the CPU is labeled as *GOOD*, the average CPU utilization is less than 80 % as well as the max CPU utilization. Both the average and the max CPU utilizations are greater than 80 % when the label is *OVER*. In addition, I/O and RAM utilizations under the three labels are plotted in Fig. 6(b) and (c), respectively. These results demonstrate that the labels are confident to represent the load level of servers.

5.2 Performance Comparisons

We implement and compare our mechanism namely LB-exchange with two methods, the first one is a method with no load balancing, referred to as noLB. And the other one is a migration-based load balancing method which solves a performance crisis through data movement. We evaluate the effectiveness of our mechanism in Fig. 7. The boxplots show sampled minimum, lower quartile, median, upper quartile, and sampled maximum SLO violation rates, and indicate the load balancing capacity of the methods. Since the noLB do nothing about the performance crisis, the data migration performs better than the noLB as expected. As we compare the LB-exchange with the data migration, the LB-exchange significantly contributes to the reduction of SLO violation.

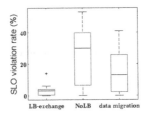

Fig. 7. The performance of each approach

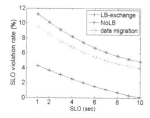

Fig. 8. SLO violation with changing SLOs

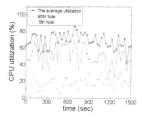

Fig. 9. Measure the balance of resource utilization across server

Figure 8 shows the SLO violation rates with a various SLOs. Focusing on 7-sec SLO, it can be observed that the data migration method violates SLO for 5.2 % of queries and the NoLB method violates SLO for 6.2 % of queries. The LB-exchange method effectively lowers the SLO violation down to 1.2 % with 7-sec SLO, achieving 4.3 times and 5.2 times reduction of SLO violation when compared to the data migration and the NoLB, respectively. Across all SLOs, we can see the noLB performs worse than the LB-exchange in general.

Since our mechanism dynamically allocates queries based on the resource consumption and balances the load to mitigate a performance crisis. So we measure the balance of resource utilization across servers. Figure 9 shows the average, 95th and 5th percentile of CPU utilizations for the nine servers in the experiment. The X-axis shows the real time since the beginning of the experiment, during which we keep sending requests to servers with different rates. The figure shows that the 95th and the 5th percentile of CPU utilizations are close to the average CPU utilization, which suggesting that we have achieved good balance. The average CPU utilization is far from the upper CPU threshold (Y), which indicating low risk of overload. Note that a performance crisis occurs at the 720 s, but it is quickly resolved by the mechanism.

6   Conclusions

In this paper, a dynamic load balancing mechanism has been developed to solve a performance crisis in multitenant databases by exchanging the roles between tenants' primary replicas and secondary replicas. When detecting a performance crisis, the ISAA algorithm can be applied to identify an optimal subset of tenants on hotspots. Then, a suitable secondary replica can be selected to exchange roles with the primary replica for each tenant in the set according to the mechanism. By doing this, a desired load balancing effect is achieved. The experimental results suggested correctness and effectiveness of our load mechanism. In addition, the comparisons with the no load balancing method and the migration-based load balancing method showed the significant reduction of SLO violation.

Acknowledgments. This work is partially supported by NSFC under Grant No. 61272241, No. 61572295, No. 61303085; Taishan industry leader talent of Shandong province; Natural Science Foundation of Shandong Province of China under Grant No. ZR2013FQ014; Science and Technology Development Plan Project of Shandong Province No. 2014GGX101047, No. ZR2014FM031; Fundamental Research Funds of Shandong University No. 2014JC025, No. 2015JC031; Shandong Province Independent Innovation Major Special Project No. 2015ZDXX0201B03, 2015ZDXX0201A04, 2015ZDJQ01002; Shandong Province key research and development plan No. 2015GGX101015, 2015GGX101007; Innovation Method Fund of China No. 2015IM010200.

References

1. Aulbach, S., Seibold, M., Jacobs, D., Kemper, A.: Extensibility and data sharing in evolving multi-tenant databases. In: 2011 IEEE 27th International Conference on Data Engineering (ICDE), pp. 99–110. IEEE (2011)
2. Jacobs, D., Aulbach, S., et al.: Ruminations on multi-tenant databases. In: BTW, vol. 103, pp. 514–521 (2007)
3. Mishima, T., Fujiwara, Y.: Madeus: database live migration middleware under heavy workloads for cloud environment. In: Proceedings of the 2015 ACM SIGMOD International Conference on Management of Data, pp. 315–329. ACM (2015)
4. Elmore, A.J., Das, S., Pucher, A., Agrawal, D., El Abbadi, A., Yan, X.: Characterizing tenant behavior for placement and crisis mitigation in multitenant DBMSS. In: Proceedings of the 2013 ACM SIGMOD International Conference on Management of Data, pp. 517–528. ACM (2013)
5. Curino, C., Jones, E.P., Madden, S., Balakrishnan, H.: Workload-aware database monitoring and consolidation. In: Proceedings of the 2011 ACM SIGMOD International Conference on Management of Data, pp. 313–324. ACM (2011)
6. Luo, X., Xin, G., Wang, Y., et al.: Superset: a non-uniform replica placement strategy towards perfect load balance and fine-grained power proportionality. Cluster Comput. **18**(3), 1127–1140 (2015)
7. Moon, H.J., Hacıgümüş, H., Chi, Y., Hsiung, W.P.: Swat: a lightweight load balancing method for multitenant databases. In: Proceedings of the 16th International Conference on Extending Database Technology, pp. 65–76. ACM (2013)
8. Curino, C., Jones, E.P.C., Popa, R.A., et al.: Relational cloud: a database-as-a-service for the cloud, 235–240 (2011)
9. Elmore, A.J., Das, S., Agrawal, D., El Abbadi, A.: Zephyr: live migration in shared nothing databases for elastic cloud platforms. In: Proceedings of the 2011 ACM SIGMOD International Conference on Management of Data, pp. 301–312. ACM (2011)

10. You, G., Hwang, S., Jain, N.: Scalable load balancing in cluster storage systems. In: Kon, F., Kermarrec, A.-M. (eds.) Middleware 2011. LNCS, vol. 7049, pp. 101–122. Springer, Heidelberg (2011)
11. Lang, W., Shankar, S., Patel, J.M., Kalhan, A.: Towards multi-tenant performance slos. IEEE Trans. Knowl. Data Eng. **26**(6), 1447–1463 (2014)
12. Campbell, D.G., Kakivaya, G., Ellis, N.: Extreme scale with full SQL language support in microsoft SQL azure. In: Proceedings of the 2010 ACM SIGMOD International Conference on Management of Data, pp. 1021–1024. ACM (2010)
13. Bernstein, P., Cseri, I., Dani, N., Ellis, N., Kalhan, A., Kakivaya, G., Lomet, D.B., Manne, R., Novik, L., Talius, T., et al.: Adapting microsoft SQL server for cloud computing. In: 2011 IEEE 27th International Conference on Data Engineering (ICDE), pp. 1255–1263. IEEE (2011)
14. Yang, F., Shanmugasundaram, J., Yerneni, R.: A scalable data platform for a large number of small applications. In: CIDR, vol. 1, p. 11 (2009)

ERPC: An Edge-Resources Based Framework to Reduce Bandwidth Cost in the Personal Cloud

Shaoduo Gan$^{(\boxtimes)}$, Jie Yu, Xiaoling Li, Jun Ma, Lei Luo,
Qingbo Wu, and Shasha Li

School of Computer, National University of Defense Technology, Changsha, China
{ganshaoduo,jackyu,lixiaoling,majun,luolei,wuqingbo}@ubuntukylin.com
shashali@nudt.edu.cn

Abstract. Personal Cloud storage and file synchronization services, such as Dropbox, Google Drive, and Baidu Cloud, are increasingly prevalent within the Internet community. It is estimated that subscriptions of personal cloud storage are projected to hit 1.3 billion in 2017. In order to provide high rates of data retrieving, cloud providers require huge amounts of bandwidth. As an attempt to reduce their bandwidth cost and, at the same time, guarantee the quality of service, we propose a novel cloud framework based on distributed edge resources (i.e., voluntary peers in P2P Networks and edge servers in Content Delivery Networks).

P2P technique is well considered as an efficient way to distribute contents, but not all of contents are applicable for it. Thus we first present an approach to select the contents which are more profitable to be placed in P2P networks. Considering the unreliability of P2P networks, we utilize CDN servers to dynamically cache the contents which can not be well served by P2P peers. The proposed caching algorithm takes the state of P2P peers into account and considers replacement and allocation policies simultaneously. According to trace-driven simulations, the ERPC achieves an impressive performance of saving bandwidth for cloud system and guaranteeing download rates for users.

Keywords: Cloud system · Distributed system · Personal cloud · Bandwidth saving · Edge resource

1 Introduction

Personal Cloud storage and file synchronization services has been increasingly popular in recent years, for it enables individuals and communities to store, fetch, edit and synchronize data stored in "cloud" which can be assessed all over the Internet. Nowadays millions of people tend to depend on personal cloud providers, like Dropbox, Google Drive, and Baidu Cloud, to manage their data. It is estimated that subscriptions of personal cloud storage are projected to hit

B. Cui et al. (Eds.): WAIM 2016, Part II, LNCS 9659, pp. 444–456, 2016.
DOI: 10.1007/978-3-319-39958-4_35

1.3 billion in 2017 [1]. As personal cloud providers have already been able to provide enough storage space, they now pay more attention to enlarging their bandwidth, which straight influences the experience of users. With the rapid ascent of users, the cost of bandwidth is increasingly higher.

It is common that most public contents in people's cloud come from others sharing and will be shared with others, like movies, music, etc. This will result in repeated distributions of the same content, which consumes a large part of bandwidth of data centers adopting server-client pattern. To save this part of bandwidth consumption and guarantee the data retrieving rates, we introduce edge resources to the personal cloud. Edge resources, including voluntary peers in P2P networks and edge servers in CDNs, are especially suitable for content distribution, because they are fully distributed among users, which enables them to serve users' requests in time and depend less on network conditions. Many former studies try to utilize edge resources to improve the performance of some cloud services, like live video stream [2], web acceleration [3], etc.

To make the Edge-Resources Based Framework of Personal Cloud (ERPC) achieve better performance, we need to deal with two challenges. The first is selecting the most profitable contents to be placed in edge resources. There are two general metrics to decide whether a certain content is applicable for edge resources: the amount of data that can be offloaded from cloud and the distribution time compared with server-client pattern. Based on these two metrics, we present an approach to make decision between HTTP and BitTorrent. The second challenge is how to overcome the unreliability of P2P networks to provide a relative stable rates of data retrieving. We utilize CDN servers to address this problem by dynamically caching the contents which can not be well served by P2P peers. A caching algorithm is proposed for CDN servers which considers replacement and allocation policies simultaneously.

We carry out series of trace-driven simulations to evaluate the performance of ERPC. The results show that our framework obviously saves bandwidth for cloud and does not degrade the rates of data-retrieving.

2   Related Work

Utilizing edge resources of networks, like volunteer peers or edge servers of CDNs, for Cloud Computing has always drawn much attentions of researchers.

In 2006, Babaoglu et al. proposed the concept of "fully decentralized P2P cloud" [4]. Then this team designed and implemented a P2P Cloud system on the infrastructure-as-a-service level [5]. Voluntary edge Clouds have been considered as a recent research trend in a survey [6] in 2010. Wu et al. designed a P2P based cloud system to improve the performance of the Map Reduce. Due to the increasing number of MapReduce jobs submitted and system scale, the master node will become the bottleneck of the system. Therefore, they used P2P Chord network to manage master node in a decentralized way [7]. Authors in [8] proposed a novel cloud framework for P2P Video on Demand (VoD) application. This new cloud framework tends to provision video contents with rapid growth in demand. However, these works all neglect the unreliability of P2P networks and can not ensure the quality of service.

CDN is a complimentary utility platforms, which make it achievable to be integrated into Cloud computing platforms. Meisong Wang et al. conducted a comprehensive survey on existing Cloud CDN solutions [9]. Their work shows that the integration of Cloud and CDN has mutual benefits allowing content to be efficiently and effectively distributed in the Internet using a pay-as-you-go model promoting the content-as-a-service model. Authors in [10] combine CDN and cloud computing to be better applied in the video business and other applications which have high traffic.

Whereas, there are rare studies integrating both of these achievable edge resources into cloud system simultaneously to save the bandwidth of personal cloud, which is the motivation of our work in this paper.

3   Overview of the ERPC

In this section, we introduce the general architecture of the ERPC. The proposed framework consists of three components: centralized cloud, CDN servers and P2P networks. Since all of them can be data sources for clients, there should exist an mechanism to guide the process of data retrieving.

To integrate these components in a smoother way, we tend to utilize the initial infrastructures of themselves, such as CDN DNS servers and P2P trackers. DNS servers can redirect the client's requests to the CDN edge servers near the client based on specialized routing, load monitoring and Internet mapping mechanism. P2P trackers are used to return the client a set of peers possessing the requested contents. In our framework, DNS servers not only forward requests to CDN servers, but also to P2P trackers with corresponding seeds. If the demanded content does not exist in P2P networks, the request will be forwarded to centralized cloud directly. Figure 1 illustrates the process of data retrieving. Note Step (2) and Step (2') are alternative step; if Step (2) is selected, and then the next step is Step (3), otherwise, it is step (3').

(1) The user firstly sends his HTTP request to the CDN DNS server as usual, then DNS servers undertake to check if the requested content has already been published in Bittorrent network;

(2) If not, the request is forwarded to centralized cloud to retrieve content directly;

(3) Centralized cloud serves the request and decides whether the requested content should be published into P2P networks;

(2') Otherwise, the request is redirected to selected CDN edge servers and P2P tracker with corresponding seed;

(3') The client retrieve data from edge resources, and CDN servers fetch or evict contents based on caching algorithm;

In the process presented above, there are two most crucial steps that influence the performance of the entire framework: step (3) and step (3'). Step (3) decides

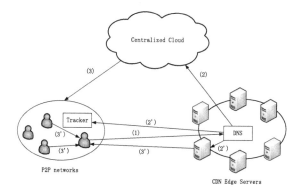

Fig. 1. Overview of the ERPC

which contents should be placed in edge resources that can save more bandwidth for cloud and not degrade the data retrieving rates. Scale of swarms, upload capacity of leechers and the content size are all conditions that should be taken into account. Due to dynamic departure and heterogeneous bandwidth of peers, P2P networks can not guarantee quality of service. To make up this drawback, CDN servers in step $(3')$ should precisely cache the contents which can not be well served by P2P networks. These two steps are quite important and sophisticated, so we specify them in next two sections.

4 Selection Between HTTP and BitTorrent

In this section, we propose two general metrics to decide whether a certain content should be placed in edge resources: the amount of data that can be offloaded from cloud and the distribution time compared with server-client pattern. We first build up a model to describe these two metrics. Although the model can not be extremely quantitatively accurate compared with real-world data, it still reveals the most important characteristics. Based on the results, we propose a practical approach to make the most optimal decision.

4.1 The Distribution Model

We start with the BitTorrent protocol. Suppose that there are a total of S swarms (all peers that distribute the same content called a swarm), which can be represented by $\mathcal{S} = \{1, 2, ..., S\}$. For a certain swarm s whose content size is η_s, the concurrent leechers (peers only possessing a part of the content) and seeders (peers possessing the entire content) are X_s and Y_s. The upload and download capacity of a certain peer i are denoted by μ_i and c_i. Table 1 summarizes the main notations used in this paper.

Apparently, if put the content in swarm s to P2P networks, the amount of data that can be offloaded from cloud is $\eta_s \cdot X_s$. We set constraints on both η_s and X_s. For the content size η_s, it is known that BitTorrent is not convenient for the distribution of small files, because overheads of BitTorrent (e.g., making seed file, searching neighbors) will exceed the download time. Thus we decide that if the requested content's size is smaller than 20 MB, centralized cloud will serve it directly with HTTP protocol. For the number of leechers X_s, since our framework intends to reduce the bandwidth consumed by widely repeat-requested contents, the number of leechers should be large enough. We set 30 as the minimum threshold.

Table 1. Frequently used notation

Parameter	Definition
S	Set of swarms
η_s	Size of content in swarm s
X_s, Y_s	Set of leechers and seeders of swarm s, respectively
μ_i, c_i	Upload and download capacity of a certain peer i
U_s, D_s	Total upload and download capacity of swarm s
U_s^{CDN}	Upload capacity provided by a CDN server to swarm s
γ_s	Average percentage of content possessed by leechers in swarm s
a_s	Percentage of the content in swarm s cached by a CDN server
b_s	Bandwidth allocated to swarm s by a CDN server
θ	Size of every single segment
n_s	The number of segments a CDN server allocate to content in swarm s
Δt	The time window of refreshing allocation state in CDN servers

To model the distribution time, we need to get the total download and upload capacities of swarm s. Only leechers have download process, so the total download capacities of swarm s is the sum of leechers download rate. Since seeders have the entire content, they can upload at their full rate. Whereas leechers' capacity of uploading is decided by how much percentage of content they have stored. We use γ_s to denote average percentage of content possessed by leechers within swarm s. Consequently, we can get the total download capacities D_s and upload capacities U_s of swarm s as:

$$D_s = \sum_{i \in X_s} c_i$$

$$U_s = \gamma_s \cdot \sum_{i \in X_s} \mu_i + \sum_{i \in Y_s} \mu_i \tag{1}$$

The distribution time can be limited by either D_s or U_s, which are called download-rate-limited case and upload-rate-limited case. Therefore, the average distribution time of swarm s with BitTorrent protocol can be formulated as:

$$T_s^{bt} = \frac{\eta_s}{\min(\frac{D_s}{|X_s|}, \frac{U_s}{|X_s|})} \tag{2}$$

This model is originally proven by Kumar et al. in [11]. We also make some modification to make it more applicable for our work.

Then we consider server-client pattern with HTTP protocol. The number of requests for the content in swarm s can be represented by the number of leechers $|X_s|$. So the download capacity are same as BitTorrent scenario. However, the upload capacity in server-client pattern can only be provided by centralized cloud servers. We denote the upload capacity cloud servers assign to swarm s as U_s^{cs}. Hence, we get the expression of distribution time of swarm s with HTTP protocol:

$$T_s^{cs} = \frac{\eta_s}{\min(\frac{D_s}{|X_s|}, \frac{U_s^{cs}}{|X_s|})} \tag{3}$$

4.2 Selection Approach

When a request from swarm s arrives in cloud, we first check its size and leechers number. If it is more than 20 M and requested by more than 30 clients, it have the potentiality to be placed in edge resources.

Then we compare the distribution time. To measure the difference between the client-server and P2P systems, we introduce the gain ratio as follows:

$$Gain = \frac{T_s^{cs} - T_s^{bt}}{T_s^{cs}} \tag{4}$$

If the $Gain$ exceeds the threshold τ we set initially, centralized cloud will create a torrent meta-data file and publish the corresponding seed to BitTorrent networks. At the same time, DNS servers are informed to add a tuple in its forward table. Thus the following requests for this content will be redirected to P2P trackers for data retrieving.

5 Caching Algorithm of CDN Edge Servers

Due to dynamic departure and heterogeneous bandwidth of peers, P2P networks can not guarantee the stable rates of data retrieving. Therefore, we use CDN edge servers, which are reliable and stable, to make up drawbacks of P2P networks. The main point of our design is that we regard CDN servers as the cache of P2P networks to store contents which can not be well served by P2P peers.

Note that only the upload-rate-limited case is considered in the caching model because in the cloud providers' point of view, download-rate-limited case caused by users' insufficient download bandwidth can not be improved. Based on the model, we propose a caching algorithm for CDN edge servers which takes the state of P2P peers into account.

5.1 Caching Model of CDN Servers

In the BitTorrent protocol, the shared content is split into pieces to be distributed, which means cache can also only store a part of the content. We assume

---

Algorithm 1. Refreshing Allocation State

// Every Δt time
1 $\mathcal{U} \leftarrow$ contents stored in cache now;
2 **for** $s \in \mathcal{U}$ **do**
3 | recalculate uV_s;
4 **end**
5 **for** $s \in \mathcal{U}$ **do**
6 | $\Delta n_s \leftarrow n_s$;
7 | recalculate n_s;
8 | $\Delta n_s = n_s - \Delta n_s$;
9 **end**
10 refresh the priority queue of unit value;

---

that a certain CDN server stores a_s percentage of the content distributed in swarm s and allocates b_s bandwidth to it. Then $a_s \cdot (D_s - U_s)$ demanded rate can be provided by this CDN server. But remember, this rate can not exceed the allocated bandwidth b_s. Thus the upload capacity this CDN server can provide to swarm s is $U_s^{CDN} = \min(a_s \cdot (D_s - U_s), b_s)$. Based on this expression, the best situation is:

$$a_s \cdot (D_s - U_s) = b_s \tag{5}$$

Otherwise, either storage or bandwidth of CDN server will be wasted.

Now we consider the value of caching a certain content. Since U_s^{CDN} is the upload capacity provided by a CDN server, thus we use:

$$V_s = \frac{U_s^{CDN}}{U_s}$$
$$= a_s \cdot \frac{D_s - U_s}{U_s} \tag{6}$$

to represent the contribution made by the CDN server to enlarging the upload bandwidth. As we mentioned before, each content is split into small segments to be cached, so we assume that the size of every single segment is θ and the number of cached segments of the content in swarm s is n_s. Then there exists the equation:

$$a_s = \frac{n_s \cdot \theta}{\eta_s} \tag{7}$$

Now our problem is to decide which contents should be cached and how many segments they deserve. To compare the value of caching different contents, we calculate the value of caching one segment of a certain content, which is called the unit value. Substituting $n_s = 1$, Eq. 7 into Eq. 6, the unit value of a certain content is formulated as:

$$uV_s = \frac{\theta}{\eta_s} \cdot \frac{D_s - U_s}{U_s} \tag{8}$$

Algorithm 2. Serving Requests

// a request for content s comes

1    $\mathcal{Q} \leftarrow$ priority queue of unit value;

2   **if** *content s is not in the cache* **then**

3   **if** *there is free space in the cache* **then**

4   cache content s as much as possible;

5   **else**

6   calculate uV_s and n_s;

7   replaceStrategy($\mathcal{Q}, uV_s, n_s$);

8   **end**

9   **else**

10   **if** *requested range $\subset$ cached range* **then**

11   serve the request directly;

12   **if** $\Delta n_s > 0$ **then**

13   replaceStrategy($\mathcal{Q}, uV_s, n_s$);

14   **end**

15   **end**

16   **Function** replaceStrategy($\mathcal{Q}, uV_s, n_s$)

17      $q \leftarrow$ number of the lowest valued content in $\mathcal{Q}$;

18   **while** $uV_s \geqslant uV_{s_q}$ *and* $n_s > 0$ *and* $q \geqslant 1$ **do**

19   **if** $n_{s_q} < 0$ **then**

20   CDN server evicts $\min(n_s, |n_{s_q}|)$ segments of content s_q and adds corresponding segments of content s;

21          $n_s - = \min(n_s, |n_{s_q}|)$;

22   **end**

23        $q - = 1$;

24   **end**

25   **end**

5.2 Caching Algorithm of CDN Servers

There are two parts of the proposed algorithm: *Refreshing Allocation State* and *Serving Requests*. We first introduce *Refreshing Allocation State* part. For a certain CDN server whose size is Λ, the total number of segments it can store is Λ/θ. It is reasonable that the higher unit value a certain content has, the more segments it deserve. But remember, n_s can not exceed η_s/θ. Then we can calculate n_s as:

$$n_{i,s} = \min(\frac{\eta_s}{\theta}, \frac{uV_s}{\sum_{s \in \text{contents in CDN server}} uV_s} \cdot \frac{\Lambda}{\theta}) \tag{9}$$

Apparently, uV_s is time-related, so it need to be refreshed periodically. We set Δt as the time window of refreshing allocation state. The pseudo-code of *Refreshing Allocation State* is presented in Algorithm 1. It is noteworthy that we maintain a value called Δn_s for every content in CDN server. It is the difference of new calculated n_s and its former counterpart. Δn_s is a crucial reference for adding and evicting contents in *Serving Requests* part. In addition, we rank contents according to their unit value, such that for contents ranked from 1 to n, we get $uV_{s_1} \geqslant uV_{s_2} \geqslant \ldots \geqslant uV_{s_n}$. All of these information is managed in a priority queue.

Now we present *Serving Requests* part of the algorithm. This part is triggered when a request for a certain content comes. If there are no segments of this content stored in CDN servers, it will decide whether store this new content. The new content will be cached only if there are free spaces or other contents in CDN servers whose Δn is negative and unit value is lower than the requested content's.

Then we discuss the situation that the requested content is stored in CDN server. Since contents are partially stored in CDN server, the requested content may not be completely cached. CDN servers can only provide the cached range. At the same time, they check Δn_s of the requested content. If it is positive, more segments of requested content should be cached. So CDN servers will evict corresponding segments of contents whose Δn is negative and unit value is lower than the requested content's. The pseudo-code of *Serving Requests* part is presented in Algorithm 2.

6   Evaluation

In this section, trace-driven simulations are presented to evaluate the performance of ERPC in two aspects: the bandwidth saved for centralized cloud and the data retrieving rates improved for clients.

6.1   Simulation Setup

We integrate the BitTorrent protocol into an open source personal cloud storage client, which uses OpenStack Swift as a storage back end. Then we use Peer-Sim [12] and the corresponding libraries to simulate edge resources. To make our simulations more close to the reality, we utilize statistical results of measurement studies on personal cloud conducted by Gracia-Tinedo et al. [13], and on BitTorrent networks conducted by T. Hossfeld et al. [14]. The former one has been performed from May 10, 2012, to July 15, 2012 using their university network (Universitat Rovira i Virgili, Spain) and PlanetLab. They measured the performance of three of the major Personal Cloud services in the market: DropBox, Box and SugarSync. The latter one has been performed for almost one year using the PlanetLab.

Based on these statistical results, we consider $S = 300$ swarms and the file size of each swarm varies from 1000 MB to 7 MB, which corresponds to the mean size range of a movie and a music. We use Zipf-Mandelbrot distribution presented in [15] to describe the probability of accessing a content ranked by popularity. Peers upload rate is in the range of 300 kbps to 50 kbps and download rate is in the range of 1500 kbps to 300 kbps. The arrival process of peers is modeled as a Poisson process.

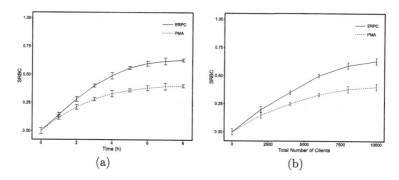

Fig. 2. Saved ratio of bandwidth for cloud

Every content is split into pieces of 256 KB to be cached, which is consistent with the BitTorrent protocol. CDN servers are set as a special kind of peers, with larger storage and bandwidth, and they can provide stable service. The time window Δt is set as 30 min. Every simulation runs for 8 h. The illustrated results are averages of 10 simulation runs with the 95 %-confidence intervals.

6.2   Bandwidth Saved for Centralized Cloud

To evaluate the performance of the ERPC, we choose a relatively similar work with ours presented in [16] to compare with. In [16], authors present a Protocol Management Algorithm (PMA) to select the distributing way between HTTP and BitTorrent in personal cloud. We set the scenario without any edge resources as the benchmark to calculate the Saved Ratio of Bandwidth for Cloud (SRBC) for these two schemes. Figure 2a and b show the ERPC can always save more bandwidth for cloud by contrast with PMA. With the increase of the clients number and simulation time, the advantage of ERPC is becoming more obvious.

6.3   Data Retrieving Rates Improved for Clients

Then we discuss the Gain Ratio of Data Retrieving Rate (GRDR). As is shown in Fig. 3a, the number of CDN servers is quite significant for the data retrieving rates, which prove the efficiency of the proposed caching algorithm. Figure 3b shows the gain ratio of data retrieving rate of each swarm. Note that the index of each swarm is assigned by following the reverse order of the corresponding contents popularity. As is illustrated in the figure, all of the swarms are beneficiaries in our framework, even the swarms whose contents are not placed in edge resources also enjoy enhancement because compared with the scenario without edge resources, more bandwidth of centralized cloud are saved for them. It is notable that popular swarms achieve more rise of data retrieving rate, which is consistent with the feature of P2P networks.

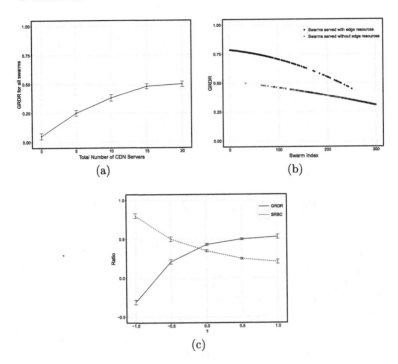

Fig. 3. Gain ratio of data retrieving rate and the impact of τ

6.4   The Impact of τ

As we mentioned before, τ is a crucial factor in selecting protocols for every content. The higher value the τ has, the fewer contents will be placed in edge resources, which means less bandwidth can be saved for cloud. But for Gain Ratio of Data Retrieving Rate (GRDR), higher τ indicates better rates of data retrieving. Figure 3c illustrates this trade-off process. In order to reduce the bandwidth cost and, at the same time, guarantee the quality of service, the value of τ should be set seriously in practice.

7   Conclusion

In this paper, we utilize voluntary peers in P2P networks and edge servers in CDNs to compose a novel framework to reduce the bandwidth cost in personal cloud without degrading the data retrieving rates. To achieve an impressive performance, we specifically deal with two problems: selecting the more profitable contents to be placed in edge resources and making up the unreliability of edge resources.

For the former problem, we take content size, request frequency, and distribution time into consideration and present an efficient approach to select serve type for each content. For the latter problem, we propose a caching algorithm for CDN servers to make up the instability of P2P networks. The proposed algorithm can appropriately cache the contents which can not be well provided by P2P networks. At last, we conduct series of trace-driven simulations to evaluate the performance of ERPC. The results show that the proposed framework achieves an impressive performance of saving bandwidth for cloud system and guaranteeing data retrieving rates for users. In the future, we will make our framework to be suitable for more cloud services.

Acknowledgment. The authors thank National University of Defense Technology for providing essential conditions to accomplish this paper. This work is supported by the NSFC under Grant 61103015, 61303191, 61402504 and 61303190.

References

1. iSuppli Research, I.: Subscriptions to cloud storage services to reach half-billion level this year (2012). https://technology.ihs.com/410084/
2. Payberah, A.H., Kavalionak, H., Kumaresan, V., Montresor, A., Haridi, S.: Clive: Cloud-assisted P2P live streaming. In: 2012 IEEE 12th International Conference on Peer-to-Peer Computing (P2P), pp. 79–90. IEEE (2012)
3. Yu, J., Lu, L., Li, Z., Wang, X., Su, J.: A simple effective scheme to enhance the capability of web servers using P2P networks. In: 2010 39th International Conference on Parallel Processing (ICPP), pp. 680–689. IEEE (2010)
4. Babaoglu, O., Jelasity, M., Kermarrec, A.M., Montresor, A., Van Steen, M.: Managing clouds: a case for a fresh look at large unreliable dynamic networks. ACM SIGOPS Operat. Syst. Rev. **40**(3), 9–13 (2006)
5. Babaoglu, O., Marzolla, M., Tamburini, M.: Design and implementation of a P2P cloud system. In: Proceedings of the 27th Annual ACM Symposium on Applied Computing, pp. 412–417. ACM (2012)
6. Zhang, Q., Cheng, L., Boutaba, R.: Cloud computing: state-of-the-art and research challenges. J. Internet Serv. Appl. **1**(1), 7–18 (2010)
7. Wu, J., Yuan, H., He, Y., Zou, Z.: Chordmr: a P2P-based job management scheme in cloud. J. Network. **9**(3), 541–548 (2014)
8. Riad, A., Elmogy, M., Shehab, A., et al.: A framework for cloud P2P VoD system based on user's behavior analysis. Int. J. Comput. Appl. **76**(6), 20–26 (2013)
9. Wang, M., Jayaraman, P.P., Ranjan, R., Mitra, K., Zhang, M., Li, E., Khan, S., Pathan, M., Georgeakopoulos, D.: An overview of cloud based content delivery networks: research dimensions and state-of-the-art. In: Hameurlain, A., Küng, J., Wagner, R., Sakr, S., Wang, L., Zomaya, A. (eds.) TLDKS XX. LNCS, vol. 9070, pp. 131–158. Springer, Heidelberg (2015)
10. Ling, L., Xiaozhen, M., Yulan, H.: Cdn cloud: a novel scheme for combining cdn and cloud computing. In: 2013 International Conference on Measurement, Information and Control (ICMIC), vol. 1, pp. 687–690. IEEE (2013)
11. Kumar, R., Ross, K.W.: Peer-assisted file distribution: the minimum distribution time. In: 1st IEEE Workshop on Hot Topics in Web Systems and Technologies, 2006. HOTWEB 2006, pp. 1–11. IEEE (2006)
12. Peersim. http://peersim.sourceforge.net/
13. Gracia-Tinedo, R., Artigas, S.: Actively measuring personal cloud storage. In: 2013 IEEE Sixth International Conference on Cloud Computing (CLOUD), pp. 301–308. IEEE (2013)

14. Hoßfeld, T., Lehrieder, F., Hock, D., Oechsner, S.: Characterization of bittorrent swarms and their distribution in the internet. Comput. Network. **55**(5), 1197–1215 (2011)
15. Hefeeda, M., Saleh, O.: Traffic modeling and proportional partial caching for Peer-to-Peer systems. IEEE/ACM Trans. Network. **16**(6), 1447–1460 (2008)
16. Chaabouni, R., Sánchez-Artigas, M., Garcia-Lopez, P.: Reducing costs in the personal cloud: Is bittorrent a better bet? In: 14-th IEEE International Conference on Peer-to-Peer Computing (P2P), pp. 1–10. IEEE (2014)

Multidimensional Similarity Join Using MapReduce

Ye Li, Jian Wang, and Leong Hou U$^{(\boxtimes)}$

Zhuhai Research Institute, University of Macau, Macau, China
{yb47438,ryanlhu}@umac.mo, wjcqu@hotmail.com

Abstract. Similarity join is arguably one of the most important operators in multidimensional data analysis tasks. However, processing a similarity join is costly especially for large volume and high dimensional data. In this work, we attempt to process the similarity join on MapReduce such that the join computation can be scaled horizontally. In order to make the workload balancing among all MapReduce nodes, we systemically select the most profitable feature based on a novel data selectivity approach. Given the selected feature, we develop the partitioning scheme for MapReduce processing based on two different optimization goals. Our proposed techniques are extensively evaluated on real datasets.

1 Introduction

With the development of information network in recent years, tremendous amount of data is being generated everywhere and anytime. Along with the rapid growth of the large-scale data volume, the variety of data also increases dramatically as a prospering result of online applications which ubiquitously record different types of data and provide various services. The variety of data types dwells in real-world applications [1]. As an example in social networks, spatio-temporal information and user preferences can play important roles in providing high-quality recommendation to users.

While the volume and variety of data contains valuable information for various potential applications, the complexity of data incurs data management and query processing problems. Massive Multidimensional Similarity Join (MDSJ) operation is one of the crucial problems in large-scale data processing. Given two sets of data with multiple data features, and the corresponding similarity functions on each features, MDSJ operation identifies a subset of pairs that satisfy the similarity thresholds. The curse of dimensionality and diverse behaviors in the join operations render difficulties in large data sets.

To process large-scale data, MapReduce [2], a high-scalability computation model in a distributed environment, has been introduced to overcome the efficiency in massive join operations [3–5]. Although computing the join on multiple machines in parallel can boost the efficiency, the performance gain is very sensitive to the data skewness. Naive partitioning schemes on MapReduce may suffer from data load balancing problem.

© Springer International Publishing Switzerland 2016
B. Cui et al. (Eds.): WAIM 2016, Part II, LNCS 9659, pp. 457–468, 2016.
DOI: 10.1007/978-3-319-39958-4_36

Recent developments of the similarity join on MapReduce, such as SAND [6] and SALA [7], have been proposed to remedy the performance degradation of the skewed data. The partitioning schemes in these methods are based on the data distribution estimated by a pre-sampling process. However, these methods focus on one single data feature and none of them is readily support the join operation on multi-features.

In this work, we propose a MapReduce framework to process the similarity join on multidimensional feature spaces. Our framework first estimates the selectivity ratio of each feature by a sampling method. Based on the selectivity ratio of each feature, we identify the most profitable feature to partition the data for the MapReduce processing. In addition, we further study two algorithms to optimize the join processing subject to different optimization goal.

In summary, our major contributions are listed as followings:

1. We propose a MapReduce-based similarity join framework for multidimensional feature spaces. We attempt to identify the profitability of each feature for data partitioning by our selectivity ratio estimation;
2. With the optimization goal (e.g., the data-oriented and the computation-oriented), we propose a dynamic programming based algorithm to minimize the skewness among the partitions;
3. We conduct extensive experiments to demonstrate the efficiency of our approaches for the multidimensional similarity join.

The remainder of this paper is organized as follows. The definition and the preliminary of the problem will be introduced in Sect. 2 and prior solutions will be discussed in Sect. 3. Section 4 will show the detail of our proposed methods and the experimental results are demonstrated in Sect. 5. Literature reviews are given in Sect. 6. We will conclude our techniques and discuss future work in the last section.

2 Preliminary

In this section, we formalize the problem of multidimensional similarity join and discuss three different data features that generally capture the characteristic of multidimensional data involved in typical similarity join.

Problem 1 (Multidimensional Similarity Join (MDSJ)). Given two multidimensional datasets R and S with k feature spaces $f_1, f_2, ..., f_k$ and the threshold of each feature θ_i, MDSJ returns a subset $\{(r,s)|r \in R, s \in S\}$ subject to the feature thresholds, i.e.,

$$sim(r.f_1, s.f_1) < \theta_1, \quad sim(r.f_2, s.f_2) < \theta_2, \quad ..., \quad sim(r.f_k, s.f_k) < \theta_k \quad (1)$$

As the similarity definition is varying on different features, in this work we discuss three main types of data features (including single value data, d-dimensional data, and multi-value data) where these data are frequently employed in multidimensional analytical tasks. The single value data is the most common feature.

For instance, the temporal data records a specific activity or an event for each timestamp. The d-dimensional data can represent the spatial locations when $d = 2$ which are widely adopted in spatial analytical tasks. The last data type can be viewed as categorical data which can indicate the preference of a user. The similarity function of these three data types is defined respectively as followings:

Definition 1 (Single Value Similarity). The single value data similarity $sim_{f_i}(r.f_i, s.f_i)$ between $r.f_i$ and $s.f_i$ is defined as their absolute value difference.

$$sim_{f_i}(r.f_i, s.f_i) = |r.f_i - s.f_i| \tag{2}$$

Definition 2 (d-dimensional Similarity). The d-dimensional data similarity of two objects $r.f_j$ and $s.f_j$ is defined by their Euclidean distance.

$$sim_{f_j}(r.f_j, s.f_j) = \sqrt{(r.f_{j_1} - s.f_{j_1})^2 + ... + (r.f_{j_d} - s.f_{j_d})^2} \tag{3}$$

Definition 3 (Multi-Value Similarity). The multi-value data similarity between $r.f_m$ and $s.f_m$ is defined by the weighted Jaccard coefficient of the two token sets.

$$sim_{f_m}(r.f_m, s.f_m) = \frac{\sum_{t \in r.f_m \cap s.f_m} w(t)}{\sum_{t \in r.f_m \cup s.f_m} w(t)} \tag{4}$$

where $w(t)$ is the weight of token t.

3   Existing Solutions

In this section, we will discuss the state-of-the-art join processing on MapReduce for single value feature, d-dimensional feature [8] and multi-value feature [9]. Besides, we introduce the MapReduce adaption of a standalone algorithm, EGO [10,11] as a baseline algorithm in this work, which is the state-of-the-art method for the multidimensional similarity join to the best of our knowledge.

3.1   Single Value Join on MapReduce

For the single value join, a typical MapReduce-based join algorithm divides the data into buckets based on their values, where each bucket is assigned a value range (based on the similarity threshold) and every data falls into this range will be assigned to this bucket. The data within the same bucket will be processed by the same processing node.

3.2 d-Dimensional Data Join on MapReduce

For the d-dimensional data (e.g., spatial objects), we use the method introduced in the paper [8]. For simplicity, we take 2-dimensional objects as an example. Given the similarity threshold (Euclidean distance), the 2-dimensional space is divided into regular cells and the objects are projected into their corresponding cells based on their values. In the MapReduce processing, we locally join the objects of a cell by processing node. To secure the correctness of the join, each processing node is required to collect the objects of the neighbor cells (e.g., 8 cells in total for 2-dimensional space).

3.3   Multi-value Data Join on MapReduce

For the multi-value data (e.g., the textual tags), we use the set similarity join approach first proposed in [9]. Their approach first builds a total order of the tokens (e.g., the tags) based on their frequency, where the frequency of the tokens can be calculated by a MapReduce process (e.g., similar to WordCount). Then, it extracts the top n tokens as the prefix tokens for subsequent computation, where n is designed based on the threshold of the Jaccard similarity. Each data object is then assigned to these n buckets based on their own prefix. The join process is then computed locally in each Mapper and the union of these results is the final result of the similarity join.

3.4   EGO Algorithm Using MapReduce

The methods introduced above use the similarity threshold to partition the data. However, their performance is highly sensitive to the data skewness of these features. For instance, the spatial objects are always very dense in city center but are very sparse in widepark areas. To address this problem, EGO [10,11] proposes a join algorithm that attempts to make the workload balancing among all sub-join processes. Their idea is to build a virtual grid with small cell length ϵ and try to merge these ϵ-cells to form a set of groups having identical data size. Afterwards, each group in the first dataset join with the corresponding groups in the second dataset in order to secure the similarity result. The EGO algorithm achieves good workload balancing for the join processing as every group pair computation has identical number of objects. This motivates us to extend their work on MapReduce.

In MapReduce, the EGO algorithm first employs a MapReduce job to sort the data and then emit the same number of objects to different reducers. Each reducer joins the received objects locally and report the join result. Compared with the basic join solutions in MapReduce, the EGO algorithm achieves better data balance and incurs less replication. However, the sorting and the partitioning strategy in EGO algorithm are only based on one of the dimensional features (e.g., single value, d-dimensional, or multi-value) which may degrade the performance significantly if the selected feature is not well distributed (e.g., very skewed for a portion of data).

4   Our Approach

4.1   Overview

Our framework includes two major components, data sampling and partitioning scheme, to address the feature selection problems discussed in Sect. 3. In order to avoid selecting a bad feature for data partitioning, we consider all possible features by our data sampling technique and estimate the most profitable feature (having the lowest selectivity) for data partitioning. We summarize our proposed techniques in the following steps.

1. Step 1: An off-line sampling process is initiated for the entire data sets and thereafter we can estimate the selectivity ratio of the features from the sampled data. The most profitable partition feature is then identified based on the selectivity ratios.
2. Step 2: Based on the chosen partition feature and the optimization goal (e.g., data balance or computation balance), the partitioning scheme is then decided by our partitioning algorithms.

4.2   Sampling and Selectivity

In this section, we introduce our general sampling algorithm which estimates the selectivity ratio of different features by only sampling on the raw data.

General Sampling Algorithm. Suppose there are two data sets R and S of F features $f_1, f_2, ..., f_F$ in the MDSJ, the goal of the general sampling algorithm is to estimate the selectivity of each features and then output the most profitable partition feature. For a feature f_k, we denote ρ_k as its selectivity factor, $R \bowtie_{f_k} S$ as the join of R and S on f_k and $|R \bowtie_{f_k} S|$ as the number of intermediate results. Therefore, the selectivity ratio can be defined as:

$$\rho_k = \frac{|R \bowtie_{f_k} S|}{|R||S|} \tag{5}$$

The selectivity ratio indicates the percentage of all possible pairs $|R||S|$ remained in the join result. The lower the selectivity a feature f_k is, the more profitable f_k is if we select f_k as the partition feature. However, estimating $|R \bowtie_{f_k} S|$ for a feature f_k requires exhausted join operations based on f_k and this overhead is obviously prohibitive for the MDSJ processing. To overcome this challenge, we propose our general sampling process which can sample a pair of record (r_i, s_j) from $|R \bowtie_{f_k} S|$ subject to a uniform probability $\frac{1}{|R \bowtie_{f_k} S|}$. Accordingly, we can estimate the selectivity of the feature f_k based on the sampling information and the estimation accuracy is increased with the sample size.

Before discussing the detail of our method, we firstly introduce a concept, *candidate set*, that is thoroughly used in our subsequent discussions. Given a specific partitioning scheme and a data object $r_i \in R.$, we say a data object $s_j \in S$ is in the candidate set if and only if s_j is assigned to the same cell (based on the partitioning scheme) or the neighbor cells of r_i. Given the feature space f_k and the candidate set of r_i denoted as $C_{f_k}(r_i)$, we show the procedures of our General Sampling Algorithm in Algorithm 1 which estimate the selectivity of the join result by a sampling processing. The correctness of our sampling processing is shown in Theorem 1.

Theorem 1. *The General Sampling Algorithm samples a pair of record (r_i, s_j) from $R \bowtie_{f_k} S$ with probability $\frac{1}{|R \bowtie_{f_k} S|}$.*

Algorithm 1. General Sampling Algorithm

Definition:
1: R, S: Data Sets
2: $|r|$: the sample size
Algorithm:
3: Scan both data sets to derive $C_{f_k}(r_i)$ and $w_{r_i} = |C_{f_k}(r_i)|$ for all $r_i \in R$, let w_{max} be the maximum among all w_{r_i}
4: **repeat**
5:      Sample a random record $r_i \in R$ from R with uniform distribution
6:      Sample a random record $s_j \in S$ from $C_{f_k}(r_i)$ with uniform distribution
7:      If $(r_i, s_j) \in R \bowtie S$, then go to next step, otherwise reject the sample
8:      Output (r_i, s_j) as a accepted sample with a probability $\frac{w_{r_i}}{w_{max}}$
9: **until** $|r|$ records have been accepted

Proof. For each iteration, Algorithm 1 attempts to sample a pair of objects (r_i, s_j), where r_i is sampled from R with probability $\frac{1}{|R|}$ and s_j is sampled from $C_{f_k}(r_i)$ with a probability $\frac{|r_i \bowtie_{f_k} S|}{w_{r_i}}$ where $w_{r_i} = |C_{f_k}(r_i)|$. The sampled pair (r_i, s_j) is accepted only if it passes a check with $\frac{w_{r_i}}{w_{max}}$ probability.

Given a specific data r_i, the probability p_{r_i} of accepting a pair (r_i, s_j) is:

$$p_{r_i} = \frac{1}{|R|} \times \frac{|r_i \bowtie_{f_k} S|}{w_{r_i}} \times \frac{w_{r_i}}{w_{max}} = \frac{|r_i \bowtie_{f_k} S|}{|R| w_{max}} \tag{6}$$

Thereby, the probability of accepting a generated pair for every $r_i \in R$ is:

$$P = \sum_{r_i \in R} p_{r_i} = \sum_{r_i \in R} \frac{|r_i \bowtie_{f_k} S|}{|R| w_{max}} = \frac{\sum_{r_i \in R} |r_i \bowtie_{f_k} S|}{|R| w_{max}} = \frac{|R \bowtie_{f_k} S|}{|R| w_{max}} \tag{7}$$

Given a specific pair (r_i, s_j), the probability p of accepting this pair is:

$$p_{(r_i, s_j)} = \frac{p_{r_i}}{|r_i \bowtie_{f_k} S|} = \frac{|r_i \bowtie_{f_k} S|}{|R| w_{max}} \times \frac{1}{|r_i \bowtie_{f_k} S|} = \frac{1}{|R| w_{max}} \tag{8}$$

Let Q be the probability of rejecting a record in an iteration, i.e., $Q = 1 - P$. The probability of a sampled record (r_i, s_j) is accepted at the m-th iteration is $Q^{m-1} p_{(r_i, s_j)}$. Accordingly, we have

$$\sum_{m=1}^{\infty} Q^{m-1} p_{(r_i, s_j)} = p_{(r_i, s_j)} \frac{1 - Q^{\infty}}{1 - Q} = p_{(r_i, s_j)} \frac{1}{1 - Q} = \frac{p_{(r_i, s_j)}}{P} = \frac{1}{|R \bowtie_{f_k} S|} \tag{9}$$

Hence, a single pair of record $(r_i, s_j) \in R \bowtie_{f_k} S$ is sampled by our general sampling algorithm with the probability of $\frac{1}{|R \bowtie_{f_k} S|}$. $\square$

Selectivity Estimation. So far we have discussed how we generate the samples with a proper probability $\frac{1}{|R\bowtie_{f_k} S|}$. In this subsection, we introduce how to estimate the selectivity ρ_k of a feature f_k based on the sampled records.

As analysis in the last section, we already have the probability of accepting a record in an iteration. Therefore, the expected number of iterations to obtain one random sample (e.g., (r_i, s_j)), denoted by E, is:

$$E = |R \bowtie_{f_k} S| \times \sum_{m=1}^{\infty} mQ^{m-1} p_{(r_i, s_j)}$$

$$= \frac{|R \bowtie_{f_k} S|}{|R|w_{max}(1 - Q)^2}$$

use Eq. 7 to replace $(1 - Q)^2$

$$= \frac{w_{max} \times |R|}{|R \bowtie_{f_k} S|} = \frac{\frac{w_{max} \times |R|}{|R||S|}}{\frac{|R\bowtie_{f_k} S|}{|R||S|}} = \frac{w_{max}}{\rho_k |S|}$$

Thereby, the relation between ρ_k and E can be established by $\rho_k = w_{max}/(E|S|)$. Hence the estimation of selectivity for a feature f_k is identical to the estimation of the expected number of iterations to obtain a sample. From the general sampling process, the estimation of E can be estimated as:

$$E \approx \frac{1}{|r|} \sum_{n=1}^{|r|} I_n \qquad (10)$$

where I_n is the number of iterations for sampling the n-th record. Thus ρ_k can be estimated by:

$$\rho_k \approx \frac{|r|w_{max}}{\sum\limits_{n=1}^{|r|} I_n |S|} \qquad (11)$$

As the sampling size $|r|$ increases, the estimation of E will approach to the expected value which results in the increasing of the estimation accuracy of ρ_k.

So far we have proposed our general sampling algorithm and the estimation of selectivity for a single feature f_k. To evaluate the selectivity of all F features, our general sampling algorithm will be run for F times picking each feature as input. Afterwards, we select the feature with the lowest selectivity ratio as the most profitable partition feature to partition the data.

4.3   Partitioning Algorithm

Given the selected feature f_k and the corresponding partitioning scheme, we assign data buckets of R and S into partitions based on the similarity threshold θ_k. Our goal is to make all partitions even in terms of their data size or their computation cost. To archive this goal, we propose a dynamic programming algorithm, which finds an evenly distributed partitioning scheme based on the sub-partitioning schemes. Suppose there are a sorted list of M buckets and the cost (e.g., data size or computation cost) of the i-th bucket is $|b_i|$. We partition the sequence of M buckets into N processing nodes evenly, where the objective is to minimize the cost of the maximum (i.e., the most costly) partition. Hence we have our dynamic programming function as follows.

$$P(m, k) = \begin{cases} \sum_{i=m}^{M} |b_i| & \text{if } k = 1 \\ \min_{m \leq j < M-k+1} [\max(\sum_{i=j}^{j} |b_i|, P(j+1, k-1))] & \text{otherwise.} \end{cases}$$

where $P(m, k)$ indicates the minimum cost (either the data size or the computation cost) of assigning the m-th bucket to the last buckets evenly into the k partitions. Our dynamic programming algorithm starts by $P(m = 1, k = N)$. In the first iteration, it assigns j buckets into the first partition and then go to a subproblem $P(m = j+1, k = N-1)$ that attempts to partition $M - j$ buckets into $N - 1$ partitions evenly.

5   Empirical Studies

5.1   Datasets and System Environments

The experiments are conducted on two real datasets.

- Paris dataset (3 features, 796,000 records) is an image collection of Paris that we collected from Flickr by its API. The dataset contains 796,427 images that contains temporal (single value), spatial (2-dimensional locations), textual (multi-value).
- Flickr dataset (20 features, 100,000,000 records) is a collection from Flickr that provides by Yahoo Lab Webscope. It contains 100 M images/videos with 23 attributes. We extract two subsets, $|R| = 7,147,956$ and $S = 7,223,493$, by removing the records with missing values. These two subsets contain 3 single value features (e.g., date token, date uploaded, and accuracy), a 2-dimensional feature (e.g., geo-location) and 3 multi-value features (e.g., user tags, capture device, and the photo description).

For the empirical studies, we use our in-house cluster with one master and four slave nodes running with Hadoop 2.6.0. Each data node has 8 cores and 50 GB memory so there are 32 cores and 200 GB memory in total.

5.2   Experiments

To verify the effectiveness of our general sampling algorithm, we compare the estimated selectivity ratio with the ground truth. There are two relational databases R and S and each of them contains 10,000 record respectively. As demonstrated in Fig. 1(a), with the increase of sampling size, the estimated

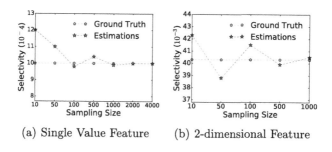

(a) Single Value Feature   (b) 2-dimensional Feature

Fig. 1. Effect of the general sampling algorithm

value progressively approaches to the true selectivity ratio that verifies the effectiveness of our sampling algorithm. We also verify the effectiveness for 2-dimensional data, where R and S contains 10000 records of geo-locations, respectively. Figure 1(b) shows similar trend to that of Fig. 1(a).

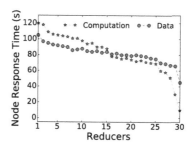

Fig. 2. Effect of two optimizations

In Fig. 2, we demonstrate the effectiveness of two optimization goals (i.e., data size and computation cost denoted by Data and Computation respectively) for our partitioning scheme on the Flickr dataset, where the execution time of the reducers (30 in total) are reported. As expected, the computation cost optimization provides better partitions in terms of the computation cost.

Next, we investigate the overall performance of the framework compared with the baseline algorithms (including all four methods discussed in Sect. 3) on the Paris and the Flickr data set in Fig. 3. We denote the basic MapReduce partitioning schemes on temporal, spatial and textual features as Temporal, Spatial and Textual respectively. Also, we denote EGO algorithm using MapReduce as EGO. Noted that for each experiment, the EGO algorithm decides its partitioning scheme based on the varied feature, e.g., Fig. 3(a) and (b) based on the temporal feature. Similarly, we denote our proposed methods in terms of data size and computation cost as Data and Computation respectively. We measure the performance by the overall running time, including the off-line sampling process, off-line partitioning scheme computation and MapReduce process. In this set of

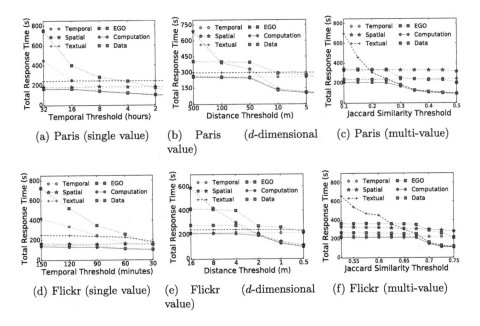

(a) Paris (single value)   (b) Paris (d-dimensional value)   (c) Paris (multi-value)

(d) Flickr (single value)   (e) Flickr (d-dimensional value)   (f) Flickr (multi-value)

Fig. 3. Effect of varying thresholds

experiments, we vary the similarity threshold to demonstrate the robustness of our partitioning approaches. In Fig. 3(a) and (b), we vary the threshold of the temporal (i.e., the single value) feature. The temporal-based algorithm (Sect. 3.1) and the EGO algorithm (Sect. 3.4) are very sensitive to the similarity threshold due to their partitioning scheme. Note that the other two competitors, spatial (Sect. 3.2) and textual (Sect. 3.3) are not sensitive to the temporal threshold as they use spatial and textual feature as their partitioning scheme, respectively. Our methods outperform all these baseline algorithms as we can identify the most profitable feature and use it in the partitioning scheme. Similar results can be found in Fig. 3(c), (d), (e), and (f) where we vary the spatial threshold and Jacard similarity threshold. In summary, our partitioning algorithm based on the selectivity ratio estimation is effective to find the most profitable feature for the partitioning. This result is beneficial to the MapReduce processing.

6 Related Work

MapReduce-based join has been intensively studied in recent years. [3] is the first attempt to address the join problem using the MapReduce framework. A Merger phase is appended after the Reducer phase, which can efficiently merge the join result already partitioned and sorted. [4] presents a comprehensive analysis on a series of MapReduce join algorithms. [5] specifically compares the Map-side join and Reduce-side join and analyzes these two algorithms thoroughly. The paper also proposes an Improved Map-side Join which adds a partition job to make full use of the Map-side join.

Generally, hash based partition schemes suffer from skewed data distribution. Partition schemes accounting for skewed distribution can generate even partitions resulting in less total running time. Such techniques require prior knowledge on the skewed data distributions which can be obtained by sampling algorithm. SAND [6] and SALA [7] take advantage of the estimated data distributions and then design their partition schemes based on balanced data partition or additionally accounting for the locality of the data. However, most of previous research on MapReduce-based join restricts in single feature. Our work has demonstrated that the selectivity of features has great impact on the overall performance and we can benefit greatly from our sampling technique that can identify the most profitable partition feature.

Prior studies have designed algorithms for sampling records uniformly from the join results [12,13]. [13] is the first work to sample records from the original data sets without the joins operation based on a uniform distribution assumption. [12] proposes a more efficient weighted sampling method on a dataset R based on the frequency statistics of another dataset S. However, their algorithms are designed in the relational database. Our general sampling algorithm shares similar ideas but we extend it to support multi feature spaces.

Sampling techniques have been introduced to process MapReduce-based join for high dimensional objects [14,15]. Specifically, these work employ both sampling and clustering techniques to balance the data distribution at the partitioning phrase. However, the closeness between the objects in these work is measured based on an aggregation function among all features. This is different from MDSJ as the closeness is measured by d similarity functions (cf. Eq. 1) instead of just one single function.

7   Conclusion

In this paper, we study the problem of processing multidimensional similarity join efficiently in large-scale data sets. We propose our MapReduce-based framework for the multidimensional similarity join. We propose a general sampling algorithm that can identify the most profitable partition feature and therefore our framework is robust to join on multi feature spaces. On the other hand, we study the data-size based and computation cost based optimizations for the partitioning scheme. Experiments demonstrate that our framework is superior to other baseline algorithms for dataset having multi feature spaces. Our future work includes considering the selectivity of multiple data features and designing the partitioning scheme based on multiple features at the same time.

Acknowledgements. This work was supported by grant MYRG109(Y1-L3)-FST12-ULH from UMAC Research Committee and grant NSFC 61502548 from National Natural Science Foundation of China.

References

1. Gunopulos, D., Kollios, G., Tsotras, J., Domeniconi, C.: Selectivity estimators for multidimensional range queries over real attributes. VLDB J.–Int. J. Very Large Data Bases **14**(2), 137–154 (2005)
2. Dean, J., Ghemawat, S.: Mapreduce: simplified data processing on large clusters. Commun. ACM **5**(1), 107–113 (2008)
3. Yang, H., Dasdan, A., Hsiao, R.L., Parker, D.S.: Map-reduce-merge: simplified relational data processing on large clusters. In: Proceedings of the 2007 ACM SIGMOD International Conference on Management of Data, pp. 1029–1040. ACM (2007)
4. Blanas, S., Patel, J.M., Ercegovac, V., Rao, J., Shekita, E.J., Tian, Y.: A comparison of join algorithms for log processing in mapreduce. In: Proceedings of the 2010 ACM SIGMOD International Conference on Management of data, pp. 975–986. ACM (2010)
5. Wu, F., Wu, Q., Tan, Y.: Comparison and performance analysis of join approach in mapreduce. In: Yuan, Y., Wu, X., Lu, Y. (eds.) Trustworthy Computing and Services. CCIS, vol. 320, pp. 629–636. Springer, Heidelberg (2013)
6. Atta, F., Viglas, S.D., Niazi, S.: Sand joina skew handling join algorithm for google's mapreduce framework. In: 2011 IEEE 14th International Multitopic Conference (INMIC), pp. 170–175. IEEE (2011)
7. Lin, Z., Cai, M., Huang, Z., Lai, Y.: SALA: A skew-avoiding and locality-aware algorithm for mapreduce-based join. In: Dong, X.L., Yu, X., Dong, X.L., Li, J., Sun, Y., Sun, Y. (eds.) WAIM 2015. LNCS, vol. 9098, pp. 311–323. Springer, Heidelberg (2015). doi:10.1007/978-3-319-21042-1_25
8. Zhang, S., Han, J., Liu, Z., Wang, K., Xu, Z.: SJMR: Parallelizing spatial join with mapreduce on clusters. In: 2009 IEEE International Conference on Cluster Computing and Workshops, CLUSTER 2009, pp. 1–8. IEEE (2009)
9. Vernica, R., Carey, M.J., Li, C.: Efficient parallel set-similarity joins using mapreduce. In: Proceedings of the 2010 ACM SIGMOD International Conference on Management of data, pp. 495–506. ACM (2010)
10. Böhm, C., Braunmüller, B., Krebs, F., Kriegel, H.P.: Epsilon grid order: an algorithm for the similarity join on massive high-dimensional data. In: ACM SIGMOD Record, vol. 30, pp. 379–388. ACM (2001)
11. Kalashnikov, D.V.: Super-ego: fast multi-dimensional similarity join. VLDB J. Int. J. Very Large Data Bases **22**(4), 561–585 (2013)
12. Chaudhuri, S., Motwani, R., Narasayya, V.: On random sampling over joins. In: ACM SIGMOD Record, vol. 28, pp. 263–274. ACM (1999)
13. Olken, F., Rotem, D.: Simple random sampling from relational databases. In: VLDB, vol. 86, pp. 25–28 (1986)
14. Das Sarma, A., He, Y., Chaudhuri, S.: Clusterjoin: a similarity joins framework using map-reduce. Proc. VLDB Endowment **7**(12), 1059–1070 (2014)
15. Wang, Y., Metwally, A., Parthasarathy, S.: Scalable all-pairs similarity search in metric spaces. In: Proceedings of the 19th ACM SIGKDD International Conference on Knowledge Discovery and Data Mining, pp. 829–837. ACM (2013)

Real-Time Logo Recognition from Live Video Streams Using an Elastic Cloud Platform

Jianbing Ding[1,2(✉)], Hongyang Chao[1], and Mansheng Yang[3]

[1] School of Data and Computer Science, Sun Yat-Sen University, Guangzhou, China
dingsword@gmail.com, isschhy@mail.sysu.edu.cn
[2] SYSU-CMU Shunde International Joint Research Institute, Foshan, China
[3] School of Computing, National University of Singapore, Singapore, Singapore
mansheng@u.nus.edu

Abstract. Real-time logo recognition from a live video stream has promising commercial applications. For example, a sports video website broadcasting a live soccer match could show advertisements of brands when their logos appear in the video. Although logo recognition is a well-studied problem, the vast majority of previous work focuses on recognition *accuracy*, rather than system *efficiency*. Consequently, existing methods cannot recognize logos in real-time, especially when a large number of logos appear in the video. Motivated by this, we propose a general framework that converts an offline logo detection method to a real-time one, by utilizing the massive parallel processing capabilities of an elastic cloud platform. The main challenge is to obtain high *scalability*, meaning that logo recognition efficiency keeps improving as we add more computing resources, as well as *elasticity*, meaning that the resource allocation is guided by the current workload rather than the peak load. The proposed framework achieves these by balancing workload, elastically provisioning resources, minimizing communication overhead, and eliminating performance bottlenecks in the system. Experiments using real data demonstrate the high efficiency, scalability and elasticity of the proposed solution.

Keywords: Real-time streams · Logo detection · Elastic cloud platform

1 Introduction

Live video broadcasting has become increasingly popular nowadays as more and more people have high-speed Internet access, e.g., through 4G LTE or fiber optic broadband. For instance, a sporting event, such as Olympic Games or FIFA World Cup, can be broadcast live; meanwhile, services such as Periscope (http://www.periscope.tv) allow users to create their own live broadcastings. Such live video streams create enormous potentials for real-time object recognition technologies. This paper focuses on real-time logo detection. Figure 1 shows two example live videos, one for a soccer game and another for a makeup lesson. Both videos contain sponsors' logos, and the live broadcasting service typically reaps revenue through advertising. Showing ads with the right timing, i.e., when the corresponding logos of the sponsor's brands appear in the video,

© Springer International Publishing Switzerland 2016
B. Cui et al. (Eds.): WAIM 2016, Part II, LNCS 9659, pp. 469–480, 2016.
DOI: 10.1007/978-3-319-39958-4_37

helps improve click through rates [10], which in turn leads to higher revenue. Hence, recognizing these logos in real time has significant commercial potential.

(a) Live soccer game   (b) Live makeup lesson

Fig. 1. Examples of logo detection from live video streams

As we review in Sect. 2, existing techniques for logo recognition largely focus on recognition accuracy. Until recently little attention has been paid on the efficiency of the algorithm, which is critical for obtaining real-time response, meaning that the algorithm must be fast enough to keep up with the frame rate of the video. Designing a new algorithm that achieves real-time speed without losing accuracy is difficult. Moreover, logo recognition is typically done by matching parts of a video frame with pre-registered logos; thus, the amount of computations depends upon the number of registered logos. Consequently, *the workload of logo recognition can fluctuate during a video stream*, depending on the number of registered logos. Meanwhile, different logos can be registered at different sections of a video stream. For instance, in a makeup lesson, brands for eye makeups can be very different from those for cheekbones. To ensure real-time recognition at all times, current methods require computational resources (e.g., CPU cores and memory) sufficient for the *peak* workload, which can be inefficient in terms of resource usage.

Motivated by this, we propose a novel framework that converts an existing offline logo recognition algorithm to a real-time one that runs on top of a cloud platform. The recognition accuracy of our solution is identical to that of the underlying offline logo detection algorithm, and our framework is compatible with a wide range of such offline algorithms. Further, the proposed solution is *elastic*, meaning that its efficiency is monitored constantly and adjusted on the fly to match the *current* workload (rather than the peak load), by dynamically provisioning computational resources from the cloud. Since cloud providers typically provide a pay-as-you-go pricing model, such elastic resource provisioning is both efficient and economic. Currently, existing stream processing systems provide insufficient elasticity for our purposes, because they typically involve high overhead when changing the amount of computational resources (e.g., virtual machines) for one application. Hence, we also propose an effective elastic processing module that is integrated into a popular stream processing system, Apache Storm [15]. Experiments using real video streams and logos confirm the efficiency, scalability and elasticity of the proposed solution. In the following, Sect. 2 surveys related work; Sect. 3 presents our logo detection algorithms. Section 4 describes the proposed elastic processing module that is integrated into Storm [15]. Section 5 contains an extensive set of experimental evaluations, respectively; Sect. 6 concludes with future directions.

2 Background

Offline Logo Detection. Logo recognition and related problems such as duplicate image detection (e.g., detect online videos that contain copyrighted contents) have been well studied in the computer vision literature. The basic problem can be formulated as follows: given two images, I_1 and I_2 (e.g., I_1 is a logo and I_2 is a video frame), we are to find all appearances of I_1 in I_2. Matching images is challenging because the algorithm must be robust against various transformations such as scaling, cropping, rotation, changes of lighting conditions, etc. Hence, most existing methods, e.g., [8, 16, 17], focus mainly on accuracy, and they tend to be rather inefficient. Consequently, they cannot be directly applied to real-time logo recognition. In particular, their time complexity is usually super-linear to the sizes of I_1 and I_2.

When the target image I_2 is large (e.g., a high-definition video frame), one way to obtain a tradeoff between accuracy and efficiency is to divide I_2 into smaller *patches*, and perform logo recognition on individual patches [9]. Figure 2 shows an example, in which image I_2 is split into 9 overlapping rectangular patches P_1–P_9. Specifically, I_2 is first partitioned into 4 patches P_1–P_4. Note that if we only search within these patches, we would miss appearances of I_1 that lie at the boundaries between adjacent patches. To fix this, we generate 5 additional patches P_5–P_9. Due to the super-linear complexity of image matching, searching in 9 smaller patches can be faster than in a large one. However, the former may miss appearances of I_1 that is larger than a patch. Several systems, e.g., [13], attempt to accelerate logo recognition using GPUs. These methods, however, suffer from two inherent limitations: (i) data transmission, e.g., from main memory to graphic memory, is often a major bottleneck; (ii) the GPU usually cannot perform an entire algorithm alone, and instead must cooperate with the CPU, which incurs communication overhead and (iii) there is a limit on the processing power of a single GPU, and high-end GPUs are rather expensive.

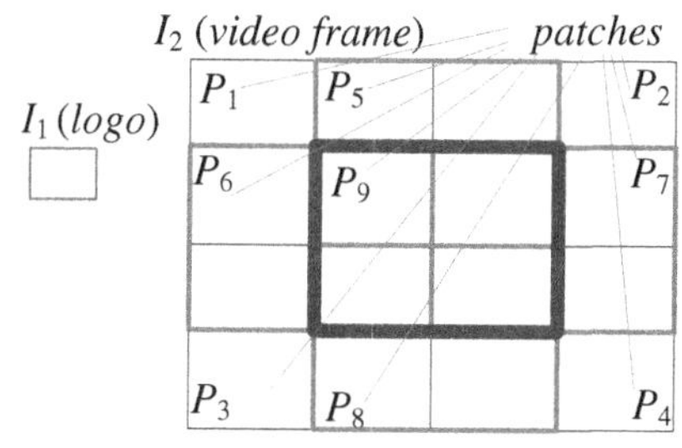

Fig. 2. Example of splitting an image into patches

Cloud-Based Stream Processing. Recently, the cloud computing community has devoted much efforts to building real-time, cloud-based stream processing engines

(SPEs) to manage fast data streams, such as Storm [15], Samza [1], S4 [11], and Spark Streaming [19]. One common characteristic of these systems is that they dynamically schedule computational resources (e.g., CPU cores and memory) to execute tasks, each of which can be understood as a fragment of work required by the application that can be performed in parallel. Several research prototypes, e.g., StreamCloud [6] and Resa [14] (which includes DRS [3]) support *elastic resource allocation*, meaning that resources can be dynamically added to or removed from an application to match its current workload, e.g., using the Abacus framework [2, 20]. Although these systems have obtained success in relatively simple streaming applications, such as text stream analytics, no existing work to our knowledge has used them for multi-media stream processing. In the next section, we identify the challenges involved in performing real-time logo detection on a cloud-based SPE, and describe the proposed solutions.

3   Cloud-Based Logo Recognition

Given a video stream consisting of frames, and a set of pre-registered logos, our goal is to recognize all appearances of these logos in every frame of the video stream. The main idea of our proposal is utilizing the cloud-based SPEs (reviewed in Sect. 2) to achieve real-time logo recognition. The fundamental requirement for our system is scalability, meaning that as we add computational resources to the system, its *throughput* (i.e., number of video frames processed per unit time) should increase, while its *result delay* (the timespan between receiving a new frame and finishing processing it) should decrease. Besides scalability, another important objective is elasticity, meaning that resources should be provisioned based on the current workload. In the following, Sect. 3.1 introduces a basic solution that helps to identify the main challenges in designing a cloud-based logo recognition system. Section 3.2 describes the proposed solution.

3.1   Basic Solution

A naive approach that executes offline logo recognition is to use different nodes to process different frames in parallel. Figure 3 illustrates the logical operator topology of this method. At the core of the system are a number of logo recognizers running in parallel that match registered logos (according to the *logo registry*) with input video frames (fed by the *frame dispatcher*), using an existing offline algorithm. The logo registry maintains the set of logos to be detected, and the user can update it by adding new logos or removing existing ones. The frame dispatcher dissembles the input video stream into individual frames, and routes each frame to a logo recognizer in a round-robin fashion. Additionally, a *result collector* gathers the logo recognition results from the recognizers, each of which consists of a logo id, a frame number and a list of positions of the logo in the frame. These results can be optionally aggregated and sent for output.

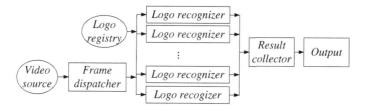

Fig. 3. Naive approach that parallelizes based on frames

The above method achieves scalability in the sense that the throughput of the system improves as more computing nodes are added. However, *it fails to scale in terms of result delay*. As an extreme case, even if we have an infinite amount of computational resources in the system, the result delay cannot be lower than the processing time of a single logo recognizer. For applications involving a large number of registered logos and high-resolution frames, the result delay can be unacceptably long. Another problem with this method is that *each logo is broadcast to all logo recognizers*, which may lead to a temporarily congested network. Consequently, the user expects a sudden increase in result delay whenever a new logo is registered, exacerbating the problem.

One improvement of the above approach is to also dispatch different logos to different recognizers. As shown in Fig. 4, a *logo dispatcher* is placed between the logo registry and the logo recognizers, which dispatches different logos to different recognizer nodes. Other parts of the system remain the same as in Fig. 3. This improvement alleviates the result delay problem as the minimum delay is reduced to the running time of the logo recognizer with one input frame and one input logo. However, the result delay can still be too long for large frames and/or expensive but accurate recognition algorithms.

Fig. 4. An improvement to the naive approach

Discussions. The basic solution treats the logo recognizer as a black box; hence, its result delay is inherently limited to the efficiency of the underlying logo recognition algorithm, regardless of the amount of provisioned computational resources. To break this limit, we must parallelize the internals of logo recognition algorithm without losing accuracy. Next we show how this is done in the proposed solution.

3.2 Proposed Solution

The proposed solution is based on the basic one presented in Sect. 3.1: the framework that dispatches logos and video frames remain the same. In order to achieve scalability in terms of result delay, we further parallelize the logo recognizer by splitting the input frame into patches, as described in Sect. 2. Figure 5 shows a simple architecture, in which a *frame broadcaster* sends the input frame to all processing nodes. Unlike the basic approach, however, the frame broadcaster also sends different patch IDs to different nodes (*patch matchers* in the figure), each of which only performs logo recognition of the patch specified by the ID. Hence, result delay can be reduced by generating a larger number of patches, and using different nodes to process them in parallel. Note that due to the use of patches, the same logo appearances may be reported by different patch matcher nodes. Hence, the result collector needs to remove these duplicates.

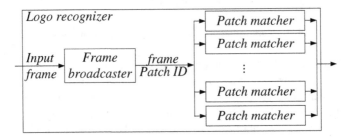

Fig. 5. Parallelized logo recognizer

The frame broadcaster in the above architecture avoids the work of splitting the frame and materializing the patches. However, when there are a large number of patch matchers, the frame broadcaster may become a bottleneck, as we show in the experiments. The reason is that broadcasting a frame to many nodes leads to network congestions, which, in turn, leads to starved patch matchers waiting for inputs. To address this problem, we propose an alternative that explicitly generates the patches and dispatches them, shown in Fig. 6.

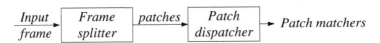

Fig. 6. Alternative design that materializes patches

Next we further optimize the patch matcher, whose performance is critical: when the patch matcher is slow, the only way to reduce result delay is to increase the number of patches generated from the frame, leading to smaller patches. However, as discussed in Sect. 2, logo appearances that are larger than the logo may not be detected; consequently, the result accuracy may decrease. Figure 7 shows how the proposed solution accelerates the patch matcher without splitting the frame into smaller patches. The main idea is to separate the processing of the patch (performed by the *patch indexer*) and the matching with processed logos (done by the *matcher*). For example, the patch indexer can extract features from the input patch, index the features, and match them against multiple logos which have been similarly processed by a *logo indexer*. Note that each patch or logo is only indexed once. For many base logo recognition algorithms, such a design can lead to considerable savings in terms of computation costs. Finally, for specific algorithms we can further parallelize the patch/logo indexers as well as the matcher.

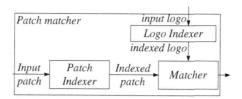

Fig. 7. Optimized patch matcher.

Discussions. Unlike the basic solution presented in Sect. 3.1, the proposed solution does not treat the base logo recognition algorithm as a black box. Thus, we need to examine its compatibility with the base algorithm. Firstly, the use of patches (Figs. 3 and 4) is standard in existing solutions, and the patch matcher can still execute the base recognition algorithm as a black box. The optimizations presented in Fig. 7 implicitly assumes that there is a pre-processing step before matching is performed, which is also common in existing offline logo recognition algorithms. Therefore, the proposed algorithms only rely on mild assumptions on the internals of the base logo recognition algorithm; hence, it is compatible with a wide range of base algorithms.

4   Elastic Streaming Logo Detection

The proposed logo detection algorithms require real-time elastic scaling, i.e., the underlying stream processing system must be able to dynamically add or remove computing nodes (e.g., virtual machines) and redistribute tasks. To scale the logo detection application, the system computes a new task configuration, reallocates computing resources, and performs *migration* to redistribute the tasks. Today most elastic stream processing systems still follow a "stop the world" approach, which first stops the running application altogether, and then restarts it with the new configuration. Clearly, this approach completely interrupts the application during the restart; further, it may cause the loss of all states information and intermediate results, unless the users manually perform additional bookkeeping. The state-of-the-art strategy adopted by prior work [5, 7, 12] is to identify the tasks to move and pause the tasks that communicate with them until the migration completes. Still, this approach often leads to pausing many components for considerable time; consequently, no jobs can be completed during the migration. In our logo detection applications, this may lead to a noticeable freeze in the output video, which is clearly unacceptable.

Hence, in the following we describe a novel elasticity module (integrated into Apache Storm [15]) that performs *smooth task migration* in an elastic streaming system, which minimizes performance degradation during migrations. A preliminary version of our results have been presented in a poster [18]. Specifically, Storm runs an application on a number of *workers*, which are OS processes running on a cluster of nodes. Every worker contains a number of *executors*, each of which resides in an independent OS thread and performs the actual computation. When scaling an application, the system redistributes the executors among a new set of workers. Currently, Storm uses the simple method described above to perform migration: it shuts down all workers of the application and then starts new ones with the new configuration. This scheme leads to a lengthy migration process (usually more than 20 s), during which no output is produced and many tasks fail and trigger expensive fault recovery. We present three migration methods to solve these issues: (i) worker level (WL) migration, (ii) executor level (EL) migration, (iii) executor level migration with reliable messaging (ELR). EL is built on top of WL; ELR in turn is built on top of EL. Our system implementation and experiments use the most efficient solution ELR.

Worker Level Migration (WL). Worker level migration eliminates the restarting of workers in Storm by reusing the existing workers. Instead of killing the workers, WL keep them alive, and only kills the executors and restarts them on the workers that they are newly assigned to. When scaling up the system, WL only starts the newly added workers; when scaling down the system, WL only shuts down the removed workers.

Executor Level Migration (EL). WL still disrupts the running application in the migration as it restarts all executors. By managing migration at the executor level, EL runs the application without disruption during the migration, as follows. (i) As in WL, workers are kept alive and added or removed only when necessary. (ii) If an executor does not need to move to a different worker after a migration, EL keeps it running without interruption; otherwise, EL kills the executor and restarts it on the destination worker. (iii) EL contains a migration-friendly task scheduler to minimize the number of moving executors in migration.

Reliable Messaging (ELR). EL may still incur messaging failures during the migration, which leads to task failures and expensive fault recovery processes. To address this issue, ELR improves the messaging system in Storm as follows. (i) ELR captures all failed messages at various places in the system, which are usually caused by inconsistent states among workers when the workers are updating their states in a migration. Using a novel message retry mechanism, ELR resends these messages later at a scheduled time. (ii) ELR saves messages from executors that are being shut down, and sends them to the correct workers using the same message retry mechanism.

5   Experiments

We have implemented the proposed live logo recognition solution using Apache Storm [15], and carried out all experiments using a cluster consisting of 8 nodes. Elastic resource provisioning is implemented using DRS [3] (available at http://github.com/ADSC-Cloud/resa). Following common practice in streaming SPEs, we dedicate one node as the master node (Zookeeper and Nimbus). Meanwhile, we run up to 5 threads (executors) on each physical node, totally 30 executors. The underlying offline logo recognition package is SIFT [8], a popular computer vision library. The dataset consists of a high-definition (1024 × 768) video of a real soccer game, and logo images that appear in the video.

The parameters investigated in the experiments include the number of executors, number of registered logos and input video resolution. In each experiment, we vary one parameter and fix all others to their default values, i.e., 5 registered logos, 20 executors and 640 × 480 frame resolution. We compare the proposed method against the alternative called *frame broadcasting* that broadcasts each frame to all patch matchers (Sect. 3.2).

Figure 8 evaluates the effect of number of executors, and reports the average throughput and result latency obtained by adding 5 more executors. Clearly, the proposed solution obtains the best throughput, result latency and scalability.

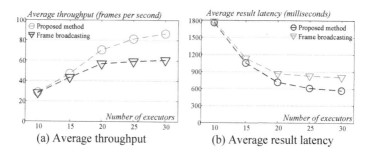

(a) Average throughput   (b) Average result latency

Fig. 8. Effect of number of executors (5 logos, 640 × 480 frame resolution)

Figure 9 plots the average throughput and result latency as functions of the number of registered logos. When the number of registered logos is low, the proposed method clearly outperforms frame broadcasting. However, the performance gap gradually closes as the number of logos increases. This is because for frame broadcasting, network transmissions is the bottleneck. As the amount of computations grows the networking overhead is no longer the limiting factor, leading to comparable performance as the proposed method.

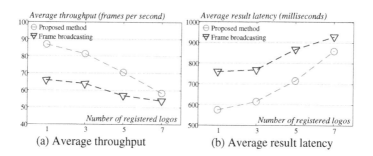

(a) Average throughput   (b) Average result latency

Fig. 9. Effect of No. of logos (20 executors, 640 × 480 frame resolution)

Figure 10 reports the effect of varying input video resolutions. As expected, the proposed solution outperforms its competitor in all settings. The performance of both methods decreases with increasing resolution, suggesting that more resources are required to reach real-time recognition efficiency.

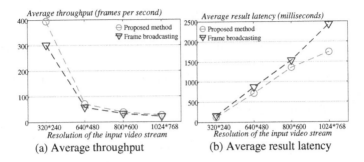

Fig. 10. Effect of input video resolution (20 executors, 5 logos)

Figures 11 and 12 present the elasticity properties of our streaming logo detection system. Specifically, in Fig. 11, we first run the system for 12 min as a warmup, with default settings, i.e., 20 executors, 5 registered logos, 640 × 480 resolution. Then, we register a new query logo in the system at the end of the 12th min. According to Fig. 11, after adding the new query logo, system performance drops (i.e., throughput decreases and result latency increases). The performance monitoring module detects the change of workload soon after the addition of the new logo; yet, in order to better present the results, we configured the system so that new executors are added at the 22nd min. Specifically, the proposed elasticity module adds 5 more executors to the application, and performs the necessary operator states migrations. As shown in the figure, the migration process happens almost instantly (i.e., within one minute), and its overhead is negligible compared to that of the main logo detection tasks. Within a couple of minutes after adding new executors, system performance in terms of result throughput and latency returns to the same level before adding the new query logo at the 12th min.

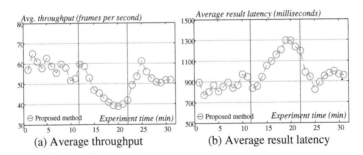

Fig. 11. Effect of registering a new logo to the system

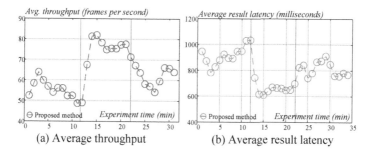

Fig. 12. Effect of removing a registered logo

Similarly, in Fig. 12, we run the system for 12 min and then remove one registered logo. Consequently, the elasticity module removes executors from the application at the 22nd min. The results again confirm two observations: (i) the migration process incurs little overhead and (ii) system performance returns to the level before the workload change, after freeing excess system resources, i.e., executors.

6   Conclusion

This paper studies real-time logo recognition from a live video stream, and proposes a generic framework that converts an offline logo recognition algorithm to a real-time one running on a cloud platform in order to obtain high efficiency, scalability and elasticity. Compared to baseline solutions, the proposed framework is carefully designed to enable high parallelism, eliminate performance bottlenecks, and reduce inter-node communication overhead. Experimental evaluations using real data validate our performance claims. Regarding future work, an interesting direction is to apply tracking techniques on the recognized logos, which are generally more efficient than recognizing logos in every frame. Meanwhile, we intend to further optimize the stream processing system to exploit shared executions for multiple registered logos.

Acknowledgments. This work is supported by the NSF of China under Grant 61173081, and the Guangzhou Science and Technology Program, China, under Grant 201510010165.

References

1. Apache Samza. http://samza.apache.org
2. Ding, J., Zhang, Z., Ma, R., Yang, Y.: Auction-based cloud service differentiation with service level objectives. Elsevier Comput. Netw. **94**, 231–249 (2016)
3. Fu, T., Ding, J., Ma, R., Winslett, M., Yang, Y., Zhang, Z.: DRS: dynamic resource scheduling for real-time analytics over fast streams. In: Proceedings of the IEEE ICDCS (2015)
4. Fu, T., Ding, J., Ma, R., Winslett, M., Yang, Y., Zhang, Z., Pei, Y., Ni, B.: LiveTraj: real-time trajectory tracking over live video streams. In: Proceedings of the ACM MM (2015). video program

5. Fernandez, R.C., Migliavacca, M., Kalyvianaki, E., Pietzuch, P.: Integrating scale out and fault tolerance in stream processing using operator state management. In: Proceedings of the ACM SIGMOD (2013)
6. Gulisano, V., Jiménez-Peris, R., Patiño-Martínez, M., Soriente, C., Valduriez, P.: StreamCloud: an elastic and scalable data stream system. IEEE Trans. Parallel Distrib. Syst. 23(12), 2351–2365 (2012)
7. Heinze, T., Jerzak, Z., Hackenbroich, G., Fetzer, C.: Latency-aware elastic scaling for distributed data stream processing systems. In: Proceedings of the ACM DEBS (2014)
8. Lampert, C., Blaschko, M., Hofmann, T.: Efficient subwindow search: a branch and bound framework for object localization. IEEE Trans. Pattern Anal. Mach. Intell. 31(12), 2129–2142 (2009)
9. Lindeberg, T.: Scale invariant feature transform. Scholarpedia 7(5), 10491 (2012)
10. Li, Y., Wan, K. Yan, X., Xu, C.: Real-time advertisement insertion in baseball video based on advertisement effect. In: Proceedings of the ACM Multimedia (2005)
11. Neumeyer, L., Robbins, B., Nair, A., Kesari, A.: S4: distributed stream computing platform. In: Proceedings of the IEEE ICDM KDCloud Workshop (2010)
12. Schneider, S., Andrade, H., Gedik, B., Biem, A., Wu, K.-L.: Elastic scaling of data parallel operators in stream processing. In: Proceedings of the IEEE IPDPS (2009)
13. Sinha, S., Frahm, J., Pollefeys, M., Genc, Y.: GPU-based video feature tracking and matching. In: Proceedings of the EDGE (2006)
14. Tan, T., Ma, R., Winslett, M., Yang, Y., Yu, Y., Zhang, Z.: Resa: realtime elastic streaming analytics in the cloud. In: Proceedings of the ACM SIGMOD (2013). undergraduate poster
15. Toshniwal, A., Taneja, S., Shukla, A., Ramasamy, K., Patel, J.M., Kulkarni, S., Jackson, J., Gade, K., Fu, M., Donham, J., et al.: Storm@Twitter. In: Proceedings of the ACM SIGMOD (2014)
16. Wu, Z., Ke, Q., Isard, M., Sun, J.: Bundling features for large scale partial-duplicate web image search. In: Proceedings of the IEEE CVPR (2009)
17. Xie, H., Gao, K., Zhang, Y., Tang, S., Li, J., Liu, Y.: Efficient feature detection and effective post-verification for large scale near-duplicate image search. IEEE Trans. Multimedia 13(6), 1319–1332 (2011)
18. Yang, M., Ma, R.: Smooth task migration in Apache storm. In: Proceedings of the ACM SIGMOD (2015). undergraduate poster
19. Zaharia, M., Das, T., Li, H., Shenker, S., Stoica, I.: Discretize streams: an efficient and fault-tolerant model for stream processing on large clusters. In: Proceedings of the ACM SOSP (2013)
20. Zhang, Z., Ma, R., Ding, J., Yang, Y.: ABACUS: an auction-based approach to cloud service differentiation. In: Proceedings of the IEEE IC2E (2013)

Profit Based Two-Step Job Scheduling in Clouds

Shuo Zhang, Li Pan(✉), Shijun Liu, Lei Wu, and Xiangxu Meng

School of Computer Science and Technology, Shandong Uiversity,
Jinan 250101, People's Republic of China
zs_sduzz@sina.com, {panli,lsj,i_lily,mxx}@sdu.edu.cn

Abstract. One of the critical challenges facing the cloud computing industry today is to increase the profitability of cloud services. In this paper, we deal with the problem of scheduling parallelizable batch type jobs in commercial data centers to maximize cloud providers' profit. We propose a novel and efficient two-step on-line scheduler. The first step is to rank the arrival jobs to decide an eligible set based on their inherent profitability and pre-allocate resources to them; and the second step is to re-allocate resources between the waiting jobs from the eligible set, based on threshold profit-effectiveness ratio as a cut-off point, which is decided dynamically by solving an aggregated revenue maximization problem. The results of numerical experiments and simulations show that our approach are efficient in scheduling parallelizable batch type jobs in clouds and our scheduler can outperform other scheduling algorithms used for comparison based on classical heuristics from literature.

Keywords: Cloud · Resource allocation · Scheduling · Profit maximization

1 Introduction

Cloud computing is now significantly changing the way people use resources such as computation hardware, storage, software applications and so on. Users can obtain on-demand cloud services to host their jobs and applications in a pay-per-use way. With virtualization as a key enabler, cloud providers can run multiple isolated Virtual Machines (VMs) simultaneously in their data centers to host users' jobs. By maximizing server consolidation, cloud providers can achieve efficient utilization of resources and thus obtain profits through economy of scale [1].

In an open and dynamic cloud environment, users have divergent requirements on services for their jobs. They usually need to pay more to get better Quality of Service (QoS), while in order to deliver better services cloud providers need to provision more resources for hosting jobs and thus incur more cost. More specifically, there exists resource competition between users' jobs submitted to cloud data centers. How to schedule these concurrent service requests efficiently on the physical resources in cloud data centers to maximize a cloud provider's profit is far from trivial.

In this paper, as an example application, we consider scheduling parallelizable batch type jobs into cloud data centers. One characteristic of these jobs is that their execution can be speeded up to an extent by allocating more resources to them. Besides, for these types of jobs, response time is a main QoS criterion. Under this circumstance, to solve

© Springer International Publishing Switzerland 2016
B. Cui et al. (Eds.): WAIM 2016, Part II, LNCS 9659, pp. 481–492, 2016.
DOI: 10.1007/978-3-319-39958-4_38

the provider's profit maximization problem, we propose a new and efficient online scheduler with admission control. The scheduling algorithm is divided into two phases. At first step, the arrived jobs are sorted by their *inherent profitability* and pre-allocated resources by the sequence of this sorted queue. And in the second step, we re-allocate resources between the accepted jobs based on the threshold profit-effectiveness ratio. This threshold is calculated dynamically based on the potential profit of the accepted jobs. And we test the effectiveness and efficiency of our proposed approach through experiments and simulations. A significant contribution of this paper is the idea of resource re-allocation based on the concept of potential profit, which can improve providers' profits by adjusting the scheduling solution based on local optimum to the global optimal solution. To our knowledge, we are the first to leverage the idea of potential profit to do resource reallocation for scheduling jobs in distributed systems such as clusters and clouds.

The rest of this paper is organized as follows: Sect. 2 lists some related work about cloud scheduling algorithms and resource allocation strategies. Section 3 gives an overview of the system model. The problem description and its formalization are presented in Sect. 4. Section 5 presents our two-step profit based scheduler. Section 6 outlines the experiment results on synthetic dataset and real dataset, respectively. Section 7 gives conclusion and future directions for this work.

2 Related Work

Economy driven approach is not new for resource allocation and job scheduling in distributed systems such as grids and clouds [2]. We will discuss those most related to our work in this section. Jiayin Li et al. (2012) proposed two online dynamic resource allocation algorithms [3]. The execution of tasks in this cloud system can be preempted. Their algorithm adjusts the resource allocation dynamically based on the real-time information of the tasks executions in the cloud system.

Gunho Lee (2012) addressed the cloud resource management problem in a way that service providers achieve high resource utilization and users meet their SLA with minimum expenditure [4]. Maria Alejandra Rodriguez et al. (2014) proposed a resource allocation and scheduling algorithm which was based on the meta-heuristic optimization technique for scientific workflows on Infrastructure as a Service (IaaS) clouds [5]. The objective of this method is to minimize the overall workflow execution cost and meeting deadline constraints of them.

Hong Wei Zhao et al. (2014) proposed a scheduling algorithm in clouds based on Artificial Fish Swarm Optimization (AFSA) [6]. The main idea of this algorithm is extending Fish Swarm Optimization to the interacting swarm model by cooperative models, and thus it becomes more efficient and can converge to global optimum faster.

The idea of leveraging economic theory to improve scheduling strategy is not scarce. There are lots of scheduling algorithms based on job value or the profit. David E. Irwin et al. proposed a scheduling strategy based on the balance between the yield and the risk of a task [7]. They proposed some heuristics as the priority of a job and investigated their parameters through experiments.

Compared to the above previous work, the main contributions of our work lies within an efficient and effective scheduling mechanism to maximize a cloud provider's profit, with the resource re-allocation strategy based on the idea of *profit-effectiveness ratio* to maximize all the overall potential profit.

3   System Model

The system contains a number of clients who submit multiple batch type jobs and a single cloud service provider, who possesses a huge amount of resources, as shown in Fig. 1. When faced with a large number of service requests for hosting jobs submitted by clients, the total demand for resources can exceed the capacity of the available resources. Admission control is employed as a general approach to avoid such over-loading of resources, in order to guarantee that the QoS requirements of the hosted jobs can be met. The cloud provider uses the admission control mechanism to decide whether to accept a client's job service request, based on the present availability and price of resources, the payment for completing a job, and the other jobs concurrently submitted to the cloud system. Then, the *scheduler* component schedules the jobs for execution by allocating resources to jobs which pass the admission control component. The operations of admission control and job scheduling can be triggered periodically or real-timely.

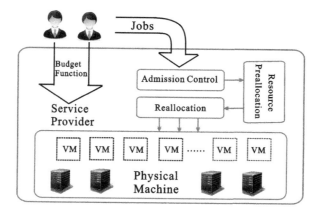

Fig. 1. The system model

If a client's job service request is accepted by the provider, a service contract establishes between them, which is usually called Service Level Agreements (SLAs) [8]. A SLA specifies the QoS expectations that the provider agrees to deliver, usually in measurable terms, as well as the payment the consumer agrees to make for the service delivered, and the penalty against violations of SLA. Since different types of jobs, such as non-interactive batch jobs and transactional applications, have different QoS and resource demand characteristics, schedulers for different types of jobs may have varying policies. In this paper, for sake of clarity and simplicity, we only deal with the scheduling problems of batch type jobs.

We assume each job submitted to the cloud data center is associated with a budget function, which reflects a user's willingness-to-pay for the job execution service, considering the quality of service delivered by the cloud provider. Generally, a user would like to pay more for a better QoS delivered. We believe that for batch type jobs, response time, i.e., expected job completion time, is a most important QoS parameter which is used to measure how a job service request is fulfilled by cloud date centers. Thus, in this paper, we assume a user's budget function holds a mapping relationship between the payment the user would make and expected response time delivered. This function should be monotonically decreasing because the user needs to pay more to get the job finished earlier and we assume the payment is inversely proportional to the service response time in this paper. We assume that the budget function is given by:

$$B_i(t_j) = \alpha_i / t_j \tag{1}$$

where α_i is a parameter given by user i which reflects the ability of the user i to pay and t_j is the response time of batch job j submitted by the user i. Each job has a rigid deadline and if the response time exceeds its deadline, penalties will be incurred. The penalty is generally assumed to be proportional to the duration of the delay for finishing the job [9]. Thus if Δt is the length of time that exceeds a job's deadline and Q is the penalty factor, the penalty for delaying a job's completion by Δt can be calculated as penalty = $Q * \Delta t$.

4   Problem Statement

We model our cloud data center to have a certain number of available physical resources with total capacity c, which are running in clusters of physical machines maintained by a provider. Within a period of time t, let J be the number of jobs accepted by the service provider and v_j be the amount of resources the job j required. For the sake of clarity and simplicity, in this paper we assume that each VM is homogeneous. i.e., they are all allocated the same number of CPU cores, memories, storages and so on. Thus each VM here has the same performance. We use r_j to denote the revenue that job j can contribute. Let K be the cost of owning and operating a unit of resource and Q is the overall penalty incurred for not completing jobs before their deadlines. Assume that all these J jobs can be finished in the time period t, and t_j is the response time of job j.

As described above, the objective of a cloud provider in scheduling jobs is to maximize its total profit, which can be formulated as:

Maximize

$$\sum_{j=1}^{J} r_j - K * \sum_{j=1}^{J} v_j - Q$$

Subject to

$$\sum_{j=1}^{J} v_j \leq c; \ max\{t_j\} < t \ \forall j \in [1, J]$$

Since in real-world commercial cloud data centers jobs are submitted online and there is usually no prior knowledge about the workload distribution, it is impossible to find an optimal scheduling considering the workload conditions of both current arrived jobs and future arriving jobs. A perfect prediction of job arrivals in such an open and dynamic cloud environment is a complex problem, which is beyond our discussion. Even with perfect prediction of job arrivals, the off-line scheduling problem shown above is NP-hard. Thus in this paper we consider an online scheduling approach to improve a cloud provider's revenue, based on heuristics of jobs' profitability, which we will discuss in details in the next section.

5   Profit-Based Two-Step Job Scheduling

When there are lots of jobs arrived in the system, how to schedule these jobs efficiently with the limited available resources is a significant question. In current literature, there are some commonly employed scheduling algorithms such as First Come First Served (FCFS), Shortest Remaining Processing Time (SRPT) and so on. These classical algorithms do not take the profit a service provider can get into account. In this paper, we first propose a new heuristic for scheduling jobs in cloud data centers. This novel heuristic is based on the *inherent profitability* of a job. We define a job's inherent profitability as the profit gained from running this job per unit time and per unit resource. Using this profitability heuristic, we first do resource pre-allocation before scheduling these jobs for execution. After this step, some adjustment about the number of resources which are pre-allocated in the first step will be made based on the potential profit between the submitted jobs. This two-step job scheduling model is based on the following premises:

- The capacity of resources is recalculated at the end of every scheduling interval. And the priority of each approved job is calculated before the beginning of scheduling.
- Scheduling only occurs at the beginning of every scheduling interval. We do not execute scheduling during an interval of execution.
- The scheduling algorithm presented in this paper is non-preemptive.

After the steps of resource pre-allocation and adjustment later, these admitted jobs can be scheduled into the cloud system for execution.

5.1 Admission Control

Before allocating resources to the arrived jobs, a critical problem needed to be solved is to determine how many and which jobs should be accepted. To solve this problem, we need to design a strategy of admission control. The job accepted by this strategy should be featured that higher profitability and the workload of our system can't exceed the capacity of our system too much. But this problem doesn't belong to our research field, so, we use the admission control strategy studied previously.

5.2 Resource Pre-allocation

As stated in Sect. 2, every job submitted to the cloud system is associated with a budget function, as given by (1), representing the profitability of a job. We also assume that the cost of maintaining a unit of virtual resource is a constant K. Figure 2 presents the varying tendency of the revenue, cost and the profit, given the number of virtual resources allocated. The profit P is calculated as [10]:

$$P = R - C \tag{2}$$

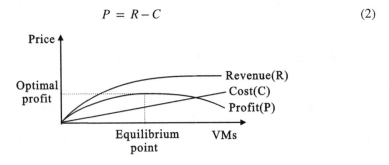

Fig. 2. Equilibrium point of a user's profit

Then, we can calculate the amount of the resources should be allocated to a job for achieving a specific service response time according to assumption [11]:

$$T(v) = M/(1 + ln(v)) \tag{3}$$

$T(v)$ is the response time of a job and v is the amount of virtual resources allocated to this job. M is the response time for running this job on a unit virtual resource.

For a specific job j which is submitted by user i, its cost and revenue are:

$$C = K * v_j \tag{4}$$

$$R = \left(\alpha_i\left(1 + \ln(v_j)\right)\right)/M_j \tag{5}$$

By differentiating the profit function (2) with respect to v_j, we can get the equilibrium point of v_j as (6). Then, we get the theoretical amount of resources which can maximize the profit obtained from job j and leverage it to measure a job's inherent profitability.

$$v_j = \alpha_j / (M_j * K) \tag{6}$$

Using this inherent profitability, we can sort all arrived jobs and put them into a sorted queue. Based on the strategy of admission control, we can determine how many jobs should be accepted. And these jobs are most profitable because we accept them by the queue sorted by the inherent profitability. In most cases, the total workload we accept often exceeds the capacity of system. Thus there are some jobs will wait for scheduling if we schedule them directly no matter what strategies we use. We call jobs which can get resources if we schedule them directly ready-to-run jobs and the rest jobs are called waiting jobs. On the other hand, although we can get the theoretically optimal amount of virtual machines allocated to a single specific job, the cost of this optimal value is too high while considering the other jobs co-arrived in the system because all jobs compete for the limited sharing resource. So, we employ the idea of economic analysis to discuss whether it is necessary to do the resource re-allocation and how to do it.

5.3 Resource Re-allocation

As stated above, the optimal amount of resource of a job may not be the best for the global resource allocation. To the best of our knowledge, there is no previous study providing quantification analysis to prove this situation indeed exists or not. In this paper we first employ the idea of an economic analysis approach, called cost-benefit analysis, to examine if we adjust the resource allocation decision made in the step of resource pre-allocation, the total profit will present what kind trends. The calculation of variation of total profit is based on a concept named *potential profit*.

If we separate some resources allocated to the job *i* to host a job *j* which is waiting for resources allocation, the profit of job *i* will decrease while new profit will be achieved by job *j*. So, we may obtain more profits after adjusting the resource allocation and we call this extra profit as *potential profit*. There will be a difference between the amount of resources of a task which is calculated by the equilibrium point and the amount of resources after being cut off by a restriction threshold. These extra resources beyond the threshold can be provisioned to host other waiting jobs. Of course, cutting off the amount of resources from a given job will decline the profit gained from this job. We also take these declined values into account.

We define the resource cutting threshold as *profit-effectiveness ratio*, which is a proportional value of the pre-allocated amount of resources, instead of an absolute amount value. It means that when you cut off resources from the pre-allocated resources by this ratio, the total profit will be influenced at the same time. And we present how to optimally decide the *profit-effectiveness ratio* for per resource allocation scenario below. According to the idea of *potential profit*, we design a computational formula and we find that there exists an extreme point of this function. Through a differentiation process we can get the expression of the extreme value point. Then we evaluate these outgoing costs against the generated revenue from the resource re-allocation to decide whether this *profit-effectiveness ratio* will result in a profit at all, and if so, it is necessary to take the step of resource re-allocation.

The computing method of *potential profit* is presented in (7) where Rc is the extra profit contributed by the waiting jobs after allocating the resources separated from ready-to-run jobs and Rd is the loss after cutting the allocated resources. Now the problem is how to calculate Rc and Rd present in (8) and (10).

$$PP = Rc - Rd \tag{7}$$

Let x be the value of the cutting proportion of the resource distribution, I be the number of waiting jobs, and $Perc_i$ be the proportion of the separated resources which the waiting job i can be allocated. J is the number of ready-to-run jobs, and $Exnum_j$ is the amount of resources of a ready-to-run job j which is calculated by the equilibrium point. $Perc_i$ is presented in (9) where $Exnum_t$ is the amount of resources of a waiting job t which calculated by the equilibrium point, the Rc as

$$Rc = \sum_{i=1}^{I} (\alpha_i/M_i) * \left(1 + \ln\left(Perc_i * x * \sum_{j=1}^{J} Exnum_j\right)\right) - K * x * \sum_{j=1}^{J} Exnum_j \tag{8}$$

$$Perc_t = Exnum_t / \sum_{t=1}^{T} Exnum_t \tag{9}$$

Rd is the loss after cutting off the allocated resources from the ready-to-run jobs. Re_j is the profit of task j obtained by allocating the amount of resources calculated by the equilibrium point.

$$Rd = \sum_{j=1}^{J} \left(Re_j - \left(\frac{\alpha_j}{M_j} * (1 + \ln((1-x) * Exnum_j)) - (1-x) * K * Exnum_j\right)\right) \tag{10}$$

We now, differentiate the potential-profit function with respect to x as (11), and then we can compute the extreme value point of PP as (12).

$$\frac{\partial PP}{\partial x} = \frac{1}{x} \sum_{i=1}^{I} \frac{\alpha_i}{M_i} - \frac{1}{(1-x)} \sum_{j=1}^{J} \frac{\alpha_j}{M_j} = 0 \tag{11}$$

$$x = \frac{\sum_{i=1}^{I} \alpha_i/M_i}{\sum_{i=1}^{I} \alpha_i/M_i + \sum_{j=1}^{J} \alpha_j/M_j} \tag{12}$$

Through this approach, we get the theoretical optimal value of the cutting proportion that maximizes the improvement of the total profit. In the next experiment and simulation section, we will examine the *practical effect* of this threshold point.

6   Experiments

In order to evaluate of the efficiency and effectiveness of our proposed two-step scheduling approach (TPRD), we have conducted extensive experiments and simulations based on both synthetic dataset and real-world dataset. The experimental results show that our scheduling model performs better than the others.

6.1 Experimental Methodology

We now explain how the synthetic dataset are generated. We assume the number of arrived jobs is a stochastic process which is i.i.d. across time slots and they are evenly submitted by users. The ability to pay of every user obeys the normal distribution. The experiment parameters are shown in Table 1.

Table 1. Values of the experiment parameters.

Name	Value	Remark
UN	10	The number of users
RT	100 days	The total operation time of the system
K	0.3$	The cost of owning a unit of virtual resource.
RN	800	The whole number of resources
L	100	The parameter of Poisson distribution
mu	5	The parameter of normal distribution
sigma	1.5	The parameter of normal distribution
PR	3	The penalty rate

For real-world dataset, we import the Grid Workload datasets recorded by Delft University of Technology to test our scheduling model. We sort some jobs in the dataset whose TraceID is GWA-T-1 DAS2 and GWA-T-2 Grid5000 to test our proposed method. In order to test the proposed methodology, we consider the runtime of a job is its execution time of running on only one machine. Without loss of generality, we assume the selected jobs are submitted by 10 users evenly. The payment capability of all users obeys normal distribution.

6.2 Efficiency

We use the execution time of our algorithm to test the efficiency of the proposed method. Under different experimental parameter configurations, we test the execution durations of step 1 and step 2 respectively. And the result is the average of 50 times of experiments. The variables are listed in Table 2 and other parameters are the same as Table 1. It is obvious that the cost of re-allocation is much smaller than pre-allocation. This is because the stage of re-allocation only needs to calculate the profit and resources of accepted jobs, but the pre-allocation stage includes the calculation of the priority of all submitted jobs and sorting them with admission control.

Figure 3 presents the comparison of efficiency between four schedulers. It shows the average execution time of these schedulers in 50 scheduling time slots. It can be seen that the execution time of our proposed approach (TPRD) is sometimes a little longer than the others because of the runtime of step 2. The average execution time of TPRD in these five different experiments is about 4 % longer than other schedule strategies. Besides, the efficiency of our proposed method is better than others in some situations.

Table 2. Experimental result of efficiency test.

Total resource	Jobs arrived per hour	Execution time in milliseconds	
		Step 1	Step 2
800	60	1.06	0.25
800	110	2.33	0.58
800	160	3.32	0.69
1000	120	8.28	1.39
1000	170	10	1.72
1000	220	11.5	1.88
1200	180	18.7	2.84
1200	230	21	3.21
1200	280	23.4	3.56
1400	340	40.9	5.99
1400	390	44.5	6.36
1400	440	48.4	6.71

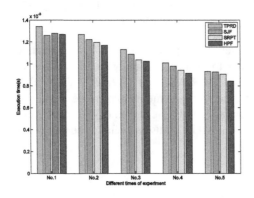

Fig. 3. Comparison of efficiency

6.3 Effectiveness

In order to test the effectiveness of our scheduling approach (TPRD), we compare its performance with some other classical scheduling algorithms such as Shortest Remaining Processing Time (SRPT), High priority First (HPF) and Short Job First (SFJ). We inspect the total profits of these scheduling strategies in the same experimental environments. Since the SRPT, SJF and HPF don't have the resource re-allocation phase, there are some jobs waiting in the queue for next scheduling period. The response time is used as the scheduling priority in SRPT. The time of running a job on only one machine is calculated as the priority of this job in SJF. And the HPF is the step one of our proposed scheduling method without step two. So the priority of a job in HPF is its inherent

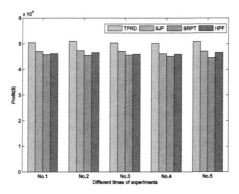

Fig. 4. Total profits of 4 scheduling strategies under the synthetic dataset

profitability. All these scheduling algorithms schedule jobs based on priorities. And we run the simulations a lot of times to exclude some accidental circumstances.

Figure 4 presents the total value obtained by the four different scheduling algorithms in 100 h under the synthetic data. It is evident that the profit obtained by our scheduling algorithm is higher than the other strategies by 9.1 %, 11.2 %, 10.7 %, respectively.

Figure 5 presents the total profit obtained by 4 scheduling algorithms in 100 h under the workload data derived from the real-world systems. The first figure is the result of the experiment based on the GWA-T-1 DAS2 and the second figure based on the GWA-T-2 Grid5000. The results show that our proposed two-step scheduling mechanism has the best performance among the four algorithms. The profit of TPRD is higher than others by 10.2 %, 12.7 %, and 9.8 % under the dataset GWA-T-1 DAS2, respectively. The profit of TPRD is higher than other three strategies by 18.3 %, 13.4 %, and 21.7 % under the dataset GWA-T-2 Grid5000, respectively.

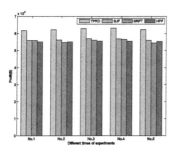 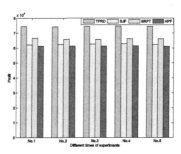

Fig. 5. Total profit of 4 scheduling strategies under the real world dataset

7 Conclusions and Future Work

In this paper we propose a novel two-step on-line scheduler for maximizing a cloud provider's revenue. The first step is to pre-allocate resources to submitted jobs based on each job's *inherent profitability* proposed in this paper; and the second step is to adjust the allocation of resources between accepted jobs, based on threshold *profit-effectiveness ratio* as a cut-off point for per-job resource allocation. We have conducted experiments based on both real dataset and synthetic dataset and the results show that the proposed two-step scheduler can obtain better performance than other traditional scheduling strategies. It suggests that our proposed two-step scheduling algorithm is effective in improving cloud providers' profit. We also test the efficiency of our two-step scheduling algorithm with very large instances and it completes scheduling in just a few milliseconds. Thus it can provide real-time performance guarantees in cloud environments.

For the future, we plan to extend our scheduling mechanism for cloud data centers in which multiple types of applications coexist with batch type jobs, each having different QoS considerations and SLAs. And we also plan to combine our scheduling algorithm with workload forecasting model to get optimal scheduling plans.

Acknowledgments. This work is supported in part by the National Natural Science Foundation of China (61402263), and the Science & Technology Development Projects of Shandong Province (2014GGX101028, 2014GGH20100).

References

1. Volker, C.E., Hamscher, V., Yahyapour, R.: Economic scheduling in grid computing. In: Feitelson, D.G., Rudolph, L., Schwiegelshohn, U. (eds.) JSSPP 2002. LNCS, vol. 2537, pp. 128–152. Springer, Heidelberg (2002)
2. Lartigau, J., Nie, L., Xu, X., Zhan, D., Mou., T.: Scheduling methodology for production services in cloud manufacturing. In: International Joint Conference on Service Sciences, pp. 34–39. IEEE Press, New York (2012)
3. Li, J., Qiu, M., Ming, Z., Quan, G., Qin, X., Gu, Z.: Online optimization for scheduling preempt able tasks on IaaS cloud systems. J. Parallel Distrib. Comput. **72**, 666–677 (2012)
4. Lee, G.: Resource allocation and scheduling in heterogeneous cloud environments. Dissertations and Theses-Grad works, University of California, Berkeley (2012)
5. Rodriguez, M.A., Buyya, R.: Deadline based resource provisioning and scheduling algorithm for scientific workflows on clouds. IEEE Trans. Cloud Comput. **2**, 222–235 (2014)
6. Zhao, H., Tian, L.: Resource schedule algorithm based on artificial fish swarm in cloud computing environment. In: 4th International Conference on Advanced Design and Manufacturing Engineering, pp. 1614–1617. Trans Tech Publications, Switzerland (2014)
7. Irwin, D.E., Grit, L.E., Chase, J.S.: Balancing risk and reward in a market-based task service. In: 13th IEEE International Symposium on High Performance Distributed Computing, pp. 160–169. IEEE Press, New York (2004)
8. Yeo, C., Buyya, R.: Service level agreement based allocation of cluster resources: handling penalty to enhance utility. In: IEEE International Conference on Cluster Computing. IEEE Press, New York (2005)
9. Garg, S.K., Toosi, A.N., Gopalaiyengar, S.K., Buyya, R.: SLA-based virtual machine management for heterogeneous workloads in a cloud datacenter. J. Netw. Comput. Appl. **45**, 108–120 (2014)
10. Tsakalozos, K., Kllapi, H., Sitaridi, E., Roussopoulos, M., Paparas, D., Delis, A.: Flexible use of cloud resources through profit maximization and price discrimination. In: 27th International Conference on Data Engineering, pp. 75–86. IEEE Computer Society, US (2011)
11. Eager, D.L., Zahorjan, J., Lozowska, E.D.: Speedup versus efficiency in parallel systems. IEEE Trans. Comput. **38**, 408–423 (1989)

A Join Optimization Method for CPU/MIC Heterogeneous Systems

Kailai Zhou[1,2], Hong Chen[1(✉)], Hui Sun[1], Cuiping Li[1], and Tianzhen Wu[1]

[1] Key Lab of Data Engineering and Knowledge Engineering of MOE, and School of Information,
Renmin University of China, Beijing, China
zkl2@163.com, {chong,sun_h,licuiping}@ruc.edu.cn,
wutianzhen@foxmail.com
[2] School of Computer and Information, Southwest Forestry University, Kunming, China

Abstract. In recent years, heterogeneous systems consisting of general CPUs and many-core coprocessors have become the main trend in the high-performance computing area due to their powerful parallel computing capabilities and superior energy efficiencies. Join is one of the most important operations in database system. In order to effectively exploit each hardware's advantages in heterogeneous systems, in this paper we focus on how to optimize the join algorithm in hybrid CPU/MIC system. We design a join method with CPU and MIC working collaboratively when implementing the join operation. In order to fully utilize the MIC's parallel computing power, we also propose a Sort-Scatter-Join (SSJ) algorithm for MIC to generate the join index. Through turning the traditional process of comparison and matching into the process of computing and scattering, the SSJ gains more beneficial from thread-level parallelism and SIMD data parallelism. Experiment results show that, compared with the traditional parallel sort-merge join algorithm, the peak performance of the SSJ running on MIC is improved by around 26 %.

Keywords: Join · Optimization · CPU-MIC · Sort-merge join

1 Introduction

With the development of computer hardware, heterogeneous systems consisting of general CPUs and various accelerators, such as GPGPUs and FPGAs, have become the main trend for high-performance computing [1]. In recent years, a new many-core architecture named Many Integrated Cores (MIC) which usually serves as accelerator

This work is supported by National Basic Research Program of China (973) (No. 2014CB340403, No. 2012CB316205), National High Technology Research and Development Program of China (863) (No. 2014AA015204) and NSFC under the grant No. 61272137, 61033010, 61202114 and NSSFC (No. 12\&ZD220), and the Fundamental Research Funds for the Central Universities, and the Research Funds of Renmin University of China (15XNH113, 15XNLQ06). It is also supported by Huawei Innovation Research Program (No. HIRP 20140507).

B. Cui et al. (Eds.): WAIM 2016, Part II, LNCS 9659, pp. 493–505, 2016.
DOI: 10.1007/978-3-319-39958-4_39

has emerged. Due to the excellent computing abilities and lower power consumption, MICs have widely used in many compute-intensive applications [2].

Join is one of the most important operations in database system. In most queries, the performance are determined by the efficiency of join operation. Sort-merge join and hash join are two popular join algorithms in most database systems. The debate over which is the best join algorithm has been going on for decades [3–6]. Alt-hough the hash join algorithm currently has been shown to outperform sort-merge join in many cases [3, 4], this situation is likely to change on the future many cores architecture, because sort-based join algorithm has a better scalability potential with the architectural trends of wider SIMD and more cores [5]. Additionally, hash join needs large memory to store a hash table, which limits its application to MICs to some extent due to their rather small on-board memory. Accordingly, in this paper, we focus on how to optimize the sort-based join algorithm on the MIC platform.

Over the past few years, researchers have proposed many high-efficiency variants of sort-merge join running on modern CPUs, such as *MPSM*, *m-way* and *m-pass* [4]. However, it is hard to seamlessly transplant these algorithms based on a homogeneous architecture into a heterogeneous system due to the following reasons.

(1) On platforms consisting of general CPUs and MICs, different computing devices may have significantly different hardware features, which leads to extremely asymmetrical parallel computing power. This asymmetry makes the existing join algorithms not well suited for task distribution or data layout.

(2) The communication between MICs and CPUs is implemented by PCIe bus. But the low-bandwidth and long-latency PCIe channel might be performance bottlenecks that need to be addressed.

(3) The available memory space for applications is generally very limited, a few gigabytes at best, thus, the traditional algorithms that rely on large memory is not suitable for MIC.

To overcome the above limitations, we propose a new join method that fits the MIC/CPU heterogeneous systems. We summarize our contributions as follows.

(1) We design a collaborative join method for the hybrid CPU/MIC system, mainly including tasks assignment, data layout and communication between CPUs and MICs. We also propose a Sort-Scatter-Join (SSJ) algorithm for MIC to generate the join index. Through turning the traditional process of comparison and matching into the process of computing and scattering, the SSJ algorithm gains more beneficial from thread-level parallelism and SIMD data parallelism provided by MIC.

(2) We have performed extensive experiments to evaluate and compare the performance of our algorithm. The experiment results show that, compared with the traditional parallel sort-merge join algorithm, the peak performance of the SSJ running on MIC is improved by around 26 %.

The remainder of this paper is organized as follows. After a review of related work in Sect. 2, we provide a brief introduction on MIC/CPU architecture in Sect. 3, followed by our algorithm design and implementation in Sect. 4. Then in Sect. 5, we present our experimental results. Finally, we conclude in Sect. 6.

2   Related Work

Recently, heterogeneous systems consisting of general CPUs and many-core coprocessors have become the main trend in the high-performance computing area due to their powerful parallel computing capabilities and superior energy efficiencies. At the same time, the database query optimization technology using many-core coprocessors has also become a new emerging research topic. Most of the studies focus on GPU so far. For example, in 2007, He et al. [7, 8] have optimized the relational operators for GPU. They design a set of data-parallel primitives, and use these primitives to implement indexed or non-indexed nested-loop, sort-merge and hash joins. Following their research, Kaldewey et al. [9] indicate that their acceleration performance analysis ignored the PCIe data transfer time, but low-speed PCIe data transmission is actually the bottleneck preventing improvement of the performance in heterogeneous systems. In order to fully utilize PCIe bandwidth, Kaldewey uses the UVA provided by modern GPUs to perform efficient memory management. Pirk [10] and Karnagel [11] studied the problem of dividing processing workloads to fit both CPU and GPU in database management modules, and the problem of utilization of limited memory capacity. They also proposed a task allocation scheme based on a cost estimation model.

Although the GPU-oriented query optimization has made great achievements,it is hard to transplant these optimization methods directly to MICs due to the obvious architectural differences between GPUs and MICs [12]. MICs are new generation accelerator devices. Currently, the study on query optimization for MICs is still in its infancy, and the literature and achievements regarding this topic are also scarce. In order to fully utilize MIC's computing power, Stuart et al. [2] design a SQL-based database engine that can execute SQL queries on the Xeon Phi. The first researchers who studied join optimization for MIC are Jha et al. [13]. They compared optimization effects of the NPO and Radix Join on both MICs and CPUs in their experiments. Their experiment results show that the join algorithm optimization for MIC must account for its special hardware features.

3   Join Processing on Hybrid CPU/MIC Architecture

This section provides a brief introduction on MIC/CPU architecture because it is the foundation for our join optimization. The hybrid CPU/MIC architecture model is illustrated in Fig. 1. On such architecture, each component has many different hardware features. A CPU commonly owns no more than a dozen of cores and supports tens of hardware threads at the best. While on a MIC where the number of active threads can be easily extended to hundreds (e.g. new generation "Knight Landing", owns 72 cores and can support 288 hardware threads). In addition, the CPU can directly access a large amount of host memory (e.g. terabytes-scale), while the MIC has only several gigabytes device memory can be accessed directly. Moreover, each core of CPU support 256-bit SIMD at the best, while each core of MIC can support 512-bit vector operation. On hybrid CPU/MIC architecture, data is transferred between the host memory and the device memory over a PCIe link, which puts an upper bound on the throughput of data-intensive operations.

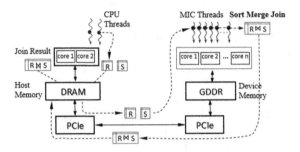

Fig. 1. Work flow of join processing on hybrid CPU/MIC architecture

In order to fully utilize every component, we design the workflow of join processing on hybrid CPU/MIC architecture, as presented in Fig. 1. Firstly, the CPU threads of a query application read relations to be joined (R and S) from the host memory. Next, R and S are transferred from the host to the MIC over the PCIe link. Then, the MIC threads join R and S using a join algorithm (e.g. sort-merge join) in parallel and generate the join result. Finally, the join result is transferred back to the host via PCIe.

4 The Collaborative Join Method for Hybrid MIC/CPU Platform

Section 3 introduces the workflow of join processing on hybrid architecture. In the following section, we will further describe how MICs and CPUs work collaboratively to finish the join operation. The main phase we design is given as follows.

(1) Join attributes are mapped to integer vectors using the method of dictionary encoding on the host, and then transferred to MIC via PCIe (Sect. 4.1).

(2) Using our join algorithm, namely Sort-Scatter-Join, to generate the join index. Then, the join index is asynchronously sent back to the host (Sect. 4.2).

(3) The host outputs the join results according to the join index (Sect. 4.3).

4.1 Data Preparation on the Host

The communication overheads generally become the bottleneck for performance improvement due to the low-bandwidth and high-latency PCIe channel. Thus, in order to minimize the cost of communication, we first reduce the data volume to be transferred and then choose the high-efficient communication mode.

Relation join is also a data-intensive operation, requiring large volume of data support. In order to reduce the data volume to be transferred, we only transfer the join attributes from the host to the MIC, whereas other attributes are hidden in the form of rows index. Similarly, the MIC only returns the join index to the host, and lets the host generate join results.

There are several communication channels between the host and the MIC, such as the Symmetric Communications Interface (SCIF), InfiniBand, and virtual network. Among these channels, the SCIF has highest communication bandwidth according to literature [15]. Hence, in this paper we use the SCIF for data communication.

Now, assume that relations R and S are to be joined with $R.x = S.y$. We first reconstruct $R.x$ and $S.y$ as two integer array vectors, V_L and V_R. If join attribute x or y is non-integers, we first need to conduct the dictionary encoding, and map them to integer array vectors. Then, the vectors V_L and V_R are transferred to the on-board memory of MIC using SCIF API functions via PCIe bus.

4.2   Sort-Scatter-Join Algorithm on the MIC

After the join attributes are transferred to the MIC, the next key issue should to be addressed is how to make full use the highly parallel computing capabilities of MIC to implement the join operation. For this, we propose a novel join algorithm named Sort-Scatter-Join (SSJ).

Algorithm 1. Sort-Scatter-Join algorithm

input: left relation R, right relation S
output: join index array *JoinRs*
01. $\{ V_L, V_R \}$ ←*acceptFromHost(R, S)*
02. $\{ C_L, C_R \}$ ←*constructComObj(V_L, V_R)*
03. C_{L+R} ←*joint(C_L, C_R)*
04. Q ← *ParallelSort(C_{L+R})*
05. $\{Q_1, Q_2, \cdots, Q_n\}$ ←*Partitioning(Q)*
06. for all threads T_i where $1 \leqslant i \leqslant n$ parallel-do
07. $\{LG_1, LG_2, \cdots, LG_m\}$ ←*ScanAndGroup(Q_i, key, 0)*
08. $\{RG_1, RG_2, \cdots, RG_m\}$ ←*ScanAndGroup(Q_i, key, 1)*
09.   for j ←0 to m
11.   if $\| LG_i \|$!=0 && $\| RG_i \|$!=0 then
12.   for each *tup* ∈ LG_i
13.   *LeftJoinIndex* ← *scatterIndex($\| RG_i \|$, tup.RowIdx)*
14.   end for
15.   *clearItems(tmpRIdxs)*
16.   for each *tup* ∈ RG_i
17.   addItems(*tup.RowIdx*, *tmpRIdxs*)
18.   end for
19.   *RightJoinIndex* ←*scatterIndex($\| LG_i \|$, tmpRIdxs)*
20.   end if
21.   end for
22.   *JoinRs*←*MergeResults(LeftJoinIndex, RightJoinIndex)*
23.   *SendbackToHost(JoinRs)*
24. end for

The pseudo-code of the SSJ algorithm is described in Algorithm 1. Line 1 shows the initialization operation, including accepting join attributes of input relations R and S from the host. In line 2, we package each data item of V_L using the two-tuple form of <*Key*, *RowIdx*>, and construct a new vector array C_L. Where, *Key* is each element of V_L and *RowIdx* is the position index of *Key* in V_L. Vector C_R is created in the same way. The next step we conduct sequence joints for C_L and C_R to construct a long sequence C_{L+R} (line 3). Then, in line 4 we sort C_{L+R} according to *Key* to obtain an ordered sequence Q in parallel. The following line we partition the Q as evenly as possible into n non-intersecting subsequences (supposing there are n threads). Lines 6-24 show the generation process of join index for each thread. In line 6, n worker threads are launched and each thread is assigned a subsequence. For each subsequence Q_i, we scan it and group its items into m groups (in lines 7–8). For each group, we first generate its partial join index using *RowIdx* scattering and then merge the results to the output array *JoinRs* (in lines 9–22). Finally, when a thread finishes the processing for a subsequence, this thread can transfer its join result *JoinRs* to the host using SCIF functions via PCIe (line 23).

4.3 Output Join Results on the Host

In order to receive the join index from the MIC, we start one or more threads on the host to listen and receive the join index. According to the left and right *RowIdx* of the join index (using *RowIdx's* sign: if it is positive, then item comes from the left table R, otherwise, the item comes from the right table S), we extract project attributes from the left and right tables and then construct the output tuples, as shown in Fig. 2.

	RA	RB	RC	RD
1	112301	6	455	2012/2/23
2	112302	2	768	2013/5/12
3	112303	3	456	2014/2/8
4	112305	7	456	2013/11/21
5	112306	6	321	2014/9/12
6	112308	2	643	2014/4/16
7	112309	9	768	2013/3/9
8	112310	8	345	2013/3/6
9	112312	3	356	2012/10/12
10	112313	2	467	2014/9/14
11	112314	6	345	2014/2/24
12	112315	6	656	2013/8/9

R

	SA	SB	SC
1	1021	1	Wuhan
2	1022	3	Beijing
3	1023	6	Kunming
4	1024	2	Guangzhou
5	1025	4	Changsha
6	1026	3	Taiyuan
7	1027	6	Jinan
8	1028	4	Zhangji
9	1029	9	FengHuang
10	1030	9	Xiangyang
11	1032	7	Wabang
12	1033	1	Chongqing
13	1034	9	Guiyang

S

RowIdx	RowIdx
1	-3
1	-7
5	-3
5	-7
11	-3
11	-7
12	-3
12	-7
4	-11

Join Index

Thread 2 MIC

PCIe

RA	SC
112301	Kunming
112301	Jinan
112306	Kunming
112306	Jinan
112314	Kunming
112314	Jinan
112315	Kunming
112315	Jinan
112305	Wabang

$\pi_{RA,SC}((R \bowtie S)_{RB=SB})$

Join results

Fig. 2. Output join results according to the join index and project attributes

4.4 Sort-Scatter-Join Algorithm Example

For further understanding our algorithm, in this section we provide an example to describe the whole process of SSJ. Suppose the input relations R and S are the same as illustrated in Fig. 2, and their join attributes (RB and SB) have been transferred to the MIC's memory via PCIe and stored in arrays V_L and V_R. We start three worker threads, namely, T_1, T_2 and T_3. The following steps describe the working process of SSJ in detail, as shown in Fig. 3.

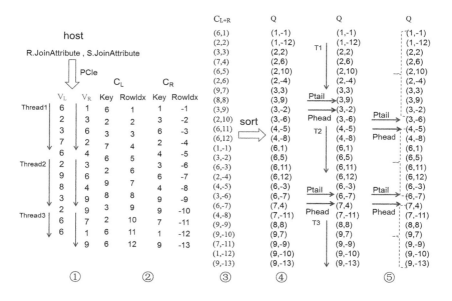

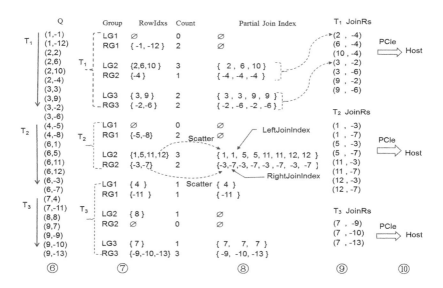

Fig. 3. Sort-Scatter-Join algorithm

Step ①: Start the threads T_1, T_2 and T_3, each thread scans one part of V_L and V_R.

Step ②: Package each item of V_L and V_R using the two-tuple form of (Key, $RowIdx$), and construct new sequences C_L and C_R. Where, $RowIdx$ from V_L is expressed by a positive number, and $RowIdx$ from V_R by a negative.

Step ③: Conduct sequence joints for C_L and C_R to construct a long sequence C_{L+R}.

Step ④: Adopt parallel sorting algorithm to sort C_{L+R} according to *Key* to obtain an ordered sequence Q.

Step ⑤: Partition Q into three subsequences as evenly as possible. Initially, Q is divided into three subsequences of the same size, where each subsequence has two pointers *Phead* and *Ptail*, which point to the head and end of the subsequence. And then constantly adjust the two pointers of each subsequence until the end item's *Key* of one subsequence is not equal to the head item's *Key* of the next subsequence.

Step ⑥: Scan each subsequence and group items by *Key*, i.e. the ones with the same *Key* are in the same group.

Step ⑦: For each group, count the number of items, LG_i, from the left table, and RG_i from the right table.

Step ⑧: Generate the join index. For each group, if neither LG_i nor RG_i of the ith group is null, then conduct the following processing: duplicate each *RowIdx* of the LG_i n times into the array *LeftJoinIndex*, where n is the number of items in RG_i. Meanwhile, consider all *RowIdxs* in RG_i as a whole and duplicate it m copies to the array *RighJoinIndex*, where m is the number of items in LG_i. The above operations can be accomplished with the SIMD instructions "gather/scatter" provided by MIC.

Step ⑨: Merge the *LeftJoinIndex* and *RighJoinIndex* to the output array *JoinRs*.

Step ⑩: Transfer the join index array *JoinRs* to the host via PCIe.

4.5   Algorithm Analysis

Similar to traditional parallel sort-merge join (PSMJ) (see Sect. 5.2), the SSJ is also a variant of sort-merge join. Thus the cost of sorting has a critical effect on performance of the join algorithm. The following are the main differences between the two algorithms. First, the PSMJ sorts and scans two lists using multithreads, while SSJ only deals with a single list, which leads our algorithm has less sync steps than the PSMJ. Second, after sorting, in order to get the joined tuples, PSMJ use the "if-then-else" instructions to control the pointers movements of the two ordered lists and then to find the matched tuples; while our approach just use the computing and scattering operations to generate the joined tuples, which reduce lots of branch instructions. This is very important, because not like CPU cores to support out-of-order execution, the MIC only arms with in-order cores. Third, in the merge-join phase, each thread in SSJ always sequentially and contiguously accesses memory, even for "many-to-many" joins; while the PSMJ algorithm may cause many pointer roll-backs when meeting with m to n joins. Thus, our method can get more benefit from SIMD "gather/scatter" instruction provided by the MIC [14], and also can reduce many cache misses caused by pointer jumping occurred in PSMJ algorithm.

5   Experimental Evaluation

5.1   Experimental Setup

We conduct our experiments on a server equipped with two MIC coprocessors. The hardware specifications of CPU and MIC are shown in Table 1. Utilizing these features is the key to achieve high performance on heterogeneous platforms.

Table 1. Experimental platforms

	Xeon Phi 5110P (MIC)	Xeon E5-2650 v3 (CPU)
# Cores × SMT	60 × 4	10 × 2
Core architecture	P54C	Haswell
Clock frequency	1.05 GHz /core	2.3 GHz /core
L1 cache size	32 Instr. + 32 Data KB/core	32 Instr. + 32 Data KB/core
L2 cache size	512 KB/Core	256 KB /Core
L3 cache size	-	25 MB
SIMD width	512-bit	256-bit
Memory capacity	8 GB	512 GB
Memory bandwidth	320 GB/s	34 GB/s

Our codes are developed using C and Pthreads, and compiled with optimization level 3 using the Intel compiler ICC 15. Xeon Phi runs Linux 2.6 as an embedded OS, and the Haswell machine has Linux CentOS (kernel version 2.6.32).

5.2   Evaluation Method

The MIC is a typical Symmetrical Multi Processing (SMP) architecture. According to the previous study [8], parallel sort-merge join algorithm (PSMJ) is more suitable to SMP. Thus, to verify the effectiveness of our algorithm, we consider the PSMJ algorithm as a benchmark for comparison. The general idea and the individual phases of the PSMJ algorithm are presented in Fig. 4.

Phase I: Sorting Phase

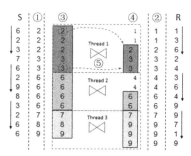

Fig. 4.   An example for parallel sort merge join

Step ① ②: Sort relation S, R in parallel.

Phase II: Range-partition Phase

Step ③: Divide the smaller relation S to be n chunks as evenly as possible, n is the number of worker threads.

Step ④: Use the key values of the first and the last tuples of each chunk in S to identify the start and the end positions of its matching chunks in R.

Phase III: Merge Join Phase

Step ⑤: Merge each pair of the chunk in S and its matching chunk in R in parallel. Each thread is responsible for a pair.

parallel sort merge join of threads on the SCIF bandwidth

In the above steps, threads not work independently for each step, thus we need to synchronize all threads via barrier before we can start the next step. This means when the number of threads increases, the synchronization overheads would also enlarge.

5.3   Data Transfer Between the MIC and the Host

The low-bandwidth and long-latency PCIe link is the main performance bottleneck for heterogeneous system, especially to join operation that rely on large volume of data transmission. Among many channels of data transfer, the SCIF has the highest bandwidth [15]. Although it has a slightly longer latency, it suits better for large volume data transmission. In order to prevent thread blocking, we select the asynchronous transfer mode when the SSJ algorithm returns the join index to the host.

Figure 5 shows the result of SCIF unidirectional data transmission experiment. Where "H ≫ M" means Host transfer to MIC and "M ≫ H" is the opposite situation. It can be seen from Fig. 5 that the upper bound of SCIF data transfer speed is 6.5 GB/s. The size of the message has a great effect on the transmission bandwidth. When the number of threads is small, the bandwidth grows along with the message size. However, when the thread number is large enough, the bigger message size has a negative impact. This indicates that when designing the parameters of data transmission, we need to consider both the message size and the number of threads to choose an appropriate send/receive buffer size. Conclusion drawn from our experiment is that the transmission is most effective for PCIe 2.0 with 60 threads and 8 MB message. At this time, the bandwidth can achieve 6.37 GB/s.

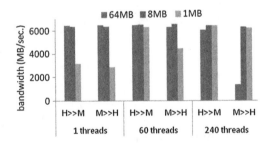

Fig. 5. Effects of message sizes and number of threads on the SCIF bandwidth

5.4   Performance Evaluation of the Sort-Scatter-Join Algorithm

The most significant advantage for MIC lies in its massive cores, which can support more than hundreds of hardware threads. Therefore, the focus of MIC-oriented optimization is how to improve the thread-level parallelism. To better observe parallel scalabilities of the SSJ algorithm, we carry out experiments to test the impact of the number of threads on the join performance. In order to avoid the interference of data communication, the following implementations all run on MIC as native programs. We perform the equal-join on two relations *R* and *S*, both with 128 million tuples. All tuples are randomly generated in memory but not unique, and each tuple consists of 4 bytes integer *key* and 4 bytes *payload*. Since the *parallel radix sort* proposed by Satish [16] is the fastest sort algorithm for many-core GPU, we use the same method in sorting phase.

As can be seen in Fig. 6, the parallel scalability is relatively good until the SSJ algorithm reaches the number of physical cores owned by a coprocessor, and the performance improves almost linearly with the number of threads. However, when the number of worker threads exceed the number of hardware cores (e.g. 60 for Xeon Phi 5110), the acceleration slows down significantly. The optimal performance is achieved when launching 180 threads. After that, the performance begins to decline due to increasing competition for shared resources. It follows that our algorithm can make full use of the computing resources offered by the MIC to a larger extent.

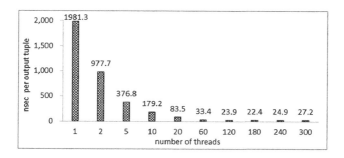

Fig. 6. Impact of the number of threads on the SSJ performance

To further examine the performance of our algorithm running on the MIC, we choose the PSMJ algorithm as a benchmark for comparison. The results of the performance comparison of the two algorithms are shown in Fig. 7.

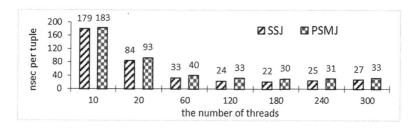

Fig. 7. Performance comparison of the PSMJ and SSJ with the number of threads varied

As shown in Fig. 7, our algorithm can achieve better performance than the PSMJ, and the more the number of threads is the better the profits we get. Wherein our algorithm achieve its optimal performance at 180 threads, and its peak performance reaches 22 ns per tuple, while at this scale parallelism, the peak performance of the PSMJ is only 30 ns per tuple. Thus, our algorithm performs better than the PSMJ algorithm by around 26 % in high concurrent environment. This is because our algorithm has less synchronization steps than the PSMJ. Moreover, our method is more adaptive to utilizing the MIC's hardware features, as described in Sect. 4.5.

5.5   Time Consumption Profiling in Heterogeneous Systems

In order to find out, under different degrees of parallelism, which stage of our join method has the higher time consumption when running on hybrid MIC/CPU system, we further perform detailed profiling for each program stages. Table 2 show the time consuming percentage for each program stage in our algorithm.

Table 2. The time consumption profiling for main program stages

Program stage	1 threads	60 threads	180 threads	240 threads
Parallel sort	90.4 %	76.0 %	63.6 %	60.2 %
Join initialization	2.7 %	8.1 %	16.5 %	18.3 %
SCIF data transfer	2.1 %	8.6 %	11.6 %	11.7 %
Scatter Join	3.8 %	4.3 %	4.5 %	4.0 %
Other operations	1.0 %	2.9 %	3.8 %	5.8 %

Table 2 shows that the *parallel sort* is always the most time-consuming operation, especially when the degree of parallelism is low. Sorting can take more than 76 % of the total time. But with the increase of number of threads, the ratio decreases. When at 240 threads, sorting only accounts for 60.2 %. The second most time-consuming operation is the *join initialization*. At this stage, it needs to create arrays: V_L, V_R, C_L, C_R, and to copy data items among different arrays. Memory allocation and lots of memory read/write operations make the time consumption relatively high. The *SCIF data transfer* takes up the third position. Although we reduce the amount of data to be transferred as possible as we can, in a high concurrent environment, it is still one of the main factors affecting the performance. With 180 threads, it can account for 11.7 % of the total time. However, the *scatter join* stage, as the degree of concurrency increases, it always maintains at a low level of about 4 %. Based on the above facts, in a highly concurrent environment, the hotspots of the SSJ algorithm distribute more uniformly, but the main hotspots are still related to sorting. Therefore, searching for a more efficient sorting algorithm is our further optimization focus.

6   Conclusion

On heterogeneous platforms, CPUs and MICs have different hardware features, and their nonsymmetrical computing abilities and storage models make the traditional join algorithms cannot make full use of each hardware. Thus, we design a collaborative join method for hybrid MIC/CPU systems, including a comprehensive study of the data layout, communication mode, tasks assignment between the MIC and CPU. In order to fully utilize the MIC's advantages, we also propose a Sort-Scatter-Join algorithm for MIC to efficiently generate the join index. Through turning the process of comparison and matching into the process of computing and scattering, the SSJ algorithm gains more beneficial from thread-level parallelism and SIMD data parallelism. Experimental results demonstrate that our join optimization on the MIC is effective.

References

1. Casper, J., Olukotun, K.: Hardware acceleration of database operations. In: Proceedings of the ACM/SIGDA International Symposium on FPGA, pp. 151–160. ACM, New York (2014)
2. Stuart, O., Brian, R., Ziliang, Z.: SQLPhi: a SQL-based database engine for Intel Xeon Phi coprocessors. In: Proceedings of the 2014 International Conference on Big Data Science and Computing, pp. 1–6. ACM Press, New York (2014)
3. Blanas, S., Li, Y., Patel, J.M.: Design and evaluation of main memory hash join algorithms for multi-core CPUs. In: Proceedings of the 2011 ACM SIGMOD International Conference on Management of Data, pp. 37–48, New York (2011)
4. Balkesen, C., Alonso, G., Teubner, J. et al.: Multi-core, main-memory joins: sort vs. hash revisited. In: The 40th International Conference on Very Large Data Bases, pp. 85–96, Hangzhou (2014)
5. Kim, C., Sedlar, E., Chhugani, J., et al.: Sort vs. hash revisited fast join implementation on modern multi-core CPUs. VLDB Endow. 2(2), 1378–1389 (2009)
6. Albutiu, M.C., Kemper, A., Neumann, T.: Massively parallel sort-merge joins in main memory multi-core database systems. VLDB Endow. 5(10), 1064–1075 (2012)
7. He, B., Lu, M., Yang, K.: Relational query co-processing on graphics processors. Trans. Database Syst. ACM 34(4), 23–32 (2009)
8. He, B., Yang, K., et al.: Relational joins on graphics processors. In: ACM SIGMOD International Conference on Management of Data, pp. 511–524. ACM, New York (2008)
9. Kaldewey, T., Lohman, G., et al.: GPU join processing revisited. In: Proceedings of the 18th International Workshop on Data Management on New Hardware, pp. 55–62 (2012)
10. Pirk, H., Kersten, M., Manegold, S.: Accelerating foreign-key joins using asymmetric memory channels. In: The 2nd International Conference on Accelerating Data Management Systems (2011)
11. Karnagel, T., Habich, D., Schlegel, B., et al.: Heterogeneity-aware operator placement in column-store DBMS. Datenbank-Spektrum 14(3), 211–221 (2014)
12. Jim, J., James, R.: Intel Xeon Phi Coprocessor High Performance Programming. Morgan Kaufmann, San Francisco (2013)
13. Jha, S., He, B., Lu, M., et al.: Improving main memory hash joins on Intel Xeon Phi processors: an experimental approach. VLDB Endow. 8(6), 642–653 (2015)
14. Tian, X., Saito, H., Preis, S.V., et al.: Effective SIMD vectorization for Intel Xeon Phi coprocessors. Sci. Program. 2015, 1–14 (2015)
15. Potluri, S., Venkatesh, A., et al.: Efficient intra-node communication on Intel-MIC clusters. In: The 13th IEEE/ACM Cluster, Cloud and Grid Computing, pp. 128–135 (2013)
16. Satish, N., Harris, M., Garland, M.: Designing efficient sorting algorithms for manycore GPUs. In: The 23rd IEEE International Symposium on Parallel and Distributed Processing, pp. 1–10 (2009)

GFSF: A Novel Similarity Join Method Based on Frequency Vector

Ziyu Lin[1($\boxtimes$)], Daowen Luo[1], and Yongxuan Lai[2]

[1] Department of Computer Science, Xiamen University, Xiamen, China
{ziyulin,luodw}@xmu.edu.cn
[2] School of Software, Xiamen University, Xiamen, China
laiyx@xmu.edu.cn

Abstract. String similarity join is widely used in many fields, e.g. data cleaning, web search, pattern recognition and DNA sequence matching. During the recent years, many similarity join methods have been proposed, for example Pass-Join, Ed-Join, Trie-Join, and so on, among which the Pass-Join algorithm based on edit distance can achieve much better overall performance than the others. But Pass-Join can not effectively filter those candidate pairs which are partially similar. Here a novel algorithm called GFSF is proposed, which introduces two additional filtering steps based on character frequency vector. Through this way, the number of pairs which are only partially similar are greatly reduced, thus greatly reducing the total time of string similarity join process. The experimental results show that the overall performance of the proposed method is better than Pass-Join.

1 Introduction

Similarity join is to find out string pairs satisfying a certain similarity threshold. It has a wide range of applications, such as coalition detect [1], fuzzy keyword matching [2], data integration [3], data cleaning [4], and so on. The classical similarity join algorithms include Ed-Join [5], Trie-Join [6], Pass-Join [7], and so on. They mainly consist of two process, namely, filtering and verification. In the verification process, at present, the widely used measurement function is edit distance [5–7], Jaccard distance [8–10], Cosine [11] and the variants [12,13] of the functions above. In the filtering process, the Ed-Join algorithm uses the n-grams-based [14] method to cut down the original string sets. Then the prefix filtering is used to filter the original candidate string pairs. Finally, the ultimate candidate pairs are verified. Trie-Join uses the form of Trie-tree structure. If the edit distance of the prefix of two branches is more than the threshold, there are no similarity strings in the descendants of the two branches. Among similarity join algorithms, the Pass-Join's overall performance is most desirable. The Pass-Join algorithm divides the string into some segments, and then restricts the

Supported by the Natural Science Foundation of China (61303004), the National Key Technology Support Program (2015BAH16F00/F01) and the Key Technology Program of Xiamen City (3502Z20151016).

B. Cui et al. (Eds.): WAIM 2016, Part II, LNCS 9659, pp. 506–518, 2016.
DOI: 10.1007/978-3-319-39958-4_40

number of segments. If there are a segment matching between two strings, the two strings are added into candidate set and finally verified. These algorithms are able to get correct results and reduce the number of candidate pairs so as to reduce the amount of edit distance calculation.

However, Pass-Join is still with some shortcomings. If there is a common segment in two strings, Pass-Join takes the two strings as a candidate pair. Evidently, many string pairs which can not be similar may also be added into the candidate sets. Because although a lot of strings have a common prefix or suffix, they may not be similar strings. For example, assuming that the string $s_1 = $ "*through*", $s_2 = $ "*thing*" and the threshold of edit distance is 2. Although s_1 and s_2 have the common prefix "th", they are not similar strings because of their edit distance being more than 2. So the candidate set which is produced by Pass-Join can be further filtered and the amount of candidate set can be further reduced.

In order to address the problem discussed above, we propose GFSF (Global-Filtering and Segment- Filtering) similarity join algorithm, which is based on Pass-Join algorithm. Compared to the original Pass-Join algorithm, here two additional filtering steps are introduced in GFSF. First, when there is a common segment in two strings, we calculate the difference of their character frequency vector which represents the degree of difference between the two strings. If the difference is relatively large, it means that it needs more editing steps to make the two strings matching. So, the two strings are mismatched and can be eliminated. Second, after the first step, we take off the common segment from the two strings and respectively calculate the difference of the character frequency vector between the two remaining segments just as the first step. By comparing the difference with a given threshold, we can further filter the two strings which have a common segment and the same characters but the order of characters of which are greatly different. It must be pointed out that although the two additional filtering steps may increases the time of filtering process, the candidate set is cut down greatly, thus greatly decreasing the time of verification. Overall, the reduction of the verification time is more than the increase of filtering time, and therefore our method can achieve much better overall performance than Pass-Join.

In a sum, the contributions of this paper can be summarized as follows:

- We propose GFSF similarity joins algorithm, which can greatly reduce the amount of candidate pairs and therefore reduce the time cost of the verification process.
- For the second filtering step in GFSF, a novel segment filtering method is used to greatly speed up the process of GFSF.
- We conduct extensive experiments to compare our algorithm with the Pass-Join algorithm, and the results show that the overall performance of our algorithm is better than Pass-Join.

The rest of the paper is organized as follows. Section 2 introduces the Pass-Join similarity join algorithm and discusses the deficiency of pass-Join algorithm. The proposed GFSF similarity join algorithm is described in detail in Sect. 3.

Section 4 gives the experimental results. Finally, we review related work in Sect. 5 and conclude in Sect. 6.

2   Pass-Join Algorithm

2.1   The Description of Pass-Join

Assuming that a string is denoted as s and the edit distance is denoted as τ. $|s|$ is the length of s. Pass-Join splits s into $\tau+1$ segments, and then compare s with another string whose length is within the range of $[|s| - \tau, |s|]$. If the string s has a segment matching with another string whose length is within the range of $[|s| - \tau, |s|]$, the two strings will be taken as a candidate pair. Finally, we calculate the edit distance and determine whether the two strings are similar.

Figure 1 shows an example of Pass-Join framework. There are two strings s_1 = "kaushik chakrab", s_2 = "kaushuk chadhui" and the edit distance threshold $\tau = 3$. First, s_1 is spilt into $\tau + 1$ segments using the split method that Pass-Join proposes. Next, we find out whether s_2 contains a substring that matching one of the four segments of s_1. In Fig. 1, s_2 has segments "kau" and "_cha" that matching one of the four segments of s_1, so s_1 and s_2 are taken as a candidate pair. Finally we calculate their edit distance. Their edit distance is greater than τ, so they are not similar strings. Since s_1 and s_2 have the same length denoted as $len = 15$, so s_2 is also split into $\tau + 1$ segments and is added into the inverted index.

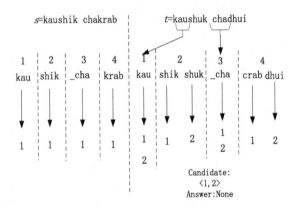

Fig. 1. An example of Pass-Join framework

String Splitting . According to the string split method of Pass-Join, we divide the string into $\tau+1$ segments, but there are many different kinds of methods for segmentation. Pass-Join's segmentation strategy is to split the string into $\tau+1$ fragments whose length are roughly equal. Assuming that a string s, $k = |s| - \lfloor \frac{|s|}{\tau+1} \rfloor * (\tau + 1)$, and then the length of the last k fragments is $\lceil \frac{|s|}{\tau+1} \rceil$, the length of the first $\tau + 1 - k$ fragments is $\lfloor \frac{|s|}{\tau+1} \rfloor$. For a fixed length string, the length of each segment is fixed. Thus we build the inverted index L, $L_l^i(w)$ represents string sets in which the length of string is l and the *ith* segment is w.

The main steps of the Pass-Join algorithm are as follows:

1. Sort all of the strings according to the length of the sequence, if the length is equal, sorted according to the lexicographical order;
2. Split strings and build inverted index;
3. Start from the first string to do the join operation: set the length of string s is $|s|$. First list all the segments w of s, then calculate the edit distance between any one of the $L_l^i(w)$ and s, finally get the similar string pairs, l and i are limited as $|s| - \tau \le l \le |s|, 1 \le i \le \tau + 1$.

Segment Selection. The third step of the Pass-Join is a segments selection process. Let $W(s, L_l^i)$ represents all the strings that s find in index L_l^i, $W(s,l)$ represents all the segments that s find in $L_l^i (1 \le i \le \tau + 1)$. If we enumerate all the segments that s has, then $|W(s, L_l^i)| = \sum_{i=1}^{|s|} (|s| - i + 1)$. That is equal to $\frac{|s|*(|s|+1)}{2}$. This is vary large if s is a long string. And it will greatly influence the efficiency of the algorithm. Therefore, Pass-Join designs four methods for segments selection.

1. **Length-based Method.** In L_l^i, the segments have the same length, so $|W(s, l)| = (\tau + 1)(|s| + 1) - l$;
2. **Shift-based Method.** As for L_l^i, s only selects the segments whose offset to *ith* segment is in $[-\tau, \tau]$. So $|W(s, l)| = (\tau + 1)(2\tau + 1)$;
3. **Position-aware Substring Selection.** As for L_l^i, s only select the segments whose offset to *ith* segment is in $[-\lfloor \frac{\tau - \triangle}{2} \rfloor, \lfloor \frac{\tau + \triangle}{2} \rfloor]$. So the $|W(s, l)| = (\tau + 1)^2$;
4. **Multi-match-aware Substring Selection.** This method selects the count of segments is $|W(s, l)| = (\tau + 1)^2$, the detail derivation see paper [6];

2.2 The Disadvantage of Pass-Join

Pass-Join designs four methods to select segments, among which, the fourth is the optimal. However, inefficient filtering is an evident shortcoming of Pass-Join, namely, many string pairs which can not be similar and should be cut, still exist in the resulting candidate set.

Firstly, when the two strings have a segment matching with each other, it is arbitrary for Pass-Join to take the two strings as similar candidate string pair. The two strings are likely to be only partially similar, and then the rest is completely different. So for Pass-Join, to determine whether the candidate pair is similar is too restrictive. For example, there are two stings $s_1 = $ "vasdlym" and $s_2 = $ "vahijkx", the edit distance threshold $\tau = 2$. The two strings contain "va" segment, so the Pass-Join algorithm will take the two strings as a candidate pair. However, the edit distance of the two strings is $6 > \tau$, so the two strings are not similar. Thus, Whether two strings are similar is related to the all the characters that the two strings contain.

Secondly, if two strings have the same characters and the order of the characters is also the same, they are not always similar. For example, there are the two string s_1 = "vasdlym" and s_2 = "myldsva". Although they have the same characters and the count of character is also the same, it's obvious that they are not similar. They only have a character matching, but they don't have a segment matching whose length is more than 2. So the edit distance of the two strings are more than 2, they are not similar strings. Therefore whether two strings are similar is also related to the order of the characters.

3   GFSF Algorithm

In order to address the problem of the Pass-Join algorithm, here a novel algorithm called GFSF(Global-Filtering and Segment-Filtering) is proposed to further filter candidate set generated by Pass-Join Selection Algorithm. In the following, the overview of GFSF is described in Sect. 3.1. Then the two additional filtering steps, namely, Global Filtering and Segment Filtering, are discussed in Sects. 3.2 and 3.3 respectively.

3.1   An Overview of GFSF

Compared with Pass-Join, two additional filtering steps are introduced in GFSF, namely, Global Filtering and Segments Filtering. Global Filtering mainly filters the candidate pairs which are only partially similar. Segment Filtering filters the two strings which have the same characters but the order of characters are different. Figure 2 shows where we add the two filtering steps in Pass-Join.

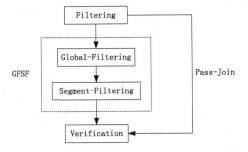

Fig. 2. The overview of GFSF

3.2   Global Filtering

The main idea of the Global Filtering is that when two strings have a segment matching, we do not take them as candidate pair like Pass-Join. Instead we calculate their γ-distance introduced next. If their γ-distance is less than a threshold discussed next, it shows that the difference of their character frequency is not very large. Thus they are likely to become similar strings. Next we will introduce character frequency vector and discussed the γ-distance threshold.

Character Frequency Vector : α_i is any character of alphabet Σ, and how many times the α_i appears is called the frequency of α_i, denoted as $f_i(s)$. $<f_1(s), f_2(s), ..., f_{|\Sigma|}(s)>$ is called s's character frequency vector, denoted as $f(s)$, which is ordered by lexicographical order.

γ-distance: There are two vectors, $U = (u_1, u_2, ..., u_m), V = (v_1, v_2, ..., v_m)$, their γ- distance is $||U - V||_\gamma = \sum_{i=1}^{m} |u_i - v_i|$.

Theorem 1. *There are two strings s and t, if $||f(s) - f(t)||_\gamma > 2\tau$, the edit distance between s and t is greater than τ.*

We know that every edit operation will result in 2 differences of frequency vector between two strings, thus if the edit distance between s and t is less than τ, then $||f(s) - f(t)|| \le 2\tau$. Subsequently, if $||f(s) - f(t)||_\gamma > 2\tau$, the edit distance between s and t is more than τ.

γ-distance threshold: If s's length is equal to t's length, the difference of γ-distance between s and t is completely resulted by the kind of character. So we can get the following theorem.

Theorem 2. *If the γ-distance between s and t is L, then the edit distance between the two is at least $L/2$, which is the number of edit operations required to convert s to t.*

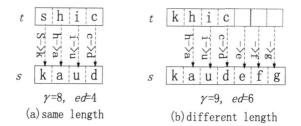

Fig. 3. The conversion process between two strings

Here, we do not take the order of character into count. For example, there are two strings $s =$ "kaud" and $t =$ "shic". The two strings are completely different, so the γ-distance between the two strings is 8, then the edit distance between them is at least 4. Figure 3(a) shows the conversion process.

If s's length is not equal to t's length, the length difference between the two strings must satisfy $||s| - |t|| \le \tau$. Since Pass-Join compares the current string with other strings whose length are in the range of $[|s| - \tau, |s|]$. If the length difference is greater than τ, the two strings must not be similar strings, because the operation to offset length difference is greater than the edit distance threshold.

Theorem 3. *Given two strings s and t with different length, assuming that $|s| > |t|$ and their γ-distance is L. If the length difference is equal to len, then the edit distance is at least $\frac{(L-len)}{2} + len$.*

Algorithm $Global - Filtering(s, L_l^i(w))$
Input: a string s;
            a inverted list $L_l^i(w)$ that contains the strings which contain segment w;
Output: a new inverted list $L\_new_l^i(w)$;

1.**Begin**
2.   // s_vector is s's character frequency vector;
3.   **for** each $r \in L_l^i(w)$ **do**
4.   //r_vector is r's character frequency vector;
5.   //v_dist is the difference between s_vector and r_vector;
6.   **if**($|s| \doteq |r|$ and $\frac{V\_dist}{2} \leq \tau$)
7. $L\_new_l^i(w) \rightarrow add(r)$
8.   **endif**
9.   **if**($|s| \neq |r|$ and $\frac{V\_dist - (||s| - |r||)}{2} + (||s| - |r||) \leq \tau$)
10. $L\_new_l^i(w) \rightarrow add(r)$
11.   **endif**
12.   **endfor**
13.**End**

Fig. 4. The algorithm of Global Filtering

Since the length difference between the two strings is *len*, it needs at least *len* edit operations to offset the length difference. *s* deducts the length difference and the γ-distance of the rest with *t* is resulted by the kind of character just as the above discussion. So the edit distance between the *s* and *t* is at least $\frac{(L-len)}{2} + len$. Here we do not take the order of character into account. For example, two strings $s =$ "kaudefg" and $t =$ "khic". Their γ-distance is 9 and length difference is 3, then $\frac{(L-len)}{2} + len = 3$, so their edit distance is at least 6. Figure 3(b) shows the conversion process.

The global filtering mainly uses the γ-distance and γ-distance threshold to filter candidate sets. The main steps of GFSF is as follows:

1. Select all the segments of *s*, denoted as $W(s, L_i^l)$, by the means of Pass-Join's selection algorithm.
2. Match all the segments of $W(s, L_i^l)$ with the segments in inverted index, if not matched, iterate next string, or go to step 3.
3. Calculate the character frequency vector of the two strings.
4. Calculate the γ-distance between the two strings, if the γ-distance is less than a threshold τ, it can judge that the two strings are candidate pairs.

From the above discussion, we know that we can further filter the candidate set by γ-distance. So we use γ-distance to filter the candidate set produced by Pass-Join.

Algorithm 1 in Fig. 4 formally describes global filtering. When Pass-Join selection method gets the segment sets of *s* and the inverted list $L_l^i(w)$ which contains the strings that also contains segment w, we do not calculate the edit distance between s and strings in $L_l^i(w)$. Instead, we iterate every strings denoted as r in $L_l^i(w)$ and calculate γ-distance between s and r. Then we continue the process as follows:

1. The lengths of the two strings are equal. If $\frac{\gamma}{2} \leq \tau$, we reserve them, else discard them;
2. The length of the two strings are not equal. If $\frac{\gamma-(||s|-|t||)}{2} + ||s| - |t|| \leq \tau$, we reserve them, else discard them.

3.3   Segment Filtering

Although global filtering can filter some candidate pairs which are taken as candidate similar string pairs because of their partial similarity, it does not take the character order into count. Two strings with completely same characters and with a segment matching are either not likely to be similar, although Pass-Join will take them as candidate pair. For example, two strings s = "abcdexyz" and t = "xyzdeabc" and assuming that $\tau \doteq 2$. Although they have completely same characters and have a common segment "de", they are not similar strings. Pass-Join will take them as a candidate pair because of common segment "de". But it can be filtered by a certain method. Next we will introduce the method and call it segment filtering.

Segment filtering is the further filtering process after global filtering. The main idea of segment filtering is that if two strings have a common segment, then the first segment of the two strings will calculate their γ-distance respectively and judge whether they meet the threshold. The second segment of the two strings will also be processed as the first segment. If they meet the threshold, then we will take the two strings as candidate pair ultimately. Global filtering analyzes the problem from the global perspective, but will ignore the order of character. Segment filtering analyzes the problem from local perspective and will well solve the problem of character order. Next we show the steps of segment filters:

1. If s's first segment and t's first segment do not meet the threshold, then they must not be similar and there is no need to calculate the γ-distance of the second segment of the two strings.
2. If s's first segment with the t's first segment meet the threshold, we denoted their γ-distance as v_1, so we will calculate the γ-distance of s's second segment and t's second segment, denoted as v_2. If $v_1 + v_2 \leq threshold$, then we will take them as candidate pair ultimately. If $v_1 + v_2 > threshold$, we will discard them.

Here we use the example previously to explain the segment filtering. Given two strings s = "abcdexyz" and t = "xyzdeabc", and assuming that $\tau \doteq 2$. They have a common segment "de", So we calculate the γ-distance between the first segment of the two strings. Because the γ-distance between the "abc" and "xyz" is 6, the edit distance is at least 3. Since $3 > \tau$, s and t must not

Algorithm *Segment − Filtering(s,$L_l^i(w)$)*
Input: a string s;
              a inverted list $L_l^i(w)$ that contains the strings which contain segment w;
Output: a new inverted list $L\_new_l^i(w)$;

14.**Begin**
15.   s_seg[2]={$se|se \in s$ and $se \neq w$} // se is the segment of s except segment w;
16.   **for each** $r \in L_l^i(w)$ **do**
17.   r_seg[2]={$re|re \in r$ and $re \neq w$} // re is the segment of r except segment w;
18.   // v1_dist is the γ-distance between s_seg[1] and r_seg[1];
19.   **if**(v1_dist is not meet the threshold)
20.   **continue**;
21.   **endif**
22.   **if**(v1_dist is meet the threshold)
23.   // v2_dist is the γ-distance between s_seg[2] and r_seg[2];
24. $v\_dist = v1\_dist + v2\_dist$;
25.   **if**(v_dist is meet the threshold)
26. $L\_new_l^i(w) \rightarrow add(r)$
27.   **endif**
28.   **endif**
29.   **endfor**
30.**End**

Fig. 5. The algorithm of Segment Filtering

be similar strings. Now we can discard s and t. If the two strings are changed to s = "abmdenxy" and t = "xymdenab". Because the γ-distance between the first segment "ab" and "xy" is $v_1 = 4$, edit distance is at least 2. Since $2 \leq \tau$, we calculate γ-distance of the second segment. The γ-distance between "xy" and "ab" is $v_2 = 4$, so the edit distance is at least 2. Since $\frac{v_1+v_2}{2} \doteq 4 > \tau$, the two strings are not similar strings, and therefore we discard them.

Algorithm 2 in Fig. 5 formally describes the segment filtering. First, we get the two segments of string s except the common segment w. And then, we iterate every string in $L_l^i(w)$, denoted as r. We also get the two segments of string r except the common segment w. Next we calculate the γ-distance between s_seg [1] and r_seg [1], denoted as *v1_dist* . If *v1_dist* does not meet the threshold discussed in global filtering, we discard the two strings. If *v1_dist* meets the threshold discussed in global filtering, we calculate the γ-distance between s_seg [2] and r_seg [2], denoted as *v2_dist*. We let *v_dist* be equal to the sum of *v1_dist* and *v2_dist*. If *v_dist* meets the threshold, we take them as candidate pair ultimately, else we discard them.

3.4   Performance Analysis

The calculation of edit distance is a time-consuming work, so many algorithms including Pass-Join for similarity join pay much attention to the filtering process, so as to greatly reduce the amount of string pairs and decrease the time of the verification process. Although GFSF introduces two additional filtering steps, the decrease of verification is more than the increase of the time resulting from the two filtering steps. So the overall performance of GFSF is much better than Pass-Join, which can be shown in experimental results in Sect. 4.

4   Empirical Study

In this section, we conduct experiments to verify the efficiency of our approach. We mainly use the number of candidate pairs to demonstrate performance difference. We compare the Pass-Join, Pass-Join with Global-Filtering and GFSF. If the number of the candidate pairs of GFSF is the least, it means that GFSF is effective for filtering candidate set. We also compare the elapsed time of the GFSF with Pass-Join. We will show that GFSF algorithm can not only filter candidate set but also speed up Pass-Join algorithm.

4.1   Environmental Setup

All the algorithms are implemented in C++ and compiled using GCC 3.4 with -O3 flag. All the experiments run on a Ubuntu machine with an Intel(R) Core(TM) i3 3.10GHZ processor and 6G memory. We use the same datasets with PASS-Join, but some details are different. We also use strings with three kinds of length, so as to show that our algorithm is effective for Pass-Join in different cases. Table 1 shows the detail of datasets.

Table 1. Datasets

Datasets	Cardinality	Avg Len	MAX Len	Min Len
Author	423178	13.348	42	6
QueryLog	469427	46.742	497	28
Author + Title	642094	110.482	893	23

4.2   Evaluating Filtering Efficiency

We first compared the number of candidate pairs. As Fig. 6 shows, when the threshold increases, the number of candidate pairs also increases. The reason is that the larger the threshold is, the more difference the candidate pairs can contain. It can also be seen from Fig. 6 that, compared with the size of candidate set produced by Pass-Join, the size of candidate sets produced by Pass-Join with global filtering and GFSF, are reduced by 19 percent and 33 percent respectively. So GFSF has a good effect for filtering string pairs and can filter many string pairs that Pass-Join take as candidate pairs.

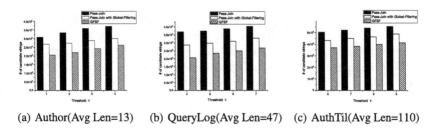

(a) Author(Avg Len=13)   (b) QueryLog(Avg Len=47)   (c) AuthTil(Avg Len=110)

Fig. 6. The number of candidate pairs in three cases

4.3   Evaluating Elapsed Time

Now we show that GFSF algorithm can not only filter many string pairs, but also can speed up the whole similarity join process than Pass-Join. We conduct three different experiments with three different experimental datasets, namely, *Author*, *QueryLog*, *Author + Title*. From the Fig. 7, we know that as the edit distance threshold becomes larger, the elapsed time of Pass-Join and GFSF also become larger because of the increase of candidate set. And compared with the elapsed time of Pass-Join, the elapsed time of GFSF is reduced by 23 percent. So we can get that GFSF can speed up the similarity join process much better than Pass-Join.

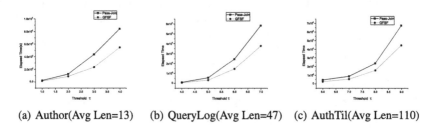

(a) Author(Avg Len=13)   (b) QueryLog(Avg Len=47)   (c) AuthTil(Avg Len=110)

Fig. 7. The elapsed time in three cases

5   Related Work

In recently years, various approaches have been proposed to deal with similarity joins. Most of the them take the Filter-And-Refine as their framework. The main idea of this framework is that in the filter step, they will use a special index structure and generate a small number of candidate pairs that may be similar. In the refine step, they will use edit distance algorithm or other measure functions to verify the candidate string pairs and then find the similar string pairs. Recently, ED-Join, Trie-Join and Pass-Join are the three mainstream approaches with high efficiency to deal with the similarity strings.

ED-Join is based on q-gram. If two q-grams are matching, it means that they have the common token and the position offsets of their positions in string do not exceed the edit distance threshold. Based on q-gram matching, ED-Join algorithm proposed three kinds of string similar filtering methods, namely, Prefix Filtering, Location-based Filtering and Content-based Filtering. Through the three filter methods, ED-Join can filter many string pairs which can not be similar.

Trie-Join has completely different filtering algorithm compared with ED-Join. Trie-Join use trie structure to store every string. Every path from root to leaf represents a string, so every node represents a character of string. Trie-Join uses active node to filter strings. If a node is not an active node of a certain string, then we can filter the descendants of active node of the string. This is the filtering principle of Trie-Join. Trie-Join is efficient for short strings.

Recently, MapReduce is introduced to improve the efficiency of similar join, such as [15–17]. First, strings are divided into groups. Second, in the map process, every group are calculated and candidate sets are found. In the reduce process, every candidate set are merged to be a ultimate candidate set. By using the parallel computation of MapReduce, it can greatly speed up the process of similarity join. In addition, Simrank [18] and SPB-tree [19] structure are also used to process similarity join, they can also get a good result.

6   Conclusion

In this paper, we propose GFSF algorithm to overcome the Pass-Join's shortcoming. On one hand, global filtering filters the string pairs that are only partially similar; on the other hand, segment filtering filters the string pairs which have the same characters but the order of characters are different. By the additional filtering steps, GFSF can filter more strings and speed up the process of verification. The experimental results show that GFSF algorithm is more efficient than Pass-Join algorithm. Our future work includes further improving the performance of GFSF algorithm and using GFSF in some applications.

References

1. Metwally, A., Agrawal, D., Abbadi, A.E.: Detectives: Detecting coalition hit inflation attacks in advertising networks streams. In: Proceedings of 16th International Conference on World Wide Web, pp. 241–250. ACM Press, New York (2007)
2. Ji, S., Li, G., Li, C., et al.: Efficient interactive fuzzy keyword search. In: Proceedings of the 18th International Conference on World Wide Web, pp. 371–380. ACM Press, New York (2009)
3. Dong, X., Halevy, A., Yu, C.: Data integration with uncertainty. Int. J. Very Large Data Bases **18**(2), 469–500 (2009)
4. Chaudhuri, S., Ganti, V., Kaushik, R.: A primitive operator for similarity joins in data cleaning. In: Proceedings of the 22nd International Conference on Data Engineering, p. 5. IEEE Press (2006)
5. Xiao, C., Wang, W., Lin, X.: Ed-Join: an efficient algorithm for similarity joins with edit distance constraints. PVLDB **1**(1), 933–944 (2008)

6. Wang, J., Li, G., Feng, J.: Trie-Join: efficient trie-based string similarity joins with edit-distance constraints. PVLDB **3**(1), 1219–1230 (2010)

7. Li, G., Deng, D., Wang, J., et al.: Pass-Join: a partition-based method for similarity joins. Proc. VLDB Endow. **5**(3), 253–264 (2011)

8. Sarwagi, S., Kirpal, A.: Efficient set joins on similarity predicates. In: Proceedings of ACM SIGMOD International Conference on Management of data, pp. 743–754. ACM Press, New York (2004)

9. Xiao, C., Wang, W., Lin, X., et al.: Efficient similarity joins for near-duplicate detection. ACM Trans. Database Syst. **36**(3), 15 (2011)

10. Vernica, R., Carey, M.J., Li, C.: Efficient parallel set-similarity joins using MapReduce. In: Proceedings of ACM SIGMOD International Conference on Management of Data, pp. 495–506. ACM Press, New York (2010)

11. Bayardo, R.J., Ma, Y., Srikant, R.: Scaling up all pairs similarity search. In: Proceedings of the 16th International WWW Conference, pp. 131–140 (2007)

12. Wang, J., Li, G., Fe, J.: Fast-join: an efficient method for fuzzy token matching based string similarity join. In: Proceedings of the 27th IEEE International Conference on Data Engineering, pp. 458–469. IEEE Press (2011)

13. Chaudhuri, S., Ganjam, K., Ganti, V., et al.: Robust, efficient fuzzy match for online data cleaning. In: Proceedings of ACM SIGMOD International Conference on Management of Data, pp. 313–324. ACM Press, New York (2003)

14. Gravano, L., Ipeirotis, P., Jagadish, H., et al.: Approximate string joins in a database (almost) for free. In: Proceedings of the International Conference on Very Large Databases, pp. 491–500 (2001)

15. Metwally, A., Faloutsos, C.: V-SMART-Join: A scalable MapReduce framework for all-pair similarity joins of multisets and vectors. PVLDB **5**(8), 704–715 (2012)

16. Deng, D., Li, G., Hao, S., Wang, J., Feng, J.: MassJoin: A MapReduce-based method for scalable string similarity joins. In: ICDE 2014, pp. 340–351 (2014)

17. Huang, J., Zhang, R., Buyya, R., Chen, J.: MELODY-JOIN: Efficient Earth Mover's Distance similarity joins using MapReduce. In: ICDE 2014, pp. 808–819 (2014)

18. Chen, L., Gao, Y., Li, X., Jensen, C.S., Chen, G.: Effcient metric indexing for similarity search. In: Proceedings of IEEE 31st International Conference on Data Engineering, pp. 591–602, April 2015

19. Maehara, T., Kusumoto, M., Kawarabayashi, K.: Scalable SimRank join algorithm. In: 2015 IEEE 31st International Conference on Data Engineering (ICDE), pp. 603–614 (2015)

Erratum to: Web-Age Information Management (Part I and II)

Bin Cui[1(✉)], Nan Zhang[2], Jianliang Xu[3], Xiang Lian[4], and Dexi Liu[5]

[1] Peking University, Beijing, China
[2] The George Washington University, Washington, D.C., USA
[3] Hong Kong Baptist University, Kowloon Tong, Hong Kong, SAR China
[4] University of Texas Rio Grande Valley, Edinburg, TX, USA
[5] Jiangxi University of Finance and Economics, Nanchang, China

Erratum to:
B. Cui et al. (Eds.)
Web-Age Information Management (Part I and II)
DOI: 10.1007/978-3-319-39958-4

In an older version of the paper starting on page 441 of the first volume of the WAIM proceedings (LNCS 9658), the name and email address of the second author (Qin Liu) were missing. This has been corrected. Consequently, the Contents and the Author Index has also been updated in this volume (LNCS 9659).

In an older version of the paper starting on p. 521 of the WAIM proceedings (LNCS 9659), the corresponding author was given incorrectly. Jia Zhu is the correct corresponding author.

The updated original online version for this Book can be found at 10.1007/978-3-319-39958-4

© Springer International Publishing Switzerland 2016
B. Cui et al. (Eds.): WAIM 2016, Part II, LNCS 9659, p. E1, 2016.
DOI: 10.1007/978-3-319-39958-4_41

Demo Papers

SHMS: A Smart Phone Self-health Management System Using Data Mining

Chuanhua Xu[1], Jia Zhu[1(✉)], Zhixu Li[2], Jing Xiao[1], Changqin Huang[1], and Yong Tang[1]

[1] School of Computer Science, South China Normal University, Guangzhou, China
{chxu,jzhu,xiaojing,cqhuang,ytang}@m.scnu.edu.cn
[2] School of Computer Science and Technology, Soochow University, Soochow, China
zhixuli@suda.edu.cn

Abstract. This work presents a smart phone-based self-health management system that aims at recording and interpreting users' physiological data so as to ensure the early detection of complex diseases, such as Diabetes. On the basis of all recorded information, the proposed system can estimate the probability of several diseases inflicting users using a multiple classifiers model. The system also provides patients with health recommendations based on their diet. In the demonstration, users may interact with the SHMS to gain a wide variation of results, which include the following: (i) self diagnosis results based on current and historical physiological data; (ii) analysis results based on daily meals; and (iii) health recommendations based on the results of (i) and (ii).

1 Introduction

The specific causes of complex diseases, such as Type-2 diabetes mellitus (T2DM), have not yet been identified. The early detection of such diseases can facilitate their prevention and treatment. Considering the greatly increased amount of data in medical databases and the availability of historical data on complex diseases, such as patients' blood glucose, traditional manual analysis has become inadequate. This inadequacy has naturally led to the application of data mining techniques to identify interesting patterns and thus make early disease detection and successful recommendations possible [3].

Self-health management is critical for patients with complex diseases as it allows them to assess the interplay among nutrition therapy, physical activity, emotional, and medications. However, various limitations include the fact that health care providers lack the time to provide continuous education, the high cost of evidence-based lifestyle interventions, and the difficulty in providing patients with access to health care providers. By combining the analysis results and daily meal results, the SHMS can further provide health recommendations [1] to achieve the purpose of self-health management.

The original version of this bookbackmatter was revised. An erratum to this book-backmatter can be found at 10.1007/978-3-319-39958-4_41

© Springer International Publishing Switzerland 2016
B. Cui et al. (Eds.): WAIM 2016, Part II, LNCS 9659, pp. 521–523, 2016.
DOI: 10.1007/978-3-319-39958-4

2   System Architecture

The main contribution of this work includes self diagnosis and health recommendations functions as shown in Fig. 1(a). The smart phone collects information from patients sends to database,then the data mining module using a Multiple Classifiers Model (MCM) [4] described below to calculate probabilities of disease and coming up with advice combined with diet analysis as shown in Fig. 1(b).

MCM uses a dynamic weighted voting scheme to calculate the weight to be used later for classification based on multiple factors, including localized generalization error bound (LGEB). Our key contribution is the calculation of the LGEB, which is the generalization error of a classifier that measures the performance of a classifier generalized to unseen samples. We should assign more weight to the classifier if the classifier has smaller generalization error bound.

Given vector X transformed from user data, we first find the greatest distance D^M between X and its K neighborhoods $(Y_1, Y_2, \ldots, Y_K)$ from training samples $D^M = max(d(X, Y_i^K))$. We then calculate the LGEB as follows: $LGEB = \sqrt{\frac{1}{K} \sum_i^K err(f, Y_i)} + \sqrt{\frac{(D^M)^2}{3K} \sum_i^K (\frac{\partial f}{\partial Y_i})^T (\frac{\partial f}{\partial Y_i})}$, where $err(f, Y_i) = f(Y_i) - F(Y_i)$ and $\frac{\partial f}{\partial Y_i} = [\frac{df}{dy_{i1}}, \frac{df}{dy_{i2}}, \ldots, \frac{df}{dy_{in}}]^T$. $f(Y_i)$ is the function for calculating the confidence for each decision to Y_i between 0 to 1, and $F(Y_i)$ is the final decision for Y_i,which is 0 or 1. $\frac{\partial f}{\partial Y_i}$ is the sensitivity term of the classifier, and $(y_{i1}, y_{i2}, \ldots, y_{in})$ are the features of Y_i.

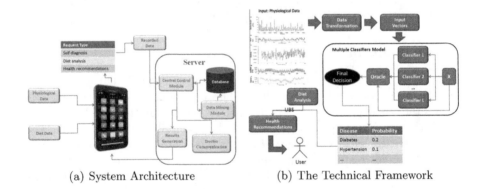

(a) System Architecture   (b) The Technical Framework

Fig. 1.

3   System Demonstration

Self Diagnosis Results. By using MCM to analyze, the estimated probability of several diseases are showing in the pie chart inflicting the user "Pavel Korchagin" in the future, as shown in Fig. 2(a). The user has 35.1 % probability of developing Diabetes and 27.2 % probability of developing high cholesterol.

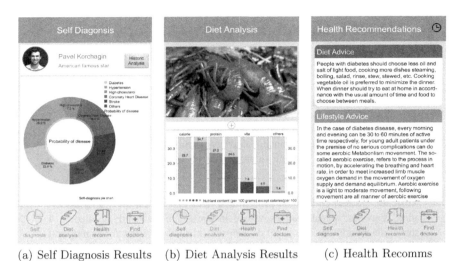

(a) Self Diagnosis Results   (b) Diet Analysis Results   (c) Health Recomms

Fig. 2.

Diet Analysis Results. Users can upload several photos of their meals using SHMS. The system then analyzes the photos and estimates the elements' quantity, as shown in Fig. 2(b).

Health Recommendations Results. The health recommendations function combines the results of the self diagnosis and diet analysis functions. By analyzing the results, the system can determine the problems in users' daily meals using UBS [2]. Recommendations are generated and shown in Fig. 2(c).

Acknowledgments. This work was supported by the Youth Teacher Startup Fund of South China Normal University (No. 14KJ18), the Natural Science Foundation of Guangdong Province, China (No. 2015A030310509), the National Natural Science Foundation of China (61370229,61272067), and the S&T Projects of Guangdong Province (No. 2014B010103004, No. 2014B010117007).

References

1. Cheplygina, V., Tax, D.M.J., Loog, M.: Combining instance information to classify bags. In: Zhou, Z.-H., Roli, F., Kittler, J. (eds.) MCS 2013. LNCS, vol. 7872, pp. 13–24. Springer, Heidelberg (2013)
2. Dong, H., Zhu, J., Tang, Y., Xu, C., Ding, R., Chen, L.: UBS: a novel news recommendation system based on user behavior sequence. In: Zhang, S., Wirsing, M., Zhang, Z. (eds.) KSEM 2015. LNCS, vol. 9403, pp. 738–750. Springer, Heidelberg (2015). doi:10.1007/978-3-319-25159-2_68
3. Homme, B.M., Reynolds, K.K., Valdes, R., Linder, M.W.: Dynamic pharmacogenetic models in anticoagulation therapy. Clin. Lab. Med. **28**(4), 539–552 (2008)
4. Zhu, J., Xie, Q., Zheng, K.: An improved early detection method of type-2 diabetes mellitus using multiple classifier system. Info. Sci. **292**, 1–14 (2015)

MVUC: An Interactive System for Mining and Visualizing Urban Co-locations

Xiao Wang, Hongmei Chen$^{(\boxtimes)}$, and Qing Xiao

Department of Computer Science and Engineering, School of Information
Science and Engineering, Yunnan University, Kunming 650091, Yunnan, China
hmchen@ynu.edu.cn

Abstract. Spatial co-location patterns and rules may reveal the spatial associations between spatial features whose instances are frequently located in a spatial neighborhood. This paper develops an interactive system for Mining and Visualizing Urban Co-locations (MVUC) from the Point Of Interest (POI) datasets of cities. According to user-specified thresholds, MVUC efficiently mines maximal patterns based on instance-trees. Then MVUC concisely demonstrates the table-instances of patterns on map based on the convex hull algorithm, intuitively illustrates areas in which the consequents' instances of rules possibly appear on map based on the intersection of areas, and visually figures patterns by graph. MVUC also allows users to compare co-locations in different regions. MVUC can help users with analyzing the functional regions of cities, selecting the locations for public services, and other urban computing.

Keywords: Spatial data mining · Co-location mining · Co-location visualization · Urban co-location

1 Introduction

Mining spatial co-location patterns and rules is one of the most important tasks of spatial data mining. A co-location pattern is a subset of spatial features whose instances are frequently located in a spatial neighborhood. A co-location rule reflects that the table-instance of the consequent pattern is frequently co-located with the table-instance of the antecedent pattern. So co-locations may reveal the spatial associations among spatial features, and have attracted interest from researchers. However, existing works, such as join-based approach, partial-join approach, join-less approach and order-clique-based approach [1], mainly focus on reducing the cost of mining co-locations. Distinct from the above, we aim to visualize co-locations, and help users to understand and utilize co-locations.

On the other hand, with the development of modern technologies such as GIS and GPS, an amount of spatial data, such as Point of Interest (POI) datasets and road networks in cities, have been collected and need to be analyzed by using spatial data mining technologies such as co-location mining. Motivated by the above, we develop an interactive system for Mining and Visualizing Urban Co-locations (MVUC) from the POI datasets of cities.

© Springer International Publishing Switzerland 2016
B. Cui et al. (Eds.): WAIM 2016, Part II, LNCS 9659, pp. 524–526, 2016.
DOI: 10.1007/978-3-319-39958-4

Generally, the main contributions of MVUC can be summarized as:

- Firstly, in order to efficiently mine urban maximal patterns, the order-clique-based approach which is based on instance-trees is modified [1]. Maximal patterns are mined from bottom to top by Aprior-like method instead of checking candidate patterns generated from 2-size patterns from top to bottom.
- Secondly, MVUC visualizes co-locations including concisely demonstrating the table-instances of patterns on map based on the convex hull algorithm [2, 5], intuitively illustrating areas in which the consequents' instances of rules possibly appear on map based on the intersection of areas, and visually figuring patterns by graph [3]. And MVUC also allows users to compare co-locations in different regions.

2 System Overview

The interactive system for Mining and Visualizing Urban Co-locations (MVUC) mainly includes 4 modules shown in Fig. 1.

- **Input Module:** In this module, users can select the POI datasets of cities or regions, set the thresholds including the distance threshold d, the participation index threshold pi, and the conditional probability threshold cp.

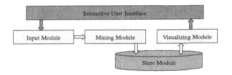

Fig. 1. System overview of MVUC

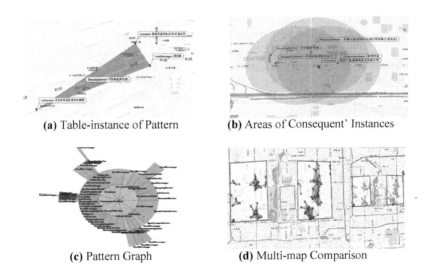

(a) Table-instance of Pattern   (b) Areas of Consequent' Instances

(c) Pattern Graph   (d) Multi-map Comparison

Fig. 2. Demonstration scenario of MVUC

- **Mining Module:** This module mines urban maximal patterns and rules. We modify the order-clique-based approach which is based on instance-trees in order to efficiently mine patterns. The order-clique-based approach generates initial candidate patterns which are maximal cliques of features in 2-size patterns, and then generates successive candidate patterns from the top down [1]. However, our approach adopts Aprior-like method to find urban maximal patterns from the bottom up.
- **Store Module:** In this module, patterns and rules are stored in files.
- **Visualizing Module:** This module includes 4 sub-modules. **(a) Demonstrating the table-instances of patterns on map.** A row-instance in the table-instance is a clique in which any two nodes have edges. In order to concisely demonstrate a row-instance, we use the convex hull algorithm to reduce unnecessary edges of a clique and get an area as Fig. 2(a) shows [2, 5]; **(b) Illustrating areas in which the consequents' instances of rules possibly appear on map.** The consequents' instances and the antecedents' instances should be located in a spatial neighborhood, so we compute the areas by the intersection of areas which are the neighborhoods of the antecedents' instances for intuitively illustrating the areas as Fig. 2(b) shows; **(c) Figuring patterns by graph.** We adopt the radiate hierarchy graph to visually figure patterns as Fig. 2(c) shows [3]; **(d) Co-locations comparison.** This is a multi-map comparison sub-module as Fig. 2(d). It allows users to explore commonness and divergences among co-locations from different regions.

3   Demonstration Scenario

In this section, we select the POI dataset of Beijing as the demonstration scenario of MVUC [4]. In Fig. 2(a) shows a row-instance in the table-instance of 4-size pattern {*company*; *shopping service*; *food service*; *life service*}; (b) shows the areas in which the consequent of rule {*company life service*}→{*shopping service*} possibly appear; (c) figures all patterns in Beijing; and (d) shows exploration for differences in spatial distribution of pattern{*life service; shopping service; food service*} between areas in the west and east of the Forbidden City by multi-map comparison.

Acknowledgments. This work is supported by the National Natural Science Foundation of China (61262069, 61472346), the Natural Science Foundation of Yunnan Province (2015FB114), and Innovation and Entrepreneurship Training Program for College Students in Yunnan Province (106732015010).

References

1. Wang, L.Z., Zhou, L.H., Lu, J., Yip, J.: An order-clique-based approach for mining maximal co-locations. Inf. Sci. **179**, 3370–3382 (2009)
2. Graham, R.L.: An efficient algorithm for determining the convex hull of a finite planar set. Inf. Process. Lett. **1**, 132–133 (1972)
3. Heim, D.A., Schneidewind, J., Sips, M.: FP-Viz: visual frequent pattern mining. In: IEEE Symposium on Information Visualization. Minneapolis, MN, USA (2005)
4. Shujutang (the POI dataset of Beijing). http://www.datatang.com/data/44484
5. CodePlex: Project-Hosting for Open Source Software. http://greatmaps.codeplex.com

SNExtractor: A Prototype for Extracting Semantic Networks from Web Documents

Chi Zhang$^{(\boxtimes)}$, Yanhua Wang, Chengyu Wang, Wenliang Cheng, and Xiaofeng He

Institute for Data Science and Engineering, East China Normal University, Shanghai, China
{51131500049,51141500045}@ecnu.cn

Abstract. Algorithms for extracting entities and relations from the Web heavily rely on semi-structured data sources or large-scale Web pages. It is difficult to extend these techniques to arbitrary Web documents in different domains. In this demonstration, we present SNExtractor, a prototype system for extracting semantic networks from documents related to any hot topics on the Web. Given a user query, it provides a comprehensive overview of relevant documents, including entities, a semantic network and a timeline summary. In the following, we will present internal mechanisms of SNExtractor and its application in the education domain.

1 Introduction

In recent years, the problem that how to harvest knowledge from the Web has attracted lots of attention. In knowledge graphs, the facts are either extracted from semi-structured data sources (in YAGO and DBPedia), or textual patterns based on the redundancy of Web data (in Probase and NELL). However, when it comes to a hot topic in a specific domain, only a (relatively small) collection of relevant Web documents can be retrieved from the Web. It is difficult to use existing techniques to extract knowledge from these documents.

Therefore, the natural question is *how to automatically extract semantic networks from Web documents related to any topics*. However, there are still several challenges to this work, illustrated as follows: (i) Entities in Web documents are appeared in the form of free text, and need to be recognized and normalized. (ii) The low redundancy nature of Web documents in a specified domain makes it difficult to apply traditional relation extraction methods. (iii) Additionally, given a query topic, a user is better served by a timeline summary of the topic, rather than having a collection of relevant documents.

In this demonstration, we present SNExtractor, a prototype system for extracting semantic networks from Web documents. SNExtractor has three novel features. (i) To improve the performance of NER, it employs a statistical method to remove noisy entities. It uses sub-string matching and entity disambiguation to resolve entity normalization. (ii) It detects potential relation instances based on linguistic and statistical features. In order to label the extracted relations,

© Springer International Publishing Switzerland 2016
B. Cui et al. (Eds.): WAIM 2016, Part II, LNCS 9659, pp. 527–530, 2016.
DOI: 10.1007/978-3-319-39958-4

we use keywords extracted from context with pre-defined patterns. (iii) For online summary generation, we detect the different stages of a hot topic by topic drift, and generate a timeline as the topic summary.

There are two parts in our demonstration. The first part is the backend of the system, including offline data analysis and online query processing modules. In the second part, we present an application based on SNExtractor for educational data analysis, which automatically collects and processes education-related Web pages and provides overviews of hot educational topics in China.

2   System Overview

As Fig. 1 illustrates, SNExtractor contains two major parts: (i) offline knowledge acquisition and (ii) online query processing. Given a user query, SNExtractor returns a semantic network relevant to the query with entities and relations, and a timeline as the summarization of relevant documents as output. The semantic network is extracted offline and stored in a knowledge repository. The timeline is generated online by detecting the topic drift phenomenon from the result of the topic modeling computed offline.

In the offline module, the data collector consists several distributed Web crawlers. Each crawler crawls Web pages of a certain website, parses the page structures, and stores the semi-structured data contents in databases. There are two stages after the data collection step. (i) We utilize NER techniques to recognize entities of various types in documents. Additionally, to improve the data quality of entities, we normalize entities by mapping surface forms to unambiguous references. Semantic relations are extracted and labeled with keywords of the context. (ii) We employ the Latent Dirichlet Allocation (LDA) to generate topic distributions for all the documents.

In the online part, a collection of relevant documents w.r.t a user query are returned by a search engine. Entities and semantic relations extracted from these documents are retrieved from the knowledge repository to form the semantic network. The timeline is generated online according to topic distributions of these documents.

In the following, we discuss our implementation details coping with some major challenges in SNExtractor.

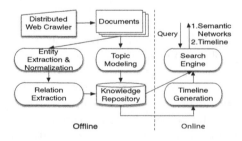

Fig. 1. System workflow of SNExtractor

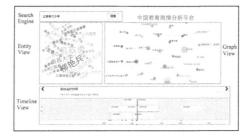

Fig. 2. Screenshot of educational public opinion analysis platform

Entity Extraction and Normalization. Given a collection of unnormalized entities M recognized by NER models, we filter out noisy or incorrect entities. We design a mapping function $f : m \rightarrow e$ such that for each entity m in the remaining entity set M' ($M' \subseteq M$), it maps m to its normalized, unambiguous form e. The techniques for entity normalization include sub-string matching and entity disambiguation, introduced in [1].

Semantic Relation Extraction. Web documents related to a certain topic and in a specific domain are relatively sparse, making it hard to apply traditional pattern-based relation extraction methods. To solve this problem, SNExtractor analyzes linguistic and statistical features to identify candidate relation tuples [2], in the form of $(e_i, e_j, C_{i,j})$, where e_i and e_j are normalized entities, and $C_{i,j}$ are the contexts of e_i and e_j. In order to label the extracted candidate relations, we cluster entity pairs which have similar contexts together as a raw relation, i.e., $R = \{(e_i, e_j, C_{i,j})\}$. The keywords for the raw relation R are labeled by extracting the frequent keywords in $C_{i,j}$ for all e_i and e_j pairs.

Online Timeline Generation. We observe that when a hot topic is reported online, the issue of topic drift arises in these news articles, which gives us a signal to detect different stages of the topic in an efficient manner. For two documents d_i and d_j, the topic drift $t(i,j)$ can be measured by the change in topic distributions θ_i and θ_j, such as self-normalized KL divergence [3]. Given a collection of documents D, we order them chronically, based on publication time. We select top-k documents $D' \subseteq D$ that indicates the drift of topics. The headlines of documents in D' are taken as the timeline.

3 Demonstration Scenario

Our demonstration includes the following two parts:

SNExtractor Backend. We will show how Web data are collected, processed, analyzed and managed in the system. The following three components will be illustrated: (i) collecting and processing data in a distributed environment, (ii) extraction of semantic network, and (iii) process of query processing, including relevant document search, semantic network retrieval and timeline generation.

SNExtractor Application. In this part, we will show an application of SNExtractor for educational public opinion analysis in China. A screenshot of this application is shown in Fig. 2.

Acknowledgment. This work is partially supported by Shanghai Agriculture Science Program (2015) Number 3-2.

References

1. Jijkoun, V., Khalid, M.A., Marx, M., de Rijke, M.: Named entity normalization in user generated content. In: AND 2008, pp. 23–30 (2008)
2. Shen, W., Wang, J., Luo, P., Wang, M., Yao, C.: REACTOR: a framework for semantic relation extraction and tagging over enterprise data. In: WWW 2011, pp. 121–122 (2011)
3. Knights, D., Mozer, M.C., Nicolov, N.: Detecting topic drift with compound topic models. In: ICWSM 2009 (2009)

Crowd-PANDA: Using Crowdsourcing Method for Academic Knowledge Acquisition

Zhaoan Dong[1], Jiaheng Lu[1,2(✉)], and Tok Wang Ling[3]

[1] DEKE, MOE and School of Information, Renmin University of China,
Beijing, China
jiahenglu@gmail.com
[2] Department of Computer Science, University of Helsinki, Helsinki, Finland
[3] School of Computing, National University of Singapore, Singapore, Singapore

Abstract. Crowdsourcing is currently being used and explored in a number of areas. Since some automatic algorithms are extremely hard to handle diverse academic documents with different quality and layouts, we present Crowd-PANDA, a **Crowd**sourcing sub-module in the **P**latform for **A**cademic k**N**owledge **D**iscovery and **A**cquisition, which combines algorithms and human workers to identify and extract meaningful information objects within academic documents named Knowledge Cells (e.g. *Figures, Tables, Definitions*, etc.) as well as their relevant key information and relationships. The extracted Knowledge Cells and their relationships can be used to build an **Academic Knowledge Graph**, which could provide a fine-grained search and deep-level information explore over academic literature.

Keywords: Crowdsourcing · Academic knowledge acquisition

1 Introduction

With an exponential growth of the volume of scientific publications, tremendous interests have been spent on extraction and management of research data within scientific literature. One example is **Digital Curation (*DC*)** [1] which indicates the activities including *selection, preservation, maintenance, collection* and *archiving* of digital assets and the process of *extraction* of important information from scientific literature. Another example is **Deep Indexing (*DI*)** [2] by which *ProQuest* [3] indexes the research data within scholarly articles that are often invisible to traditional bibliographic searches. Similar capabilities are available in *CiteSeerX* [4] and *ScienceDirect* [5] et al. However, they only focus on *Figures* and *Tables*, none of them have pay attention to other kind of objects like *Algorithms, Theroms, Lemmas*, etc., and the relationships among them. In PandaSearch [6–8], the information units within academic literature including all mentioned above are defined as **Knowledge Cells**. Knowledge Cells and

This work is partially supported by NSFC (No. 61472427), NSSFC (No. 12&ZD220) and RUC Fund (No. 11XNJ003).

B. Cui et al. (Eds.): WAIM 2016, Part II, LNCS 9659, pp. 531–533, 2016.
DOI: 10.1007/978-3-319-39958-4

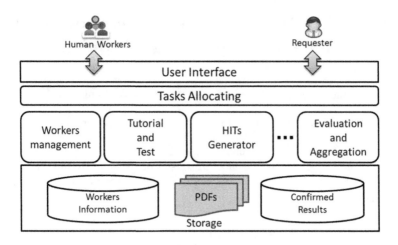

Fig. 1. The system architecture of Crowd-PANDA.

their relationships can be used to build an**Academic Knowledge Graph** for a fine-grained search and deep-level information explore over academic literature.

However, the most important prerequisite is to correctly identify and extract the Knowledge Cells and their relationships from huge amount of documents [8]. Existing automatic computer algorithms can extremely hard to handle diverse PDF documents that published in different years, conferences or journals with different quality and layouts.

As *crowdsourcing* has become a powerful paradigm for large scale problem-solving especially for those tasks that are difficult to computers but easy to human [9, 10], we make use of crowdsourcing to identify and extract those Knowledge Cells as well as their relevant key information and relationships from huge amount of PDFs. It is notable that during the process of identification and extraction some activities can generally be broken into small tasks which are often repetitive and do not require any specific expertise. For example, a human worker can firstly locate the content of a **Figure** by browsing the PDF pages and then crop the content only by *"drag and draw"*.

Obviously, if all the documents are crowdsourced to anonymous workers, the crowdsourcing cost will be extremely high. Therefore, the natural alternative is to combine automatic algorithms with crowdsourcing, which is the most challenging work mentioned in [8]. In this demo, we only focus on the crowdsourcing sub-module for identifying and extracting Knowledge Cells.

2   System Design and Implementation

2.1   System Architecture

The architecture of Crowd-PANDA is illustrated in Fig. 1. The system is implemented in Python and use PostgreSQL to store the data.

2.2   Crowdsourcing Tasks

There are two basic kinds of Human Intelligence Tasks (HITs): *identifying* and *review*. **Identifying tasks** ask workers to identify the knowledge cells from those PDF pages that are difficult for automatic extraction algorithms [8]. **Review tasks** ask workers to check the answers of other workers for the sake of quality control. To guarantee the effectiveness and the accuracy, human workers have to accomplish the tutorial tasks and some tests to learn how to perform the tasks. Additionally, each HITs should be assigned a time limit.

2.3   Results Aggregations

Majority Vote is the common-used strategy for results evaluation and aggregation where an answer will be accepted if it is confirmed by most of the reviewers. For more complex tasks, we can devise a weighted voting strategy based on the confidences and historical performances of human workers.

3   Demonstration Scenarios

We plan to demonstrate the Crowd-PANDA system with the following scenarios:

Issuing Crowdsourcing Tasks: The requester can set some parameters of the crowdsourcing tasks before publishing them including the type of extracted Knowledge Cells, time limit and the number of tasks, etc.

Identifying Tasks: Human workers undertake the identifying tasks by browsing the PDF pages and crop the contents of the Knowledge Cells such as Figures, Definitions, Tables, Algorithms, etc.

Reviewing Tasks: Human workers are asked to confirm or reject the answers from other workers.

User Performance: Human workers can view the performance of their completed tasks and see their performance rank among all the workers.

References

1. Digital Curation. http://en.wikipedia.org/wiki/Digital_curation
2. Illustrata Deep Indexing. http://proquest.libguides.com/deepindexing
3. ProQuest. http://search.proquest.com/
4. Citeseer. http://citeseer.ist.psu.edu/
5. ScienceDirect. http://www.sciencedirect.com/
6. PandaSearch. http://pandasearch.ruc.edu.cn/
7. Huang, F., Li, J., Lu, J., Ling, T.W., Dong, Z.: PandaSearch: a fine-grained academic search engine for research documents. In: 2015 IEEE 31st International Conference on Data Engineering (ICDE 2015), pp. 1408–1411. IEEE Press (2015)
8. Dong, Z., Lu, J., Ling, T.W.: PANDA: a platform for academic knowledge discovery and acquisition. In: 2016 3rd International Conference on Big Data and Smart Computing (BigComp 2016). IEEE Press (2016)
9. Howe, J.: The rise of crowdsourcing. Wired Mag. **14**, 1–4 (2006)
10. Luz, N., Silva, N., Novais, P.: A survey of task-oriented crowdsourcing. Artif. Intell. Rev. **44**, 1–27 (2014)

LPSMon: A Stream-Based Live Public Sentiment Monitoring System

Kun Ma[1(✉)], Zijie Tang[1], Jialin Zhong[2], and Bo Yang[1]

[1] Shandong Provincial Key Laboratory of Network Based Intelligent Computing,
University of Jinan, Jinan 250022, China
{ise_mak,yangbo}@ujn.edu.cn, zijietang@mail.ujn.edu.cn
[2] School of Automation and Electrical Engineering,
University of Jinan, Jinan 250022, China
ygkszjl@sina.com

Abstract. Classical public sentiment monitoring systems are based on the text sentiment analysis which has limitation in short text without clear sentiment words. In our demonstration, we present a stream-based live public sentiment monitoring system (LPSMon). This system takes the social media such as forum, microblog, and WeChat as the input, and takes sentiment alerts as the output. Innovations of LPSMon are stream processing paradigm and latent topic matching.

Keywords: Public sentiment · Stream processing · Topic model · Latent dirichlet allocation · Message processing · Crawler

1 Introduction

With the explosive growth of social media such as forum, microblog, and WeChat, automatic sentiment analysis and monitoring on media data has provided critical decision making in various domains [1]. For instance, a news media can study the public sentiment to obtain the latest news material towards its products, while a politician can adjust his/her position with respect to the sentiment change of the public. However, traditional sentiment monitoring systems are usually based on text matching without latent topic, the relevance ranking of the result indexing is not always correct. To overcome the lack of real-time monitoring of sentiment monitoring systems, this paper has proposed a new stream-based live public sentiment monitoring system called LPSMon. The innovations of LPSMon come in stream processing paradigm and latent topic matching.

2 Architecture

The architecture of LPSMon is shown in Fig. 1, which is composed of four layers: collector layer, messaging layer, topic acquisition layer, and topic alert layer. The spider crawlers firstly collect the social media data from the Internet to push them to the messaging layer and asynchronously store in the search engine. Next,

© Springer International Publishing Switzerland 2016
B. Cui et al. (Eds.): WAIM 2016, Part II, LNCS 9659, pp. 534–536, 2016.
DOI: 10.1007/978-3-319-39958-4

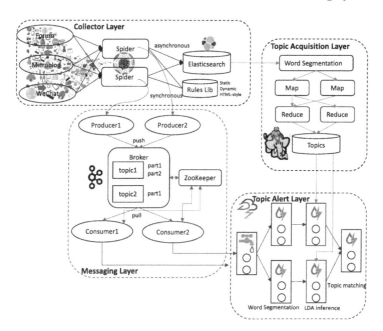

Fig. 1. Architecture of LPSMon.

the messaging layer timely emits the processed tuples of RESTful media data to the topic alert layer. Finally, the topic alert layer based on stream computing infers to sentiment alerts to end users. The training and estimation of topic model comes from the MapReduce-based topic acquisition layer. Some open source softwares are used in this architecture such as Elasticsearch, Apache Kafka [2], Apache Storm [3], and Apache Mahout [4].

Collector Layer. The collector layer collects the media data using spider crawlers. The designed crawlers depends on the rule library, including static hard coding, dynamic configuration and HTML-style template. The output of crawler results are timely pushed to the Messaging Layer, as asynchronously stored in the search engine for further retrieval.

Messaging Layer. The messaging layer is designed to spread the collected feeds of data streams from the collector layer over a cluster of machines to make data transmission timely and elastically. We use publish-subscribe messaging framework Kafka [2] to achieve this goal. At a high level, producers send messages over Kafka clusters, and consumers emit the processing results to the topic alert layer. A topic within a broker is a category or feed name to which messages are published, and each partition in a topic is an ordered, immutable sequence of messages that is continually appended to a commit log. The ZooKeeper is a fault tolerant coordination service for group membership, coordinated workflow, as well as generalized distributed data structures. The messaging layer is mainly to guarantee the high availability with good modularity in failure of stream-based topic alert.

Topic Acquisition Layer. To optimize the ranking of searching results, we adopt Latent Dirichlet Allocation (LDA) model [5, 6] to jointly clustering collected media data into mixtures of topics. We use Apache mahout [4] to train and estimate the media data to find latent topics from the Elasticsearch storage. All the topics are used to infer the topics of timely streaming tuples in the topic alert layer.

Topic Alert Layer. To find timely latest sentiment, we use the stream computing framework Apache Storm to design the stream topology. First, the streaming tuples emitting from the spout are split into words and then remove (a) non-alphabetic terms, (b) short terms with <4 characters, and (c) the most common stopwords. Second, the words are segmented in the word segmentation bolt if they are Chinese. Third, the topics are inferred for previously new data in the LDA inference bolt. Finally, the last bolt sends the alert if the topics of new data match the alerting topics.

3   Demonstration Scenarios

We will first use a poster and several slides to introduce the motivation of LPSMon and highlight how it works. After that, we will show the audience the live demonstration system hosted on our personal website. Finally, we invite them to participate in the interactions by their smart mobile phones or laptops.

Acknowledgments. This work was supported by the Shandong Provincial Natural Science Foundation (ZR2014FQ029), and the Shandong Provincial Key R&D Program (2015GGX106007).

References

1. Cao, D., Ji, R., Lin, D., Li, S.: A cross-media public sentiment analysis system for microblog. Multimedia Syst. 1–8 (2014)
2. Wang, G., Koshy, J., Subramanian, S., Paramasivam, K., Zadeh, M., Narkhede, N., Rao, J., Kreps, J., Stein, J.: Building a replicated logging system with Apache Kafka. Proc. VLDB Endowment **8**(12), 1654–1655 (2015)
3. Toshniwal, A., Taneja, S., Shukla, A., Ramasamy, K., Patel, J.M., Kulkarni, S., Jackson, J., Gade, K., Fu, M., Donham, J., et al.: Storm@ twitter. In: Proceedings of the 2014 ACM SIGMOD International Conference on Management of Data, pp. 147–156. ACM (2014)
4. Landset, S., Khoshgoftaar, T.M., Richter, A.N., Hasanin, T.: A survey of open source tools for machine learning with big data in the Hadoop ecosystem. J. Big Data **2**(1), 1–36 (2015)
5. Blei, D.M., Ng, A.Y., Jordan, M.I.: Latent dirichlet allocation. J. Mach. Learn. Res. **3**, 993–1022 (2003)
6. Ma, K., Lu, T., Abraham, A.: Hybrid parallel approach for personalized literature recommendation system. In: 2014 6th International Conference on Computational Aspects of Social Networks (CASoN), pp. 31–36. IEEE (2014)

DPBT: A System for Detecting Pacemakers in Burst Topics

Guozhong Dong[1], Wu Yang[1(✉)], Feida Zhu[2], and Wei Wang[1]

[1] Information Security Research Center, Harbin Engineering University,
Harbin, China
yangwu@hrbeu.edu.cn
[2] Singapore Management University, Singapore, Singapore

Abstract. Influential users usually have a large number of followers and play an important role in the diffusion of burst topic. In this paper, pacemakers are defined as the influential users that promote topic diffusion in the early stages of burst topic. Traditional influential users detection approaches have largely ignored pacemakers in burst topics. To solve this problem, we present DPBT, a system that can detect pacemakers in burst topics. In DPBT, we construct burst topic user graph for each burst topic and propose a pacemakers detection algorithm to detect pacemakers in Twitter. The demonstration shows that DPBT is effective to detect pacemakers in burst topics, such that the historical detection results can effectively help to detect and predict burst topics in the early stages.

1 Introduction

With the development of social media, Twitter has been an important medium for providing the rapid spread of burst topic. Influential users play an important part in the process of burst topic diffusion, especially in the burst topics related to Internet public opinion. So far, plenty of works focus on the influential users who are popular or famous in burst topics. However, in quite a lot of scenario, these famous influential users are not the early adopters of burst topic. Unlike previous works [1, 2], we focus on detecting the influential users that can promote the topic diffusion in the early stages of burst topic, which are defined as pacemakers. In this paper, we develop DPBT(Detecting Pacemakers in Burst Topics) system that contains three functional layers. DPBT implements a polling based method to monitor new burst topics detected by CLEar system[1] [3]. When new burst topics are detected, DPBT can construct burst topic user graph for each burst topic and apply pacemakers detection algorithm to detect pacemakers.

2 DPBT Design and Implementation

The system architecture of DPBT is shown in Fig. 1, which contains three functional layers. The Data Layer provides two databases for efficient data storage

---

[1] http://research.pinnacle.smu.edu.sg/clear/.

© Springer International Publishing Switzerland 2016
B. Cui et al. (Eds.): WAIM 2016, Part II, LNCS 9659, pp. 537–539, 2016.
DOI: 10.1007/978-3-319-39958-4

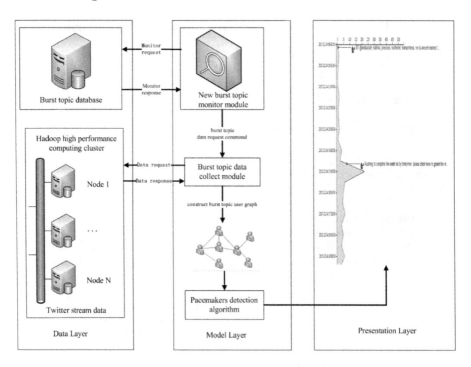

Fig. 1. The system architecture of DPBT.

and data query. The first one is to store burst topics detected by CLEar system and provide query operation for new burst topic monitor module (NBTM) in Model Layer. The second one is to store Twitter stream data, which stores necessary data involved in burst topics. The Model Layer utilizes several important modules to detect pacemakers in burst topics. NBTM monitors new burst topics via polling burst topic database. Once new burst topics are detected, NBTM sends burst topic data collect command to burst topic data collect module (BTDC). BTDC retrieves burst topic data from Hadoop cluster, constructs burst topic user graph for further processing. In order to detect pacemakers in burst topic, pacemakers detection algorithm based on burst topic user graph is proposed. The Presentation Layer presents the life cycle of burst topic and pacemakers detected by DPBT with a user-friendly interface.

The burst topic user graph of burst topic k can be formally defined as $G_k =< V_k, E_k, T_k >$. In detail, $V_k = \{u, \cdots, v, \cdots\}$ is the set of Twitter users over burst topic k, E_k represents the set of edges among Twitter users, in which a directed edge (u, v) means that u is the follower of v. $T_k = \{t(u), \cdots, t(v), \cdots\}$ is the earliest post time set of users over burst topic k. By considering time information, the directed edges in topic user graph model can represent the direction of information flow and play a key role in detecting pacemakers, so we include time information in the edge weight. For each $(u, v) \in E_k$, the edge weight $w(u, v)$ and the normalization of edge weight $W(u, v)$ can be defined as follows:

$$w(u, v) = \begin{cases} e^{-\frac{t(u)-t(v)}{\alpha}}, & if \ t(v)>0 \ and \ t(v)<t(u), \ \alpha>0 \\ 0, & otherwise \end{cases} \quad (1)$$

$$W(u, v) = \frac{w(u, v)}{\sum\limits_{p \in Out(u)} w(u, p)} \quad (2)$$

where $Out(u)$ is the following set of user u in burst topic k. The pacemaker weight of user v in G_k, denoted by $PM(v)$, is given in Eq. 3.

$$PM(v) = dD(v) + (1-d) \sum\limits_{p \in In(v)} PM(p)W(p, v), \ 0 \le d \le 1 \quad (3)$$

where d is the damping factor, $In(v)$ is the follower set of user v in burst topic k and $D(v)$ is a probability distribution over V_k. The distribution is topic dependent and is set to $1/|V_k|$ for all $v \in V_k$. The pacemakers of burst topic k are the top N pacemaker weight of users in V_k and the value of N can be adjusted adaptively in the system.

3   Demonstration Scenarios

Here we use burst topics detected by CLEar system as an example scenario. The Presentation Layer in Fig. 1 shows the results of DPBT. Assume that there are burst topics detected by CLEar system, DPBT can monitor them real-time and detect pacemakers by pacemakers detection algorithm. When a user is interested in a burst topic and click the burst topic link, the presentation interface presents the life cycle of burst topic and the pacemakers detected by DPBT.

Acknowledgment. This paper is funded by the International Exchange Program of Harbin Engineering University for Innovation-oriented Talents Cultivation, China Scholarship Council, the Fundamental Research Funds for the Central Universities (no. HEUCF100605), the National High Technology Research and Development Program of China (no. 2012AA012802) and the National Natural Science Foundation of China (no. 61170242, no. 61572459).

References

1. Weng, J., Lim, E.P., Jiang, J., He, Q.: Twitterrank: finding topic-sensitive influential twitterers. In: Proceedings of the third ACM international conference on Web search and data mining, pp. 261–270. ACM, New York (2010)
2. Chang, B., Zhu, F., Chen, E., Liu, Q.: Information source detection via maximum a posteriori estimation. In: 2015 IEEE International Conference on Data Mining (ICDM), pp. 21–30. IEEE, Atlantic (2015)
3. Xie, R., Zhu, F., Ma, H., Xie, W., Lin, C.: CLEar: a real-time online observatory for bursty and viral events. Proc. VLDB Endowment **7**(13), 1–4 (2014)

CEQA - An Open Source Chinese Question Answer System Based on Linked Knowledge

Zeyu Du[1,2], Yan Yang[1,2(✉)], Qinming Hu[1,2], and Liang He[1,2(✉)]

[1] Shanghai Key Laboratory of Multidimensional Information Processing,
East China Normal University, Shanghai 200241, China
zydu@ica.stc.sh.cn, {yanyang,lhe,qmhu}@cs.ecnu.edu.cn
[2] Department of Computer Science and Technology, East China Normal University,
Shanghai 200241, China

Abstract. In this paper, we proposed an open source Chinese question answer system (CEQA), which focuses on the field of E-commerce question answer system. Particularly, CEQA is able to improve the performance in these aspects compared with the traditional method: Chinese-English mixed commodity name identification, semantic link and dependency parsing of complex questions, linking knowledge to natural language. Specifically, the key algorithm utilized in CEQA includes Conditional Random field and Word2Vec. Finally, we evaluate the CEQA on the open data sets, where our experimental results show that our approach is promising and superior. A demonstration video has been published at http://t.cn/RbkKklF.

1 Introduction and Motivation

Most of the question answer (QA) systems focus on English language. However, in the field of the QA system based on Chinese language, there are more challenges: (1) diversity in Chinese spoken language. (2) spoken language does not conform to the rules of grammar. (3) E-commerce domain contains several terminologies, for example, Chinese-English mixed commodity name.

CEQA provides a framework to solve these problems. First, it can distinguish complex commodity name by utilizing Conditional Random field (CRF). Then, based on Word2Vec algorithm, CEQA can easily train a great linking model to link the knowledge to natural language. Besides, by Semantic Dependency Parsing [1], we propose a reducing algorithm to transform the complex parsing result into three triples.

Figure 1 illustrates an example of CEQA system, the question is "What is the size of JL78?". It is noteworthy that CEQA can not only give us the answer, but also the Lists, Charts, Groups of the result.

2 System Architecture

The main contribution of CEQA is to improve the performance of transforming Chinese natural language into SPARQL query.

© Springer International Publishing Switzerland 2016
B. Cui et al. (Eds.): WAIM 2016, Part II, LNCS 9659, pp. 540–543, 2016.
DOI: 10.1007/978-3-319-39958-4

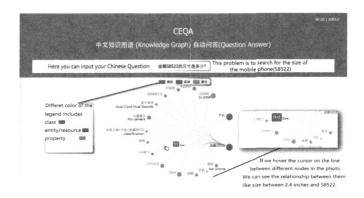

Fig. 1. Example question: "What is the size of JL78?".

Classifier. First, CEQA classifies the questions into 8 question types: UNKNOWN, COUNT, MAX, MIN, NUMERIC, BOOL, FACT, LIST.

Identification. Second, CEQA identifies the named entity in the questions. Based on the CRF++ (http://crfpp.sourceforge.net), CEQA incorporates more features, such as: n-gram feature template, word type and single product features in the field of commodity title data.

Reducing. Third, CEQA reduces the complex parsing result to simple three triples. This method is referred as Semantic Dependency Parsing (SDP) Reducing [1]. Given reduction operations for F, for any semantic dependency s, CEQA will certainly be able to transform the semantic dependency into an operation.

SPARQL. Fourth, CEQA transforms the triples to none-linked SPARQL temples. For different question types, we define different temples and strategies to transform them.

Linking. Finally, CEQA creates the truly SPARQL query. CEQA builds a dictionary of the triples and class URL resource based on the RDF files. Most of the triples can be linked to the URL resources according to the dictionary. For the other triples, we first classify the natural language expressions according to Eq. 1:

$$Score_{class}W_{(i)} = PARS_{W_{(i)}} + SDP_{W_{(i)}} + CLASS_{W_{(i)}} \tag{1}$$

where the $PARS_{W_{(i)}}$ is part of speech score, $SDP_{W_{(i)}}$ is the semantic score, and $CLASS_{W_{(i)}}$ means the score of word exist in the RDF dictionary. After that, based on the initial classification results, CEQA builds a set of candidate URL resources for each natural language. At last, CEQA takes advantage of Word2Vec [2] to calculate the similarity between the candidate URL resources and select the optimal one.

3   Experiments

In order to evaluate the effect of our CEQA system, we conduct the experiments on the data sets clawed from JD, TaoBao, Amazon.cn and DangDang etc. The data sets are made up of phone class (103137 triples), computer class (103137 triples), and 200 problems crawled from http://zhidao.baidu.com. Our evaluation metrics are F-measure (F), precision (P) and recall (R) [3]. In Table 1, TBSL++ incorporates Chinese named entity recognition algorithm into basic TBSL algorithm [4]. CEQA-Link improves the performance by using Word2Vec to linking. The last CEQA-Link-R algorithm used SDP-Reduce algorithm and has the best result.

Table 1. Experimental Results

Algorithm	SPARQL nums	P	R	F
TBSL++	42	0.33	0.21	0.25
CEQA-Link	56	0.35	0.28	0.31
CEQA-Link-R	62	0.51	0.31	0.38

4   Conclusions and Future Work

With the Word2Vec, CRF and SDP-Reduce, we create a novel open source Chinese question answer system. Furthermore, with more question answer log data, we can focus on some session problem and provide a more efficient way to parsing reducing. With the visualized results, CEQA provides an insight into the data and knowledge. In the future, we will extend CEQA to deal with the common field problems and keep it open source.

Acknowledgement. This research is funded by National Key Technology Support Program (No. 2015BAH01F02), Science and Technology Commission of Shanghai Municipality (No. 14DZ2260800), Shanghai Zhangjiang National Innovation Demonstration Zone Development Fund (No. 201411-JA-B108-002) and Minhang talent development special funds.

References

1. Che, W., Li, Z., Liu, T.: Ltp: a chinese language technology platform. In: Proceedings of the 23rd International Conference on Computational Linguistics: Demonstrations, pp. 13–16. Association for Computational Linguistics (2010)
2. Goldberg, Y., Levy, O.: Word2vec explained: Deriving mikolov et al.'s negative-sampling word-embedding method. arXiv preprint (2014). arXiv:1402.3722

3. Lopez, V., Unger, C., Cimiano, P., Motta, E.: Evaluating question answering over linked data. Web Semant.: Sci. Serv. Agents World Wide Web **21**, 3–13 (2013)
4. Unger, C., Bühmann, L., Lehmann, J., Ngonga Ngomo, A.-C., Gerber, D., Cimiano, P.: Template-based question answering over rdf data. In: Proceedings of the 21st international conference on World Wide Web, pp. 639–648. ACM (2012)

OSSRec: An Open Source Software Recommendation System Based on Wisdom of Crowds

Mengwen Chen[✉], Gang Yin, Chenxi Song, Tao Wang,
Cheng Yang, and Huaimin Wang

National Laboratory for Parallel and Distributed Processing,
National University of Defense Technology, Changsha, China
{chenmengwen1991,delpiero710}@126.com,
songchenxi92@163.com,
{hmwang,yingang,taowang2005}@nudt.edu.cn

Abstract. The massive amounts of OSS provide abundant resources for software reuse, while introducing great challenges for finding the desired ones. In this paper, we propose OSSRec, an Open Source Software Recommendation System, which leverages the wisdom of crowds in both collaborative development communities and knowledge sharing communities to do recommendation. OSSRec can recommend proper candidates with high precision, whose results are much better than existing OSS communities. In this demonstration, we present the architecture and the recommendation process of OSSRec.

Keywords: Open source software · Recommendation system

1 Introduction

Open Source Software (OSS for brief) has exploded incredibly in recent years. A lot of well-known companies have made their efforts in the open source movement. To name but a few, Google and Apple have opened the source code of deep learning system and Swift respectively. Alibaba and Twitter have made publicly available of their programs by GitHub. Meanwhile, more and more companies attempt to reuse OSS, which improves the efficiency of software development.

Since there exist a plenty of OSS communities, e.g., GitHub, SourceForge. And these OSS community platforms are constructed in diverse ways. Thus, to reuse OSS, the crucial task is to find out the most suitable OSS across heterogeneous platforms to meet a specific user's requirement both in efficiency and effectiveness, which is a great challenge.

Although the search and recommendation technologies have been well studied in E-Commerce, these methods are not suitable for OSS. This is because recommending software greatly differ from recommending commodities. On the one hand, the input keywords representing users' requirements on the objected OSS are not as clearly as on commodities in E-Commerce. On the other hand, conventional recommendation

© Springer International Publishing Switzerland 2016
B. Cui et al. (Eds.): WAIM 2016, Part II, LNCS 9659, pp. 544–547, 2016.
DOI: 10.1007/978-3-319-39958-4

methods such as collaborative filtering or content-based methods rely on people's information and commodities' attributes [1], which are often unavailable for OSS.

Motivated by this, in this paper, we for the first time propose **OSSRec**, an open source software recommendation system based on wisdom of crowds. This system is able to recommend the proper OSS to meet users' requirements with two main innovations. The first one is that OSSRec provides a unified data platform for users by extracting and merging more than 10^5 software projects and discussion posts from representative communities such as SourceForge, StackOverflow and so on. The second one is that OSSRec recommend the objected OSS by taking consideration of wisdom of crowds including users' discussion posts and emotion. By combining them, we implement this system with a friendly interface and powerful functions.

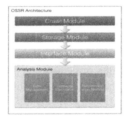

Fig. 1. System architecture

(a) Recommendation result   (b) Feedback adjust

Fig. 2. The user interface of OSSRec

2   System Architecture Overview

In this section, we overview on the architecture of OSSRec, which is shown in Fig. 1. The functions of the 4 main modules in OSSRec are listed as follows.

Crawl Module: This module crawls information of software projects and discussion posts from heterogeneous community websites in an online manner, including names, descriptions, tags and etc.

Storage Module: This module stores the information crawled from the Web.

Interface Module: This module provides a friendly interface for users. It accepts the users' input in form of keywords and shows the recommendation results. It also records the users' feedbacks on the results.

Analysis Module: This module is the core in OSSRec, which is responsible for managing a large amount of crawled web data and recommending the proper software. The details of the techniques will be discussed in Sect. 3.

3 Data Management and Recommendation Method

In this section, we describe the detailed techniques in the analysis module. The analysis module mainly has three important functions: data management, similarity calculation and recommend model.

Data Management. We have crawled a large number of software projects and discussion posts from heterogeneous online websites. There exists duplication and incompleteness within the crawled data. So we use the attributes of the projects to clean the data. In details, we use tags of the projects to recognize duplicate projects. And we use the matching posts to delete the projects with low hot degree.

Similarity Calculation. The similarity between two projects is calculated based on three aspects, namely tags, descriptions and posts, denoted as s_t, s_d, s_s, respectively. We transform the set of tags for each project into 0-1 vectors by *vector space model*. We use TF-IDF to represent the descriptions on each project as weighted vectors. Since we observe that projects on the same post are often similar, we also consider the set of matching posts of each project as wisdom of crowds in similarity estimation [3]. After all, we use *Cosine Similarity* to compute s_t, s_d, s_s, respectively.

Recommendation Model. For any two projects, we have three similarity measures, s_t, s_d and s_s. We estimate the similarity by gathering s_t, s_d and s_s together. For a specific user's requirement, we recommend 5 projects that are in a list of the intersection of higher similar software and higher user rated software. We also take consideration of users' feedbacks as their emotions to adjust recommendation results.

4 Demonstration

We demonstrate the OSSRec in the following two parts, as illustrated in Fig. 2.

Recommendation Result: A user inputs his requirements in form of keywords, such as software name or tag, for example "MySQL". The recommend results about database software are shown on the right part of the web page in Fig. 2(a).

Feedback Adjust: OSSRec allows users give feedbacks about the results of recommendation. The user clicks up or down button on the specific OSS express positive or negative emotion to adjust the results shown in Fig. 2(b), which is the result of a user clicking 'up' in "MariaDB".

5   Conclusion

To recommend proper OSS across heterogeneous platforms, we develop OSSRec, an open source software recommendation system based on wisdom of crowds, which has served in Trustie-OSSEAN, a platform for evaluating, analyzing and networking of entities in global open source communities. The system gathers the wisdom and emotion of crowds and recommend on the project level. Demonstration has shown the powerful function on recommending software and adjusting results.

Acknowledgments. The demonstration is supported by the National Natural Science Foundation of China (Grant No. 61432020 and 61472430).

References

1. Kompan, M., Bieliková, M.: Content-based news recommendation. In: Buccafurri, F., Semeraro, G. (eds.) EC-Web 2010. LNBIP, vol. 61, pp. 61–72. Springer, Heidelberg (2010)
2. Wang, H., Wang, T., Yin, G., et al.: Linking issue tracker with q&a sites for knowledge sharing across communities

Author Index

Aksoy, Cem II-171
Aljohani, Naif Radi I-479
An, Yunzhe II-40
Aslam, Muhammad Ahtisham I-479

Bao, Jinling I-164

Cai, Lijun II-319
Cai, Peng II-159
Cai, Xiangrui I-454
Cai, Zhongmin II-3
Cao, Jiannong I-109
Chao, Han-Chieh I-3
Chao, Hongyang II-469
Chen, Bole I-124
Chen, Hong I-219, II-493
Chen, Hongmei II-524
Chen, Lei I-191, II-265, II-319
Chen, Mengwen II-544
Chen, Siding I-31, I-43
Chen, Wei II-380
Chen, Xinyuan II-3
Chen, Yan I-257
Chen, Yipeng I-402
Cheng, Ce I-178
Cheng, Wenliang II-527
Cui, Lizhen II-430
Cui, Ningning II-15
Cui, Yanling II-356
Cui, Zhiming II-132

Dai, Chengcheng I-151
Dai, Jianhua II-392, II-403
Dai, Xinyuan II-304
Dass, Ananya II-171
Deng, Bailong I-31, I-43
Ding, Jianbing II-469
Ding, Junmei I-257
Ding, Yue I-286
Dong, Guozhong II-537
Dong, Guozhu I-82

Dong, Zhaoan II-531
Dou, Peng I-325
Du, Cuilan I-219
Du, Sizhen I-325
Du, Zeyu II-540
Duan, Lei I-82

Fan, Lilue II-65
Feng, Chong I-390
Feng, Jianhua I-124
Feng, Shi I-272
Feng, Shuo I-338
Fournier-Viger, Philippe I-3, I-17
Fu, Guoqiang II-417
Fu, Shaojing II-28

Gan, Shaoduo II-444
Gan, Wensheng I-3, I-17
Gao, Chao I-82
Gao, Dawei I-191
Gao, Ming I-517
Gao, Yifan II-78
Gu, Jiahui I-390
Guan, Jihong I-109
Guo, Jinwei II-159

Han, Huifeng II-392, II-403
Han, Shupeng I-454
Han, Yutong II-184
He, Ben I-491
He, Liang II-540
He, Tingqin II-319
He, Xiaofeng II-527
Hong, Tzung-Pei I-17
Hu, Hu II-392, II-403
Hu, Qinghua II-392, II-403
Hu, Qinming II-540
Hu, Xuegang II-369
Huang, Changqi II-521
Huang, Chunlan I-429
Huang, Heyan I-390

Huang, Jiajia I-415
Huang, Jiajin I-178
Huang, Jiayi I-504
Huang, Jimin I-415
Huang, Junming I-244
Huang, Kai II-28
Huang, Xinyu II-380

Ji, Anming I-353
Ji, Donghong I-310
Ji, Qiang II-356
Jiang, Bo II-291
Jin, Beihong II-356
Jin, Cheqing I-137
Jin, Songchang I-353

Kong, Lanju II-430
Kou, Yue I-338
Kuang, Zhixin II-291

Lai, Yongxuan II-107, II-506
Lee, JooYoung II-330
Lei, Kai I-378
Li, Bo II-120
Li, Chunhua II-132
Li, Cuiping I-219, II-493
Li, Fenglan I-353
Li, Guoliang I-124
Li, Hongyan I-402, II-65
Li, Jia II-279
Li, Jiajia II-40
Li, Li I-31, I-43
Li, Peipei II-369
Li, Peng I-467, II-291
Li, Qingzhong II-430
Li, Rui I-467, II-291
Li, Shasha II-444
Li, Shun II-380
Li, Wengen I-109
Li, Xiang II-213
Li, Xiaohua II-145
Li, Xiaoling II-444
Li, Xin I-257
Li, Xinchi II-417
Li, Xiu I-95
Li, Yangxi I-219
Li, Ye II-457
Li, Yuming II-78

Li, Zhixu II-132, II-521
Liang, Jiguang II-291
Liao, Guoqiong I-55
Lin, Jerry Chun-Wei I-3, I-17
Lin, Lanfen I-231
Lin, Meiqi II-319
Lin, Ziyu II-107, II-506
Liu, Chuang I-244
Liu, Guanfeng II-132
Liu, Guiquan I-257
Liu, Haichi I-298
Liu, Lei II-430
Liu, Lu II-369
Liu, Peilei I-298
Liu, Qiang I-353
Liu, Qin I-441
Liu, Shijun II-481
Liu, Teng II-430
Liu, Xiangyu II-40
Liu, Xiaoguang I-504
Liu, Xiaoyou I-378
Liu, Xingshan I-164
Liu, Xingwu II-92
Liu, Xinran II-92
Liu, Yang II-417
Liu, Zhimin II-120
Lu, Jiaheng II-531
Lu, Jie II-342
Lu, Junli I-67
Luo, Daowen II-506
Luo, Lei II-444
Luo, Shiying II-198
Luo, Zhunchen II-251

Ma, Jiansong II-52
Ma, Jun II-444
Ma, Kun II-534
Ma, Qingli I-491
Ma, Yuchi I-205
Mao, Jiali I-137, II-78
Meng, Xiangfu I-257
Meng, Xiangxu II-481
Miao, Gaoshan I-402
Mirzanurov, Damir II-330

Nawaz, Waqas II-330
Nie, Tiezheng I-338
Ning, Hong I-298

Ouyang, Shuang   I-415, II-304

Pan, Li   II-481
Pan, Yanhong   II-78
Peng, Kanggui   II-145
Peng, Min   I-415, II-304
Pu, Juhua   II-92

Qu, Qiang   II-330

Rao, Weixiong   I-441
Ren, Yafeng   I-310

Sha, Dandan   I-272
Sha, Ying   II-291
She, Jieying   I-191, II-265
Shen, Aili   I-257
Shen, Derong   I-338
Sheng, Victor S.   II-132
Song, Chenxi   II-544
Song, Guojie   I-325
Song, Qiuge   I-137
Song, Rongwei   I-43
Song, Tianshu   I-191, II-265
Stones, Rebecca   I-504
Sun, Hui   II-493
Sun, Rui   I-310
Sun, Tao   I-378

Ta, Na   I-124
Tan, Yong   II-213
Tang, Changjie   I-82
Tang, Jintao   I-298
Tang, Yong   II-521
Tang, Zijie   II-534
Theodoratos, Dimitri   II-171
Tian, Gang   I-415, II-304
Tian, Xiuxia   II-52
Tong, Qiuli   II-213
Tong, Yongxin   I-191, II-265
Tseng, Vincent S.   I-17

U, Leong Hou   II-457

Wan, Changxuan   I-55
Wang, AndWei   II-537
Wang, Bin   I-164, I-467, I-491, II-15, II-184, II-198
Wang, Chao   I-454, II-225, II-238
Wang, Chengyu   II-527

Wang, Daling   I-272
Wang, Dong   I-286
Wang, Dongsheng   II-28
Wang, Feng   II-120
Wang, Gang   I-504
Wang, Huaimin   II-544
Wang, Jian   II-457
Wang, Jianyong   I-365
Wang Ling, Tok   II-531
Wang, Lihong   II-291
Wang, Lizhen   I-67
Wang, Ning   II-145
Wang, Peipei   II-198
Wang, Qinyi   II-265
Wang, Shanshan   II-65
Wang, Shaoqing   I-219
Wang, Tao   II-544
Wang, Tengjiao   II-380
Wang, Ting   I-298
Wang, Wentao   II-392
Wang, Xiao   II-524
Wang, Xiaoling   II-52
Wang, Xiaoxuan   I-67
Wang, Yanhua   I-517, II-527
Wang, Yashen   I-390
Wang, Ying   II-251
Wang, Yuanhong   II-92
Wang, Zhenchao   I-310
Wang, Zheng   I-219
Wei, Bingjie   II-403
Wei, Dengping   I-298
Wen, Yanlong   I-454
Weng, Weitao   I-378
Wu, Jian   II-132
Wu, Lei   II-481
Wu, Qingbo   II-444
Wu, Shengli   I-429
Wu, Tianzhen   II-493
Wu, Xiaoying   II-171

Xia, Xiufeng   II-40
Xiao, Jing   II-521
Xiao, Qing   II-524
Xie, Lijuan   II-213
Xie, Qianqian   I-415, II-304
Xie, Sihong   I-55
Xie, Songxian   I-298
Xu, Chenyang   I-441
Xu, Chuanhua   II-521
Xu, Chunlin   I-429

Xu, Hua   II-279
Xu, Jungang   I-491
Xu, Ke   I-191, II-265
Xu, Kuai   I-378
Xu, Ming   II-28
Xue, Qian   II-107

Yan, Jihong   I-517
Yan, Xinwei   I-95
Yan, Yuejun   II-403
Yang, Bo   II-534
Yang, Cheng   II-544
Yang, Hao   I-82
Yang, Mansheng   II-469
Yang, Ning   I-205
Yang, Shuqiang   I-353
Yang, Wenjing   I-467
Yang, Wu   II-537
Yang, Xiaochun   II-15, II-145, II-184
Yang, Xiaoting   I-55
Yang, Yan   II-540
Yang, Zhifan   II-238
Yang, Zhihai   II-3
Yao, Yuangang   I-231
Ye, Benjun   I-504
Ye, Qi   II-120
Yin, Gang   II-544
Yin, Litian   I-286
Yin, Ning   II-65
Yiu, Man Lung   I-109
Yu, Ge   I-272, I-338, II-145
Yu, Jie   II-444
Yu, Li   II-225
Yu, Lu   I-244, II-342
Yu, Penghua   I-231
Yu, Philip S.   I-55
Yu, Tinghao   II-319
Yu, Yang   II-251
Yuan, Xiaojie   II-238

Zang, Yuda   I-365
Zeng, Guanyin   II-304
Zhang, Chendong   II-159

Zhang, Chi   II-527
Zhang, Chuxu   I-244, II-342
Zhang, Duoduo   I-205
Zhang, Haiqing   I-82
Zhang, Haiwei   I-454, II-225
Zhang, Jinbo   I-402
Zhang, Jing   II-369
Zhang, Jinghong   II-392
Zhang, Kai   II-304
Zhang, Rong   II-78
Zhang, Shuo   II-481
Zhang, Xiangliang   II-92
Zhang, Yifei   I-272
Zhang, Ying   II-225, II-238
Zhang, Yuhong   II-369
Zhang, Zeyu   I-31
Zhang, Zhaohua   I-504
Zhang, Zi-Ke   I-244, II-342
Zhao, Dapeng   II-52
Zhao, Kankan   I-219
Zhao, Pengpeng   II-132
Zhao, Xue   II-225, II-238
Zheng, Yang   I-219
Zheng, Zhejun   II-356
Zhong, Jialin   II-534
Zhong, Ning   I-178
Zhou, Aoying   I-517, II-159
Zhou, Dahai   II-40
Zhou, Ge   I-244
Zhou, Juming   I-286
Zhou, Kailai   II-493
Zhou, Lihua   I-67
Zhou, Meilin   I-467
Zhou, Minqi   II-159
Zhou, Qiang   I-390
Zhou, Rui   I-164
Zhou, Shuigeng   I-109
Zhou, Tao   II-342
Zhou, Xiangmin   I-272
Zhu, Feida   II-537
Zhu, Jia   II-521
Zhu, Jiahui   I-415, II-304
Zhu, Xuan   I-365

Printed in the United States
By Bookmasters